The Pacific Arctic Region

Bering Strait region off Cape Prince of Wales, Alaska, USA (Photo credit : Elizabeth Labunski, US Fish & Wildlife Service)

Jacqueline M. Grebmeier • Wieslaw Maslowski
Editors

The Pacific Arctic Region

Ecosystem Status and Trends in a Rapidly Changing Environment

Springer

Editors
Jacqueline M. Grebmeier
Chesapeake Biological Laboratory
University of Maryland Center
for Environmental Science
Solomons, MD, USA

Wieslaw Maslowski
Department of Oceanography
Graduate School of Engineering and
Applied Sciences
Naval Postgraduate School
Monterey, CA, USA

ISBN 978-94-017-8862-5 ISBN 978-94-017-8863-2 (eBook)
DOI 10.1007/978-94-017-8863-2
Springer Dordrecht Heidelberg New York London

Library of Congress Control Number: 2014941794

Printed on acid-free paper

Springer is part of Springer Science+Business Media (www.springer.com)

Dedication to Martin (Marty) Bergmann, February 19, 1956–August 20, 2011

Martin (Marty) Bergmann was Director of Canada's Polar Continental Shelf Program (PCSP) of Natural Resources Canada when he died at the age of 55 in the First Air B-737 plane crash on August 20, 2011, at Resolute Bay, Nunavut, Canada. He was traveling to Resolute to escort Prime Minister Stephen Harper on a tour of Canada's Arctic facilities.

Marty was a strong advocate for collaborative Arctic science and he was instrumental in developing the Pacific Arctic Group (PAG) in 2002, and he served as the forum's first Chair. PAG is an active nexus for international networking for marine programs in the Pacific Arctic region and the sponsor of this Springer volume. Marty worked diligently to provide opportunities for scientific interactions from all interested countries, including scientists, agency managers, politicians, and local community members. He facilitated many Arctic scientific programs leading up to the

2007–2008 International Polar Year (IPY), some that continue today and many of the results from which are presented in this volume. Marty will be remembered for his unbounded energy and efforts to move our understanding of the Arctic System forward.

Prior to taking on a leadership role at PCSP, Marty served as Director of the National Center for Arctic Aquatic Research Excellence in Fisheries and Oceans Canada, managing the logistics for Arctic Ocean science aboard the Canadian Coast Guard fleet. Marty had a suite of accomplishments, both in Canada through the highly successful IPY years and internationally by coordinating agreements to facilitate Arctic logistics and science activities with many countries, including Great Britain, the United States, China, Japan and the Republic of Korea.

In 2012 the Royal Canadian Geographical Society established the Martin Bergmann Medal for Excellence in Arctic Leadership and Science to recognize Marty's leadership for advancing science in the Arctic. Specifically, the award recognizes "achievement for 'excellence in Arctic leadership and science'. It celebrates "Marty" Bergmann, a public servant with an outstanding talent for networking that led him to connect scientists with resources and technology, to inspire business leaders, explorers and innovators towards new goals and to consider and attempt to meet the challenges inherent in opening up the Arctic, whether these were related to logistics, safety, resources, people, knowledge or will."

Marty was posthumously awarded the inaugural medal at the 2012 International Polar Year Conference in Montreal, Canada.

With these memories of his efforts on behalf of Arctic science in our hearts and minds, we dedicate this volume to our friend, Marty Bergmann, as we work to better our understanding of the Arctic system for future generations.

Jackie Grebmeier
Chair of the Pacific Arctic Group (PAG)
and Wieslaw Maslowski, co-editor

Contents

Contributors

Yevgeny Aksenov Marine Systems Modelling, National Oceanography Centre, Southampton, UK

Leif G. Anderson Department of Chemistry, University of Gothenburg, Gothenburg, Sweden

Carin J. Ashjian Woods Hole Oceanographic Institution, Woods Hole, MA, USA

Xuezhi Bai Cooperative Institute for Limnology and Ecosystems Research (CILER), School of Natural Resources and Environment, University of Michigan, Ann Arbor, MI, USA

Nicholas R. Bates Bermuda Institute of Ocean Sciences (BIOS), Ferry Reach, Bermuda

Bodil A. Bluhm School of Fisheries and Ocean Sciences, University of Alaska Fairbanks, Fairbanks, AK, USA

Wei-Jun Cai School of Marine Science and Policy, The University of Delaware, Newark, DE, USA

Liqi Chen Third Institute of Oceanography, The State Oceanic Administration of China, Xiamen, China

Jim Christian Institute of Ocean Sciences, Fisheries and Oceans Canada, Sidney, BC, Canada and the Canadian Centre for Climate Modeling and Analysis, Environment Canada, Victoria, BC, Canada

Kathleen E. Conlan Canadian Museum of Nature, Ottawa, ON, Canada

Jessica N. Cross School of Fisheries and Ocean Sciences, University of Alaska Fairbanks, Fairbanks, AK, USA

Beverly de Cuevas Marine Systems Modelling, National Oceanography Centre, Southampton, UK

Clara J. Deal International Arctic Research Center, University of Alaska Fairbanks, Fairbanks, AK, USA

Ken L. Denman School of Earth and Ocean Sciences, University of Victoria, Victoria, BC, Canada

Hajo Eicken Geophysical Institute, University of Alaska Fairbanks, Fairbanks, AK, USA

Lisa Eisner Alaska Fisheries Science Center, National Oceanic and Atmospheric Administration, Seattle, WA, USA

Scott M. Elliott Climate Ocean Sea Ice Modeling, Los Alamos National Laboratory, Los Alamos, NM, USA

Edward V. Farley Jr. Alaska Fisheries Science Center, National Oceanic and Atmospheric Administration, Seattle, WA, USA

Karen E. Frey Graduate School of Geography, Clark University, Worcester, MA, USA

Georgina Gibson International Arctic Research Center, University of Alaska Fairbanks, Fairbanks, AK, USA

Rolf R. Gradinger School of Fisheries and Ocean Sciences, University of Alaska Fairbanks, Fairbanks, AK, USA

Jacqueline M. Grebmeier Chesapeake Biological Laboratory, University of Maryland Center for Environmental Science, Solomons, MD, USA

Laodong Guo School of Freshwater Sciences, University of Wisconsin-Milwaukee, Milwaukee, WI, USA

Dennis A. Hansell Rosenstiel School of Marine and Atmospheric Science, University of Miami, Miami, FL, USA

Lois A. Harwood Canada Department of Fisheries and Oceans, Yellowknife, NT, Canada

Victoria J. Hill Department of Ocean, Earth and Atmospheric Sciences, Old Dominion University, Norfolk, VA, USA

Russell R. Hopcroft School of Fisheries and Ocean Sciences, University of Alaska Fairbanks, Fairbanks, AK, USA

Haoguo Hu Cooperative Institute for Limnology and Ecosystems Research (CILER), School of Natural Resources and Environment, University of Michigan, Ann Arbor, MI, USA

Brian P.V. Hunt Department of Earth, Ocean and Atmospheric Sciences, University of British Columbia, Vancouver, BC, Canada

Moto Ikeda Graduate School of Earth and Environmental Sciences, Hokkaido University, Sapporo, Japan

Motoyo Itoh Japan Agency for Marine-Earth Science and Technology, Yokosuka-city, Kanagawa, Japan

Jaromir Jakacki Institute of Oceanology, Polish Academy of Sciences, Sopot, Poland

Meibing Jin International Arctic Research Center, University of Alaska Fairbanks, Fairbanks, AK, USA

Hyoung M. Joo Korea Polar Research Institute, Incheon, South Korea

Jaclyn Clement Kinney Department of Oceanography, Graduate School of Engineering and Applied Sciences, Naval Postgraduate School, Monterey, CA, USA

David L. Kirchman School of Marine Science and Policy, University of Delaware, Lewes, DE, USA

Ksenia N. Kosobokova Shirshov Institute of Oceanology, Russian Academy of Sciences, Moscow, Russian Federation

Kathy Kuletz USF&WS, Alaska Region, Anchorage, AK, USA

Diane Lavoie Maurice Lamontagne Institute, Fisheries and Oceans Canada, Mont Joli, QC, Canada

Sang H. Lee Department of Oceanography, Pusan National University, Busan, South Korea

Warren Lee Canadian Centre for Climate Modeling and Analysis, Environment Canada, Victoria, BC, Canada

William K.W. Li Fisheries and Oceans Canada, Dartmouth, NS, Canada

Elizabeth Logerwell Alaska Fisheries Science Center, National Oceanic and Atmospheric Administration, Seattle, WA, USA

Connie Lovejoy Department of Biology, Université Laval, Quebec, QC, Canada

James Lovvorn Southern Illinois University, Carbondale, IL, USA

James A. Maslanik Colorado Center for Astrodynamics Research, University of Colorado, Boulder, CO, USA

Wieslaw Maslowski Department of Oceanography, Graduate School of Engineering and Applied Sciences, Naval Postgraduate School, Monterey, CA, USA

Jeremy T. Mathis Pacific Marine Environmental Laboratory, National Oceanic and Atmospheric Administration, Seattle, WA, USA and School of Fisheries and Ocean Sciences, University of Alaska Fairbanks, Fairbanks, AK, USA

Kohei Mizobata Department of Ocean Sciences, Tokyo University of Marine Science and Technology, Tokyo, Japan

Sue E. Moore Fisheries Office of Science and Technology, National Oceanic and Atmospheric Administration, Seattle, WA, USA

S. Bradley Moran Graduate School of Oceanography, University of Rhode Island, Narragansett, RI, USA

James R. Murphy Alaska Fisheries Science Center, National Oceanic and Atmospheric Administration, Seattle, WA, USA

R. John Nelson Department of Biology, Centre for Biomedical Research, University of Victoria, Victoria, BC, Canada

An Nguyen Program in Atmospheres, Oceans, and Climate, Massachusetts Institute of Technology, Cambridge, MA, USA

Stephen R. Okkonen School of Fisheries and Ocean Sciences, University of Alaska, Fairbanks, AK, USA

Robert Osinski Institute of Oceanology, Polish Academy of Sciences, Sopot, Poland

James E. Overland Pacific Marine Environmental Laboratory, National Oceanic and Atmospheric Administration, Seattle, WA, USA

Robert S. Pickart Woods Hole Oceanographic Institution, Woods Hole, MA, USA

Michel Poulin Canadian Museum of Nature, Ottawa, ON, Canada

Lori T. Quakenbush Alaska Department of Fish and Game, Fairbanks, AK, USA

Andrew F. Roberts Department of Oceanography, Naval Postgraduate School, Monterey, CA, USA

Igor P. Semiletov Pacific Oceanological Institute, Far Eastern Branch of the Russian, Academy of Sciences, Vladivostok, Russia and International Arctic Research Center, University of Alaska Fairbanks, Fairbanks, AK, USA

Evelyn Sherr College of Earth, Ocean, and Atmospheric Sciences, Oregon State University, Corvallis, OR, USA

Emily Shroyer College of Oceanic and Atmospheric Sciences, Oregon State University, Corvallis, OR, USA

Michael Steele Applied Physics Laboratory, University of Washington, Seattle, WA, USA

Nadja Steiner Institute of Ocean Sciences, Fisheries and Oceans Canada, Sidney, BC, Canada and the Canadian Centre for Climate Modeling and Analysis, Environment Canada, Victoria, BC, Canada

Sam VanLaningham School of Fisheries and Ocean Sciences, University of Alaska Fairbanks, Fairbanks, AK, USA

Dao-Ru Wang Hainan Marine Development and Design Institute, Hainan, China

Jia Wang Great Lakes Environmental Research Laboratory (GLERL), National Oceanic and Atmospheric Administration, Ann Arbor, MI, USA

Muyin Wang Joint Institute for the Study of the Atmosphere and Ocean, University of Washington, Seattle, WA, USA

Rik Wanninkhof Atlantic Oceanographic and Meteorological Laboratory, National Oceanic and Atmospheric Administration, Miami, FL, USA

Eiji Watanabe International Arctic Research Center, University of Alaska Fairbanks, Fairbanks, AK, USA and Research and Development Center for Global Change, Japan Agency for Marine-Earth Science and Technology, Yokohama, Japan

William J. Williams Fisheries and Oceans Canada, Institute of Ocean Sciences, Sidney, BC, Canada

Rebecca A. Woodgate Applied Physics Laboratory, University of Washington, Seattle, WA, USA

Kelly V. Young Fisheries and Oceans Canada, Sidney, BC, Canada

Yanling Yu Arctic Science Center, Applied Physics Laboratory, University of Washington, Seattle, WA, USA

Jinlun Zhang Applied Physics Laboratory, University of Washington, Seattle, WA, USA

Chapter 1
The Pacific Arctic Region: An Introduction

Jacqueline M. Grebmeier and Wieslaw Maslowski

Abstract The Pacific Arctic region (PAR) is experiencing atmospheric changes, rapid seasonal sea ice retreat, seawater warming, regional ocean acidification, along with other environmental changes and biological responses in lower to upper trophic organisms. Both physical and biogeochemical modeling indicate the potential for step-function changes to the overall ecosystem, both under current and in the projected conditions. This volume of synthesis papers was coordinated within the Pacific Arctic Group (PAG), a network of international partners undertaking and facilitating collaborative research in the Pacific influenced Arctic seas and basin. It also serves as a product of activities from the 2007–2008 International Polar Year. The topics range from atmospheric and physical sciences to chemical processing and biological response to changing environmental conditions. Physical and biogeochemical modeling results highlight the need for continued data collection together with interdisciplinary modeling activities to track and forecast the changing ecosystem of the Pacific Arctic in response to climate change.

Keywords Pacific Arctic Region • Physical forcing • Ecosystem response • Climate change

J.M. Grebmeier (✉)
Chesapeake Biological Laboratory, University of Maryland Center for Environmental Science, 146 Williams Street, Solomons, MD 20688, USA
e-mail: jgrebmei@umces.edu

W. Maslowski
Department of Oceanography, Graduate School of Engineering and Applied Sciences, Naval Postgraduate School, Dyer Road, Bldg. SP339B, Monterey, CA 93943, USA
e-mail: maslowsk@nps.edu

J.M. Grebmeier and W. Maslowski (eds.), *The Pacific Arctic Region: Ecosystem Status and Trends in a Rapidly Changing Environment*, DOI 10.1007/978-94-017-8863-2_1,

1.1 Introduction

The Pacific Arctic Region (PAR) synthesis effort presented in this volume is an ongoing contribution of the Pacific Arctic Group (PAG) to better understand the processes inherent in controlling and sustaining the marine ecosystem and projected drivers of change in the region. PAG is an international consortium of scientists, agency managers and organizations interested in scientific progress and collaboration in the Pacific Arctic. The PAG definition of the Pacific Arctic extends from the northern Bering Sea into the Chukchi Sea and adjacent Arctic seas, and extending into the deep basins of the Arctic Ocean.

The core objectives of the PAR synthesis activities within PAG are to: (1) present new synthetic results from observational and modeling activities in the Pacific-influenced Arctic region, (2) characterize current status and trends of the ecosystem to improve our understanding of the state and key processes influencing the marine system, (3) use model projections to evaluate ongoing processes and future evolution of the physical, biogeochemical and biological system to guide new science activities, and (4) to identify critical marine components for further development of observational networks in the Pacific Arctic.

As a core contribution to the PAG synthesis efforts, PAG members initiated the current synthesis book. The chapters in this book are multi-disciplinary with a systems approach, including retrospective, field, and modeling efforts. The chapter format is thematic in content, with spatial regions inside the chapters used as case studies to evaluate the Pacific Arctic region. The overarching themes are process-oriented, and consider both temporal and spatial scales. The modeling chapters cover a larger spatial domain, including the Pacific inflow through the Aleutian Islands/southern Bering Sea (upstream), while maintaining the main focus on the northern Bering, Chukchi, and Beaufort Seas, western Arctic Ocean basin as well as the downstream connectivity of the Pacific to Atlantic side, including the Canadian Arctic Archipelago.

Multiple PAG workshops and contributions set the stage for the various synthesis chapters in this book. For example, the first PAG workshop was held in Sanya, China, in January 2008, which resulted in a special issue in the Chinese Journal of Polar Science (Wang et al. 2008). The second PAG workshop, focused on biological processes, was held in May 2009 in Seattle, Washington, USA, resulting in a feature article for EOS (Grebmeier et al. 2010) and a workshop report. The third PAG workshop was focused on the marine carbon cycle, and was held in June 2009 in Xiamen, China and resulted in a special issue of Deep-Sea Research II (DSR II; Cai et al. 2012). Another special issue of scientific findings is being assembled in a DSR II special issue related to the Korean Arctic Expeditions during 2008–2010 (Sang Lee et al. 2013, personal communication). We introduce here this current book volume of synthesis chapters. The book concept was endorsed by the International Arctic Science Committee (IASC), the past Arctic Ocean Sciences Board (AOSB, now the Marine Working Group of IASC), and the IPY project office of the International Council of Science Unions (ICSU) also endorsed the volume as an IPY legacy effort.

During the course of this synthesis effort other interdisciplinary programs have been carried out in various sections of the PAR, many providing data supporting the syntheses in this volume. These programs include, but are not limited to the Bering Sea Project (Weise et al. 2012), the Bowhead Whale Feeding Ecology Study (BOWFEST) project (Ashjian et al. 2010), Canada's Three Ocean (C30) project (Carmack et al. 2010), the Chukchi Acoustic, Oceanographic, and Zooplankton (CHAOZ) study (Berchok et al. 2014), the Chukchi Sea Monitoring in Drilling Area-Chemistry and Benthos (COMIDA-CAB) project (Dunton et al. 2014), the Chukchi Sea Environmental Studies Program (CSESP; Day et al. 2013), the Distributed Biological Observatory (DBO) program (Grebmeier et al. 2010), the Impacts of Climate on the Eco-Systems and Chemistry of the Arctic Pacific Environment (ICESCAPE) project (initial results reported in Arrigo et al. 2012), and the Russian-American Long-term Census of the Arctic (RUSALCA) program (Bluhm et al. 2010), along with other multidisciplinary studies by PAG member countries and other international collaborators.

1.2 The Pacific Arctic Region

Pacific water transiting across the wide Bering, Chukchi and the eastern portion of the East Siberian shelves, and western portion of the Beaufort Sea, is a major driving force for the physical environmental state, ice extent and thickness, productivity and carbon transport in the Amerasian Arctic (Fig. 1.1). There are key physical, biogeochemical, and biological oceanographic features that distinguish the PAR from the rest of the Arctic Ocean. Hydrographic characteristics and distribution of Pacific water have important implications for shelf productivity as well as shelf-basin exchange at the continental margins, including the downstream influence that Pacific water has on the upper halocline below and sea ice above in the Arctic Basin as well as in the Canadian Arctic Archipelago.

1.3 Physical Processes, Hydrography and Sea Ice: Field and Modeling

The Bering Strait region is a major gateway from the perspective of transport of sea ice and ocean volume and properties, including heat, freshwater and nutrients, and fluxes of atmospheric heat and moisture (Cooper et al. 1997; Overland and Stabeno 2004; Shimada et al. 2006; Woodgate and Aagaard 2005). It also plays a central role as a pathway for fluxes of biological organisms and organic carbon (Grebmeier et al. 2006a; Grebmeier 2012; Walsh et al. 2004). Seasonal evaluation of time series measurements (1990–2010) from the Bering Strait indicate annual variability in salinity, temperature, and transport (Woodgate et al. 2012). Recently it has been determined that the freshwater flux in Bering Strait has likely been

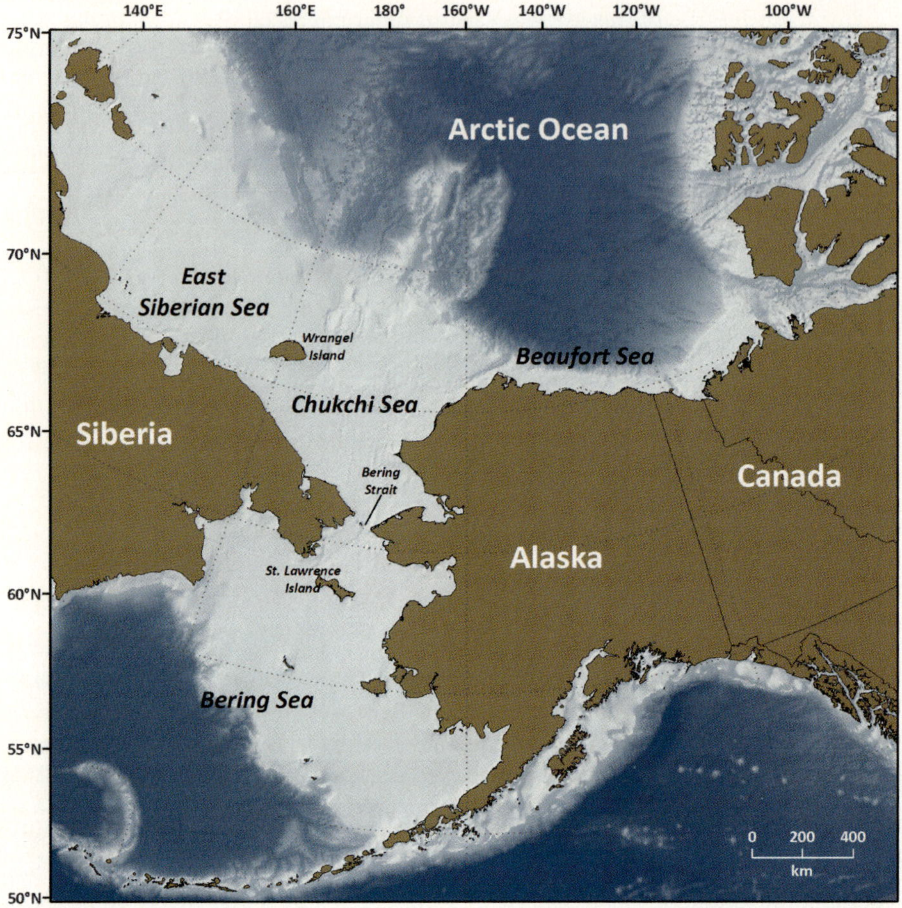

Fig. 1.1 Schematic of the Pacific Arctic Region, including the Bering, Chukchi, East Siberian and Beaufort Seas and the Arctic Ocean

underestimated and should be revised upwards (Woodgate and Aagaard 2005). While these estimates require further verification (since continuous moored measurements have been limited to deeper, more saline waters of Bering Strait) the newer flux estimates indicate that freshwater in Bering Strait may provide ~40 % of the total freshwater input to the Arctic Ocean (Woodgate and Aagaard 2005).

The nutrient-rich Pacific waters transiting through the Bering Strait are transformed seasonally by oceanographic processes over the wide continental shelves south and north of the strait, with far-reaching implications for Arctic halocline formation and basin dynamics. Changes in the freshwater flux may also potentially influence thermohaline circulation in the global ocean. Both winter and summer Pacific Water types play variable, yet distinct roles in the transport of heat, freshwater, nutrients, carbon, and biological organisms northwards through the Bering

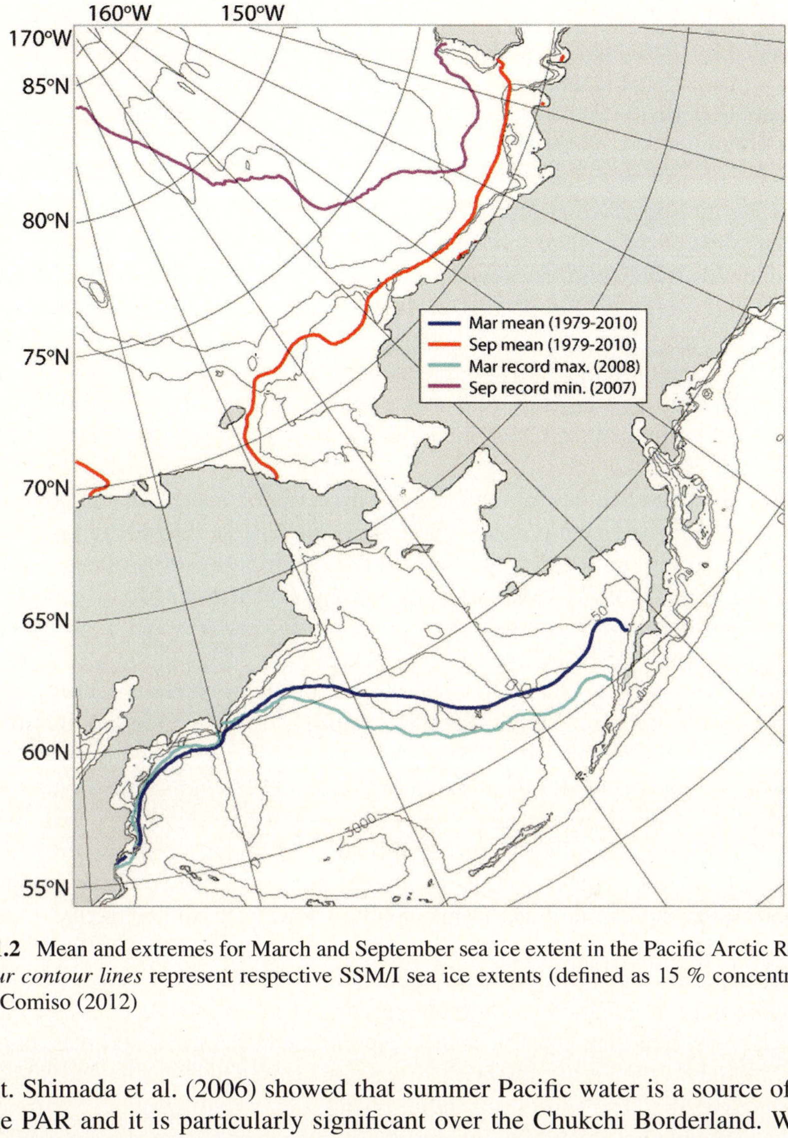

Fig. 1.2 Mean and extremes for March and September sea ice extent in the Pacific Arctic Region. *Colour contour lines* represent respective SSM/I sea ice extents (defined as 15 % concentration) from Comiso (2012)

Strait. Shimada et al. (2006) showed that summer Pacific water is a source of heat to the PAR and it is particularly significant over the Chukchi Borderland. Winter Pacific water is influenced by ice formation and brine rejection, so the timing, extent and location of these processes are intimately tied to halocline formation.

The PAR is experiencing the greatest seasonal retreat and thinning of sea ice in the Arctic (Fig. 1.2). September 2007, at the start of IPY, at that point, marked the highest seasonal sea ice retreat on record (Stroeve et al. 2007); this record was eclipsed in 2012 (Stroeve et al. 2012). Changes in sea ice formation, extent and thickness influence albedo feedback, brine formation and halocline maintenance, so atmosphere-ice-ocean interactions and dynamics are extremely critical for regulating climatic conditions in the Arctic, and can have global ramifications. Exchange of materials from shelf to slope and into the deep Arctic basin is also expected to be influenced by these changing conditions.

Key physiographic features of the shelf-slope region of the East Siberian, Chukchi and Beaufort Seas also influence shelf-basin exchange. In particular, Herald Valley/Canyon and Barrow Canyon are key conduits for transformed Pacific water and associated organisms that transit the Chukchi shelf to the deep Arctic Basin. Eddy formation and fluxes, boundary current dynamics, and advection are some of the critical transport mechanisms at the shelf-basin interface that facilitate the transfer of freshwater/salt, heat, nutrients, and various forms of organic and inorganic carbon that dictate the current state of the Arctic Ocean. Recent findings show increased northward heatflow within the Atlantic water transiting through Fram Strait into the deep Arctic Basin (Walczowski et al. 2012), which may increase advection of warmer Atlantic water and its impacts along the continental margins of the Chukchi and Beaufort seas (Pickart et al. 2011; Schulze and Pickart 2012).

Arctic systems can be rich and diverse habitats for marine life in spite of the extreme cold environment (Grebmeier et al. 2006a, 2010). Biogeochemical cycling processes and biological communities are directly influenced by changing sea ice extent, light availability, seawater hydrography (nutrients, salinity, temperature, density), currents, and water column production. Earlier seasonal sea ice melt and retreat in the Bering Sea and western Arctic will have dramatic impacts on the biological system, such as changes in overlying primary production, altering organic carbon transformation, pelagic-benthic coupling, and benthic production and community structure that could have cascading effects to higher trophic levels, which are of importance for local indigenous residents of the Arctic. Some specific examples of biological changes include time-series observations that indicate declines in bottom-dwelling clam populations and diving seaducks over the last few decades (Grebmeier et al. 2006b). In addition, a decline in benthic amphipod populations in the Chirikov Basin just south of Bering Strait has likely influenced the movement of migrating gray whales to feeding areas north of Bering Strait during this time period (Moore et al. 2003; Moore 2008).

Retreating sea ice and warming temperatures have increased coastline erosion of terrigenous materials into the coastal environment. An increased seasonal open water period in the Arctic results in increased wave fetch, leading to additional shoreline erosion. The subsequent input of old, terrestrially-produced carbon into the ocean could alter microbial transformation processes as well as dilute the labile marine carbon pool with less-usable terrigenous material, with a potential negative impact on food availability to marine organisms. The cycling of carbon (particulate, dissolved, inorganic) is a key concern in these extremely productive regions of the Arctic Ocean. Changes in these processes will have cascading impacts to all components of the ecosystem (from bacteria to man). Ocean acidification is a potential, large-scale negative impact on the Arctic marine carbon system since increased atmospheric CO_2 will reduce the buffering capability of seawater and increase corrosive trends (Mathis et al. 2007).

The following summaries highlight findings in each of the subsequent chapters of this book from atmospheric and physical forcing, biogeochemical cycling, biological processes and modeling efforts.

1.4 Atmospheric Forcing and Sea Ice

In Chap. 2, Overland et al. (2014, this volume) discuss the meteorological setting of the Pacific Arctic as the transition zone between the relatively warm and moist air mass of the North Pacific and cold and dryer air mass of the Arctic. The northern part of the PAR is dominated by high surface atmospheric pressure, but low-pressure cyclonic systems can propagate from the south, where the semi-permanent low pressure center, the Aleutian Low, is located. This latitudinal sea level pressure gradient results in predominantly easterly winds, which generate upwelling favorable conditions along the Beaufort and Chukchi continental slopes. Climate change in the Pacific Arctic is not occurring in a single phase, as the western Arctic Ocean proper has experienced warming and retreating sea ice cover in summer that has accelerated since the late 1990s, while the Bering Sea has had extended sea ice cover and much colder temperatures in winter and spring since 2006. Overland et al. also review coupled climate model predictions of climate change, which suggest increased interannual variability but relatively slow changes in seasonal sea ice cover trends and temperature in the Bering Sea during its sea ice covered season. This is in contrast to model predictions of a significant reduction of, or no, seasonal sea ice cover in the Chukchi Sea from July to November within the next couple decades. These changes will have major ecological impacts and could also lead to increased commercial activities that will take advantage of longer open-water seasons.

In Chap. 3, Frey et al. (2014, this volume) continue this theme of understanding the variability in sea ice conditions in the PAR, by synthesizing observational data and model results, and discuss the range of effects from the minimal change in the northern Bering Sea to the rapid and dramatic decline of sea ice cover in the Bering Strait region north of St. Lawrence Island, the Chukchi Sea and Canada Basin. Both earlier spring breakup and later sea ice formation occur in these seasonally changing sea ice regions, with portions of the Chukchi Sea experiencing the most extreme trends. Another common characteristic of the region is the widespread transition from multi-year ice to thinner, first year ice. Frey et al. discuss the implications of reduced sea ice areal coverage, age, and thickness on the overall productivity of the ecosystem that are discussed in more detail in later chapters of this book.

In Chap. 4, Wang et al. (2014, this volume) consider large-scale atmospheric patterns and how these patterns influence sea ice decline and other changes in the Pacific Arctic. In particular, they argue that in addition to the Arctic Oscillation (AO), the Dipole Anomaly (DA), which is a second mode of northern hemisphere sea level pressure variability, might be driving declines of sea ice cover. Wind anomalies associated with DA are typically meridional (north-south), hence their impact on sea ice distribution (e.g. its northward retreat or export through Fram Strait) may be more significant than the larger variability associated with zonal winds, which are driven by the AO. Meridional wind anomalies can also affect northward advection of warm Pacific water across the Chukchi shelf and reduction or persistence of landfast ice along the Beaufort and Chukchi coast. Wang et al. suggest a possible air-ice-sea feedback loop in the western Arctic involving DA, however the origins and controls of these modes require further investigations, including observations and modeling of the fully coupled Arctic System at process scales.

1.5 Physical Processes and Modeling

In Chap. 5, Maslowski et al. (2014, this volume) review the regional ocean circulation over an extended PAR, starting from the northern North Pacific into the western Arctic Ocean. They argue that upstream circulation of the Alaskan Stream and variability in fluxes across the Aleutian Island passes not only regulate the hydrology of the deep Bering Sea but also influence characteristics (especially salinity) of water advected onto the Bering Shelf and crossing Bering Strait. They envision the Bering Slope Current as more of a system of eddies generated along the slope rather than a continuous current. Year-round high eddy kinetic energy along the slope and in Anadyr and Bering straits is simulated just upstream of highly productive regions along the outer Bering shelf, in the Chirikov Basin that is north of St. Lawrence Island and again north of Bering Strait. This suggests strong regional biophysical coupling and indicates needs for development of improved modeling capabilities. Further downstream, in the Chukchi Sea, Maslowski et al. emphasize the importance of the Alaska Coastal Current in transporting heat via warm shelf waters into the Arctic Ocean basin. They find little or no correlation between modeled heat fluxes through Bering Strait and those entering the Beaufort Sea, which implies that most of the heat transiting across Bering Strait exchanges heat by melting ice or by transfer to the atmosphere within the Chukchi shelf. The water exported into the Beaufort Sea is instead warmed locally due to insolation, which increases with earlier retreat of sea ice cover. Winds and eddies distribute warm water further north and often under the ice cover. The authors hypothesize that some of the heat associated with this water (in particular that available below the shallow mixed layer) is not removed to the atmosphere during freeze up and instead it has been accumulating below the mixed layer in the western Arctic Ocean. This 'new' source of heat is available year around for entrainment into the mixed layer to reduce sea ice growth in winter and to accelerate its melt in summer.

In Chap. 6, Williams et al. (2014, this volume) focuses on shelf-break exchanges and associated processes within the Pacific Arctic. In particular, the authors consider ocean dynamics along the continental slope in the Bering, and in the Chukchi and Beaufort Seas, which define the southern and northern edges of the shelf system, respectively, that separate the North Pacific from the Arctic Ocean. Ocean on-shelf transport of volume and properties in the Bering Sea is concentrated in submarine canyons, especially in Zhemchug and Bering canyons, where mesoscale eddies within the Bering Slope Current are commonly observed (via remote sensing) and can also be modeled. Similarly on the northern end of the Bering-Chukchi shelf system, Herald and Barrow Canyons control the outflow into the deep basin. This outflow interacts with the shelf-break jets, resulting in instabilities and eddies, which in addition to wind-driven cross-slope fluxes in the Ekman (surface and bottom) boundary layers determine shelf-basin exchange along the Chukchi and Beaufort slopes. In addition, on-shelf upwelling of Atlantic water due to easterly winds in summer and the outflow of dense shelf water formed in polynyas due to brine rejection in winter can further affect shelf-basin exchange. Even though numerical models are the primary tool in quantifying such exchanges and their variability, models need to be evaluated and constrained by observations, which is why the authors argue for new field programs focused on shelf-break and processes there.

In Chap. 7, Clement-Kinney et al. (2014, this volume) focus on the flow through Bering Strait, the only connection between the Pacific and the Arctic. The physiographic narrow and shallow features of the strait has been challenging to apply to ocean general circulation models, using relatively coarse grids, and to realistically represent volume and property fluxes. A synthesis of observations and models presented in this chapter includes quantified fluxes of heat and freshwater and their seasonal and interannual variability. The models give reasonable estimates of volume transport compared to observational estimates, but uncertainty still remains in both. It is likely the existing models do not simulate well the narrow (10 km wide) flow of Alaska Coastal Current on the eastern margin of the strait where there is significant horizontal and vertical velocity shear and may be due to the lateral and vertical boundary conditions used. However, the available continuous observations are still limited as they do not measure the upper part of the water column with higher freshwater fractions and observational flux estimates assume homogeneity of the flow at all locations in the strait. Clement-Kinney et al. expect that further refinements of both model and observational estimates of volume flux at Bering Strait are needed, including higher spatial model resolution (both horizontal and vertical) and increased numbers of moorings. Since property fluxes are dependent on the volume transport, improved volume flux estimates are expected to reduce uncertainty of property fluxes.

1.6 Carbon Transformations and Cycling

In Chap. 8, Cai et al. (2014, this volume) evaluate the inflow of dissolved inorganic carbon (DIC) from the Pacific Ocean into the Pacific Arctic and the transformation processes occurring as the water transits both into the Arctic Ocean and east into the Canadian Arctic Archipelago (CAA) and to a limited extent, west into the East Siberian Sea (ESS). Available data indicate that the Chukchi Sea is the dominant PAR region for atmospheric carbon dioxide (CO_2) uptake, while the Beaufort Sea (BS) and CAA remove comparatively less CO_2, and the CAA and ESS are weak seasonal sources of CO_2. With the extensive reduction of sea ice in the deep Canada Basin (CB), CO_2 uptake has increased in the summer period. There is a small positive-sign export of DIC, indicating that the CB is a net heterotrophic system. The marginal seas can seasonally produce labile organic carbon from primary production and biological processing of carbon, riverine input and coastal erosion that are also recycled within and in transit out of the ecosystems, with the potential of a warming climate causing an increase in the production and export of DIC. Uncertainty remains as to whether a changing climate will make the PAR a sink or source of CO_2 in the future.

In Chap. 9, Mathis et al. (2014, this volume) discuss the chemical and biological components influencing the Arctic marine carbon cycle, including primary production, carbon transformations, and export production that are sensitive to sea ice loss, warming temperatures, changes in the timing and location of primary production, freshwater inputs, and ocean acidification. Seasonal variability in dissolved organic carbon (DOC) and particulate organic carbon (POC) production are key components of the marine carbon cycle, which vary with seasonal gradients in light, ice-cover and riverine input to the Arctic Ocean and to marginal seas. Secondary production and microbial processes in the water column, along with microbial processing in the underlying sediments, are key elements of the marine carbon cycle. Regions within the PAR are experiencing seasonal ocean acidification and decreasing pH. This is reducing the saturation states of calcium carbonate, which could have detrimental impacts on calcifying fauna, both in the water column and in sediments. Since these organisms are important prey for upper trophic level predators, additional studies of key biochemical and biological processes in the marine carbon cycle are necessary in a rapidly changing Arctic.

1.7 Lower and Upper Trophic Levels and Ecosystem Modeling

In Chap. 10, Nelson et al. (2014, this volume) evaluate the changes occurring in lower trophic level productivity, composition and biomass in the PAR, including northward range extensions of certain species. Variability in species composition can impact primary production, trophic connectivity and carbon cycling that influence ecosystem dynamics in the PAR. This chapter includes information on the distribution and abundance of microbes, zooplankton, and benthic organisms, including their potential response to increasing seawater temperature, freshwater content, lower pH, and changes in sea ice dynamics. Smaller organisms, such as microbes and microzooplankton, are most susceptible to environmental changes and are expected to respond at a faster rate. Changes in the species composition and productivity of benthic organisms in the PAR will affect the roles played by this diverse fauna in carbon cycling and as a prey items. Multidisciplinary studies of the linkages between lower tropic species ecology, physics and geochemistry are enhancing our understanding and capacity to predict the future status of PAR marine ecosystems during this period of rapid change.

In Chap. 11, Moore et al. (2014, this volume) provide an overview of the key components of upper trophic level groups, including fish, seabirds and marine mammals, which are top predators in the Arctic system. These organisms can respond to environmental variability and specific examples in this chapter outline species responses to ecosystem change. For example, some fish and snow crab are being found now further north in the PAR, while seabirds and marine mammals have changed their phenology and diet to respond to variability in sea ice and lower trophic level prey. Since migrating species have the ability to identify variation in dense prey, upper trophic level organisms can reflect ecosystem shifts by changing migration routes, abundance levels, diet, reproduction and body condition. Partnerships between local indigenous arctic residents and scientists are enhancing our understanding of the biophysical links between sea ice conditions and migrating species. Continued collaborations are an important need for understanding upper trophic level resources, habitat use and ecological relationships.

In Chap. 12, Deal et al. (2014, this volume) discuss ongoing and developing marine biogeochemical and ecosystem models in the PAR. This includes the complexities associated with different scales of modelling, from one-dimensional to regional and global Earth System scales, and combining physical forcing with biological and chemical processes. Models are being used to evaluate benthic, pelagic and sea ice impacts on ecosystem dynamics with the objective of building scenarios of ecosystem response to climate impacts. The authors highlight the importance of capturing the timing and location of the ice edge and open water blooms, including the subsurface chlorophyll maximum layer, for simulating the food web structure. Deal et al. emphasize the need for continued biogeochemical observations to use as input into modeling efforts that improve our understanding and forecasting of ecosystem dynamics for policy decision making.

1.8 Summary

The PAR is currently experiencing the largest sector changes in Arctic seasonal sea ice extent and thickness, but with different responses within the ecosystem. A challenge to both the modeling and observational community is to develop workable scenarios to investigate:

1. How will changes in the flow-through dynamics of what can be termed the Bering Strait continental shelf complex affect downstream Arctic ecosystems?
2. Will changes in the timing and extent of ice formation influence halocline formation and thickness, and if so, what are the ramifications of a reduction in the density gradients across the halocline?
3. Will an increase in freshwater and heat flux through Bering Strait move the PAR to a new stable state and what consequences would this have for the influence of

nutrients, heat, and freshwater on near-field (Pacific Arctic region) ecosystems and downstream (Arctic basin and Canadian Archipelago) ecosystems?
4. How will physical and biogeochemical fluxes vary in the Pacific Arctic Region in concert with lower latitude climate variability and change?

Determining the key drivers and responders to change in this region are essential in order to determine the downstream impacts on the Arctic system, including its connectivity to the world ocean. A hierarchy of models, including process, regional Arctic and global climate and Earth System scales (GC/ESMs) are needed to address these and other PAR related questions.

Early season ice retreat influences timing of the spring bloom and associated lower trophic level consumption of organic carbon. Changes will have cascading effects to benthos and higher trophic organisms. There are indications of increased freshwater flux and summer seawater temperatures, both that influence biological processes, and changes in the timing of productivity over the shelf and slope regions that will rapidly impact trophic structure and carbon transport from shelf to basin. Sea-ice free areas will allow for biological northward expansion of fish, but could have negative impacts on competing, benthic-feeding marine mammals. Impacts on marine mammals harvested for subsistence could also be affected.

The need remains for pan-Arctic comparative studies with standardized data collections to allow compare/contrast evaluations and projection of various scenarios of Arctic change. Ultimately, the need is to determine whether observed changes are due to climate warming or natural variability. These research needs are best met by collection of long-term time-series data at key select sites, integrated with relevant process and coupled modeling studies. Newly evolving programs, such as the Distributed Biological Observatory (Grebmeier et al. 2010) will help to fill these observational gaps, but could be supplemented with efforts in different regions, seasons and climate regimes.

Modelling approaches discussed in this volume renew a call for process-level understanding of the coupled Arctic climate system and improved modeling capabilities to advance knowledge and prediction of climate change in the Pacific Arctic. There is a clear need for eddy-resolving, coupled physical and biogeochemical modeling capabilities. Success with these approaches will lead to better understanding of critical physical processes, potential feedbacks between them and their overall impact of the whole ecosystem. Current global models still have significant limitations with regard to representing the Pacific Arctic. In particular, challenges are associated with representing processes, such as surface mixed layer, eddies, cold halocline, seasonal pycnocline, near surface temperature maximum as well as sea ice thickness distribution, deformation and drift. Regional climate system models, with high spatial and temporal resolution and focus on the Arctic or Pacific Arctic region are key requirements (Maslowski et al. 2012). An Arctic System Model (ASM) should readily allow addition of new system components, such as marine biogeochemistry, ice-sheet/ocean interaction, etc. A fully coupled ASM should allow resolving processes and feedbacks between them. Along with the development of ASMs, observations to constrain and evaluate them are as critical.

Acknowledgements We thank the co-authors of this special synthesis activity for their time and effort in developing the scientific products for this volume. We especially thank Dr. John Calder for his continuous support of the Pacific Arctic Group synthesis activities through the Arctic Research Program of the US National Oceanic and Atmospheric Administration (NOAA). In addition, we thank our international colleagues in the PAG for their continued collaboration in research activities and the support of their managers. We thank Lee Cooper for his editorial efforts on the dedication and introductory chapter. Financial support related to this special synthesis volume was provided to JM Grebmeier (NOAA grant # NA09OAR4310092) and W Maslowski by the DOE Regional and Global Climate Modeling and NSF Office of Polar Programs.

References

Arrigo KR, Perovich DK, Pickart RS, Brown ZW, Van Dijken G, Lowry KE, Mills MM, Palmer M, Balch W, Bahr F, Bates N, Benitez-Nelson C, Bowler B, Brownlee E, Ehn JK, Frey KE, Garley R, Laney SR, Lubelczyk L, Mathis J, Matsuoka A, Mitchell BG, Moore GWK, Ortega-Retuerta E, Pal S, Polashenski CM, Reynolds R, Schieber B, Sosik H, Stephens M, Swift JH (2012) Massive phytoplankton blooms under Arctic sea ice. Science 336:1408

Ashjian CJ, Braund SR, Campbell RG, George JC, Kruse J, Maslowski W, Moore SE, Nicolson CR, Okkonen SR, Sherr BF, Sherr EB, Spitz Y (2010) Climate variability, oceanography, bowhead whale distribution, and Iñupiat subsistence whaling near Barrow AK. Arctic 63:179–194

Berchok CL, Clapham PJ, Crance JL, Moore SE, Napp J, Overland J, Wang M, Stabeno P, Guerra M, Clark CW (2014) Passive acoustic detection and monitoring of endangered whales in the Arctic (Beaufort, Chukchi) & Ecosystem observations in the Chukchi Sea: biophysical moorings and climate modeling. 2013 Annual Report, BOEM/NOAA Interagency Agreement Number M09PG00016. National Marine Mammal Laboratory, Alaska Fisheries Science Center, NOAA, 14 p

Bluhm BA, Iken K, Hopcroft RR (2010) Observations and exploration of the Arctic's Canada Basin and the Chukchi Sea: the Hidden Ocean and RUSALCA expeditions. Deep-Sea Res 57:1–4

Cai W-J, Guo L, Grebmeier JM, Chen L, Zhang H (2012) Biogeochemical studies from the Chinese National Arctic Research Expeditions. Deep-Sea Res 81–84:1–2

Cai W-J, Bates NR, Guo L, Anderson LG, Mathis JT, Wanninkhof R, Hansell DA, Chen L, Igor Semiletov IP (2014) Chapter 8: Carbon fluxes across boundaries in the Pacific Arctic region in a changing environment. In: Grebmeier JM, Maslowski W (eds) The Pacific Arctic region: ecosystem status and trends in a rapidly changing environment. Springer, Dordrecht, pp 199–222

Carmack EC, McLaughlin FA, Vagle S, Melling H, Williams WJ (2010) Structures and property distributions in the three oceans surrounding Canada in 2007: a basis for a long-term ocean climate monitoring strategy. Atmos Ocean 48:211–224

Clement Kinney J, Maslowski W, Aksenov Y, de Cuevas B, Jakacki J, Nguyen A, Osinski R, Steele M, Woodgate RA, Zhang J (2014) Chapter 7: On the flow through Bering Strait: a synthesis of model results and observations. In: Grebmeier JM, Maslowski W (eds) The Pacific Arctic region: ecosystem status and trends in a rapidly changing environment. Springer, Dordrecht, pp 167–198

Comiso JC (2012) Large decadal decline of the Arctic multiyear ice cover. J Clim 25:1176–1193

Cooper LW, Whitledge TT, Grebmeier JG, Weingartner T (1997) Nutrient, salinity and stable oxygen isotope composition of Bering and Chukchi Sea in and around the Bering Strait. J Geophys Res 102:12563–12574

Day RH, Weingartner TJ, Hopcroft RR, Aerts LAM, Blanchard AL, Gall AE, Gallaway BJ, Hannay DE, Holladay BA, Mathis JT, Norcross BL, Questel JM, Wisdom SS (2013) The offshore northeastern Chukchi Sea: a complex high-latitude system. Cont Shelf Res 67:147–165

Deal CJ, Steiner N, Christian J, Clement Kinney J, Denman K, Elliott S, Gibson G, Meibing J, Lavoie D, Lee S, Lee W, Maslowski W, Wang J, Watanabe E (2014) Chapter 12: Progress and challenges in biogeochemical modeling of the Pacific Arctic region. In: Grebmeier JM, Maslowski W (eds) The Pacific Arctic region: ecosystem status and trends in a rapidly changing environment. Springer, Dordrecht, pp 393–446

Dunton KE, Grebmeier JM, Trefry JH (2014) The benthic ecosystem of the northeastern Chukchi Sea: an overview of its unique biogeochemistry and biological characteristics. Deep-Sea Res II 102:1–8

Frey KE, Maslanik JA, Clement Kinney J, Maslowski W (2014) Recent variability in sea ice cover, age, and thickness in the Pacific Arctic region. In: Grebmeier JM, Maslowski W (eds) The Pacific Arctic region: ecosystem status and trends in a rapidly changing environment. Springer, Dordrecht, pp 31–64

Grebmeier JM (2012) Shifting patterns of life in the Pacific Arctic and Sub-Arctic seas. Annu Rev Mar Sci 4:63–78

Grebmeier JM, Cooper LW, Feder HM, Sirenko BI (2006a) Ecosystem dynamics of the Pacific-influenced Northern Bering and Chukchi seas in the Amerasian Arctic. Prog Oceanogr 71:331–361

Grebmeier JM, Overland JE, Moore SE, Farley EV, Carmack EC, Cooper LW, Frey KE, Helle JH, McLoughlin FA, McNutt L (2006b) A major ecosystem shift observed in the northern Bering Sea. Science 311:1461–1464

Grebmeier JM, Moore SE, Overland JE, Frey KE, Gradinger R (2010) Biological response to recent Pacific Arctic sea ice retreats. EOS Trans Am Geophys Union 91(18):161–162

Maslowski W, Clement Kinney J, Higgins Roberts AF (2012) The future of Arctic sea ice. Annu Rev Earth Planet Sci 40:625–654

Maslowski W, Clement Kinney J, Okkonen SR, Osinski R, Roberts AF, Williams W (2014) Chapter 5: The large scale ocean circulation and physical processes controlling Pacific-Arctic interactions. In: Grebmeier JM, Maslowski W (eds) The Pacific Arctic region: ecosystem status and trends in a rapidly changing environment. Springer, Dordrecht, pp 101–132

Mathis JT, Hansell DA, Kadko D, Bates NR, Cooper LW (2007) Determining net dissolved organic carbon production in the hydrographically complex western Arctic Ocean. Limnol Oceanogr 52:1789–1799

Mathis JT, Grebmeier JM, Hansell DA, Hopcroft RR, Kirchman DL, Lee SH, Moran SB, Bates NR, VanLaningham S, Cross JN, Cai W-J (2014) Chapter 9: Carbon biogeochemistry of the western Arctic: primary production, carbon export and the controls on ocean acidification. In: Grebmeier JM, Maslowski W (eds) The Pacific Arctic region: ecosystem status and trends in a rapidly changing environment. Springer, Dordrecht, pp 223–268

Moore SE (2008) Marine mammals as ecosystem sentinels. J Mammal 89(3):534–540

Moore SE, Grebmeier JM, Davies JR (2003) Gray whale distribution relative to forage habitat in the northern Bering Sea: current conditions and retrospective summary. Can J Zool 81:734–742. doi:10.1139/Z03-043

Moore SE, Logerwell E, Eisner L, Farley E, Harwood L, Kuletz K, Lovvorn J, Murphy J, Quakenbush L (2014) Chapter 11: Marine fishes, birds and mammals as sentinels of ecosystem variability and reorganization in the Pacific Arctic region. In: Grebmeier JM, Maslowski W (eds) The Pacific Arctic region: ecosystem status and trends in a rapidly changing environment. Springer, Dordrecht, pp 337–392

Nelson RJ, Ashjian C, Bluhm B, Conlan K, Gradinger R, Grebmeier J, Hill V, Hopcroft R, Hunt B, Joo H, Kirchman D, Kosobokova K, Lee S, Li WKW, Lovejoy C, Poulin M, Sherr E, Young K (2014) Chapter 10: Biodiversity and biogeography of the lower trophic taxa of the Pacific Arctic region: sensitivities to climate change. In: Grebmeier JM, Maslowski W (eds) The Pacific Arctic region: ecosystem status and trends in a rapidly changing environment. Springer, Dordrecht, pp 269–336

Overland JE, Stabeno PJ (2004) Is the climate of the Bering Sea warming and affecting the ecosystem? Eos Trans Am Geophys Union 85:309–316

Overland JE, Wang J, Pickart RS, Wang M (2014) Chapter 2: Recent and future changes in the meteorology of the Pacific Arctic. In: Grebmeier JM, Maslowski W (eds) The Pacific Arctic region: ecosystem status and trends in a rapidly changing environment. Springer, Dordrecht, pp 17–30

Pickart RS, Spall MA, Moore GWK, Weingartner TJ, Woodgate RA, Aagaard K, Shimada K (2011) Upwelling in the Alaskan Beaufort Sea: atmospheric forcing and local versus on-local response. Prog Oceanogr 88:78–100. doi:10.1016/j.pocean.2010.11.005

Schulze LM, Pickart RS (2012) Seasonal variation of upwelling in the Alaskan Beaufort Sea: impact of sea ice cover. J Geophys Res 117:C06022. doi:10.1029/2012JC007985

Shimada K, Kamoshida T, Itoh M, Nishino S, Carmack E, McLaughlin F, Zimmermann S, Proshutinsky A (2006) Pacific Ocean inflow: influence on catastrophic reduction of sea ice cover in the Arctic Ocean. Geophys Res Lett 33:L08605

Stroeve J, Holland MM, Meier W, Scrambos T, Serreze M (2007) Arctic sea ice decline: faster than forecast. Geophys Res Lett 32, L02602. doi:10.1029/2004GL021747

Stroeve JC, Serreze MC, Holland MM, Kay JE, Maslanik J, Barrett AP (2012) The Arctic's rapidly shrinking sea ice cover: a research synthesis. Clim Chang 110:1005–1027

Walczowski W, Piechura J, Goszczko I, Wieczorek P (2012) Changes in Atlantic water properties: an important factor in the European Arctic marine climate. ICES J Mar Sci 69(5):864–869. doi:10.1093/icesjms/fss068

Walsh JJ, Dieterle DA, Maslowski W, Whitledge TE (2004) Decadal shifts in biophysical forcing of Arctic marine food webs: numerical consequences. J Geophys Res 109:C05031. doi:10.1029/2003JC001945

Wang J, Mizobata K, Hu H, Jin M, Zhang S, Johnson W, Shimada K (2008) Modeling seasonal variations of ocean and sea ice circulation in the Beaufort and Chukchi Seas: a model-data fusion study. Chin J Polar Sci 19(2):168–184

Wang J et al (2014) Chapter 4: Abrupt climate changes and emerging ice-ocean processes in the Pacific Arctic region and the Bering Sea. In: Grebmeier JM, Maslowski W (eds) The Pacific Arctic region: ecosystem status and trends in a rapidly changing environment. Springer, Dordrecht, pp 65–100

Weise FK, Wiseman WJ, Van Pelt TI (2012) Bering sea linkages. Deep-Sea Res 65–70:2–5

Williams W, Shroyer E, Clement Kinney J, Itoh M, Maslowski W (2014) Chapter 6: Shelf-break exchange in the Bering, Chukchi and Beaufort seas. In: Grebmeier JM, Maslowski W (eds) The Pacific Arctic region: ecosystem status and trends in a rapidly changing environment. Springer, Dordrecht, pp 133–166

Woodgate RA, Aagaard K (2005) Revising the Bering Strait freshwater flux into the Arctic Ocean. Geophys Res Lett 32:L02602. doi:10.1029/2004GL021747

Woodgate RA, Weingartner TJ, Lindsay R (2012) Observed increases in Bering Strait oceanic fluxes from the Pacific to the Arctic from 2001 to 2011 and their impacts on the arctic Ocean water column. Geophys Res Lett 39:L24603. doi:10.1029/2012GL054092

Chapter 2
Recent and Future Changes in the Meteorology of the Pacific Arctic

James E. Overland, Jia Wang, Robert S. Pickart, and Muyin Wang

Abstract The meteorology of the Pacific Arctic (the Bering Sea through the Chukchi and southern Beaufort Seas) represents the transition zone between the moist and relatively warm maritime air mass of the Pacific Ocean to the cold and relatively dry air mass of the Arctic. The annual cycle is the dominant feature shifting from near total darkness with extensive sea ice cover in winter to solar heating in summer that is equal to that of sub-tropical latitudes. Strong north-south gradients in air temperatures and sea level pressure are typical over the Pacific Arctic giving rise to climatological polar easterly winds (blowing from the east) throughout the year. Localized storms (regions of low sea level pressure) can propagate into the region from the south but high pressure regions are typical, connected to either northeastern Siberia or the southern Beaufort Sea. The northern portion of the Pacific Arctic has participated in the general Arctic-wide warming in all seasons over the last decade while the southern Bering Sea turned to near record cold temperatures after 2006. Future climate changes in the Pacific Arctic will come from shifts in the timing and extent of seasonal sea ice. Based on climate model projections, cold and dark conditions will still dominate over a climate warming scenario in the Bering Sea of +2 °C by 2050. The northern Bering Sea will continue to have

J.E. Overland (✉)
Pacific Marine Environmental Laboratory, National Oceanic and Atmospheric Administration, 7600 Sand Point Way NE, Seattle, WA 98115, USA
e-mail: james.e.overland@noaa.gov

J. Wang
Great Lakes Environmental Research Laboratory (GLERL), National Oceanic and Atmospheric Administration, Ann Arbor, MI, USA
e-mail: jia.wang@noaa.gov

R.S. Pickart
Woods Hole Oceanographic Institution, Woods Hole, MA 02543, USA

M. Wang
Joint Institute for the Study of the Atmosphere and Ocean, University of Washington, Seattle, WA 98195, USA

J.M. Grebmeier and W. Maslowski (eds.), *The Pacific Arctic Region: Ecosystem Status and Trends in a Rapidly Changing Environment*, DOI 10.1007/978-94-017-8863-2_2,

extensive sea ice January through April, while the southern shelf will have on average less sea ice than currently observed but with large interannual variability. The largest change has the southern Chukchi Sea shifting from being sea ice free in September and October at present to becoming sea ice free for 5 months from July through November within a decade or two, impacting shipping, oil exploration, and ecosystems.

Keywords Arctic meteorology • Siberian high • Aleutian low • Arctic Oscillation • Climate change • Sea ice • Bering Sea • Chukchi Sea

2.1 Introduction

In this atmospheric review, the Pacific Arctic covers the area from the Bering Sea south of St. Lawrence Island north through Bering Strait to the southern Chukchi and Beaufort Seas. This is the transition zone between the relatively warm and moist storm tracks of the Aleutian low weather system in the south to the cold, dry and high pressure Arctic air mass to the north. This is a region of large north-south gradients in atmospheric properties such as air temperature and atmospheric sea level pressure (SLP). The region of strongest gradients moves north and south with the seasonal cycle. Maximum temperature gradients in winter are over the central Bering Sea with sub-freezing temperatures and sea ice coverage. In summer, the strongest air temperature gradients are across the Chukchi Sea and along the north slope and seaward of the coast of Alaska with temperatures above freezing. Large gradients in SLP produce an east-west region of strong wind speeds in all seasons; for example the annual mean surface wind speed at Barrow, Alaska for 1972–2007 is 5.6 m s^{-1} (Wendler et al. 2010).

The causes for the meteorology of the Pacific Arctic region are a seasonal swing from a large heat loss in winter and the dominating presence of sea ice to a gain of heat in summer. The primary determinant of this seasonal climate shift is the annual cycle of insolation from a maximum of 500 W m^{-2} near the summer solstice to darkness in winter. As summer progresses, absorption of insolation at the ocean surface increases as the albedo decreases due to snow and ice melt and increased open water areas. This annual cycle results in a change from a winter continental-like air mass similar to the adjacent land areas to a summertime marine air mass characterized by low clouds and fog.

In the recent decade, the Pacific air mass to the south and the Arctic air mass to the north appear to be on different climate change trajectories (Overland et al. 2012). To the north, the region is part of the decadal change of Arctic warming where recent sea ice losses are changing the climatology of the region, with periods of warm temperature anomalies extending through the autumn months. To the south, the Bering Sea has turned cold with extensive sea ice cover in 2007–2011, extents that had not been seen since the mid-1970s. This contrasts to a warmer than normal temperature anomaly period for the southern Bering Sea from 2001 through

2005. This north-south decoupling is due to the northern region being dominated by the thermodynamics of melting sea ice, while changes in the south are dominated by changes in North Pacific storm tracks and large-scale Pacific climate.

2.2 Climatological Fields

The basic driver of Pacific Arctic meteorology is the annual cycle of solar insolation (Fig. 2.1). Although the sun is close to the horizon during summer, the 24 h of daylight provides the same potential solar radiation (i.e., without cloud effects) in the Arctic as occurs at lower latitudes in the northern hemisphere. North of 60°N, there is substantial solar heating from mid-April through August.

Climatologically, in terms of surface air temperature (SAT) and SLP gradients, the Pacific Arctic can be considered to have two main seasons: an extended 9-month winter (September through May) and a summer (June through August). This is demonstrated by examining the SAT fields for standard seasons, fall (Sept–Nov), winter (Dec–Feb), spring (Mar–May) and summer (Jun–Aug), presented in Fig. 2.2. Strong north-south temperature gradients are present fall through spring from northern Alaska southward through the southeastern Bering Sea. In summer, more efficient heating of the continental land masses changes the orientations of the isotherms so that a sea-land temperature gradient occurs over the coastal regions of northwestern Alaska and eastern Siberia. For the SLP fields (Fig. 2.3), during fall through spring, the Aleutian low pressure center is dominant over southwestern Alaska, with the strongest pressure gradient occurring in the vicinity of Bering Strait. The strength of SLP gradient over the Pacific Arctic is greatest in

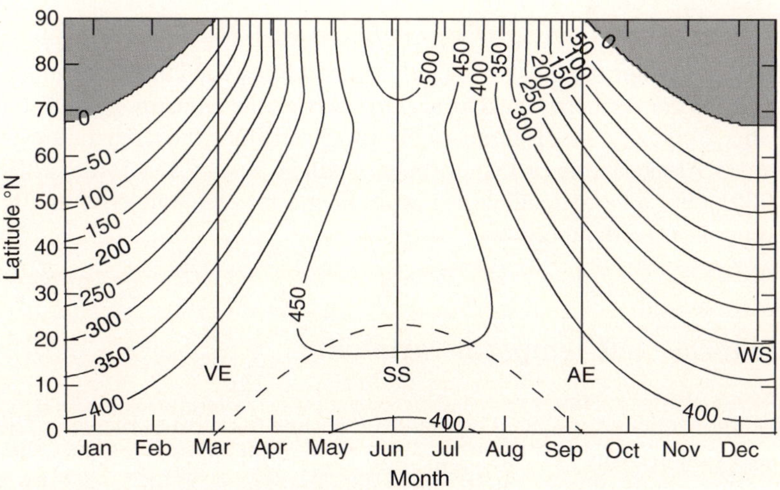

Fig. 2.1 The daily variation of solar insolation at the top of the atmosphere as a function of latitude and day of year in units of Wm^{-2} for the Northern Hemisphere (Modified from Liou 2002)

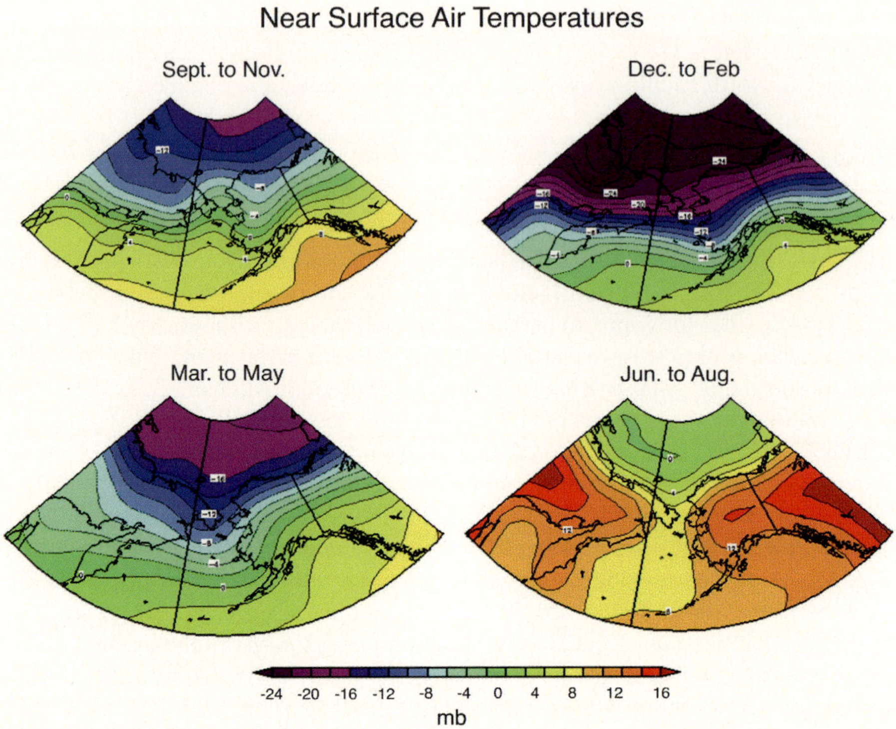

Fig. 2.2 Mean (1961–2010) near-surface temperature (°C) for the four seasons over the Pacific Arctic (Data are from the NCEP–NCAR Reanalysis through the NOAA/Earth Systems Research Laboratory, generated online at: http://www.esrl.noaa.gov/psd/cgi-bin/data/composites/printpage.pl)

winter. Climatological winds are easterly (from the east) and their magnitude drops off rapidly in the northern Chukchi Sea. In summer, the climatological SLP field is nearly flat. Typically, small storms can be produced along the temperature gradient in northern Alaska and these can drift out into the Chukchi/Beaufort Seas (Serreze et al. 2001). For a more detailed review of the marine climatology of the Beaufort Sea proper see Overland (2009).

2.3 Storms and Temporal Variability

The Pacific Arctic is dominated by high SLP regions for most of the year. These are large and slow-moving systems that persist for multiple days. The center of one of these systems is the Siberian high, far to the west. However, the edge of the high, where strong pressure gradients exist, can reside over the Pacific Arctic region, producing strong and persistent northwesterly or northeasterly winds. The other

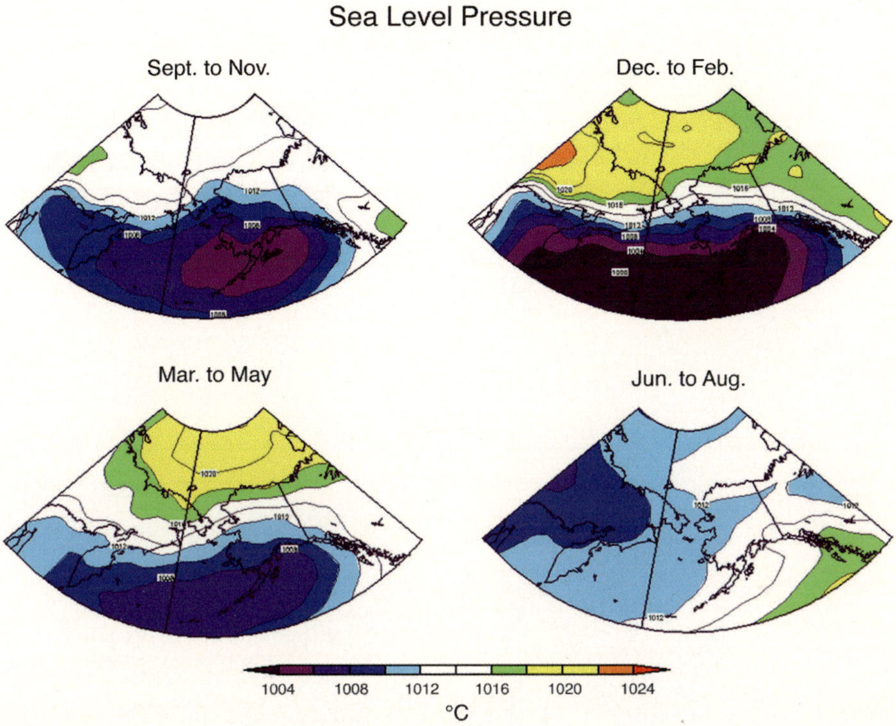

Fig. 2.3 Mean (1961–2010) sea level pressure (*SLP*) for the four seasons over the Pacific Arctic (Data are from the NCEP–NCAR Reanalysis through the NOAA/Earth Systems Research Laboratory, generated online at: http://www.esrl.noaa.gov/psd/cgi-bin/data/composites/printpage.pl)

high pressure center is over the Canadian Beaufort Sea. If there is very low pressure to the south of the region in the form of Aleutian low storm systems, there can be stronger than average easterly winds across northern Alaska and adjacent waters. These east wind events can cause coastal ocean upwelling events across the narrow northern Alaskan continental shelf (Nikolopoulos et al. 2009).

Storm track trajectories originating in the northwest Pacific and moving into the Pacific Arctic are generally of two types (Wang et al. 2004; Pickart et al. 2009). One type exhibits a west-to-east trajectory south of the region and may increase the north-south pressure gradient as the low pressure region passes to the south of the more permanent higher pressure zones to the north (Fig. 2.4). The second trajectory type curves northward from the south into the central Bering Sea. Conventional wisdom at the Anchorage NOAA Weather Service is that a sequence of storms on this latter trajectory would continue to progress farther north displacing the northern high pressure centers. The first storm would make it to the central Bering Sea, the second storm would intrude to north of Bering Strait, and the third would transit Bering Strait and then curve eastward along the northern slope of Alaska.

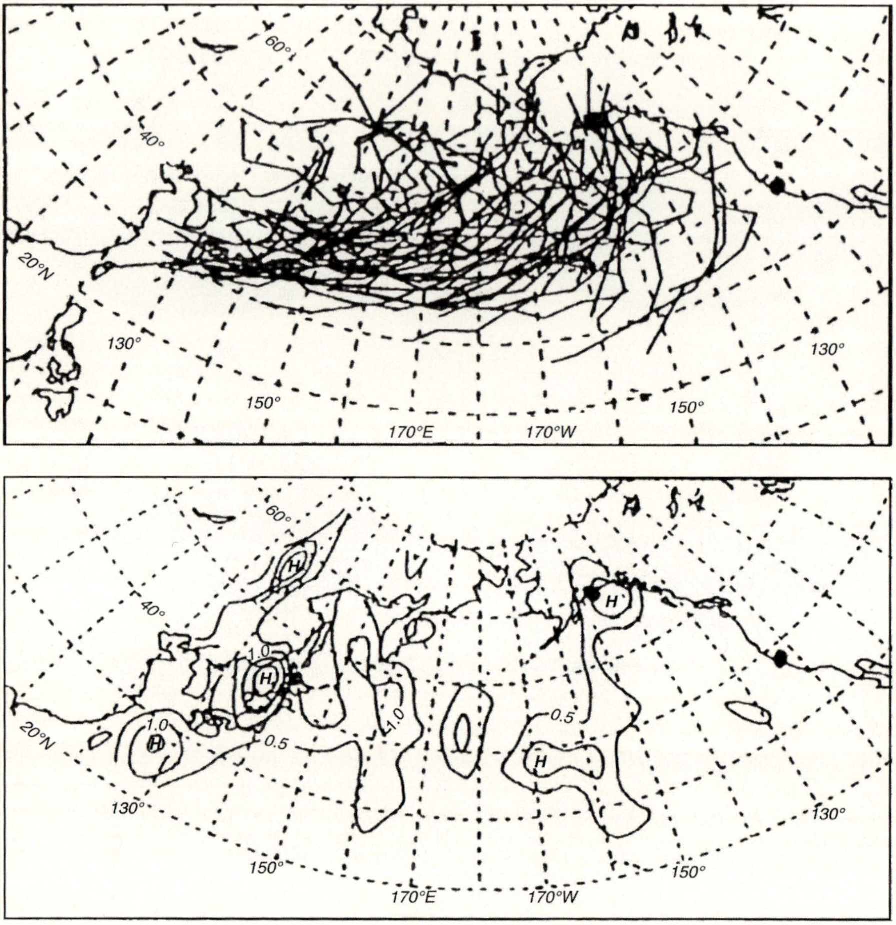

Fig. 2.4 Cyclone (storm) tracks during a typical wet season (October 1979–May 1980). A total of 57 cyclones were found during this period (*upper panel*). Cyclogenesis (storm development) location distribution is also shown for January 1980 (*lower panel*, units are in cyclones/month, with interval being 0.05) (After Wang et al. 2004)

The lower troposphere of the Pacific Arctic (eastern Siberia to northern Canada) was relatively warm during spring in the 1990s relative to the previous four decades (Overland et al. 2002). The primary difference in the 1990s was the presence of several highly episodic springtime storms and associated advection of heat compared to nearly storm-free periods in previous years. Krupnik and Jolly (2002) note that native elders in the region suggested that the weather was less dependable (predictable) starting in the 1990s. The addition of a few storms that propagated further north may be a cause of this diminished predictability relative to previous persistent easterly winds.

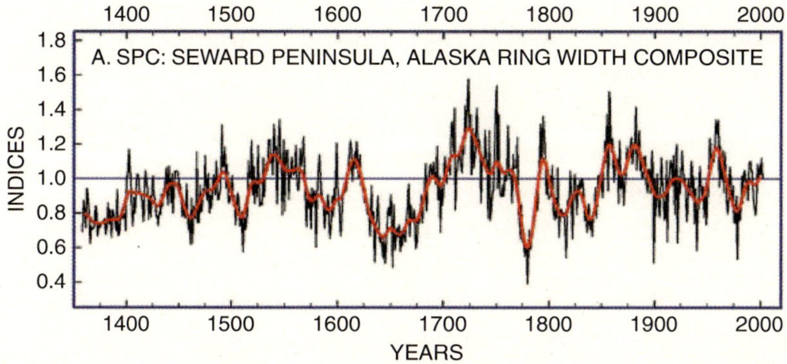

Fig. 2.5 Ring-width composite chronology extending from 1358 to 2001 A.D. A 25-year smoothing spline (*red line*) has been superimposed on this record to emphasize multi-decadal-scale fluctuations (After D'Arrigo et al. 2005)

Such changes, however, may have begun as early as the late 1960s or at least by the well-known Alaskan regime shift in the 1970s (Wang et al. 2006; Wendler et al. 2010). Before the 1960s, Barrow and Nome were dominated by Arctic air masses and St. Paul was dominated by North Pacific maritime air masses. After the 1960s, the surface air temperature correlation in winter between Barrow and St. Paul increased from 0.2 to 0.7 and between Nome and St. Paul from 0.4 to 0.8, implying greater north–south penetration of both Arctic and Pacific air masses. Relatively stable, high correlations are found among the stations in the fall, whereas correlations are low in the summer. These climatological results support the concept that the southeast Bering Sea ecosystem may have been dominated by Arctic species for most of the twentieth century, with a gradual replacement by sub-Arctic species, such as pollock, in the last 30 years (Aydin et al. 2007).

The opening date for the Prudhoe Bay shipping season depends on antecedent sea ice and weather conditions in the Bering Sea (Drobot et al. 2009). In years with early opening dates, the sea-ice cover in the southern Bering Sea is reduced in February and, as the season progresses, sea-ice concentrations in the central and northern Bering Sea remain low. Further, fewer accumulated freezing degree days (FDDs) suggests that temperatures are warmer over a broad area, ranging from the Bering Sea through the Chukchi Sea and the Beaufort Sea, in winter and spring months preceding early opening dates.

Based on tree ring widths of 14 white spruce chronologies for the Seward Peninsula, Alaska, some historical changes can be inferred for the Pacific Arctic back to 1400 (D'Arrigo et al. 2005). The chronologies correlate significantly with Bering and Chukchi Sea sea surface temperatures and with the Pacific Decadal Oscillation index weighted towards the spring and summer months. There is inferred cooling during periods within the Little Ice Age (LIA) from the early-to-middle 1600s and from the early-to-late 1700s, and warming from the middle 1600s to early 1700s (Fig. 2.5). The chronologies imply that growth conditions were somewhat below average until the middle of the twentieth century. Following a brief period of increasing growth in the 1950s, there is a decline, with some growth recovery in the 1990s.

2.4 The Differences of the Pacific Sector Relative to the Larger Arctic System

The largest climate signal for the Pacific Arctic in the recent decade is the apparent decoupling of trends in the marine air masses to the south and Arctic air masses. Major temperature increases are seen in the Arctic related to sea ice loss in the summer and subsequent positive temperature anomalies in the fall. In contrast, southern Bering Sea warming in 2001–2005 and subsequent cooling 2007–2012 are tied to sea surface temperature shifts in the North Pacific. A long-term, vertically-averaged ocean temperature record from the M2 mooring (Stabeno et al. 2012; Fig. 2.6) on the southeastern Bering shelf illustrate these warming and cooling trends. The annual average temperature anomaly chart (Fig. 2.7) for 2006 through 2012 shows extensive large positive temperature anomalies over much of the Arctic, with a cooling in the southeastern Bering Sea. The large-scale warming of the Arctic is evident in all months north of St. Lawrence Island, but is most pronounced in the autumn.

The warming in the Arctic appears to be associated with the recent predominance of a climate pattern referred to as the Arctic Dipole (AD), characterized by low SLP on one side of the Arctic and high SLP on the other (Overland and Wang 2005; Wang et al. 2009; Zhang et al. 2008). On a hemispheric basis, the AD occurs as the third most prominent pattern after the Arctic Oscillation (AO) pattern, which has low SLP over the central Arctic in its positive phase, and the Pacific North-American pattern (Overland et al. 2008). While the AD was present in spring as early as 1997, its recent occurrence began in summer 2007 when it was present in all months and contributed to 2007 record minimum summer sea ice extent (Fig. 2.8). Each year after 2007 has also seen the AD pattern for part of the summer. For example, in 2010, the AD pattern was present in May and June, but then the Arctic reverted to the more traditional climatological SLP pattern for summer of a weak, relatively small low-pressure center. However, by August 2010 the AD pattern had returned.

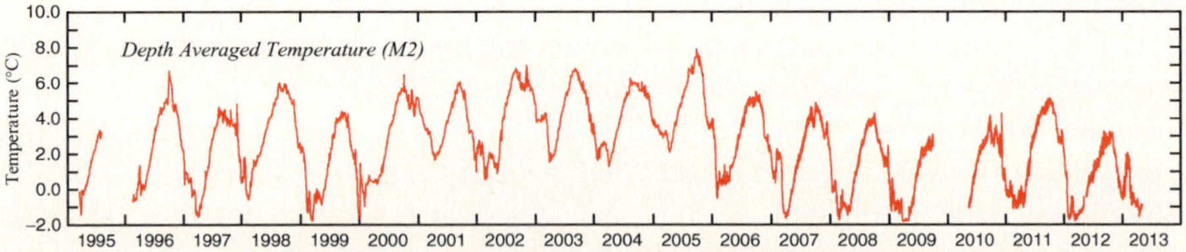

Fig. 2.6 Depth averaged ocean temperature at the M2 mooring location on the southeastern Bering Sea shelf (Updated from Stabeno et al. 2012)

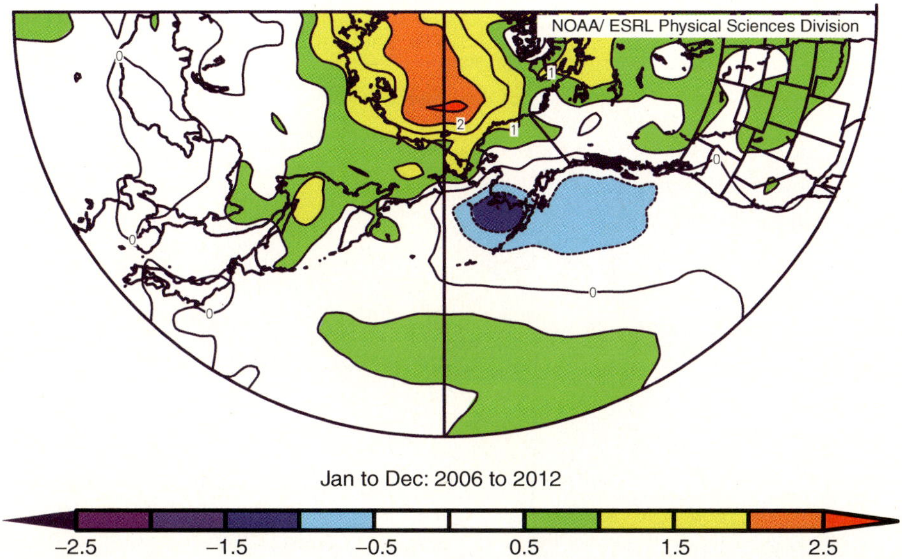

Fig. 2.7 Near surface air temperature anomaly multiyear composites (°C) for 2006–2012. Anomalies are relative to 1981–2010 mean and show a strong Arctic amplification of recent temperature trends and cooling of the southeastern Bering Sea (Data are from the NCEP–NCAR Reanalysis through the NOAA/Earth Systems Research Laboratory, generated online at: http://www.esrl.noaa.gov/psd/cgi-bin/data/composites/printpage.pl)

2.5 The Future Climate of the Pacific Arctic

The anthropogenic contribution to a warming Arctic is estimated to be 4–6 °C by 2080 (Chapman and Walsh 2007). Climate models also predict that the Arctic Ocean will be nearly sea ice free during the summer before the second half of this century, but recent data on thinning of Arctic sea ice suggest that the loss of Arctic summer sea ice may occur much sooner, within a decade or so (Stroeve et al. 2008; Wang and Overland 2009, 2012; Overland and Wang 2013). The Pacific Arctic is essentially sea ice free in summer at the present time. Such an occurrence will be limited to summer, since during the cold season, the Arctic Ocean, with its lack of sun light, will continue to be an ice covered sea.

As the Bering Sea is sea ice free in summer, it is characterized not by summer minima, but rather by winter/spring sea ice maxima. During the winters and springs of 2008–2013, there were near record maximum sea-ice extents in the southeastern Bering Sea. This contrasts with the record 2007–2012 summer Arctic sea ice minima and suggests a lack of continuity or "decoupling" between summer sea ice minima in the Arctic and the eventual winter/spring sea-ice maxima in the Bering Sea for the following season. This is a result of the main thermodynamic physics at high latitudes; it is cold and dark in winter. The northern Bering Sea (at 64°N) has only 4 h of daylight at the winter solstice. Mean monthly maximum temperatures at

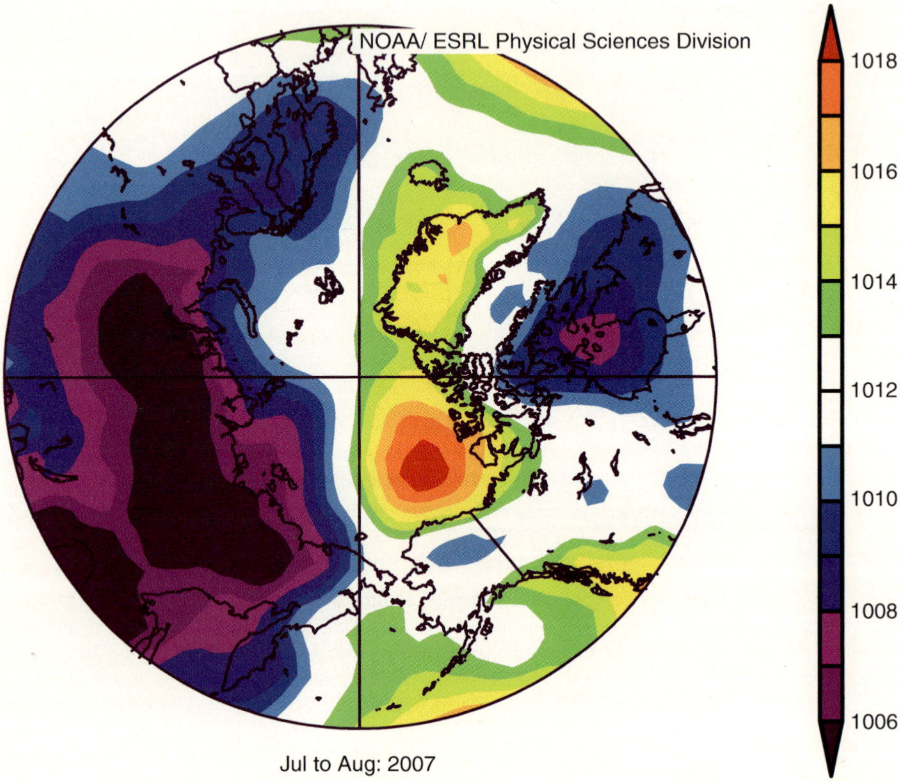

Fig. 2.8 Sea level pressure (*SLP*) field corresponding to the minimum sea ice extent in summer 2007 showing the Arctic Dipole (*AD*) pattern (Data are from the NCEP–NCAR Reanalysis through the NOAA/Earth Systems Research Laboratory, generated online at: www.cdc.noaa.gov)

Nome, Alaska, averaged from 1949 through 2006, are below −2.9 °C for all months between November and April indicating freezing conditions during these months. Mean maximum temperatures in these months will stay below freezing even with a hypothesized regional increased of 2 °C due to an overall global warming signal by 2050. Thus, winter and spring conditions in the western Arctic should not be dissimilar than current conditions for the foreseeable future.

We have some confidence for using climate models from the Intergovernmental Panel on Climate Change fourth Assessment Report (now referred to as CMIP3 and updated to CMIP5) to project April, May and November sea ice areas (Overland and Wang 2007). Figure 2.9 shows the observed area of sea ice cover (black lines) in the eastern Bering Sea for April and May for past years. May, in particular, has considerable year-to-year variability with extensive sea ice coverage in the mid-1970s. The blue line is the composite mean projection of 12 CMIP5 models (ACCESS1.0, ACCESS1.3, CCSM4, CESM1-CAM5.1, EC-EARTH, HadGEM2-AO, HadGEM2-CC, HadGEM2-ES, MIROC-ESM-CHEM, MIROC-ESM,

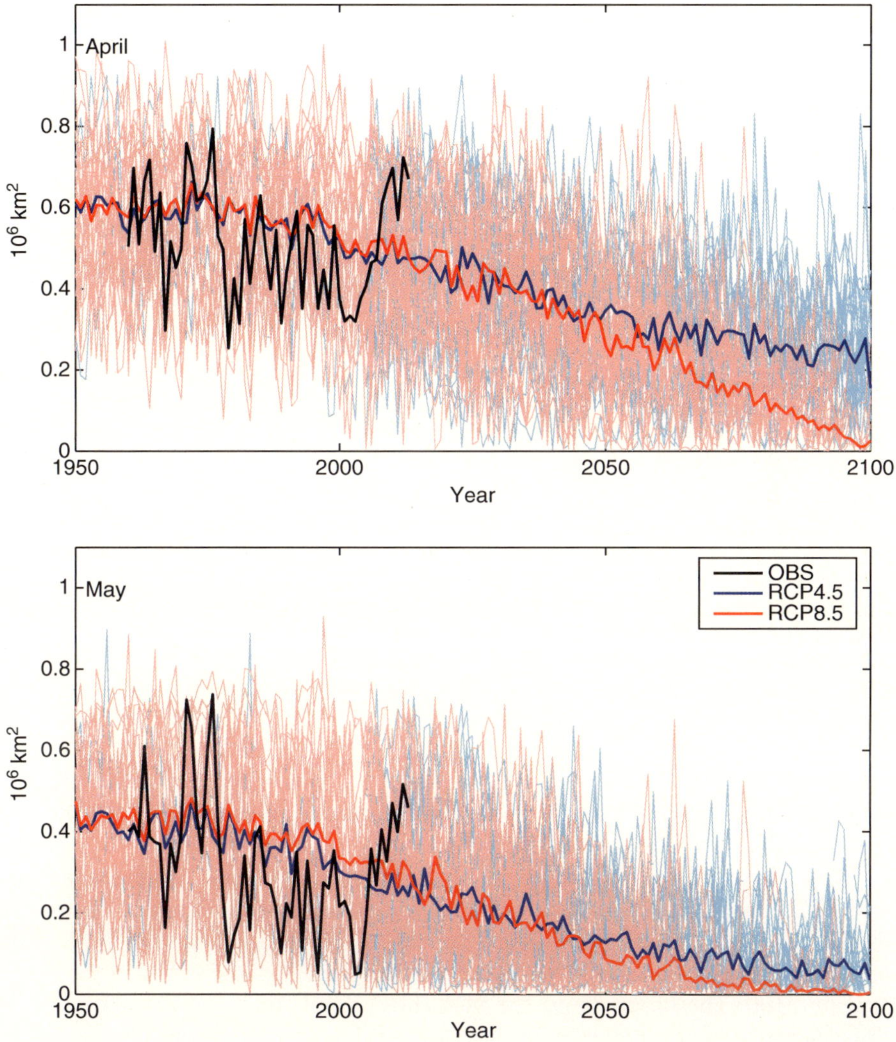

Fig. 2.9 April and May sea ice extent over the eastern Bering Sea based on HadleyISST_ice analysis (*black line*) and projections from CMIP5 models. *Thin colored lines* are individual model projections from 12 models that simulate the Arctic sea ice extent well in their historical simulations (Updated from Wang and Overland 2012). The *thick colored lines* are the ensemble mean value for each year averaged over these individual projections under RCP4.5 (*blue*) and RCP8.5 (*red*) emissions scenarios

MPI-ESM-LR, and MPI-ESM-MR.) which simulate well the mean and magnitude of seasonal cycle of sea ice coverage during 1981–2005 period. The thin blue/pink curves in the background represent multiple single runs (ensemble members) of the 12 selected models and suggest the possible range of future sea-ice areal coverage due to the influence of natural variability and to the anthropogenic

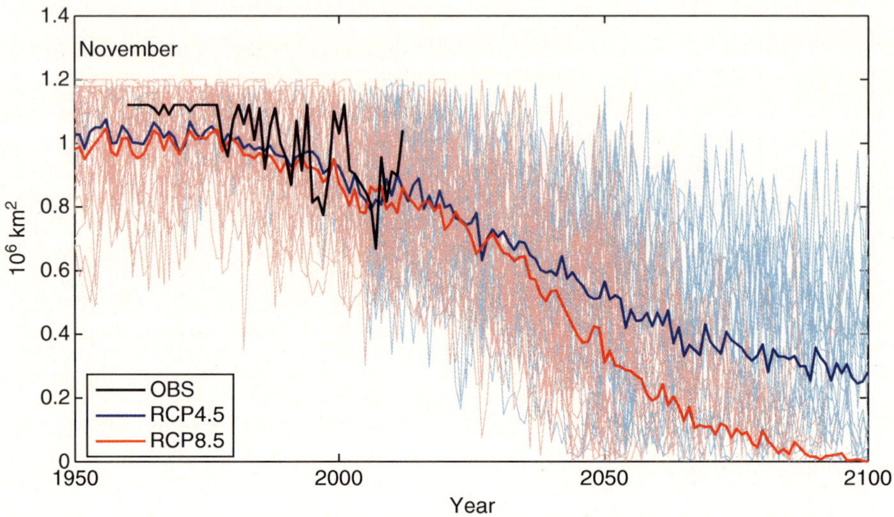

Fig. 2.10 November sea ice extent over the Chukchi Sea based on HadleyISST_ice analysis (*black line*) and projections from 12 CMIP5 models. *Light colored lines* are individual model projections from these models under RCP4.5 (*blue*) and RCP8.5 (*pink*) emissions scenarios. The *thick colored line* are the ensemble mean value for each year averaged over these individual projections under RCP4.5 (*blue*) and RCP8.5 (*red*) emissions scenarios

contribution represented by the thick blue and red lines under proposed emissions scenarios. These results from CMIP5 models show similar results to CMIP3 for the Pacific Arctic (Wang et al. 2012).

While there are measurable decreases in mean sea-ice areal estimates in April after 2010, the individual simulations (blue/pink lines) suggest that it is unlikely for the Bering Sea to be ice free in April during this century. In contrast, in May, the ensemble mean sea-ice area (blue line) decreases to 56 % of the 1960–1980 average area by 2100. The observations for May (black line) show many years in recent decades which approach a near-zero sea-ice areal coverage, but there are still individual model simulations from 2010 to 2050 where the sea-ice coverage exceeds the historical coverage from 1980 to the present. The trend toward minimal ice coverage can be expected to continue, but significant ice cover in May during any given year remains a distinct possibility through at least 2050. The main message from Fig. 2.9 is that large interannual variability in sea ice cover will continue to dominate the eastern Bering Sea in May for the next 40 years.

In contrast to spring sea ice conditions in the eastern Bering Sea, we project large changes in sea ice cover for the Chukchi Sea in autumn. Even for November, we see future reductions in sea ice coverage (Fig. 2.10). As in the Bering Sea, the range of individual ensemble members (blue/pink lines) is large, showing the uncertainty due to natural variability relative to the global warming signal. Loss of November sea ice cover has major implications for potential shipping and resource exploration, as well as overall ecosystem shifts including habitat changes for marine mammals. For the southern Chukchi Sea there is a shift from September and October being sea ice free at the present time to July through November being sea ice free within a decade or two.

2.6 Summary

The Pacific Arctic is not characterized by uniformity, but by its large north-south differences in marine habitats. Strong north-south gradients in air temperature and sea level pressure are established by latitudinal differences in the annual cycle of solar insolation. An anthropogenic temperature increase of 2 ° C by 2050 will have only a modest impact on the Bering Sea which will still be characterized by large interannual variability in addition to the weak upward temperature trend and a decreasing sea ice extent. Because of the unique thermodynamic influences of sea ice (albedo changes and ocean heat storage in newly sea-ice free areas), loss of autumnal sea ice will continue to provide major changes in the Chukchi Sea through November over the next decades with major increases in economic access and ecological impacts.

Acknowledgements We appreciate the support from NOAA Arctic Research of the NOAA Climate Program Office. This publication is partially funded by the Joint Institute for the Study of the Atmosphere and Ocean (JISAO) under NOAA Cooperative Agreement No. NA10OAR4320148 (M. Wang), Contribution No.2168. PMEL contribution #2679. We also like to thank EcoFOCI program of PMEL who maintains M2 mooring on the Bering Sea shelf.

References

Aydin K, Gaichas S, Ortiz I, Kinzey D, Friday N (2007) A comparison of the Bering Sea, Gulf of Alaska, and Aleutian Islands large marine ecosystems through food web modeling, NOAA technical memorandum 2007 NMFS-AFSC-178. U.S. Department of Commerce, Seattle, 298 pp

Chapman WL, Walsh JE (2007) Simulations of Arctic temperature and pressure by global coupled models. J Clim 20:609–632

D'Arrigo R, Mashig E, Frank D, Wilson R, Jacoby G (2005) Temperature variability over the past millennium inferred from Northwestern Alaska tree rings. Clim Dyn 24:227–236

Drobot SD, Maslanik JA, Anderson MR (2009) The Arctic frontal zone as seen in the NCEP–NCAR reanalysis. Int J Climatol 29:197–203

Krupnik I, Jolly D (2002) The earth is faster now. Arctic Research Consortium of the United States, Fairbanks

Liou KN (2002) An introduction to atmospheric radiation. Academic Press, Amsterdam/Boston

Nikolopoulos A, Pickart RS, Fratantoni PS, Shimada K, Torres DJ, Jones EP (2009) The western Arctic boundary current at 152 W: structure, variability, and transport. Deep Sea Res II Top Stud Oceanogr 56:1164–1181

Overland JE (2009) Meteorology of the Beaufort Sea. J Geophys Res 114:C00A07. doi:10.1029/2008JC004861

Overland JE, Wang M (2005) The third Arctic climate pattern: 1930s and early 2000s. Geophys Res Lett 32:L23808. doi:10.1029/2005GL024254
Overland JE, Wang M (2007) Future regional Arctic Sea ice declines. Geophys Res Lett 34:L17705. doi:10.1029/2007GL030808
Overland JE, Wang M (2013) When will the summer Arctic be nearly sea ice free? Geophys Res Lett 40:2097–2101. doi:10.1002/grl.50316
Overland JE, Bond NA, Adams JM (2002) The relation of surface forcing of the Bering Sea to large-scale climate patterns. Deep Sea Res II Top Stud Oceanogr 49:5855–5868
Overland JE, Wang M, Salo S (2008) The recent Arctic warm period. Tellus 60A:589–597
Overland JE, Wang M, Wood KR, Percival DB, Bond NA (2012) Recent Bering Sea warm and cold events in a 95-year perspective. Deep-Sea Res II 65–70:6–13. doi:10.1016/j.dsr2.2012.02.013
Pickart RS et al (2009) Seasonal evolution of Aleutian low pressure systems: implications for the North Pacific subpolar circulation. J Phys Oceanogr 39:1317–1339
Serreze MC, Lynch AH, Clark MP (2001) The Arctic frontal zone as seen in the NCEP–NCAR reanalysis. J Clim 14:1550–1567
Stabeno PJ et al (2012) Comparison of warm and cold years on the southeastern Bering Sea shelf and some implications for the ecosystem. Deep-Sea Res II 65–70:31–45. doi:10.1016/j.dsr2.2012.02.020
Stroeve JM, Serreze M, Drobot S, Gearheard S, Holland M, Maslanik J, Meier W, Scambos T (2008) Arctic sea ice extent plummets in 2007. EOS Trans Am Geophys Union 89:13–14
Wang M, Overland JE (2009) A sea ice free summer Arctic within 30 years? Geophys Res Lett 36:L07502. doi:10.1029/2009GL037820
Wang M, Overland JE (2012) A sea ice free summer Arctic within 30 years-an update from CMIP5 models. Geophys Res Lett 39:L18501. doi:10.1029/2012GL052868
Wang J, Jin M, Musgrave D, Ikeda M (2004) A numerical hydrological digital elevation model for freshwater discharge into the Gulf of Alaska. J Geophys Res 109:C07009. doi:10.1029/2002JC001430
Wang M, Overland JE, Percival D, Mofjeld HO (2006) Change in the Arctic influence on Bering Sea climate during the twentieth century. Int J Climatol 26:531–539
Wang J et al (2009) Is the Dipole Anomaly a major driver to record lows in Arctic summer sea ice extent? Geophys Res Lett 36:L05706. doi:10.1029/2008GL036706
Wang M, Overland JE, Stebeno P (2012) Future climate of the Bering and Chukchi Seas projected by global climate models. Deep-Sea Res II 65–70:46–57
Wendler G, Shulsk M, Moore B (2010) Changes in the climate of the Alaskan North Slope and the ice concentration of the adjacent Beaufort Sea. Theor Appl Climatol 99:67–74
Zhang X, Sorteberg A, Zhang J, Gerdes R, Comiso J (2008) Recent radical shifts in atmospheric circulations and rapid changes in Arctic climate system. Geophys Res Lett 35:L22701. doi:10.1029/2008GL035607

Chapter 3
Recent Variability in Sea Ice Cover, Age, and Thickness in the Pacific Arctic Region

Karen E. Frey, James A. Maslanik, Jaclyn Clement Kinney, and Wieslaw Maslowski

Abstract Over the past several decades, there has been a fundamental shift in sea ice cover, age, and thickness across the Pacific Arctic Region (PAR). Satellite data reveal that trends in sea ice cover have been spatially heterogeneous, with significant declines in the Chukchi Sea, slight declines in the Bering Strait region, yet increases in the northern Bering Sea south of St. Lawrence Island. Declines in the annual persistence of seasonal sea ice cover in the Chukchi Sea and Bering Strait region are due to both earlier sea ice breakup and later sea ice formation. However, increases in the persistence of seasonal sea ice cover south of St. Lawrence Island occur primarily owing to earlier sea ice formation during winter months. Satellite-based observations of sea ice age along with modeled sea ice thickness provide further insight into recent sea ice variability throughout the PAR, with widespread transitions towards younger, thinner ice. Investigation of sea ice cover, age, and thickness in concert provides critical insight into ongoing changes in the total volume of ice and therefore the future trajectory of sea ice throughout the PAR, as well as its likely impacts on ecosystem productivity across all trophic levels.

Keywords Sea ice • Bering Sea • Chukchi Sea • Sea ice age • Sea ice thickness • Satellite data

---

K.E. Frey (✉)
Graduate School of Geography, Clark University, Worcester, MA 01610, USA
e-mail: kfrey@clarku.edu

J.A. Maslanik
Colorado Center for Astrodynamics Research, University of Colorado, Boulder, CO 80309, USA

J. Clement Kinney • W. Maslowski
Department of Oceanography, Graduate School of Engineering and Applied Sciences, Naval Postgraduate School, Dyer Road, Bldg. SP339B, Monterey, CA 93943, USA
e-mail: jlclemen@nps.edu; maslowsk@nps.edu

J.M. Grebmeier and W. Maslowski (eds.), *The Pacific Arctic Region: Ecosystem Status and Trends in a Rapidly Changing Environment*, DOI 10.1007/978-94-017-8863-2_3,

3.1 Introduction

The Arctic Ocean has experienced significant warming (Zhang 2005; Polyakov et al. 2007; Steele et al. 2008) and dramatic declines in sea ice over the past few decades (Parkinson et al. 1999; Cavalieri et al. 2003; Serreze et al. 2003; Stroeve et al. 2005, 2008, 2012). These reductions in sea ice have facilitated a positive feedback through decreased albedo and enhanced absorption of solar insolation (e.g., Perovich et al. 2007), leading to model predictions of a near absence of summer sea ice by the year 2040 (Holland et al. 2006) and possibly sooner (Stroeve et al. 2007; Wang and Overland 2009). Although significant sea ice declines have been observed for all months, seasonal sea ice minima in September have been particularly striking (Stroeve et al. 2008; Perovich et al. 2011). Key factors in the declines in sea ice cover include long-term thinning trends of sea ice and replacement of thick, multi-year ice with thin, first-year ice (Nghiem et al. 2007; Maslanik et al. 2007a, 2011). These reductions are exacerbated by increased heat fluxes entering the Chukchi Sea through Bering Strait (Shimada et al. 2006; Woodgate et al. 2006), leading to ocean warming that in turn delays autumn sea ice re-growth (Steele et al. 2008). Reductions in Arctic sea ice cover have been most pronounced in the marginal seas of the Alaskan and Russian continental shelves, including areas in the Pacific Arctic Region (PAR) (Steele et al. 2008; Stroeve et al. 2012) that are the focus of the chapters in this book.

In the context of the overriding theme of this book, it is important to consider the potential impacts of recent sea ice variability on ecosystem productivity throughout the PAR. Sea ice melt and breakup during spring strongly impact primary production in the Arctic Ocean and its adjacent shelf seas by enhancing light availability as well as increasing stratification and stabilization of the water column. Recently observed declines in sea ice extent, thickness, and annual persistence should therefore have important consequences for primary production throughout the PAR. Recent studies indeed document significant increases in primary production in several sectors of the Arctic Ocean, in addition to significant shifts in the timing of phytoplankton blooms. For example, newly compiled satellite observations of primary production in the Arctic Ocean over a 12-year period (1998–2009) reveal a ~20 % overall increase, resulting primarily from increases in open water extent (+27 %) and duration of the open water season (+45 days) (Arrigo and van Dijken 2011). Enhanced light availability through increasingly melt-ponded sea ice surfaces (Frey et al. 2011) can also contribute to high levels of primary production underneath the ice (Arrigo et al. 2012). Furthermore, Kahru et al. (2010) found significant trends towards earlier phytoplankton blooms for ~11 % of the Arctic Ocean over the 1997–2009 period in areas roughly coincident with those experiencing earlier sea ice breakup.

Although general increases in primary production are predicted to accompany continuing losses in sea ice cover, shifts in primary production are expected to be spatially heterogeneous and dependent on several potentially confounding factors. For instance, models presented by Slagstad et al. (2011) suggest that while some

Arctic shelf areas may have significant increases in primary production with further sea ice declines, the deep central basin of the Arctic Ocean may see smaller increases in production owing to low nutrient concentrations, areas that lose all ice cover may see decreases in production owing to increased stratification with atmospheric warming, and some inner coastal shelves may see little increase in production owing to the enhanced turbidity from river runoff and coastal erosion. Empirically-based extrapolations presented by Arrigo and van Dijken (2011) suggest that when summer ice cover falls to zero, total annual primary production could reach ~730 Tg C year^{-1} total across the Arctic (a ~48 % increase over the 1998–2009 average). However, this value is highly dependent upon future distributions of nutrients, the extent of warming-induced enhanced stratification, and other limitations to primary production such as river-associated turbidity in coastal regions.

In this chapter, we investigate the variability and trends in sea ice physical characteristics across the PAR (Sect. 3.1) over the past several decades, providing a physical context for the ecosystem-centric chapters in this book. The PAR has experienced some of the most dramatic shifts in sea ice of any region across the pan-Arctic. Quantifying changes in sea ice cover, age, and thickness combined provides an assessment of overall trends sea ice volume, thereby garnering important insights into the potential future trajectory of sea ice conditions and potential impacts on ecosystem productivity. In Sect. 3.2 we focus on trends and recent interannual variability of sea ice cover (including the seasonal duration of ice cover, timing of breakup, and timing of formation) based on passive microwave satellite observations. In Sect. 3.3, we focus on changes in the age of sea ice derived from satellite observations blended with drifting-buoy vectors. In Sect. 3.4 we estimate changes in of sea ice thickness in the PAR based on a pan-Arctic coupled ice-ocean model. Lastly, in Sect. 3.5, we provide a discussion of the implications and possible future states of sea ice throughout the PAR.

3.2 Sea Ice Cover

3.2.1 Trends in Sea Ice Cover

The sea ice cover in the PAR is highly seasonally variable, with much of the northern Bering and Chukchi seas covered with first-year sea ice for several winter months of each year. Northern portions of the Chukchi Sea have also contained variable amounts of multiyear sea ice (i.e., ice that has survived at least one melt season) over the last several decades of the observed satellite record. In order to investigate trends in sea ice cover throughout the PAR over the last several decades, sea ice concentrations (spanning the years 1979–2008) derived from the Scanning Multichannel Microwave Radiometer (SMMR) and Special Sensor Microwave/ Imager (SSM/I) passive microwave instruments (Cavalieri et al. 2008) were utilized for this study. These data are available at a 25 km spatial resolution, where the

SSM/I data are daily and the SMMR data are available every other day (but were temporally interpolated to create a daily time series). As is standard in most studies, we used a 15 % ice concentration threshold to define the presence vs. absence of sea ice cover for calculations of sea ice extent.

Positions of the median sea ice edge for the last three decades (1979–1988, 1989–1998, and 1999–2008) for both March (seasonal sea ice maximum) and September (seasonal sea ice minimum) are shown in Sect. 3.1. Substantial shifts in sea ice cover throughout the PAR over the last 30 years of the satellite record are clearly evident. For March, the sea ice edge advanced slightly southward during 1989–1998 compared with its position during 1979–1998. However, the March ice edge during 1999–2008 was at its northernmost position of the three decades. In contrast, the position of the September ice edge has shifted dramatically and systematically northward over the last three decades. During 1979–1988, the median September edge was so far south that it barely retreated off the Siberian coast, was nearly coincident with the shelf break in the Beaufort Sea, and only partially exposed the Chukchi Sea shelf. Two decades later, during 1999–2008, the median September ice edge was positioned far north of the coasts, exposing the deep Arctic Ocean basin in the eastern PAR.

Trends in areal coverage of sea ice in the northern Bering and Chukchi Seas were also quantified over the satellite record from 1979 to 2008 (Figs. 3.2 and 3.3). In this case, the northern Bering Sea is defined as the shelf area north of St. Matthew Island to Bering Strait, whereas the Chukchi Sea is defined as the shelf area north of the Bering Strait between Wrangel Island and Barrow Canyon (Fig. 3.1). For the Bering Sea, statistically significant trends ($p<0.10$) are found only during the months of June, July, September, October, and November (Sect. 3.2, Table 3.1), with no trend in August owing to the lack of ice cover during that month. During winter and spring, however, sea ice cover in the northern Bering Sea has been highly variable and without trend (Fig. 3.2). In contrast, trends in sea ice cover in the Chukchi Sea have been far more dramatic and statistically significant for all months except April (Fig. 3.3, Table 3.1). During January, February, and March, slight increases in ice cover in the Chukchi Sea have been observed. For May through December, however, all trends show significant declines, where the greatest losses occurred during September: 63,840 km^2/decade or 31.8 %/decade.

Seasonal patterns and longer-term shifts in the spatial extent of sea ice were also compared over an annual time span. Daily averaged sea ice extent over 5-year increments exhibit extreme wintertime variability in the northern Bering Sea, with peak extent (~550,000 km^2) typically occurring during March and complete loss of seasonal sea ice by June–July (Sect. 3.4). Wintertime sea ice in the Chukchi Sea is comparably stable, with similar peak extent (~550,000 km^2) attained by December and spring breakup starting in May. In the Chukchi Sea, sea ice breakup has shifted earlier and ice formation has shifted later over the 30-year satellite record, with sea ice spatial extent substantially reduced during all non-winter months (Fig. 3.4).

The timing of sea ice breakup and formation is also important to consider, where we define the timing of these events as the day of year that sea ice

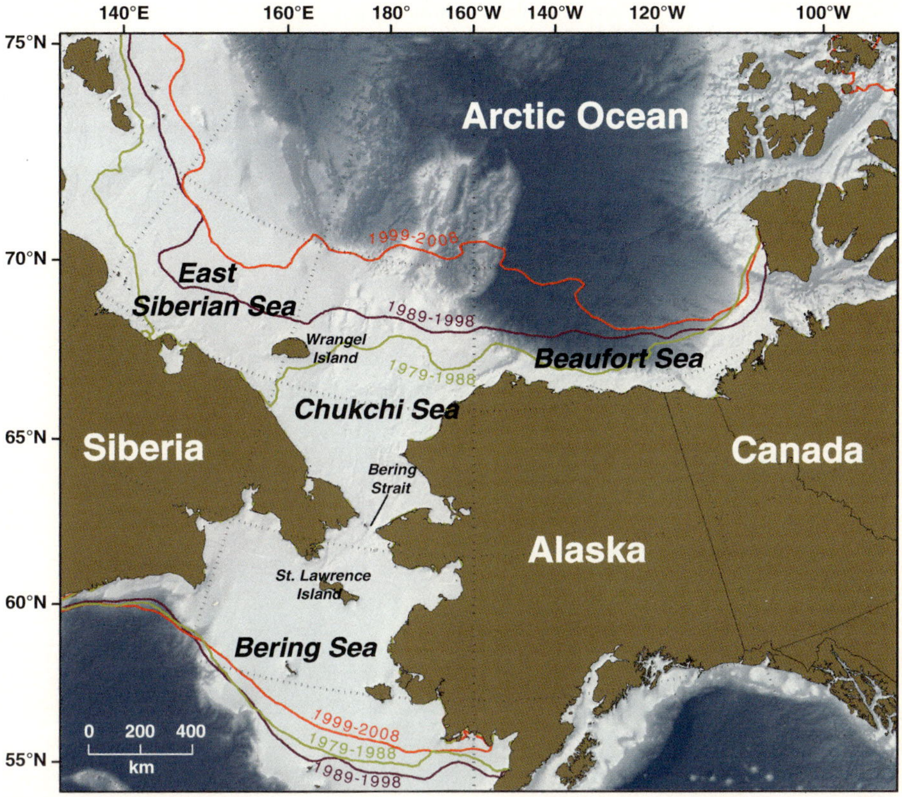

Fig. 3.1 Location of the Pacific Arctic Region (*PAR*). Isolines (based on SMMR and SSM/I sea ice concentrations) indicate the position of the median sea ice edge (defined by a 15 % sea ice concentration threshold) for 1979–1988, 1989–1998, and 1999–2008 during both March (southern isolines) and September (northern isolines)

concentrations fall below or exceed 15 %, respectively. We also define annual sea ice persistence as the total number of days per year that sea ice concentrations are greater than 15 %. Viewing the persistence and timing of breakup/formation spatially (Fig. 3.5) provides valuable insight into sea ice variability across the past three decades, particularly in the context of seasonally varying and sea ice dependent processes such as biological productivity. The PAR is highly dynamic in terms of annual sea ice persistence, ranging from 0 through >300 days per year of sea ice cover (Fig. 3.5). The timing of sea ice breakup also exhibits a broad range, with the earliest breakup in the Bering Sea (April) and the latest breakup in the northern Chukchi Sea (August–September). Similarly, the timing of sea ice formation is earliest in the northern Chukchi Sea (October–November) and latest in the Bering Sea (January–March). The timing of these events has shifted dramatically over the past three decades, which can be directly and spatially quantified by presenting linear decadal trends of annual sea ice persistence, the timing of sea ice

Table 3.1 Linear trends of sea ice decline in the northern Bering and Chukchi Seas over the 30 year record (1979–2008). Only those trends that are statistically significant ($p<0.10$) are shown

	Northern Bering Sea		Chukchi Sea	
	km²/decade	%/decade	km²/decade	%/decade
January			4,466	1.0
February			5,354	1.2
March			3,502	0.8
April				
May			−8,107	−1.8
June	−2,788	−8.9	−24,186	−6.1
July	−568	−19.9	−39,048	−13.8
August			−48,273	−25.0
September	−726	−20.0	−63,840	−31.8
October	−2,283	−15.2	−70,118	−24.7
November	−7,913	−13.9	−48,820	−12.6
December			−13,329	−2.9

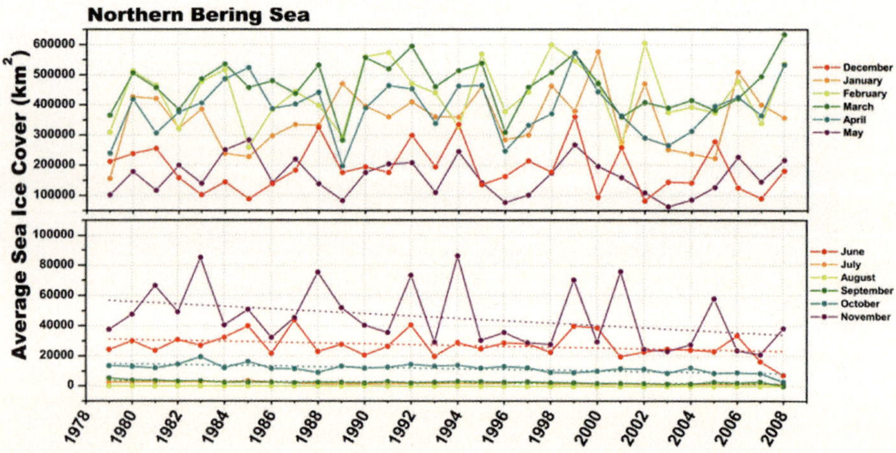

Fig. 3.2 Average sea ice cover for the northern Bering Sea for each month from 1979 to 2008. Trend lines are shown for only those months showing statistically significant ($p<0.10$) trends (Table 3.1). Results are based on SMMR and SSM/I sea ice concentrations

breakup, and the timing of sea ice formation (Fig. 3.6). The most extreme decadal shifts in sea ice cover have occurred within the central Chukchi Sea, where sea ice persistence has declined by more than 30 days/decade over the 30 year record. Decreased ice persistence was caused by not only earlier sea ice breakup (~10 days/decade earlier), but also later sea ice formation (~20 days/decade later). This reduction in sea ice cover has been pronounced in the Chukchi Sea, but trends lessen moving southward through the Bering Strait into the northern Bering Sea region surrounding St. Lawrence Island. South of St. Lawrence Island, sea ice

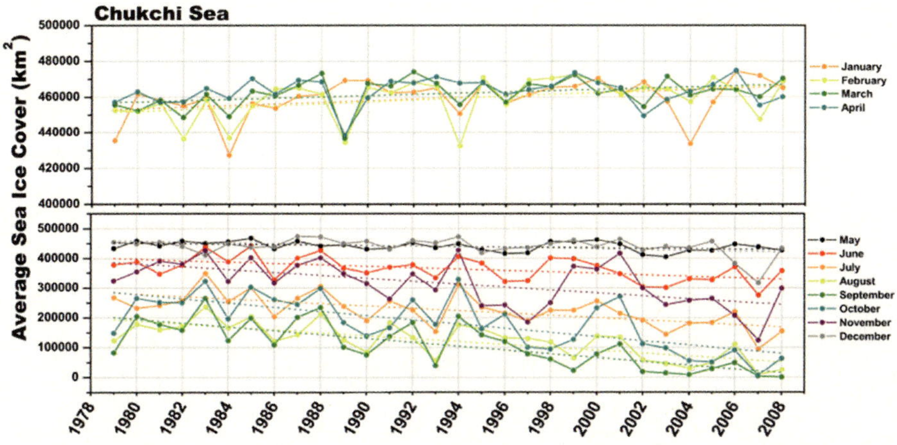

Fig. 3.3 Average sea ice cover for the Chukchi Sea for each month from 1979 to 2008. Trend lines are shown for only those months showing statistically significant ($p<0.10$) trends (Table 3.1). Results are based on SMMR and SSM/I sea ice concentrations

persistence trends show a slight increase in ice cover, driven more by earlier sea ice formation in winter rather than by later ice breakup in spring.

3.2.2 Interannual Variability in Sea Ice Cover

Sea ice cover was also investigated at a higher spatial resolution (6.25 km) with daily sea ice concentrations derived via the Advanced Microwave Scanning Radiometer for EOS (AMSR-E) sensor on the Aqua satellite platform launched in May 2002 (Spreen et al. 2008). Based on a 15 % sea ice concentration threshold, annual sea ice persistence (Fig. 3.7), the timing of sea ice breakup (Fig. 3.8), and the timing of sea ice formation (Fig. 3.9) were quantified for the PAR with AMSR-E data spanning 2002–2009. Although the time series of AMSR-E data is not long enough to determine statistically significant trends in these parameters, these data nevertheless allow for an investigation of the interannual variability of sea ice cover across the PAR with high spatial detail. Just as with the SMMR and SSM/I data, similar patterns in annual sea ice persistence emerge (Fig. 3.7), with a distinct latitudinal gradient from no sea ice cover south of the ice edge in the Bering Sea to >300 days/year of sea ice cover in the northern Chukchi Sea. Consistent with observations in other studies showing a 2007 minimum in Arctic sea ice extent (e.g., Stroeve et al. 2008), 2007 shows the fewest number of days of sea ice coverage throughout the PAR as a whole over the years 2003–2009. In contrast, however, 2003 and 2004 show the smallest sea ice cover for the Bering Sea specifically, although the position of the winter sea ice edge was farthest north in 2005.

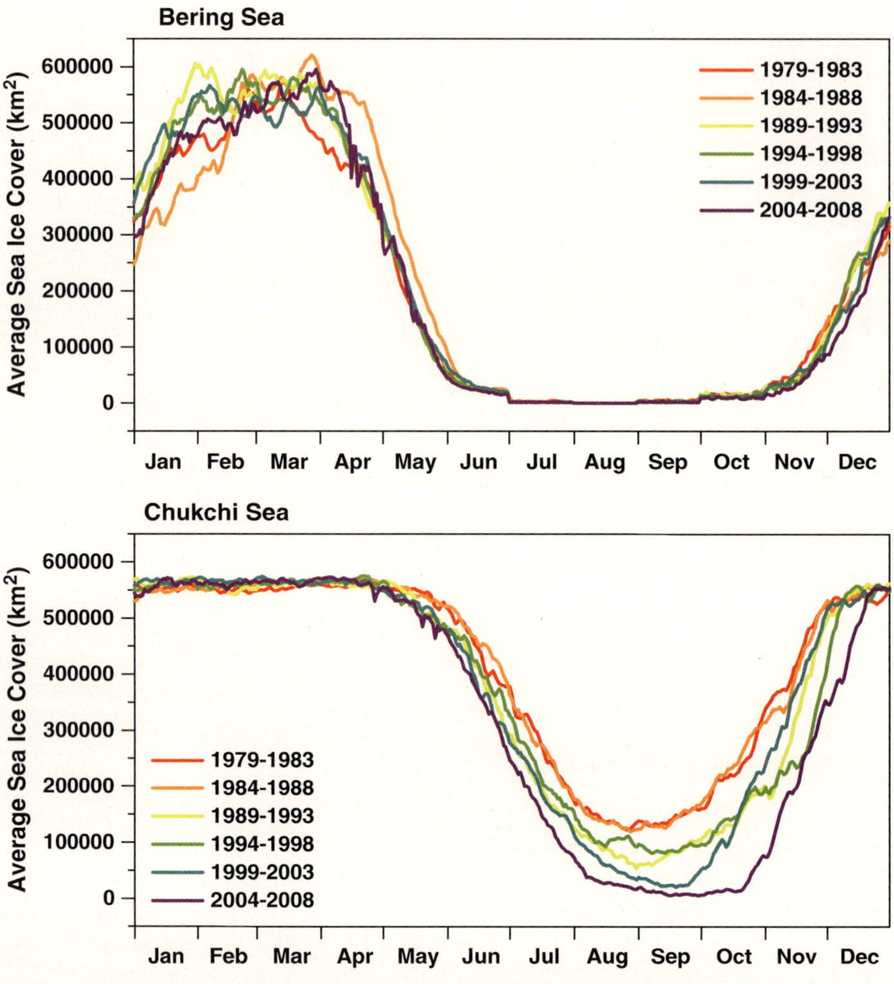

Fig. 3.4 Daily average sea ice extent for the northern Bering Sea (*top*) and Chukchi Sea (*bottom*) for 5-year periods from 1979 to 2008. Results are based on SMMR and SSM/I sea ice concentrations (using a 15 % sea ice concentration threshold). Overall, the open water season is significantly longer for the northern Bering than for the Chukchi Sea. Shifts towards an earlier seasonal sea ice breakup and later sea ice formation are particularly apparent for the Chukchi Sea

Investigation of the timing of sea ice breakup and sea ice formation additionally illustrates important spatial patterns in these events across the PAR. In general, spatial patterns in sea ice breakup were distinctly spatially heterogeneous (Fig. 3.8), where the timing of sea ice breakup did not necessarily follow an ideal latitudinal gradient. However, ice at the sea ice edge in the Bering Sea generally breaks up first (early March) and advances northward into the Bering Strait by late April-early May. Ice breakup tends to occur relatively early within polynya regions (e.g., south of St. Lawrence Island), which is particularly apparent in 2005, 2006, 2007, and 2008. Ice breakup occurs later in the Chukchi Sea, with ice in the southern Chukchi

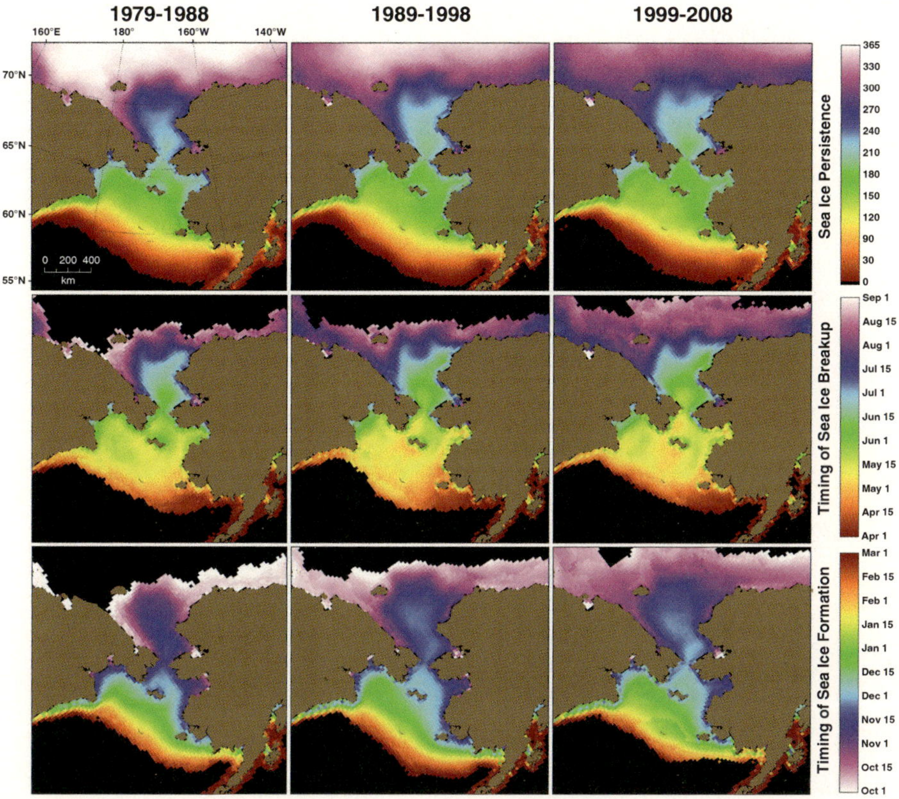

Fig. 3.5 Average annual sea ice persistence, timing of sea ice breakup, and timing of sea ice formation in the PAR during the past three decades. Results are based on SMMR and SSM/I sea ice concentrations (using a 15 % sea ice concentration threshold)

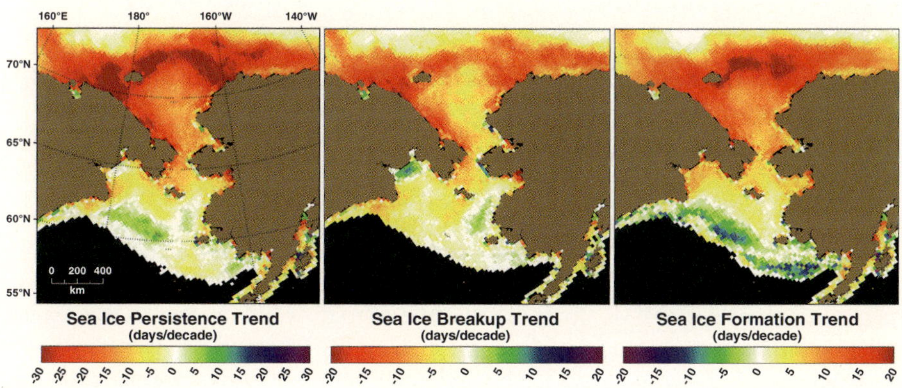

Fig. 3.6 Linear decadal trends in annual sea ice persistence, sea ice breakup, and sea ice formation over the 1979–2008 period. Results are based on SMMR and SSM/I sea ice concentrations (using a 15 % sea ice concentration threshold)

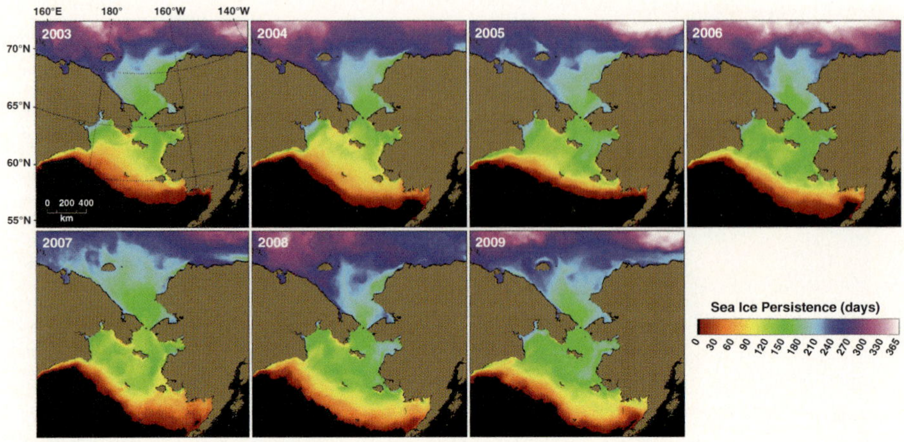

Fig. 3.7 Annual sea ice persistence for the years 2003–2009 showing recent interannual variability in sea ice cover. Results are based on AMSR-E sea ice concentrations (using a 15 % sea ice concentration threshold)

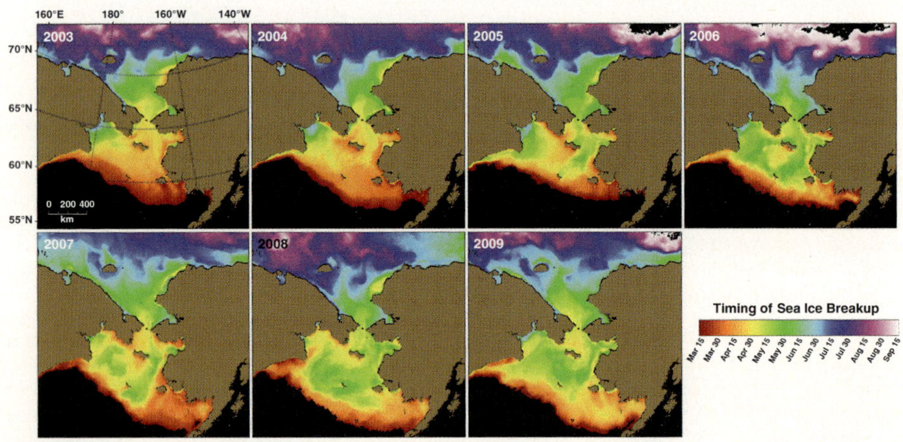

Fig. 3.8 Recent interannual variability in the timing of sea ice breakup for the years 2003–2009. Results are based on AMSR-E sea ice concentrations (using a 15 % sea ice concentration threshold)

Sea typically breaking up in late April-early May and advancing into the northern Chukchi Sea by July or August. The progression of sea ice breakup from south to north across the entire PAR (from the Bering Sea to the Chukchi Sea) occurs over a ~6 month time span (from mid-March through mid-September).

Compared to patterns in sea ice breakup, the timing of sea ice formation is distinctly more homogenous across the PAR and more closely follows an ideal latitudinal gradient (Fig. 3.9). This likely occurs in part because sea ice variability associated with polynya opening events primarily occurs during late winter and

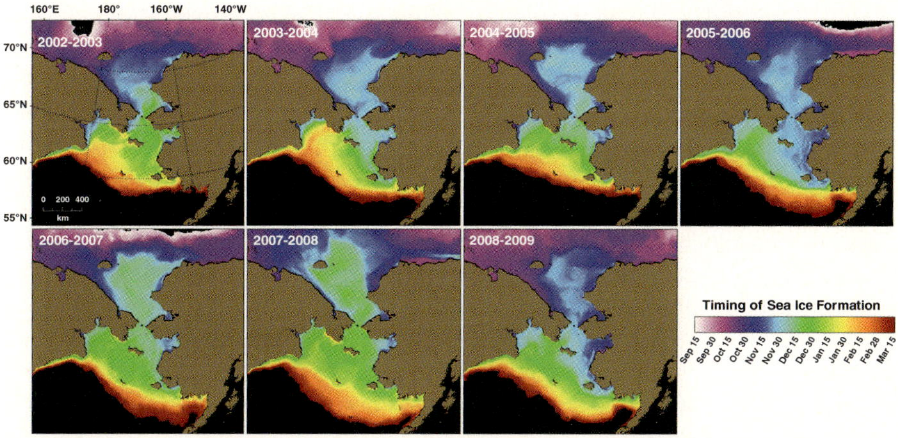

Fig. 3.9 Recent interannual variability in the timing of sea ice formation for the years 2002–2009. Results are based on AMSR-E sea ice concentrations (using a 15 % sea ice concentration threshold)

spring (e.g., Fu et al. 2012), rather than during autumn and early winter with sea ice formation. Across the PAR, sea ice forms first in the northern Chukchi Sea during September and October, whereas much of the remainder of the region (including both the Bering and Chukchi Seas) forms sea ice slightly later and relatively quickly during November and December. The most southern portion of the ice edge in the Bering Sea (in the vicinity of St. Matthew Island) then slowly advances southward over a longer timeframe (January through March) before ice breakup begins again in the spring. Over the 7 years investigated here, the 2006–2007 and 2007–2008 seasons show distinctly later seasonal sea ice formation compared with the other years (Fig. 3.9). These two seasons bound the September 2007 minimum in observed sea ice cover across the Arctic Ocean. The later sea ice formation during the 2007–2008 season is likely the direct aftermath of the shortened duration of summer 2007 sea ice cover. In addition, a shorter ice-covered period during the previous 2006–2007 season could have contributed to a younger, thinner ice cover that was more susceptible to melt during the following summer of 2007. Further discussion of the recent variability in sea ice age and thickness is found in the following two sections (Sects. 3.3 and 3.4).

3.3 Sea Ice Age

3.3.1 Sea Ice Age Data and Analysis

Recent decreases in summer sea ice extent (Sect. 3.2) not only affect total ice coverage but also have resulted in a fundamental change in the nature of the Arctic ice itself – the change from a largely perennial (multiyear) ice cover to a more seasonal coverage dominated by first-year ice (Johannessen et al. 1999; Comiso 2002; Belchansky et al. 2004; Nghiem et al. 2006; Kwok 2007; Maslanik et al. 2007a, 2011; Kwok and Cunningham 2010). There are different ways of looking at this change: (a) the amount of multiyear vs. first-year ice, (b) which areas of the Arctic now experience periods of open water instead of continuous ice coverage, and (c) the characteristics of the surviving multiyear ice pack. Each of these carries different implications in terms of understanding how the Arctic Ocean is changing. Here, we consider how the ice cover has evolved since 1979 in terms of the amount and spatial coverage of first-year and multiyear ice in the PAR, and in terms of the age distribution within the category of multiyear ice.

We use sea ice "age" data prepared by Fowler et al. (2004; updated 2010), and described further by Maslanik et al. (2007a, 2011) and Stroeve et al. (2011). In brief, using satellite data and drifting buoys, it is possible to monitor the formation, movement, and disappearance of sea ice. This history can then be used to estimate ice age, as shown by Fowler et al. (2004) and Rigor and Wallace (2004). In the Fowler et al. (2004) approach, ice movement is calculated using a cross-correlation technique applied to sequential, daily satellite images acquired by the SMMR, SSM/I, and Advanced Very High Resolution Radiometer (AVHRR) sensors. Motion vectors are then blended via optimal interpolation with International Arctic Buoy Program drifting-buoy vectors. Using the resulting 12.5 km resolution Equal-Area Scalable Earth (EASE) grid of ice motion vector fields from 1979 onward, ice age can then be estimated by treating each grid cell that contains ice as an independent Lagrangian particle and transporting the particles at weekly time steps. The datasets are therefore similar to the buoy-derived age fields of Rigor and Wallace (2004), but provide additional spatial detail. If the passive microwave-derived ice concentration remains at least 40 %, then that particle is assumed to have survived summer melt and its age is incremented by 1 year. A second version of the product uses a 15 % concentration threshold to age ice in areas of the diffuse, marginal ice zone (Maslanik et al. 2011). Unless noted otherwise (e.g., Fig. 3.10), the 40 % concentration version of the age data is used for the analyses presented here. It is important to emphasize that with a 40 % concentration threshold (and even with the 15 % threshold), some ice may still be present. In the figures presented, these areas are specifically flagged as "<40 %" or "<15 %" rather than as "open water". When age classes are aggregated into first-year and multiyear (i.e., second-year and older ice) categories, the information is comparable to the passive and active microwave satellite-derived time series of first-year and multiyear ice analyzed in other multiple studies (Johannesen et al. 1999; Nghiem et al. 2006; Comiso et al. 2008; Kwok et al. 2010). See Fowler et al. (2004), Rigor and Wallace (2004), Maslanik et al. (2007a, 2011) and Stroeve et al. (2011) for further details.

3.3.2 *Recent Variability in Sea Ice Age*

To place the changes in sea ice within the PAR into a broader context, Figs. 3.10 and 3.11 show the spatial and fractional coverage of ice age classes from 1985 through 2010 for the Arctic as a whole. The data capture the general trend seen in other

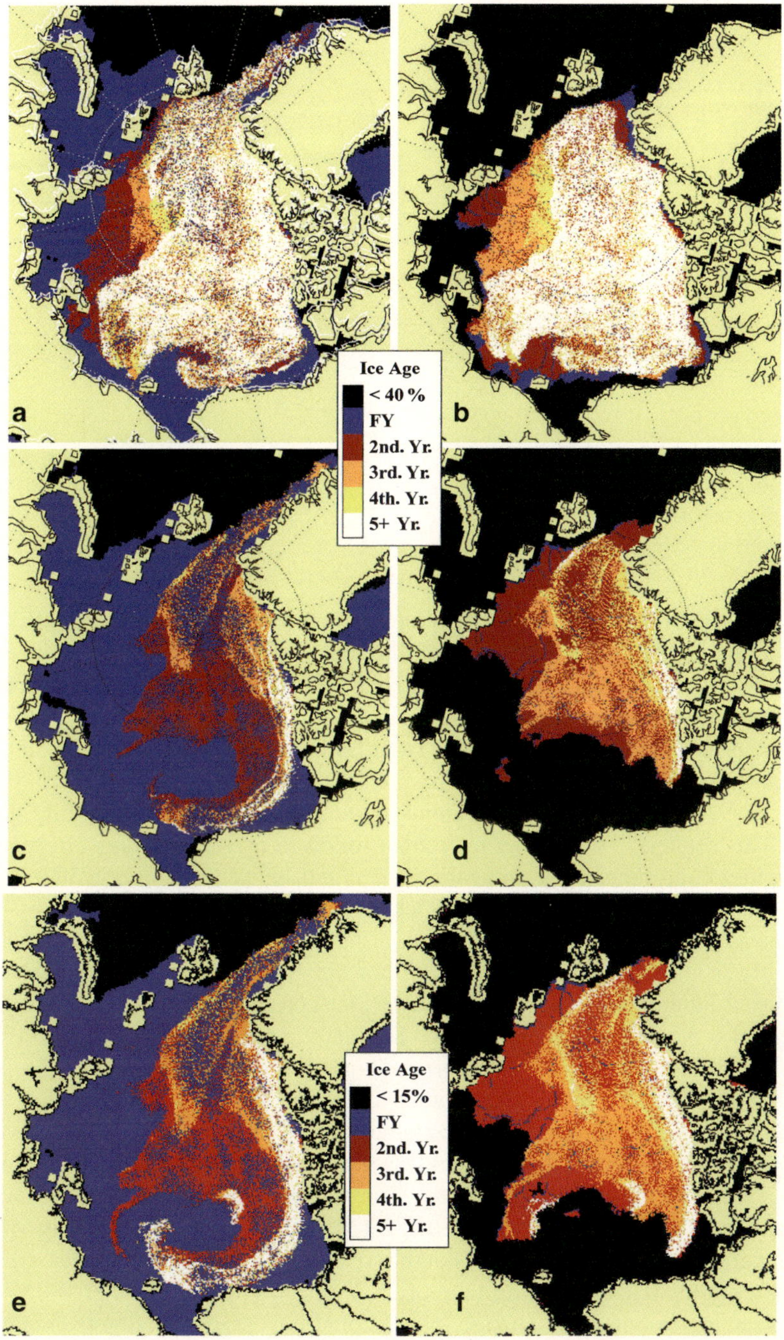

Fig. 3.10 Distribution of sea ice of different ages for (**a**) mid-May 1985, (**b**) end of September 1985, (**c**) mid-May 2010, and (**d**) end of September 2010. Panels (**e**) and (**f**) show the 2010 age maps generated using a 15 % ice concentration threshold during mid-May and September, respectively

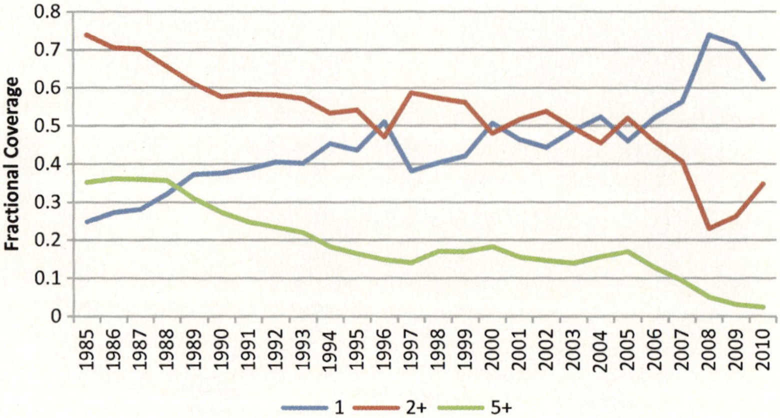

Fig. 3.11 Fractional coverage of sea ice for different age ranges for mid-May 1985–2010. The geographic domain is the Arctic Ocean proper. First-year ice (*blue*), second-year and older ice (*red*), fifth-year ice and older (*green*)

studies of the overall loss in multiyear ice extent, but as discussed by Maslanik et al. (2007a, 2011), the overall reduction of the older multiyear ice types has been particularly large. In recent years, the multiyear ice in the Beaufort and Chukchi seas and Canada Basin is not ice that persists in the region from year to year, but instead is transported into the area each year from farther north and east by the clockwise motion of the Beaufort Gyre (e.g., Proshutinsky et al. 1997) replacing first-year ice that formed in the area following extensive retreats of the pack edge during summer melt. This sequence of summer ice retreat followed by replenishment of multiyear ice typifies conditions in recent years, as seen in panels c and d of Fig. 3.10. In summary, the ice over most of the Arctic Ocean basin is no longer dominated by perennial ice as it was prior to the late 1990s.

Shifts in the age structure of ice within the Arctic Ocean basin translate into significant interannual changes for the PAR. Ice conditions prior to melt onset (mid-May) and at the end of the summer melt period (end of September) for 1985 through 2010 for the PAR and four sub regions (Figs. 3.12, 3.13, 3.14 and 3.15) show a general trend of diminishing multiyear ice extent in the western Arctic, with a greatly reduced fraction of older age classes. Using the 40 % concentration version of the age data, there was a consistent loss in the oldest ice types from 1985 to 2010 (Fig. 3.15), with the oldest ice coverage decreasing from approximately 40 % of the ice pack in 1985 to less than 4 % in 2010 for May. For multiyear ice during May, the fractional coverage remained fairly stable from 1990 to 2001 but decreased markedly after 2009, reaching a minimum of 15 % in 2009 (or a net loss of ~80 % of the multiyear ice extent). The change in fractional coverage was slightly greater for September, with the multiyear ice coverage decreasing from ~80 % in 1985 to less than 30 % in 2010. Despite the fact that relatively little multiyear ice has survived summer melt in recent years in the PAR, multiyear ice of different age categories

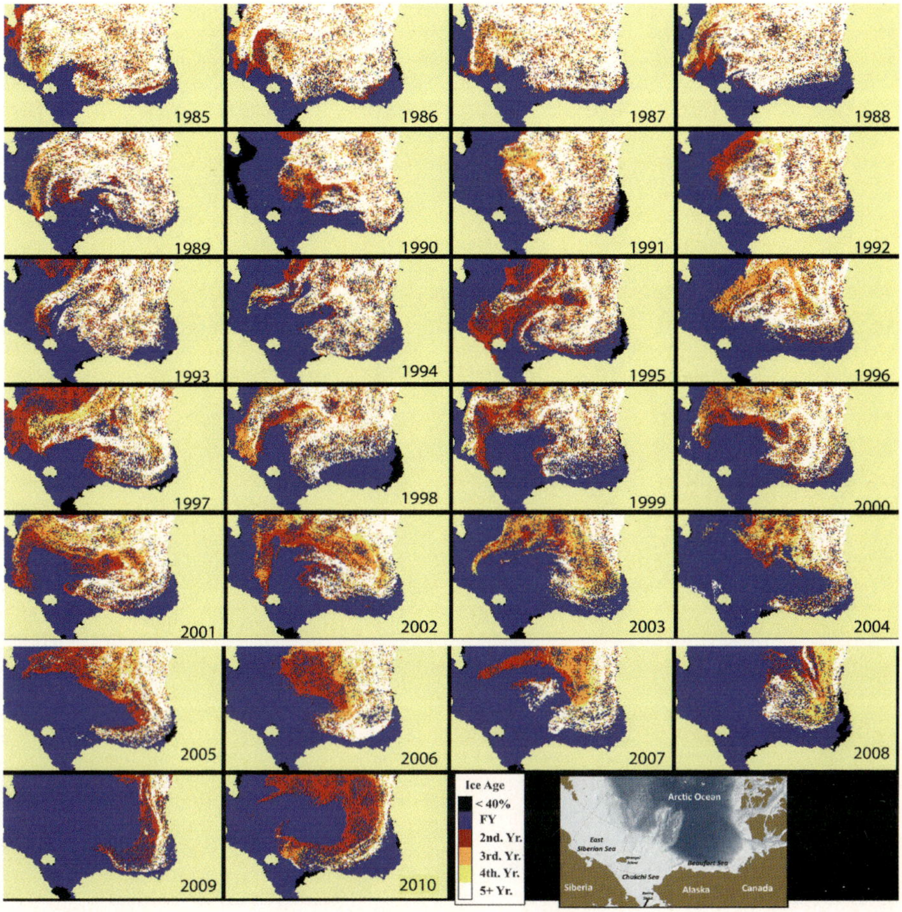

Fig. 3.12 Fractional coverage sea ice by age category for the PAR for mid-May of 1985 through 2010

continues to be present in the area at other times of the year, via transport of multiyear ice from the north and east. This continued presence is more apparent when the minimum ice concentration threshold of 15 % is used (e.g., panels e and f of Fig. 3.10), which retains more of the diffuse, marginal ice cover.

While there is considerable interannual variability in ice conditions, two periods of substantial change can be seen. The first is a pronounced loss of multiyear ice in 1989 and 1990, associated with the extremely strong positive Arctic Oscillation (AO) that favored northward transport away from the Siberian Arctic (Rigor et al. 2002). The time series of ice age composition within sub-regions 1 and 3 of the PAR depict this well, particularly sub-region 1 (Fig. 3.16). While multiyear ice recovered somewhat through the late 1990s, it was younger overall and likely thinner (e.g., Maslanik et al. 2007a, 2011; Lindsay et al. 2009) than prior to 1989. The second

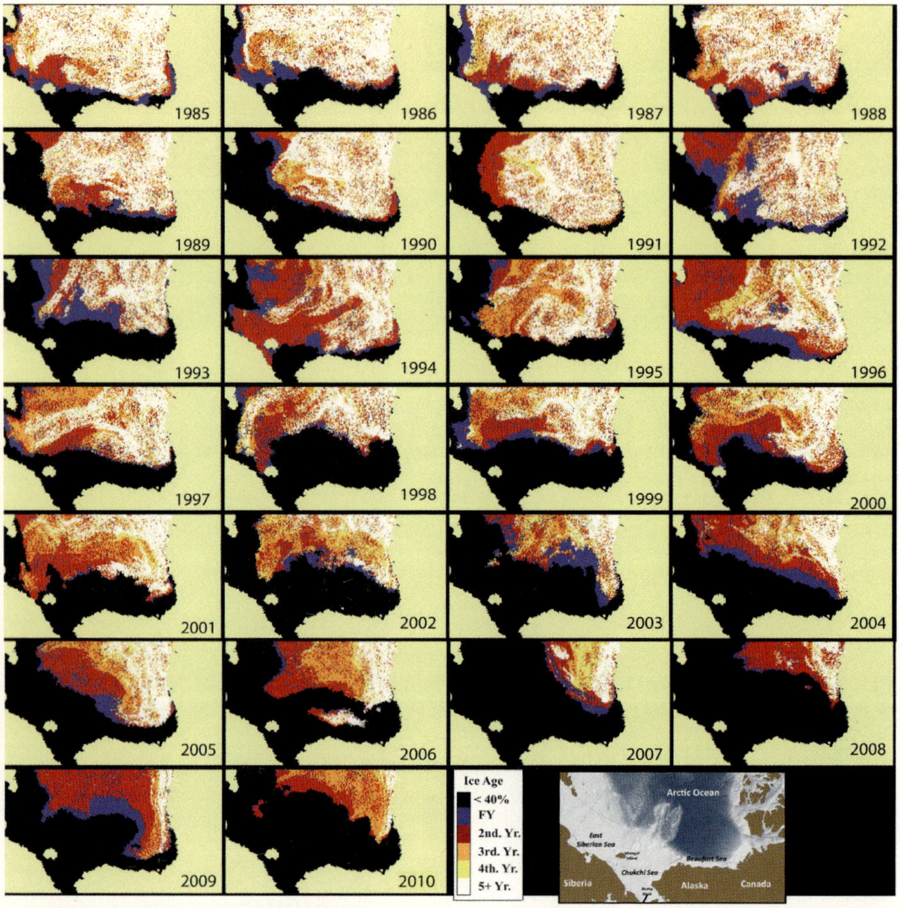

Fig. 3.13 Fractional coverage sea ice by age category for the PAR for the end of September, 1985 through 2010. Note that the *black areas* indicate areas where passive microwave-derived ice concentration is less than 40 %, rather than indicating areas of 100 % open water

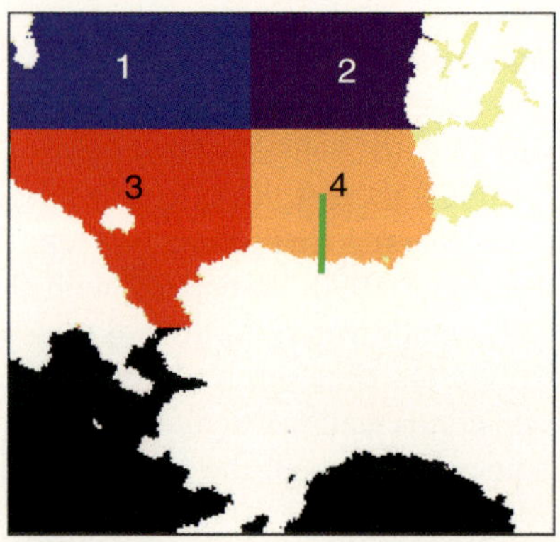

Fig. 3.14 Geographic domain used for calculations of ice coverage by age category. The four individual regions are labeled 1–4. Land (*white*) and portions of the Canadian Archipelago (*yellow*) are excluded. The transect used for transport calculations (Fig. 3.17) is in *green*

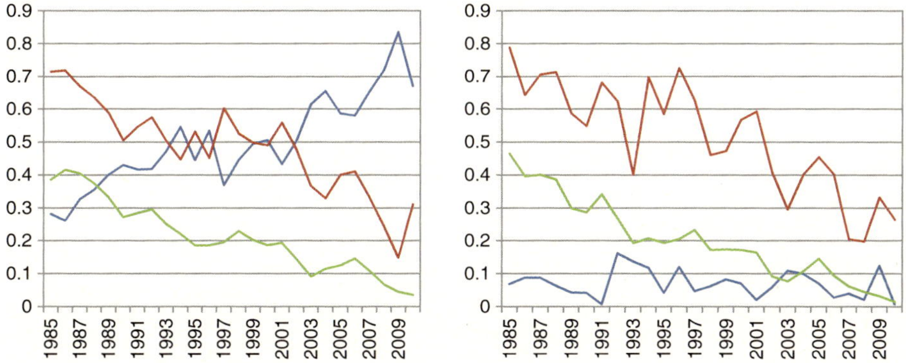

Fig. 3.15 Fractional coverage of sea ice of different age ranges summarized over the full PAR (regions 1–4 in Fig. 3.14) for mid-May (*left panel*) and end of September (*right panel*). First-year ice (*blue*), second-year and older ice (*red*), fifth-year and older ice (*green*)

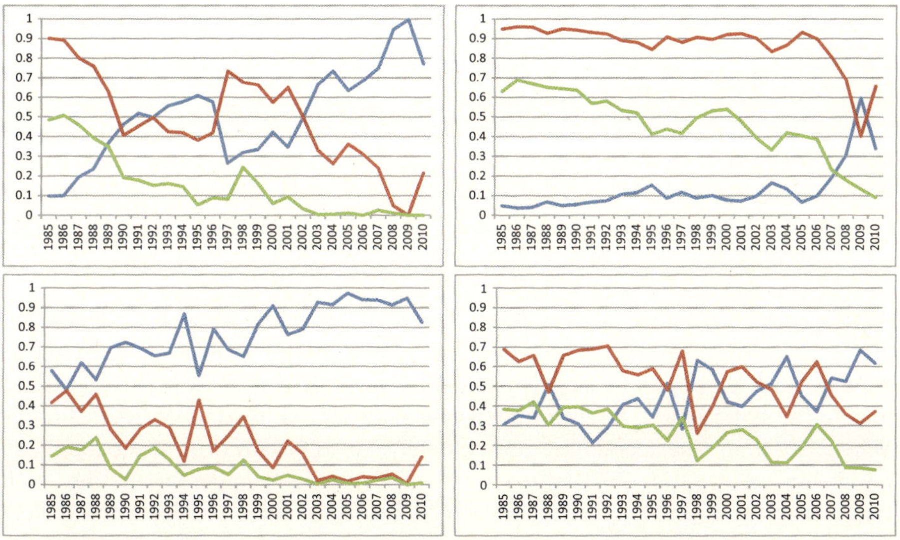

Fig. 3.16 Fractional coverage of sea ice of different age ranges in mid-May for the four regions indicated in Fig. 3.14. Region 1 (*top left panel*), Region 2 (*top right*), Region 3 (*bottom left*), Region 4 (*bottom right*). First-year ice (*blue*), second-year and older ice (*red*), fifth-year and older ice (*green*)

notable period of change began in 2003, when atmospheric circulation patterns and perhaps changes in northward ocean heat transport (Maslanik et al. 2007b; Wang et al. 2009) diminished the multiyear ice from the western Beaufort and northern Chukchi Seas (as in sub-region 3 in Fig. 3.16). This likely occurred through a

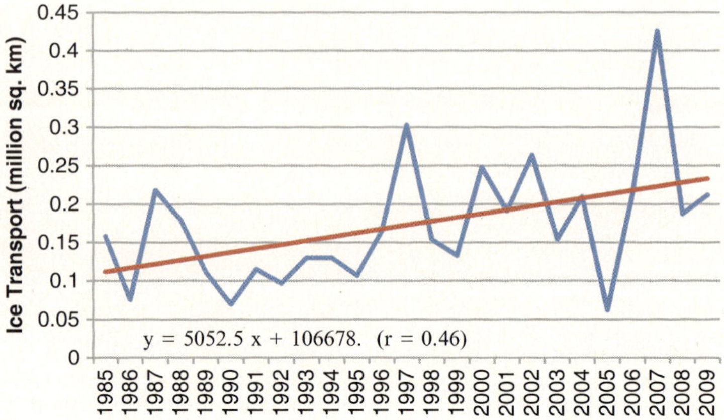

Fig. 3.17 Annual ice transport during winter October–April across a north-south transect in the Beaufort Sea as indicated in Fig. 3.14. The values are for ice years (e.g., the value for 1985 spans October 1985–April 1986)

combination of northward ice drift and greater melt, but the specific contributions of the different mechanisms remain unclear.

In sub-region 4 (eastern Beaufort Sea), the pack ice has oscillated between being predominantly first-year vs. multiyear since 1997, before which it was mostly multiyear ice. The proportion of the multiyear pack that consists of the oldest ice types has remained relatively constant. The greater variability in the mixtures of ice types since 1997 is likely related to variability in east-to-west transport prior to the melt season. Calculation of ice transport (Fig. 3.17) across the transect location indicated in Fig. 3.14, estimated using satellite derived ice motion vectors (Stroeve et al. 2011) for October through April 1985–2009, shows considerable interannual variability but with a general trend toward greater transport in the latter years. Record transport was observed in Autumn 2009 through Winter 2010, in part accounting for the extensive westward drift of the tongue of multiyear ice seen in 2010 (Stroeve et al. 2011) and apparent in Fig. 3.12. Examining the time series of ice transport across the Beaufort Sea transect suggests that the record transport arose from strong, episodic wind-driven events rather than continuous uniform ice drift.

In sub-region 2 (eastern Canada Basin), although the oldest ice types have decreased over time, the multiyear ice coverage remained constant until 2007 indicating a shift to a younger multiyear pack. The extreme ice loss in summer 2007 was the first case in the satellite record where ice was removed throughout the entire western portion of the PAR. This has been followed by some recovery of the multiyear ice extent but as with the effects from the AO in 1990, the multiyear ice is younger than it was previously and may be less able to withstand summer melt (Lindsay et al. 2009).

3.4 Sea Ice Thickness

3.4.1 Sea Ice Thickness Data and Background

While many previous studies have analyzed changes in ice extent and concentration (e.g. Comiso 2009; Stroeve et al. 2011), this section focuses on ice thickness as it gives a better indication of ice volume variability. Temporal changes in the Arctic sea ice thickness (or draft) have been noted by several investigators over the last few decades. McLaren (1989) compared 1958 ice draft measurements, obtained via acoustic profilers aboard the USS *Nautilus*, to similar measurements from the 1970 expedition of the USS *Queenfish*. The *Queenfish* resampled along the original *Nautilus* route across the Arctic Basin via the North Pole and thus provided a robust, yet snapshot, comparison between these two time periods. The mean draft decreased over the 12 years from 3.08 to 2.39 m (or by ~22 %). A reduction in the sea ice thickness north of Greenland was noted by Wadhams (1990) between 1976 and 1987. This reduction was found to be equivalent to at least a 15 % loss of volume over an area of 300,000 km^2. Rothrock et al. (1999) utilized data obtained via U.S. Navy submarines from the Scientific Ice Expeditions (SCICEX) of the 1990s and compared those with historic cruise datasets from 1958 to 1976. Geographically nearly overlapping samples from the end of the melt season averaged over five cruises during 1958–1976 (i.e. 1958, 1960, 1962, 1970, and 1976) were compared to similar averaged data from three cruises during the 1990s (i.e., 1993, 1996, and 1997) and reported a mean ice draft reduction of 1.3 m at the end of the melt season. This difference (from 3.1 m in 1958–1976 to 1.8 m in 1993–1997) represents a mean decline of 40 % over much of the deep-water portion of the Arctic Ocean.

Although many previous studies have shown reductions in the sea ice thickness of the Arctic Ocean, questions remained as to whether the true long-term trend could be distinguished from natural temporal and spatial variability using the available limited data. Rothrock et al. (2008) used all U.S. submarine data in the data release area (DRA, the region of U.S. Navy submarine data) from 1975 to 2000, along with multiple regression techniques, to separate the interannual change, the annual cycle, and the spatial field. The observed ice draft within the DRA showed declines from a maximum of 3.42 m in 1980 to 2.29 m in 2000 (and an overall 1.25 m decrease in ice thickness), with an observational error standard deviation of 0.38 m. Rothrock et al. (2008) noted a need for observations outside the DRA in order to assess the role of sea ice redistribution (i.e., with no or little change in the Arctic-wide sea ice volume; Holloway and Sou 2002) from the DRA into Russian waters or into the region south of the DRA and north of Canada. Because the DRA only encompasses ~38 % of the Arctic Ocean (Kwok and Rothrock 2009), sea ice thickness variability outside this region is still poorly known.

More recently, sea ice thickness has been estimated in the Arctic Ocean from satellite measurements of ice freeboard, which on average accounts for 10–15 % of ice thickness. Based on estimates of ice freeboard from satellite altimeter measurements

(ERS-1 and ERS-2), Laxon et al. (2003) demonstrated strong interannual thickness variability of the Arctic sea ice cover between 1993 and 2001. Haas (2004) showed a similar magnitude of interannual variability in the Transpolar Drift during 1991–2001 using independent measurements from drilling and electromagnetic (EM) sounding. Giles et al. (2008) reported on significant thinning of Arctic sea ice (0.49 m in the western Arctic) following the 2007 ice extent minimum based on the Envisat-derived sea ice thickness anomalies between 2002 and 2007. Kwok et al. (2009) compared ice thickness data estimated from the Ice, Cloud, and land Elevation Satellite (ICESat) autumn (mid-October to mid-November) and winter (late February to late March) campaigns during 2003–2008 with historic submarine sonar measurements (Rothrock et al. 1999). They showed a dramatic reduction in the wintertime sea ice thickness between 1980 (3.64 m) and 2008 (1.89 m) within the DRA (Kwok et al. 2009), which represents a thickness decline of 1.75 m during the last three decades. However, sea ice thickness estimates from ICESat depend on several not well known factors, including total freeboard, snow depth, and densities of snow, ice and seawater (e.g., Kwok et al. 2007; Kwok and Cunningham 2008).

Overall, observations of ice thickness are space and time-limited, and are not yet sufficient to diagnose long-term changes in Arctic sea ice volume. Ice thickness estimates derived from ice draft measurements by upward looking sonar from submarines and moorings are less uncertain than those derived from ice freeboard measured by satellites. However, ice draft data are not available basin-wide or long-term, while satellite data in practice are still limited to ice thickness anomalies relative to short-term means.

3.4.2 Sea Ice Thickness Model Description

In this section, we present sea ice thickness results from a high-resolution (~9 km) pan-Arctic coupled ice-ocean model (Naval Postgraduate School Arctic Modeling Effort, NAME). The NAME coupled sea ice-ocean model consists of a regional adaptation of the Parallel Ocean Program (POP) and dynamic-thermodynamic sea ice models (Maslowski et al. 2004) configured at a horizontal grid spacing of 1/12° (or ~9 km). Such horizontal resolution permits calculation of flow through the narrow passes of the Aleutian Islands (Maslowski et al. 2008a) and straits of the northern Bering Sea (Clement et al. 2005). In the vertical direction, there are 45 vertical depth layers ranging from 5 m near the surface to 300 m at depth, with eight levels in the upper 50 m. The model domain is configured in a rotated spherical coordinate system to eliminate grid singularity at the North Pole and to minimize changes in grid cell area due to convergence of meridians with latitude. It contains the sub-polar North Pacific (including the Gulf of Alaska, the Sea of Japan and the Sea of Okhotsk), the Arctic Ocean, the Canadian Arctic Archipelago (CAA), the Nordic Seas and the sub-polar North Atlantic (see Fig. 3.1a of Maslowski et al. (2004) for model domain). Model bathymetry is derived from two sources: the ETOPO5 at 5-km resolution for the region south of 64°N and the International Bathymetric

Chart of the Arctic Ocean (IBCAO; Jakobsson et al. 2000) at 2.5-km resolution for the region north of 64°N. Daily climatological runoff from the Yukon River (and all other major Arctic rivers) is included in the model as a virtual freshwater flux at the river mouth. However, in the sub-polar regions (e.g., Gulf of Alaska, Hudson Bay, Baltic Sea) the freshwater flux from runoff is introduced by restoring the surface ocean level (5 m thick) to climatological (Polar Science Center Hydrographic Climatology, PHC; Steele et al. 2001) monthly mean salinity values over a monthly time scale (as a correction term to the explicitly calculated fluxes between the ocean and underlying atmosphere or sea ice).

The ocean model was initialized with climatological, three-dimensional temperature and salinity fields (PHC) and integrated for 48 years in a spinup mode. The production run, forced with realistic daily averaged European Centre for Medium-Range Weather Forecasts (ECMWF) interannual reanalysis and operational data, covers the 26-year time period from 1979 through 2004. Additional details on the model, including sea ice, have been provided elsewhere (Maslowski and Lipscomb 2003; Maslowski et al. 2004, 2008b). Results from the production run are used for the analyses, to validate against available observations of sea ice thickness, and to provide information where and when data are missing.

3.4.3 Sea Ice Thickness Model Validation

In order to evaluate the skill of the model in representing sea ice thickness variability, its output has been compared to sea ice thickness data gathered during the last three decades (McNamara 2006; Whelan 2007). Those data sets include the draft measurements conducted by U.S. Navy submarines between 1979 and 2000, EM induction ice thickness measurements gathered using a helicopter by the Alfred Wegener Institute in April 2003, and data collected by NASA's ICESat satellite for 2003. The comparison with submarine and EM data is problematic for several reasons. The main problem is that the data and model output are at very different spatial resolutions: A typical sonar beam footprint ranges in diameter from 2.6 to 6 m and it collects data every 1 m along-track, whereas the model grid cell is about 9 km × 9 km (or ~85 km^2) and tends to remove values at either extreme. Assuming the sonar footprint diameter of 6 m, a 50 km cruise segment yields a 0.3 km^2 swath for the ULS dataset. The same 50 km segment results in a 1,390 km^2 swath when using the three grid cell wide model output comparison. Another problem is the large mismatch (up to three orders of magnitude) between the number of data points and the number of grid cells for any given distance. Finally, the time of the observations used in this work is known at least to the day, which is in marked contrast to the monthly average model output available for the analysis.

We focus here on a comparison with submarine data which was analyzed similarly to Rothrock et al. (2008), by averaging the ice thickness for each straight segment of the cruise (corresponding to a distance of 50 km or less) and assigning

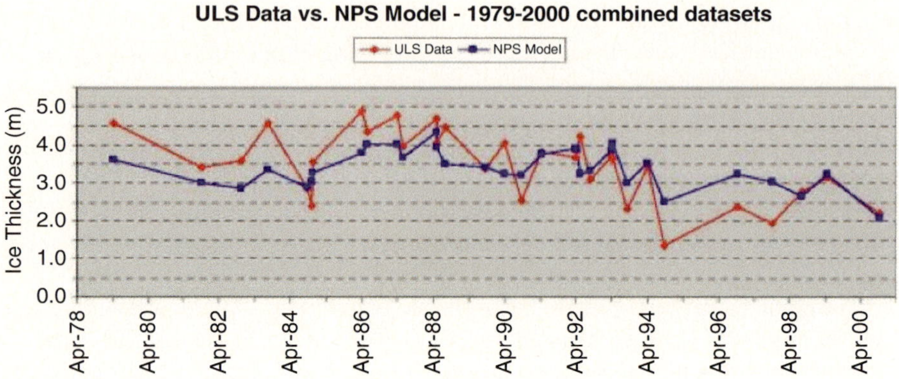

Fig. 3.18 ULS weighted mean ice thickness (*red*) vs. NPS model weighted mean ice thickness (*blue*)

weights to the segments according to their lengths (McNamara 2006). The data were then binned into 10-cm ice thickness bins and plotted together as a probability distribution function (PDF) with y-axis values representing a percentage of points with a given ice thickness relative to the total number of points. This analysis allows comparison of both modal and mean ice thickness per cruise. A summary of mean ice thickness comparison from 32 submarine cruises is shown in Fig. 3.18. The NPS model performed reasonably well when comparing its sea ice thickness output to the ULS derived sea ice thickness measurements from the analyzed 32 submarine cruises. It showed practically no bias against data when examined across the record. However, the lack of sub-grid scale variations in ice thickness effectively contributed to a limited agreement with the data on very thin and very thick ice. This is evident in comparisons where the mean model thickness is in good agreement with the mean cruise thickness from the submarines. However, the modal distribution is not in good agreement, as it shows differences of order 0.5 m and/or missing thinner/thicker secondary modal maxima.

3.4.4 Recent Variability in Modeled Sea Ice Thickness

Overall, modeled sea ice thickness has declined over the last few decades (Fig. 3.19). Mean September thickness in 1982 was in the range of 2.5–3.5 m over the central Arctic Ocean, with the thickest ice found along the northern coasts of the Canadian Archipelago and Greenland (Maslowski et al. 2007). Ten years later, the sea ice thickness in September 1992 was slightly thicker over much of the Arctic Ocean (Fig. 3.19b), although there was little change overall. In 2002, a dramatic reduction in sea ice thickness occurred with values <2 m across most of the central Arctic Ocean. Many marginal seas were ice-fee during this September time frame, in agreement with satellite observations. We also note the significant thinning of ice

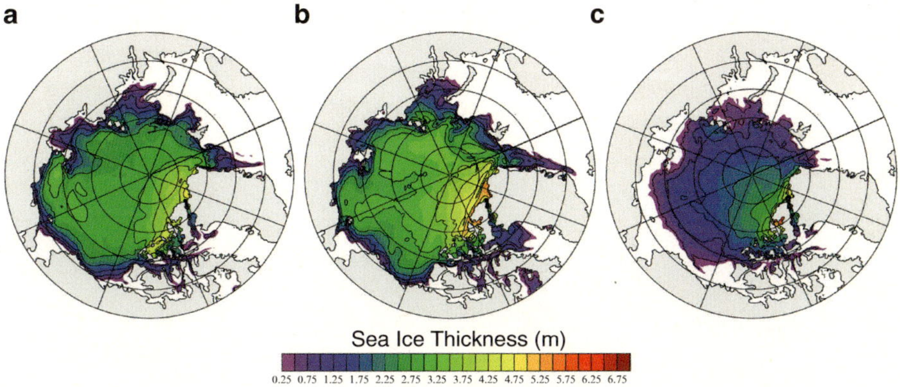

Fig. 3.19 Modeled sea ice thickness (m) during September (**a**) 1982, (**b**) 1992, and (**c**) 2002

along the northern coasts of the Archipelago and Greenland. In addition, the tongue of ice that typically extends along the east coast of Greenland (above the East Greenland Current) was not present for the most part. The dramatic thinning described here is robust and independent of the strength of surface relaxation to climatological temperature and salinity, as shown by Maslowski et al. (2007). A shift in the mode (as defined by the highest percentage of total model grid cells per ice thickness bin) of ice thickness occurred during the mid-to-late 1990s (Fig. 3.20). The PDF shows that the mode prior to 1997 was in the range of 2.5–3.5 m thick. After that time, the mode was between 1.0 and 2.5 m thick – a strong shift toward thinner ice (Stroeve and Maslowski 2007). The reduction of the modal ice thickness represents the thinning of the multi-year pack ice toward the ice thickness range representative of first-year ice.

In light of the overall changes in Arctic sea ice thickness, we now focus on the changes in the Bering and Chukchi seas. The modeled mean ice thickness in the Chukchi Sea has declined throughout the seasonal cycle in a dramatic fashion since the late 1990s (Fig. 3.21). While the late summer/early autumn differences are the highest (~1 m for 1999–2003 and ~1.4 m in 2004), the wintertime changes are still noticeable (~0.5 m for 1999–2003), particularly in 2004 (~1 m). The seasonal cycle in the Bering Sea (Fig. 3.21), in contrast, shows much smaller changes (up to 0.15 m), with thinner ice occurring during the winter through early summer time frame in 2004. The only significant trends (over the 26-year time series) (Fig. 3.22, Table 3.2) in the Bering Sea ice thickness occur during June (−3.7 cm/decade) and July (−1.6 cm/decade). The Chukchi Sea ice thickness shows significant downward trends during June–December (Fig. 3.23, Table 3.2). The most severe ice thickness trends occur in September at −51.2 cm/decade, however early summer through early winter losses are all statistically significant and noteworthy (>25 cm/decade). Although the trends for January–May are not significant (at the 90 % level), sharp declines occur for all of these months beginning in 2001 (Fig. 3.23).

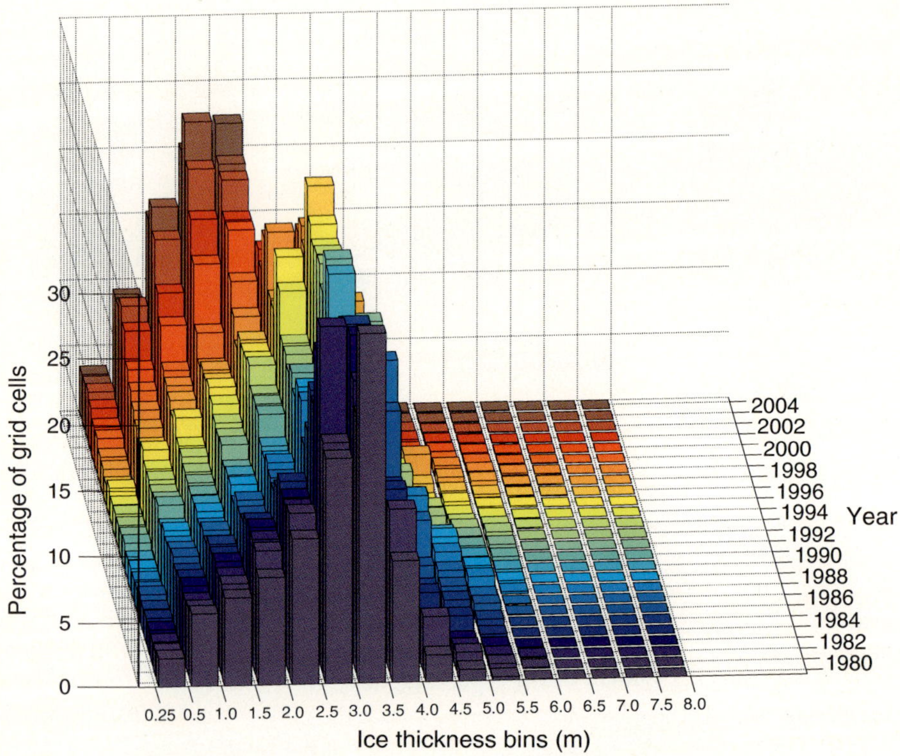

Fig. 3.20 Probability distribution function (*PDF*) of modeled annual mean binned ice thickness (m). The *z*-axis shows percentage of the total model grid cells per thickness bin defined along the *x*-axis. The *y*-axis is time in years from 1979 to 2004 with a different color histogram for each year

3.4.5 Potential Mechanisms of Sea Ice Thinning

Although many previous studies have shown reductions in the sea ice thickness of the Arctic Ocean (e.g., McLaren 1989; Wadhams 1990; Rothrock et al. 1999, 2008; Kwok et al. 2009) the direct cause(s) remain somewhat obscure. Rothrock et al. (1999) proposed three possible mechanisms for thinning the sea ice to the degree that was observed: (a) increased oceanic heat flux (additional 4 W m^{-2}), (b) increased poleward atmospheric heat transport (additional 13 W m^{-2}), or (c) increased downwelling shortwave radiation (additional 23 W m^{-2}). Still another possible cause of the thinning could be changes in precipitation and snow cover or advective processes such as sea ice export via Fram Strait. Rothrock et al. (1999) also noted all of these processes would be at the very limits of the present observational capability, and hence reaching a conclusion on the cause of the sea ice decline would prove difficult.

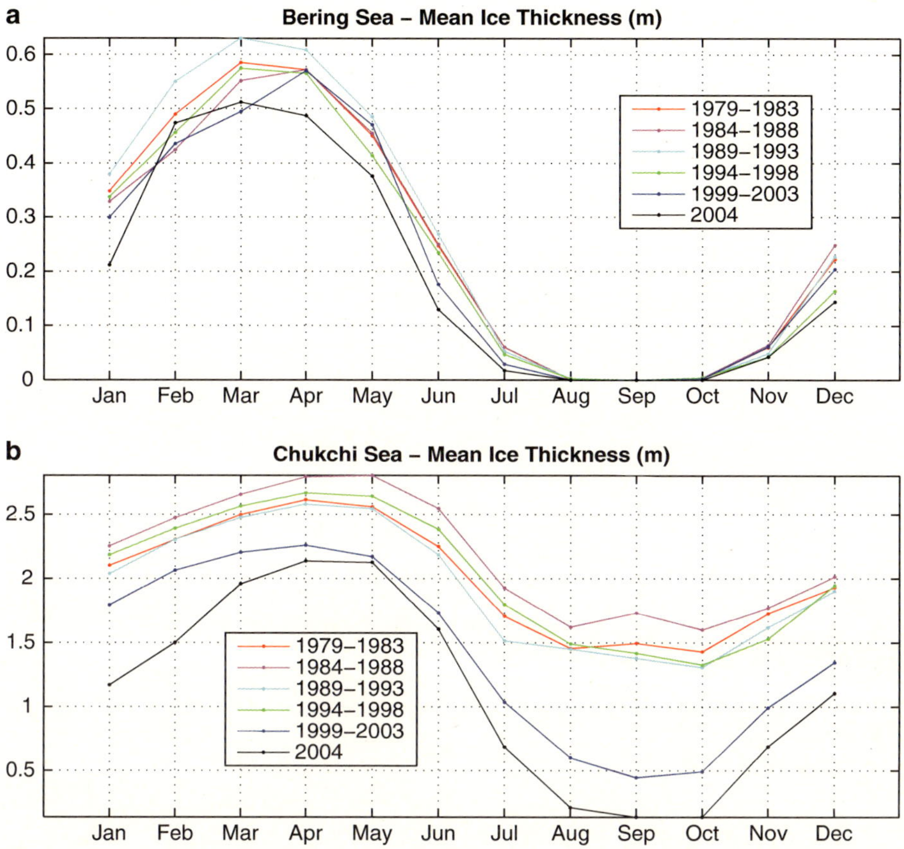

Fig. 3.21 Monthly mean modeled sea ice thickness in 5-year periods (except for 2004) as shown in the legend for the (**a**) northern Bering Sea and (**b**) Chukchi Sea

Warming in the Arctic has been typically associated with high positive AO index (Rigor and Wallace 2004), however the recent accelerated warming in the late 1990s through 2000s has occurred under a relatively neutral AO regime, which poses important questions about the actual role of the AO in sea ice variability (Overland and Wang 2005). There is a clear indication that Arctic sea ice thickness has declined at a similar, if not faster, rate than that of ice extent/area. The overall decline in sea ice cover has been most prominent in the western Arctic, which implies that some of its causes might be related to summer Pacific water advected from the Bering Sea, over the Chukchi shelf, and into the deep Canada Basin (see Maslowski et al. 2014, this volume).

In the Chukchi Sea, the modeled ice thickness trends are the most severe during September (−51.2 cm/decade or −54.8 %/decade) (Table 3.2), which is, again, consistent with the observed sea ice extent trend in the same region for September (−63,840 km^2/decade or −31.8 %/decade) (Table 3.1). An earlier summer melt in

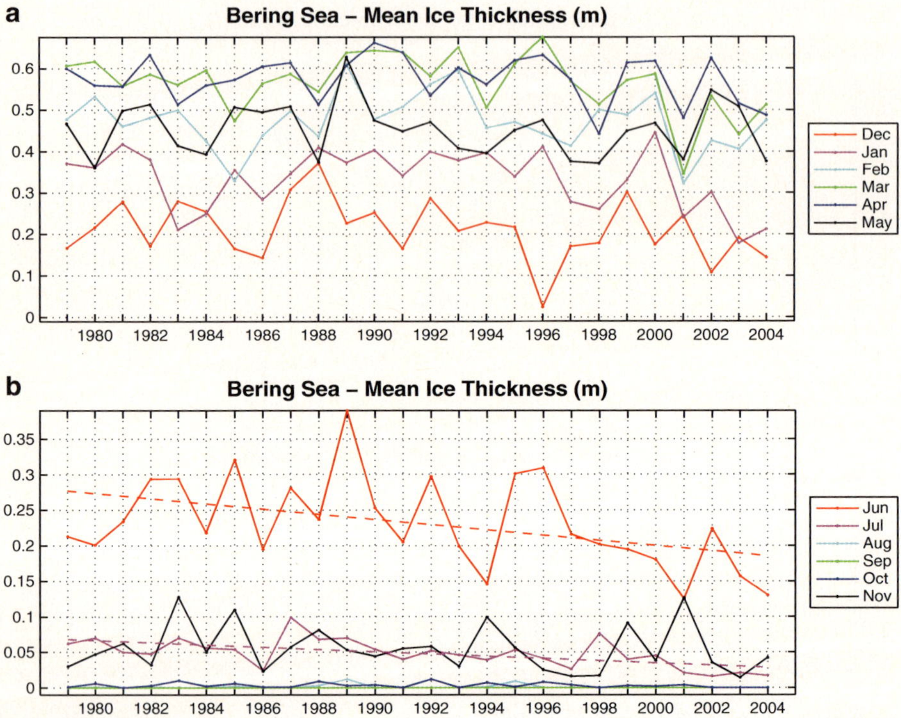

Fig. 3.22 Monthly mean modeled sea ice thickness for 1979–2004 in the northern Bering Sea. Significant ($p<0.10$) linear trends are shown as *dashed lines* in the same color as the monthly mean values

Table 3.2 Linear trends of modeled sea ice thickness decline in the Bering and Chukchi Seas over the 26 year time period (1979–2004). Only those trends that are statistically significant ($p<0.10$) are shown

	Bering Sea		Chukchi Sea	
	cm/decade	%/decade	cm/decade	%/decade
January				
February				
March				
April				
May				
June	–3.7	–17.4	–27.6	–12.0
July	–1.6	–25.8	–34.7	–20.1
August			–40.2	–39.3
September			–51.2	–54.8
October			–46.7	–60.1
November			–37.8	–26.9
December			–28.3	–16.3

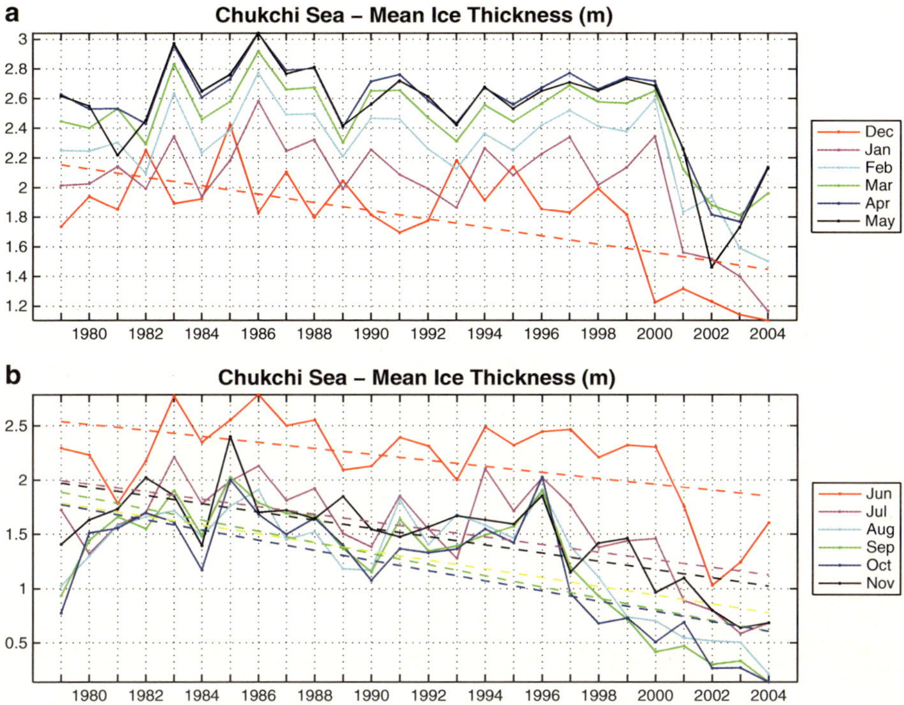

Fig. 3.23 Monthly mean modeled sea ice thickness for 1979–2004 in the Chukchi Sea. Significant ($p<0.10$) linear trends are shown as *dashed lines* in the same color as the monthly mean values

the modeled sea ice thickness trends for the Bering Sea (Table 3.2) is consistent with observations, which also show declines in ice extent during June and July (Table 3.1). However, later sea ice formation (i.e., thinner ice during autumn) is not shown in the model results through 2004.

3.5 Implications and Possible Future States

Summer retreat of the sea ice pack tends to stop at the multiyear ice edge (i.e., first-year ice typically melts out more easily than multiyear ice), as is apparent for corresponding years in Figs. 3.12 and 3.13. As such, once multiyear ice is established it helps maintain ice extent. In other words, there tends to be two relatively stable states of sea ice: a seasonal ice state and a multiyear ice state. For example, once an area is converted to first-year ice, it tends to remain an area of seasonal ice cover, as happened during the late 1980s positive AO event. But if multiyear ice survives or re-captures a region, then it favors the multiyear ice state. However, since the late 1990s, even the oldest and likely thickest (e.g., Maslanik et al. 2007a) ice transported into the Beaufort Sea has typically not survived through summer. The PAR has thus essentially become a region of multiyear ice loss, behaving more like a sub-arctic sea than the historical Arctic region dominated by perennial ice cover (Kwok and Cunningham 2010; Maslanik et al. 2011). An exception occurred in 2006, however, when a portion of multiyear ice in the Beaufort Sea consisting of the oldest ice types survived the melt season (Fig. 3.13). This suggests that while conditions have not been favorable for ice survival in recent years, they are likely close to a threshold in which favorable dynamic and thermodynamic conditions may allow extensive ice cover to survive through summer in the southern Beaufort Sea. Once a multiyear ice cover is re-established, it may act to maintain itself for several years, or until the next set of extreme conditions occurs (as in 1990 or 2007).

The tendency for the ice pack to melt back to the multiyear ice edge, along with the decreased survivability of multiyear ice in the southern Beaufort Sea, implies continued trends of more open water in the PAR. It is important to note that multiyear ice continues to be transported into the southern Beaufort Sea and occasionally farther west into the Chukchi Sea, causing some older ice types to be quite close to shore (Fig. 3.12). Unlike conditions in the 1980s and 1990s, the multiyear ice is now often confined to a relatively narrow band or tongue, with a tendency for the oldest and heaviest ice to be farthest south, with younger multiyear ice to the north transitioning to first-year ice. Some of the multiyear ice in the southern Beaufort Sea is among the oldest and thickest in the Arctic, so the potential exists to encounter multiyear ice during winter and throughout mid-summer in this region. In some cases, isolated multiyear floes can survive in the Beaufort and Chukchi seas throughout the melt season, possibly providing some isolated habitat for marine mammals and posing hazards for shipping.

While most of the ice pack might have now retreated well north of the continental shelf, it is possible that residual multiyear floes within the tongues of east-to-west transported multiyear ice might provide platforms useful for marine mammals and species across other trophic levels in relatively shallow water. However, the presence of a narrow band of heavy multiyear ice near shore might also give a false perception that heavy multiyear ice persists all the way to the North Pole. As such, any organism that assumed that ice north of the multiyear ice edge offshore of the Alaskan coast in recent years (particularly from 2004 onward) was perennial ice would in fact find itself over first-year ice and potentially a great distance from survivable ice once the summer melt season starts. In addition to ice extent, the nature of the ice itself in terms of the shift from a predominance of ice of several years of age to mainly first-year ice and young multiyear ice may have implications for biological processes and habitat. Ice characteristics such as thickness, salinity, snow cover, and surface and subsurface topography differ considerably between first-year and multiyear ice of different ages.

A decline in Arctic sea ice extent over the second half of the twentieth century was simulated by all of the global climate models participating in the Intergovernmental Panel of Climate Change Fourth Assessment Report (IPCC AR4) (Stroeve et al. 2007). However, the magnitude of satellite-observed negative trends in sea ice extent was not represented by the majority of models, implying the models were too conservative. Possible causes presented by Stroeve et al. (2007 and references therein) included lack of a parameterization of sub-grid scale ice thickness distribution, as well as poor representation of modes of atmospheric variability, ocean circulation, heat convergence, and water column vertical structure. However, in general climate projections suggest that the ice pack will continue to transition toward a nearly entirely seasonal, first-year ice cover (e.g., Overland and Wang 2007; Douglas 2010; Wang et al. 2012). Even so, conditions might be similar to those observed in the last several years, with some multiyear ice persisting along the periphery of the Arctic Ocean adjacent to the high-latitude Canadian Archipelago coast. Some of this ice will be transported into the Canada Basin and Beaufort and Chukchi seas during the winter and as noted above, it would be scattered, low-concentration, residual areas of multiyear ice may or may not survive the melt season. The same situation may hold for the Canadian Archipelago channels, where residual multiyear ice might be transported into the passages during winter. As noted above, this could be significant for habitat and shipping, so the "ice free" summers projected by models might be better termed "nearly ice free". Also, over the coming decades, there will likely be periods (such as in 2010) when there is some recovery of the multiyear ice extent. However, it is likely that these will amount to periods of natural interannual variability superimposed on the overall long-term trend toward a nearly entirely seasonal ice cover. The tendency for two stable states of ice coverage (seasonal vs. multiyear) to persist may continue, with occasional extreme conditions promoting one ice state over the other. That being said, however, it seems that the nature of the extreme events favors a shift from thicker multiyear to thinner first-year ice rather than vice versa.

3.6 Summary

Sea ice across the Pacific Arctic Region (PAR) is highly seasonally variable, with sea ice existing only a few days each year in the most southern reaches of the sea ice zone of the northern Bering Sea, to nearly year-round coverage sea ice in the northern Chukchi Sea. Over the past several decades, the sea ice of the PAR has experienced some of the most dramatic changes in areal coverage, age, and thickness of the entire Arctic Ocean basin. The most extreme trends in sea ice cover across the PAR have occurred within portions of the Chukchi Sea, where annual sea ice persistence has declined by more than 30 days/decade over the 30 year record (1979–2008) of satellite observations. This decline results from earlier sea ice breakup (~10 days/decade earlier), and even more so by later sea ice formation (~20 days/decade later). In terms of sea ice age, the PAR has become a region of multiyear ice loss, behaving more like a sub-Arctic sea than the historical Arctic region dominated by perennial cover. The loss of multiyear ice is particularly striking in September, where the fractional coverage north of Bering Strait diminished from 80 % in 1985 to less than 30 % in 2010. Changes in sea ice thickness are also most dramatic in the Chukchi Sea, with significant downward trends during June–December of 1979–2004. While September shows the highest trends in the Chukchi Sea (−51.2 cm/decade), recent sharp declines occur in all months beginning in 2001. Climate projections suggest that sea ice across the PAR will continue on a path toward a nearly entirely seasonal, first-year ice cover, which by its nature will be thinner and less seasonally persistent, with the possibility of some continued presence of multiyear ice through ridging and transport from the Arctic Basin to north of the Canadian Archipelago. As these trends continue, the changing sea ice will undoubtedly have profound impacts on ecosystem productivity across all trophic levels within the PAR.

Acknowledgements This research was in part supported by the NASA Cryospheric Sciences Program (Grant NNX10AH71G to K. Frey, and Grants NNX07AR21G and NNX11AI48G to J. Maslanik) and the NSF Arctic Sciences Division (Grants ARC-0804773 and ARC-1107645 to K. Frey, and Grant ARC-0712950 to J. Maslanik). P. Panday is thanked for assistance in satellite data processing. Funding support for the development, integration, and analyses of results from the NAME model (to W. Maslowski and J. Clement Kinney) was provided by multiple grants from the Arctic System Science (ARCSS) Program of the National Science Foundation, the Climate Change Prediction Program of the Department of Energy, and the Office of Naval Research. Computer resources were provided by the Department of Defense High Performance Computer Modernization Program (DOD/HPCMP).

References

Arrigo KR, van Dijken GL (2011) Secular trends in Arctic Ocean net primary production. J Geophys Res 116:C09011. doi:10.1029/2011JC007151

Arrigo KR, Perovich DK, Pickart RS, Brown ZW, van Dijken GL, Lowry KE, Mills MM, Palmer MA, Balch WM, Bahr F, Bates NR, Benitez-Nelson C, Bowler B, Brownlee E, Ehn JK, Frey KE, Garley R, Laney SR, Lubelczyk L, Mathis J, Matsuoka A, Mitchell BG, Moore GWK, Ortega-Retuerta E, Pal S, Polashenski CM, Reynolds RA, Scheiber B, Sosik HM, Stephens M, Swift JG (2012) Massive phytoplankton blooms under Arctic sea ice. Science 336:1408. doi:10.1126/science.1215065

Belchansky GI, Douglas DC, Alpatsky IV, Paltonov NG (2004) Spatial and temporal multiyear sea ice distributions in the Arctic: a neural network analysis of SMM/I data, 1988–2001. J Geophys Res 109:C10017. doi:10.1029/2004JC002388

Cavalieri DJ, Parkinson CL, Vinnikov KY (2003) 30-year satellite record reveals contrasting Arctic and Antarctic decadal sea ice variability. Geophys Res Lett 30:1970. doi:10.1029/2003GL018031

Cavalieri D, Parkinson C, Gloersen P, Zwally HJ (2008) Sea ice concentrations from Nimbus-7 SMMR and DMSP SSM/I passive microwave data. National Snow and Ice Data Center, Boulder. Digital media

Clement JL, Maslowski W, Cooper L, Grebmeier J, Walczowski W (2005) Ocean circulation and exchanges through the northern Bering Sea – 1979–2001 model results. Deep Sea Res II 52:3509–3540. doi:10.1016/j.dsr2.2005.09.010

Comiso JC (2002) A rapidly declining perennial sea ice cover in the Arctic. Geophys Res Lett 29. doi:10.1029/2002GL015690

Comiso J (2009) Polar oceans from space, vol 41, Atmospheric and oceanographic sciences library. Springer, New York. doi:10.1007/978-0-387-68300-3

Comiso JC, Parkinson CL, Gersten R, Stock L (2008) Accelerated decline in the Arctic sea ice cover. Geophys Res Lett 35:L01703. doi:10.1029/2007GL031972

Douglas DC (2010) Arctic sea ice decline: projected changes in timing and extent of sea ice in the Bering and Chukchi Seas, U.S. Geological Survey open-file report 2010–1176. U.S. Department of the Interior/U.S. Geological Survey, Reston, 32 pp

Fowler C, Emery WJ, Maslanik J (2004) Satellite-derived evolution of Arctic sea ice age: October 1978 to March 2003. IEEE Geosci Remote Sens Lett 1:71–74. doi:10.1109/LGRS.2004.824741

Frey KE, Perovich DK, Light B (2011) The spatial distribution of solar radiation under a melting Arctic sea ice cover. Geophys Res Lett 38:L22501. doi:10.1029/2011GL049421

Fu H, Zhao J, Frey KE (2012) Investigation of polynya dynamics in the northern Bering Sea using greyscale morphology image-processing techniques. Int J Remote Sens 33:2214–2232. doi:10.1080/01431161.2011.608088

Giles KA, Laxon SW, Ridout AL (2008) Circumpolar thinning of Arctic sea ice following the 2007 record ice extent minimum. Geophys Res Lett 35:L22502. doi:10.1029/2008GL035710

Haas C (2004) Late-summer sea ice thickness variability in the Arctic Transpolar Drift 1991–2001 derived from ground-based electromagnetic sounding. Geophys Res Lett 31:L09402. doi:10.1029/2003GL019394

Holland MM, Bitz CM, Tremblay B (2006) Future abrupt reductions in the summer Arctic sea ice. Geophys Res Lett 33:L23503. doi:10.1029/2006GL028024

Holloway G, Sou T (2002) Has Arctic sea ice rapidly thinned? J Clim 15:1691–1701

Jakobsson M, Cherkis N, Woodward J, Macnab R, Coakley B (2000) New grid of Arctic bathymetry aids scientists and mapmakers. Eos Trans Am Geophys Union 81(9):89

Johannessen OM, Shalina EV, Miles MW (1999) Satellite evidence for an arctic sea ice cover in transformation. Science 286:1937–10939. doi:10.1126/science.286.5446.1937

Kahru M, Brotas V, Manzano-Sarabia M, Mitchell BG (2010) Are phytoplankton blooms occurring earlier in the Arctic? Glob Chang Biol. doi:10.1111/j.1365-2486.2010.02312.x

Kwok R (2007) Near zero replenishment of the Arctic multiyear sea ice cover at the end of 2005 summer. Geophys Res Lett 34:L05501. doi:10.1029/2006/GL028737

Kwok R, Cunningham GF (2008) ICESat over Arctic sea ice: estimation of snow depth and ice thickness. J Geophys Res 113:C08010. doi:10.1029/2008JC004753

Kwok R, Cunningham GF (2010) Contribution of melt in the Beaufort Sea to the decline in Arctic multiyear sea ice coverage: 1993–2009. Geophys Res Lett 37:L20501. doi:10.1029/2010GL044678

Kwok R, Rothrock DA (2009) Decline in Arctic sea ice thickness from submarine and ICESat records: 1958–2008. Geophys Res Lett 36:L15501. doi:10.1029/2009GL039035

Kwok R, Cunningham GF, Zwally HJ, Yi D (2007) Ice, Cloud, and land Elevation Satellite (ICESat) over Arctic sea ice: retrieval of freeboard. J Geophys Res 112:C12013. doi:10.1029/2006JC003978

Kwok R, Cunningham GF, Wensnahan M, Rigor I, Zwally HJ, Yi D (2009) Thinning and volume loss of the Arctic Ocean sea ice cover: 2003–2008. J Geophys Res 114:C07005. doi:10.1029/2009JC005312

Kwok R, Toudal Pedersen L, Gudmandsen P, Pang SS (2010) Large sea ice outflow into the Nares Strait in 2007. Geophys Res Lett 37:L03502. doi:10.1029/2009GL041872

Laxon S, Peacock N, Smith D (2003) High interannual variability of sea ice thickness in the Arctic region. Nature 425:947–950

Lindsay RW, Zhang J, Schweiger AJ, Steele MA, Stern H (2009) Arctic sea ice retreat in 2007 follows thinning trend. J Clim 22:165–176. doi:10.1175/2008JCLI2521

Maslanik J, Drobot S, Fowler C, Emery W, Barry R (2007a) On the Arctic climate paradox and the continuing role of atmospheric circulation in affecting sea ice conditions. Geophys Res Lett 34:L03711. doi:10.1029/2006GL028269

Maslanik JA, Fowler C, Stroeve J, Drobot S, Zwally J, Yi D, Emery W (2007b) A younger, thinner Arctic ice cover: increased potential for rapid, extensive sea-ice loss. Geophys Res Lett 34:L24501. doi:10.1029/2007GL032043

Maslanik J, Stroeve J, Fowler C, Emery W (2011) Distribution and trends in Arctic sea ice age through spring 2011. Geophys Res Lett 38:L13502. doi:10.1029/2011GL047735

Maslowski W, Lipscomb WH (2003) High-resolution simulations of Arctic sea ice during 1979–1993. Polar Res 22:67–74

Maslowski W, Marble D, Walczowski W, Schauer U, Clement JL, Semtner AJ (2004) On climatological mass, heat, and salt transports through the Barents Sea and Fram Strait from a pan-Arctic coupled ice-ocean model simulation. J Geophys Res 109:C03032

Maslowski W, Clement JL, Jakacki J (2007) Towards prediction of environmental Arctic change. Comput Sci Eng 9:29–34

Maslowski W, Clement Kinney J, DMarble DC, Jakacki J (2008a) In: Hecht MW, Hasumi H (eds) Ocean modeling in an eddying regime, vol 177, Geophysical monograph series., 350 pp

Maslowski W, Roman R, Kinney JC (2008b) Effects of mesoscale eddies on the flow of the Alaskan Stream. J Geophys Res 113:C07036. doi:10.1029/2007JC004341

McLaren AS (1989) The under-ice thickness distribution of the Arctic Basin as recorded in 1958 and 1970. J Geophys Res 94:4971–4983

McNamara TP (2006) Determination of changes in the state of the Arctic Icepack using the NPS pan-Arctic coupled ice-ocean model. MS thesis, Naval Postgraduate School, Monterey, CA. http://theses.nps.navy.mil/06Mar_McNamara.pdf

Nghiem SV, Chao Y, Neumann G, Li P, Perovich DK, Street T, Clemente-Colon P (2006) Depletion of perennial sea ice in the East Arctic Ocean. Geophys Res Lett 33:L17501. doi:10.1029/2006GL027198

Nghiem SV, Rigor IG, Perovich DK, Clemente-Colón P, Weatherly JW, Neumann G (2007) Rapid reduction of Arctic perennial sea ice. Geophys Res Lett 34:L19504. doi:10.1029/2007GL031138

Overland JE, Wang M (2005) The Arctic climate paradox: the recent decreases of the Arctic Oscillation. Geophys Res Lett 32:L06701. doi:10.1029/2004GL021752

Overland JE, Wang M (2007) Future regional Arctic sea ice declines. Geophys Res Lett 34:L17705. doi:10.1029/2007GL030808

Parkinson CL, Cavalieri DJ, Gloersen P, Zwally HF, Comiso JC (1999) Arctic sea ice extents, areas, and trends, 1978–1996. J Geophys Res 104:20837–20856

Perovich DK, Light B, Eicken H, Jones KF, Runciman K, Nghiem SV (2007) Increasing solar heating of the Arctic Ocean and adjacent seas, 1979–2005: attribution and role in the ice-albedo feedback. Geophys Res Lett 34:L19505. doi:10.1029/2007GL031480

Perovich D, Meier W, Maslanik J, Richter-Menge J (2011) Sea ice [in Arctic Report Card 2011]. http://www.arctic.noaa.gov/reportcard

Polyakov I et al (2007) Observational program tracks Arctic Ocean transition to a warmer state. Eos Trans Am Geophys Union 88(40):398. doi:10.1029/2007EO400002

Proshutinsky A, Yu A, Johnson MA (1997) Two circulation regimes of the wind-driven Arctic Ocean. J Geophys Res 102(C6):12493–12514

Rigor IG, Wallace JM (2004) Variations in the age of Arctic sea-ice and summer sea-ice extent. Geophys Res Lett 31:L09401. doi:10.1029/2004GL019492

Rigor IG, Wallace JM, Colony RL (2002) On the response of sea ice to the Arctic Oscillation. J Clim 15(18):2546–2663

Rothrock DA, Yu Y, Maykut GA (1999) Thinning of the Arctic sea-ice cover. Geophys Res Lett 26:3469–3472

Rothrock DA, Percival DB, Wensnahan M (2008) The decline in arctic sea-ice thickness: separating the spatial, annual, and interannual variability in a quarter century of submarine data. J Geophys Res 113:C05003. doi:10.1029/2007JC004252

Serreze MC, Maslanik JA, Scambos TA, Fetterer F, Stroeve J, Knowles K, Fowler C, Drobot S, Barry RG, Haran TM (2003) A record minimum arctic sea ice extent and area in 2002. Geophys Res Lett 30(3):1110. doi:10.1029/2002GL016406

Shimada K, Kamoshida T, Itoh M, Nishino S, Carmack E, McLaughlin F, Zimmermann S, Proshutinsky A (2006) Pacific Ocean inflow: influence on catastrophic reduction of sea ice cover in the Arctic Ocean. Geophys Res Lett 33:L08605. doi:10.1029/2005GL025624

Slagstad D, Ellingsen IH, Wassmann P (2011) Evaluating primary and secondary production in an Arctic Ocean void of summer sea ice: an experimental simulation approach. Prog Oceanogr 90:117–131

Spreen G, Kaleschke L, Heygster G (2008) Sea ice remote sensing using AMSR-E 89 GHz channels. J Geophys Res 113:C02S03. doi:10.1029/2005JC003384

Steele M, Morley R, Ermold W (2001) PHC: a global ocean hydrography with a high quality Arctic Ocean. J Clim 14:2079–2087

Steele M, Ermold W, Zhang J (2008) Arctic Ocean surface warming trends over the past 100 years. Geophys Res Lett 35:L02614. doi:10.1029/2007GL031651

Stroeve J, Maslowski W (2007) Arctic sea ice variability during the last half century. In: Brönnimann S, Luterbacher J, Ewen T, Diaz HF, Stolarski RS, Neu U (eds) Climate variability and extremes during the past 100 years. Springer, Dordrecht, pp 143–154

Stroeve JC, Serreze MC, Fetterer F, Arbetter T, Meier W, Maslanik J, Knowles K (2005) Tracking the Arctic's shrinking ice cover: another extreme September minimum in 2004. Geophys Res Lett 32. doi:10.1029/2004GL021810

Stroeve J, Holland MM, Meier W, Scambos T, Serreze M (2007) Arctic sea ice decline: faster than forecast. Geophys Res Lett 34:L09501. doi:10.1029/2007GL029703

Stroeve J, Serreze M, Drobot S, Gearheard S, Holland M, Maslanik J, Meier W, Scambos T (2008) Arctic sea ice extent plummets in 2007. Eos Trans Am Geophys Union 89(2):13. doi:10.1029/2008EO020001

Stroeve JC, Maslanik JA, Serreze MC, Rigor I, Meier W, Fowler C (2011) Sea ice response to an extreme negative phase of the Arctic Oscillation during winter 2009/2010. Geophys Res Lett 38:L02502. doi:10.1029/2010GL045662

Stroeve JC, Serreze MC, Holland MM, Kay JE, Maslanik J, Barrett AP (2012) The Arctic's rapidly shrinking sea ice cover: a research synthesis. Clim Chang 110:1005–1027

Wadhams P (1990) Evidence for thinning of the Arctic ice cover north of Greenland. Nature 345:795–797

Wang M, Overland JE (2009) A sea ice free summer Arctic within 30 years? Geophys Res Lett 36:L07502. doi:10.1029/2009GL037820

Wang J, Zhang J, Watanabe E, Ikeda M, Mizobata K, Walsh JE, Bai X, Wu B (2009) Is the Dipole Anomaly a major driver to record lows in Arctic summer sea ice extent? Geophys Res Lett 36:L05706

Wang M, Overland JE, Stabeno P (2012) Future climate of the Bering and Chukchi Seas projected by global climate models. Deep Sea Res Part II 65–70:46–57

Whelan J (2007) Understanding recent variability in the Arctic sea ice cover – synthesis of model results and observations. MS thesis, Naval Postgraduate School, Monterey, CA. https://edocs.nps.edu/npspubs/scholarly/theses/2007/Sep/07Sep_Whelan.pdf

Woodgate RA, Aagaard K, Weingartner TJ (2006) Interannual changes in the Bering Strait fluxes of volume, heat and freshwater between 1991 and 2004. Geophys Res Lett 33:L15609. doi:10.1029/2006GL026931

Zhang J (2005) Warming of the arctic ice-ocean system is faster than the global average since the 1960s. Geophys Res Lett 32:L19602. doi:10.1029/2005GL024216

Chapter 4
Abrupt Climate Changes and Emerging Ice-Ocean Processes in the Pacific Arctic Region and the Bering Sea

Jia Wang, Hajo Eicken, Yanling Yu, Xuezhi Bai, Jinlun Zhang, Haoguo Hu, Dao-Ru Wang, Moto Ikeda, Kohei Mizobata, and James E. Overland

Abstract The purpose of this chapter is to reveal several emerging physical ice-ocean processes associated with the unprecedented sea ice retreat in the Pacific Arctic Region (PAR). These processes are closely interconnected under the scenario of diminishing sea ice, resulting in many detectable changes from physical environ-

J. Wang (✉)
Great Lakes Environmental Research Laboratory (GLERL),
National Oceanic and Atmospheric Administration, Ann Arbor, MI, USA
e-mail: jia.wang@noaa.gov

H. Eicken
Geophysical Institute, University of Alaska Fairbanks, Fairbanks, AK, USA

Y. Yu
Arctic Science Center, Applied Physics Laboratory, University of Washington, Seattle, WA, USA

J. Zhang
Applied Physics Laboratory, University of Washington, Seattle, WA, USA

X. Bai • H. Hu
Cooperative Institute for Limnology and Ecosystems Research (CILER), School of Natural Resources and Environment, University of Michigan, Ann Arbor, MI, USA

D.-R. Wang
Hainan Marine Development and Design Institute, Hainan, China

M. Ikeda
Graduate School of Earth and Environmental Sciences, Hokkaido University, Sapporo, Japan

K. Mizobata
Department of Ocean Sciences, Tokyo University of Marine Science and Technology, Tokyo, Japan

J.E. Overland
Pacific Marine Environmental Laboratory, National Oceanic and Atmospheric Administration, 7600 Sand Point Way NE, Seattle, WA 98115, USA

J.M. Grebmeier and W. Maslowski (eds.), *The Pacific Arctic Region: Ecosystem Status and Trends in a Rapidly Changing Environment*, DOI 10.1007/978-94-017-8863-2_4,

ment to ecosystems. Some of these changes are unprecedented and have drawn the attention of both scientific and societal communities. More importantly, some mechanisms responsible for the diminishing sea ice cannot be explained by the leading Arctic Oscillation (AO), which has been used to interpret most of the changes in the Arctic for the last several decades. The new challenging questions are: (1) What is the major forcing? (2) Is the AO, the DA, or their combination, contributing to the sea ice minima in recent years? How do we use models to investigate the recent changes in the PAR. Is the heat transport through the Bering Strait associated with the DA? What processes accelerate sea ice melting in the PAR?

Keywords Arctic Dipole Anomaly • Arctic summer ice minima • Landfast ice • Ice/ocean albedo feedback

4.1 Introduction

The northern North Pacific Ocean, including the Bering Sea, is among the most productive marine ecosystems in the world, as evidenced by large populations of marine fish, birds, and mammals. Fish and shellfish from these regions constitute more than 10 % of total seafood harvest of the world and about 52 % of the U.S. (PMEL 2000). As a result, the productivity is critical not only to the U.S. economy, but also to the economy of surrounding countries.

The Bering Sea (Fig. 4.1) is a complex semi-enclosed sea with a shallow, broad shelf, a shelf break, and deep basins. The ocean circulation pattern is complicated. The Alaskan Stream (AS) mainly flows along the Aleutian Peninsula and provides some of its water via Aleutian passes to the Bering Slope Current (BSC) and to the Aleutian North Slope Current (ANSC; Stabeno et al. 1999; Hu and Wang 2010). The BSC extends into two coastal currents: the Anadyr Current (AC), which flows along eastern Siberia into the Chukchi Sea through the western side of the Bering Strait, and the East Kamchatka Current (EKC), which flows southwestward along the Koryak and Kamchatka peninsula. Numerous sites of mesoscale eddy generation possibly exist within the AS and BSC, due to baroclinic instability interacting with the sloping shelfbreak (Wang and Ikeda 1997; Mizobata et al. 2002, 2006, 2008). The basic schematic surface circulation pattern, based on available observational evidence (Kinder and Coachman 1978; Stabeno and Reed 1994), is fairly well known (Fig. 4.1). The interaction or exchange between the shelf and deep basins is a typical phenomenon that significantly influences primary and secondary productivities (Mizobata et al. 2002; Mizobata and Saitoh 2004).

In the Bering Sea and the southern Chukchi Sea, seasonal sea-ice cover is an important predictor of regional climate (Niebauer 1980; Wang and Ikeda 2001). Sea-ice extent also influences ocean circulation patterns, thermal structure, water stratification, and deep convection (Wang et al. 2010) because: (1) wind stress drag is different in magnitude over the water surface with and without ice cover (Pease et al. 1983); (2) the albedo of ice differs from that of water; thus, prediction of the sea-ice extent (i.e., edge) is crucial to predicting the ocean mixed layer and ocean

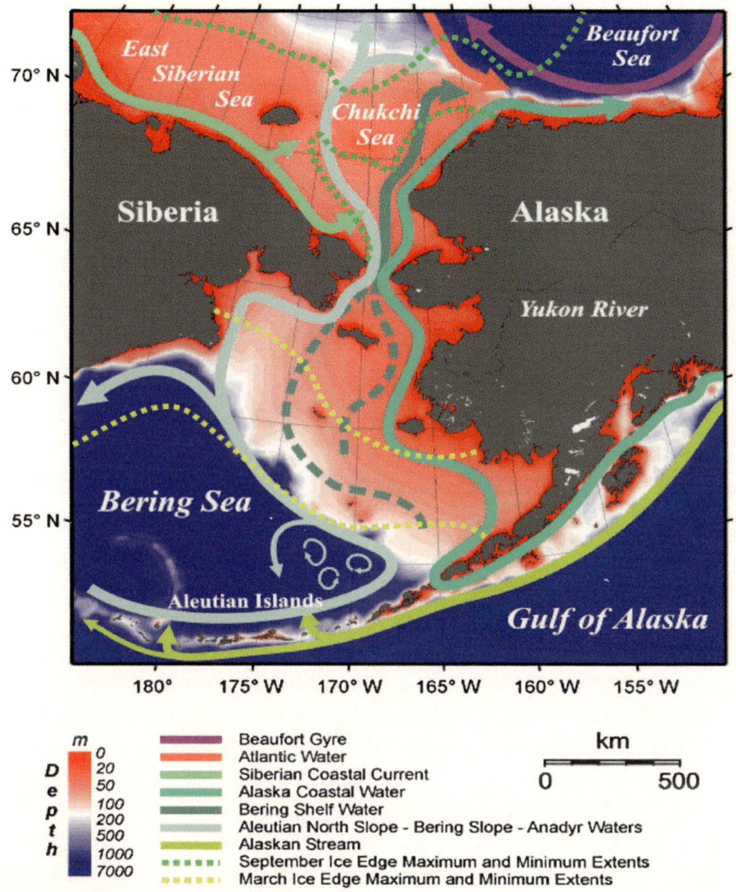

Fig. 4.1 Topography and bathymetry of the Bering Sea and the western/Pacific Arctic region (*PAR*), and schematic circulation systems (by *colored arrows*). Water depths are in meters (Courtesy of T. Weingartner, University of Alaska Fairbanks)

circulation, and thus, to predicting primary and secondary productivities (Springer et al. 1996); and (3) spring ice-melt freshwater increases stratification in the upper layer, which may enhance phytoplankton blooms. In addition, climate change, through its effect on the timing of ice melting, would determine the timing of phytoplankton and zooplankton blooms. As a result, sea-ice conditions and the ecosystem in the Bering Sea are driven mainly by atmospheric and oceanic forcing, from tidal, synoptic, and seasonal to interannual and decadal time scales.

The Western Arctic Ocean, with a freshwater and heat pathway to the Bering Sea via the Bering Strait, includes the Eastern Siberian Sea, the Chukchi Sea, and the Beaufort Sea (Fig. 4.1). There are two important current systems in the regions: the Alaska Coastal Current (ACC) and the Eastern Siberian Current (ESC), which are both driven by the freshwater. Two branches flow through the Central Channel and Herald Canyan (Woodgate et al. 2005), and join the Beaufort Slope Current

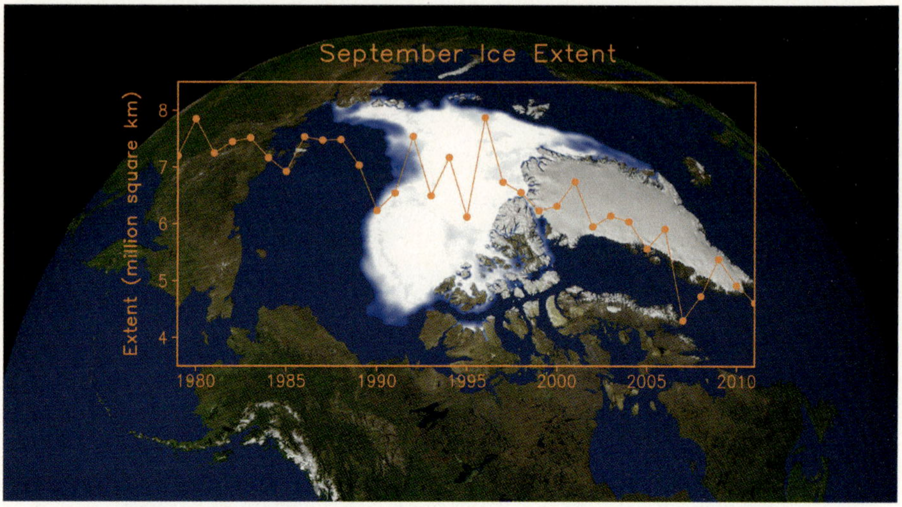

Fig. 4.2 Time series (1979–2011) of average September sea ice extent from satellite measurements. The background is the spatial distribution of the recording-breaking minimum sea ice extent on September 14, 2007 (all time low)

(Pickart 2004). The surface Beaufort Gyre is an anticyclonic circulation for both the ocean and sea ice, driven by the Beaufort high pressure system. Significant changes in sea ice cover and ice-related physical and ecological phenomena occurred in the last decade, particularly in the PAR. In particular, the extent of Arctic sea ice reached its all time low in September 2007, shattering all previous lows (Serreze et al. 2007; Comiso et al. 2008; Gascard et al. 2008; Wang et al. 2009a) since satellite record-keeping began nearly 30 years ago (Fig. 4.2). The Arctic sea ice extent in September 2007 stood at $4.3 \cdot 10^6$ km^2. Compared to the long-term (1979–2000) minimum average, the new minimum extent was lower by about $2.56 \cdot 10^6$ km^2 – an area about the size of Alaska and Texas combined, or 10 United Kingdoms. The cause of this significant ice loss was thought to be the combined effects of Arctic Oscillation (AO)-induced warming (Wang and Ikeda 2000) due to natural climate variability and exporting of multiyear ice (Rigor and Wallace 2004; Steele et al. 2004), the climate warming trend due to greenhouse gases (caused by anthropogenic activities), and the culmination of an ice/ocean-albedo positive feedback (Ikeda et al. 2001, 2003; Wang et al. 2005b).

As the Arctic environment changes at a faster rate than the rest of the world, an emerging concern is how soon the Arctic Ocean will become ice-free in summer (Wang and Overland 2009). The diminishing summer ice cover in the western Arctic can have significant impacts on the Arctic and subarctic marine ecosystems (Jin et al. 2011; Deal et al. 2014, this volume), including a lengthened algae bloom period due to increasing absorption of solar radiation (Grebmeier et al. 2006). The present Bering ecosystems and food webs have been experiencing significant changes due to an increase in water temperature (Overland and Stabeno 2004;

Grebmeier et al. 2010). The Chukchi Sea ecosystems will experience significant changes due to more open water, lengthened growth seasons, and an abundance of biomass. Seasonal landfast ice along the Alaskan Arctic coast (Eicken et al. 2006; Mahoney et al. 2007) would melt early and form late, causing a shortening of whaling and fishing seasons (autumn and spring) for the local community. These emerging consequences will alter not only the ecosystems (Jin et al. 2011; Deal et al. 2014, this volume), but also the lifestyle of the native community and commercial activity in the Arctic. Such impacts are far-reaching and inherently interdisciplinary.

Since 1995, the annual sea ice variability shows that a series of record lows in summer Arctic sea ice extent were set (Fig. 4.2). During the same time, the AO index became mostly neutral or even negative (Overland et al. 2008; Maslanik et al. 2007), suggesting a weak link between the AO and the rapid sea ice retreat. Whenever the Arctic sea ice has reached a new minimum, searching for mechanisms responsible for the individual year's event was appealing (Serreze et al. 2007; Nghiem et al. 2007; Gascard et al. 2008). However, there has been no convincing physical explanation that accounts for the complete series of such events (1995, 1999, 2002, 2005, 2007, and 2008). Wang et al. (2009a) proposed that the Arctic Dipole Anomaly (DA) pattern is the major forcing in advecting sea ice out of Arctic Ocean under the thinning ice conditions (which is due to long-term cumulative thermodynamic effect, or ice/ocean albedo feedback), causing a series of Arctic ice minima since 1995.

This chapter, in addition to Overland et al. (2014, this volume) provides a large picture of the recent and future change in meteorology in the PAR, provides an discussion and explanation of several emerging sea ice-related processes in the PAR and the Bering Sea, and is organized as follows: After the introduction, Sect. 4.2 describes data and methods used. Section 4.3 investigates the DA and summer sea ice minima. Section 4.4 shows the modeling studies of ice minima and their relationship to the dynamic (wind) and thermodynamic (including ice/ocean albedo feedback) forcings. Section 4.5 investigates the Bering Strait heat transport (Clement Kinney et al. 2009, 2014, this volume) and its possible relation to the DA. Section 4.6 discusses the modeling studies of landfast ice along Beaufort coast using a Coupled Ice-Ocean Model (CIOM). Section 4.7 investigates seasonal and interannual variability of the Bering Sea cold pool using the CIOM. A positive air-ice-sea feedback loop is proposed and discussed in Sect. 4.8. Section 4.9 summarizes the conclusions and proposes future work.

4.2 Data and Methods

The average September sea ice extent, archived at the National Snow and Ice Data Center (NSIDC), was obtained from SMMR (Scanning Multichannel Microwave Radiometer) for 1978–1987 and SSM/I (Special Sensor Microwave Imager) for the period 1987 to present based on the NASA Goddard algorithm (Comiso et al. 2008). The monthly NCEP (National Centers for Environmental Prediction) Reanalysis dataset from 1948 to 2010 were used to derive the EOF (Empirical Orthogonal

Function) modes for individual seasons: winter (DJF), spring (MAM), summer (JJA), and autumn (SON). Oceanic heat transport via the Bering Strait was calculated using in situ shipboard measurements and satellite-measured SST across the Bering Strait (Mizobata et al. 2010). A Pan-arctic Ice-Ocean Modeling and Assimilation System (PIOMAS, Zhang et al. 2008a, b) was used to simulate the sea ice and ocean circulation for the period 1978–2009 using daily NCEP forcing.

Regional CIOM was used in the Chukchi, Beaufort, and Bering Seas to investigate the ice-ocean systems and the responses to climate change in recent years. For a detailed description of development of the CIOM, readers should refer to Yao et al. (2000) and Wang et al. (2002, 2010), which was applied to the pan-Arctic Ocean (Wang et al. 2004, 2005a; Wu et al. 2004; Long et al. 2012), Chukchi-Beaufort seas (Wang et al. 2003, 2008), and the Bering Sea (Wang et al. 2009b; Hu and Wang 2010; Hu et al. 2011). The ocean model used is the Princeton Ocean Model (POM) (Blumberg and Mellor 1987), and the ice model used is a full thermodynamic and dynamics model (Hibler 1979, 1980) that prognostically simulates sea-ice thickness, sea ice concentration (SIC), ice edge, ice velocity, and heat and salt flux through sea ice into the ocean.

4.3 Leading Climate Forcing: Arctic Dipole (DA) Pattern

The EOF analysis of the NCEP reanalysis dataset (1948–2010) shows the AO as the first mode, and the DA as the second mode (Wang et al. 1995, 2009a; Wu et al. 2006; Watanabe et al. 2006) (Fig. 4.3). The AO has one annular (circled) center covering the entire Arctic, producing zonal wind anomalies. The AO variance (intensity) is strongest in winter (63 %), and decreases in magnitude from spring (61 %), summer (50 %), to autumn (49 %). The DA (Fig. 4.3) differs from the AO in all seasons because the anomalous SLP has two action centers in the Arctic. In contrast to the AO-driven wind anomaly, which is either cyclonic or anticyclonic during the positive or negative phase of the AO (Proshutinsky and Johnson 1997; Wu et al. 2006), the resulting wind anomaly driven by the DA is meridional. During a positive phase of the DA (i.e., the SLP has a positive anomaly in the Canadian Archipelago and negative one in the Barents Sea), the anomalous meridional wind blows from the western to the eastern Arctic, accelerating the Trans-polar Drift Stream (TDS; see the red-dashed arrow) that flushes more sea ice out of the Arctic into the Barents and Greenland Seas (Wu et al. 2006; Watanabe et al. 2006). During the negative phase of the DA, the opposite scenario occurs, i.e., more sea ice remains in the western Arctic (Watanabe et al. 2006).

Note that the intensity of the DA increases from winter (13 %) and spring (14 %) to summer (16 %) and autumn (16 %). The variance ratio of DA to AO is 0.21 for winter, 0.23 for spring, 0.32 for summer, and 0.33 for autumn. This pattern indicates that the DA-derived anomalous wind plays a more important role in summer and autumn during the melting and freezing seasons, or during the thinning ice season, than winter with thicker ice. The second noticeable feature is that the orientation of

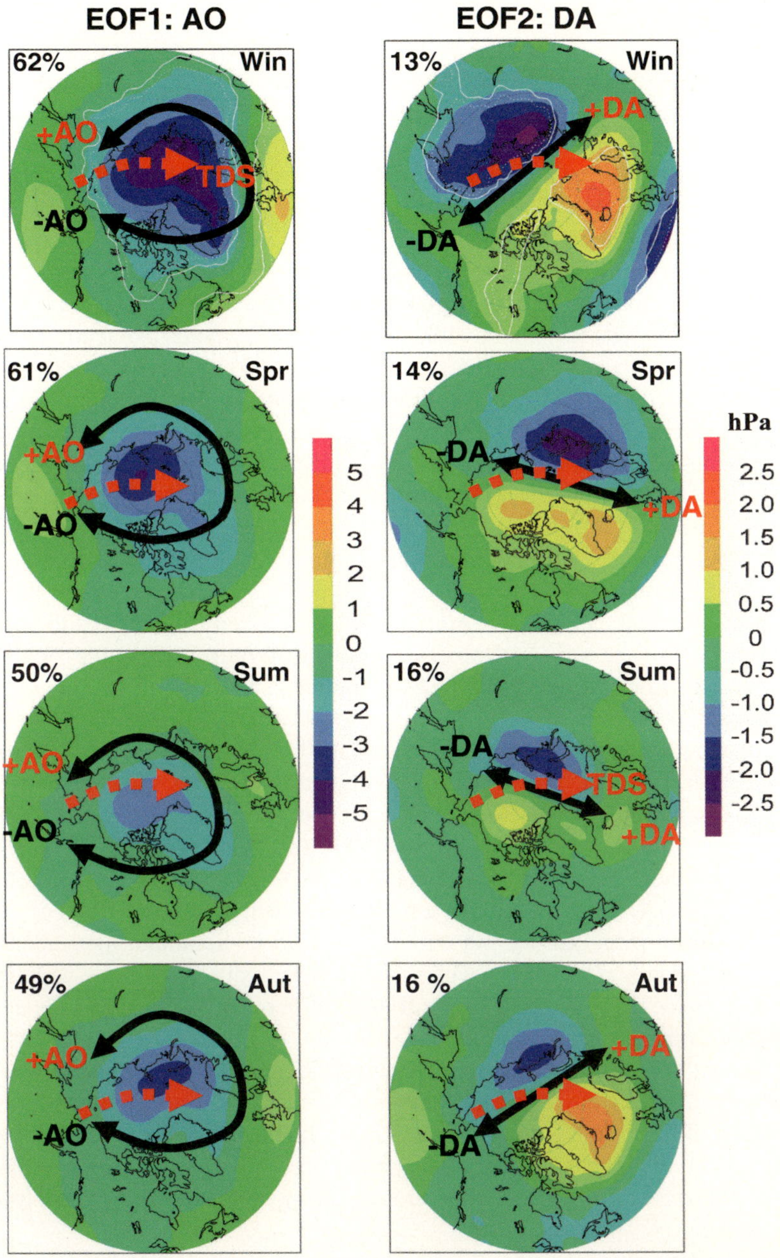

Fig. 4.3 Seasonal variations of regression maps of the first two leading modes (AO and DA) to the winter, spring, summer, and autumn mean Northern Hemisphere SLP field using the NCEP Reanalysis dataset from 1948 to 2010. Contour intervals are 0.5 hPa (see *color bars*). The *black arrows* on the *left column* indicate the cyclonic (anticlockwise, divergent) wind anomaly during the +AO phase (which promotes advection of sea ice out of Arctic via Fram Strait) and anticyclonic (clockwise, convergent) wind anomaly during the –AO phase. On *right column*, the *black arrows* indicate that the wind anomaly blows from the western to the eastern Arctic during the +DA phase that accelerates the TDS (in *red-dashed arrows*), and vice versa during the –DA phase that slows down the TDS

the maximum wind anomalies are more parallel to TDS-Fram Strait during spring and summer than winter and autumn, indicating spring and, in particular, summer DA are more effective in advecting sea ice to the Nordic seas than the winter and autumn during a positive DA phase. The third feature is that both AO and DA have seasonal variations.

During the positive/negative AO (i.e., when Arctic SLP has a negative/positive anomaly), a cyclonic/anticyclonic wind anomaly occurs, indicating a sea ice divergence/convergence. The divergence (anomalous cyclonic circulation) of sea ice leads to anomalous ice export, while the convergence results in retention of sea ice inside the Arctic Ocean (Proshutinsky and Johnson 1997; Wu et al. 2006). Thus, the +DA pattern is the more effective, direct driver of the Arctic sea ice (including multiple year ice (Maslanik et al. 2011)) export from the western Pacific Arctic to the northern Atlantic, as compared to the +AO. Kwok et al. (2004) tried to establish the relationship between the Fram Strait ice flux with the NAO index (see their Fig. 4.6e) using a 24-year time series of sea ice flux, which is not significant. A better, but not at 95 % significance level, linear relationship was obtained after removing the strong –NAO cases (5 years), with large scattering (see their Fig. 4.6f), suggesting the weak correlation. Actually, the best relationship of sea ice flux was found to be with the sea-level pressure (SLP) difference across the Fram Strait (see their Fig. 4.6c), suggesting the DA is the major forcing, rather than NAO/AO (Wang et al. 2009a).

We conducted the cross composite analysis of both phases of the DA and the AO. A matrix was constructed based on the combination of the two leading drivers that account for about 65–76 % of the total variance. Following Wang et al. (2009a), the Arctic climate patterns can be defined by the following four climate states using AO and DA indices of larger than ±0.6: (1) +AO/+DA, (2) +AO/–DA, (3) –AO/+DA, and (4) –AO/–DA. It was found that a positive DA is the key, regardless of the sign of the AO, to flushing sea ice out of Arctic due to its dominant meridional wind anomaly (Maslanik et al. 2007), while the wind anomaly driven by a negative DA can retain sea ice in the western Arctic (Fig. 4.3). In other words, the DA is dynamically more important and more effective than the AO in driving sea ice out of the Arctic Basin. These four states can represent major atmospheric circulation patterns in the Arctic, which are the major drivers of sea ice and ocean circulation. The importance of the air temperature advection by the DA is consistent with the recent findings by Overland et al. (2008). They concluded that the warming anomalies in the central Arctic during 2000–2007 are driven by meridional advection from the Pacific sector.

Figure 4.4 shows a scatter plot between the summer DA index (x-axis) and winter-spring mean DA index (y-axis) for the period since 1995. The figure examines whether the persistency of the DA may cause a series of ice minima. Most of the ice minima years fall into the first quadrant, which indicates that, when the +DA persists from winter, spring, all the way to summer, more ice would be advected out of the Arctic Ocean, leading to ice minima in 1995, 2002, 2005, 2007, 2008, and 2009; the 2007 ice minimum is the all time low (see Fig. 4.2). Although the –DA occurs in the winter-spring period when sea ice is normally retained in the Pacific Arctic, a strong +DA in summer can reverse the process, effectively advecting sea

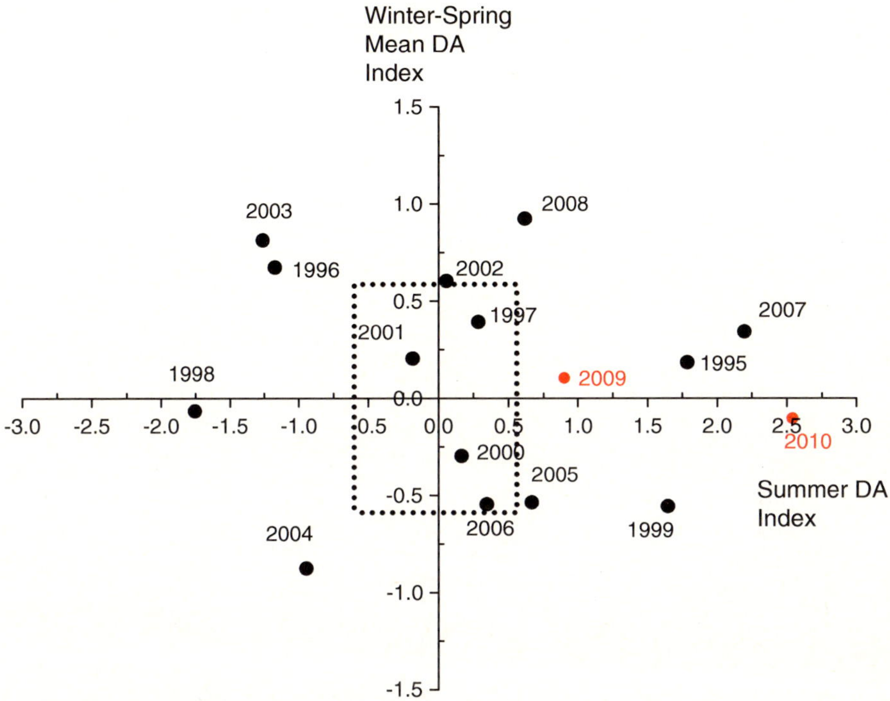

Fig. 4.4 Scatter plot between the summer DA index (x-axis) and winter-spring mean DA index (y-axis) from 1995 to 2010. The box outlined by *dotted lines* indicates the 0.6 index threshold; the indices greater than 0.6 are considered significant DA years

ice to the Nordic seas and leading to ice minima in 1999, 2005, and 2010 (4th quadrant). Therefore, the summer DA is the most critical dynamically, since summer sea ice is more mobile than winter ice because summer sea ice is about 1 m thinner than winter ice. Because the summer anomalies have weaker AO-related SLP signals, the +DA-derived wind anomalies are more effective in reducing sea ice, primarily due to two thermodynamic effects, i.e., advecting warm air from land to the ocean and enhancing melt through a positive ice-albedo feedback (Ikeda et al. 2001; Wang et al. 2005b).

Recently, Ikeda (2012) reveals an important finding that the DA (or Arctic Dipole Mode, ADM) has different frequency spectra between fall-winter (several year period) and spring-summer (biennial). In the 1980s, the most influential mode shifted from the NAM to the ADM, when the Pacific sector had low ice cover at a 1-year lag from the positive ADM, which was marked by low pressure over Siberia. In years when the spring-summer ADM was pronounced, it was responsible for distinct ice variability over the East Siberian–Laptev seas. The frequency separation in this study identifies the contributions of the ADM and spring-summer ADM. Effects of the latter are difficult to predict since it is intermittent and changes its sign biennially. The ADM and spring-summer ADM should be closely watched in relation to the on-going ice reduction in the Pacific–Siberian region.

4.4 Investigating Mechanisms Responsible for Arctic Sea Ice Minima Using PIOMAS

On the basis of the above analyses, the +DA-derived wind forcing is the key to the ice minima. To further confirm such a key mechanism in the summers since 1995, in particular from 2007 to 2010, the PIOMAS (Zhang et al. 2008a) was used to simulate the sea ice and ocean circulation for the period 1978–2009 using daily NCEP forcing. Figure 4.5 shows the August SLP and wind anomalies from 2007 to 2010.

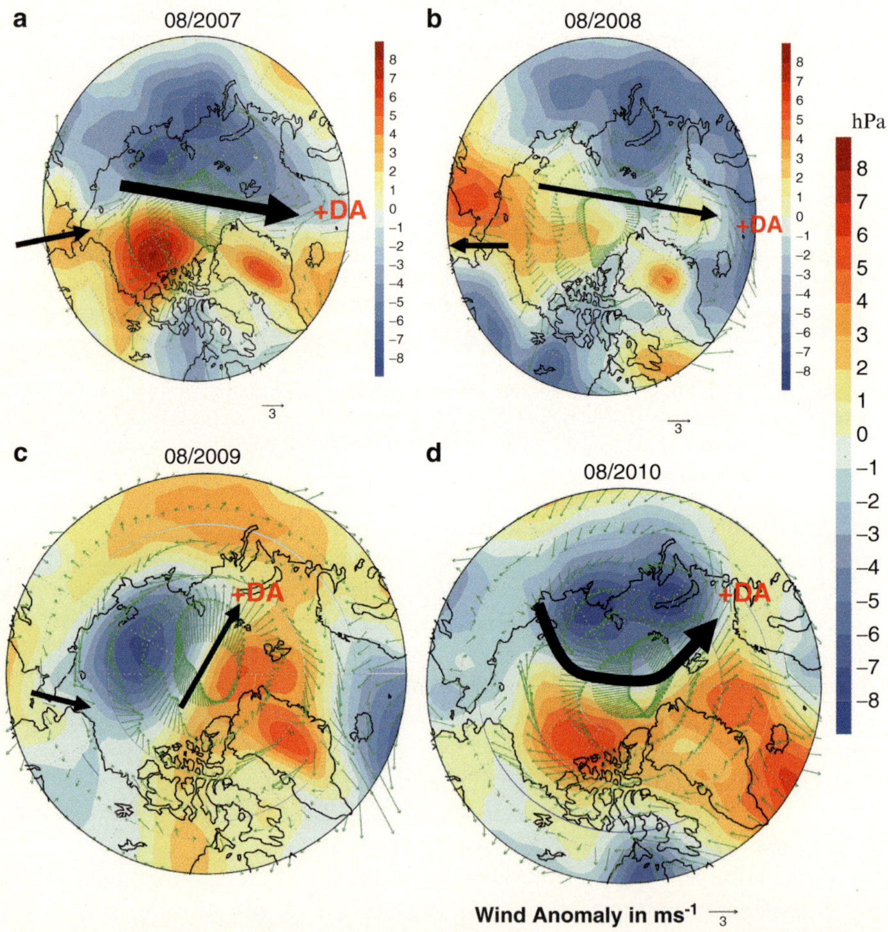

Fig. 4.5 The 2007 (**a**), 2008 (**b**), 2009 (**c**), and 2010 (**d**) August SLP (*shaded*) and wind (*vectors*) anomalies relative to the 1948–2008 mean (data from NCEP Reanalysis). *Red/blue* indicates the positive/negative anomalies in SLP. The *green arrows* indicate the +DA-derived anomalous wind velocity (in ms^{-1}). The DA-derived wind anomaly was the dominating driver for advecting sea ice toward the eastern Arctic, leading to the record minimum in the Arctic

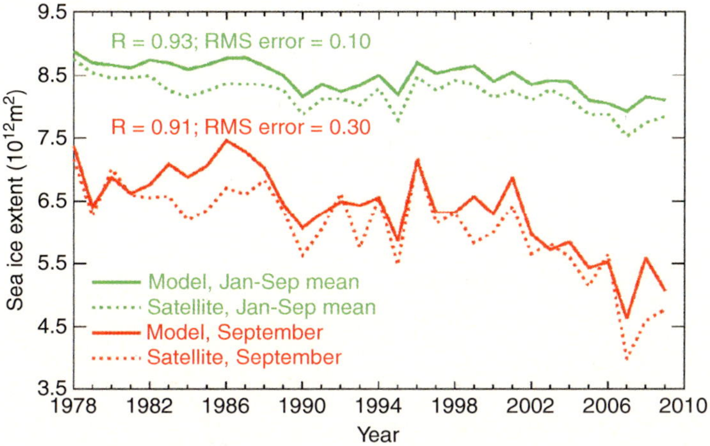

Fig. 4.6 Comparison between the PIOMAS-simulated and SSM/I-observed ice extent for the period 1978–2009

A common feature is that the DA was positive for all these summers. The differences are the magnitude and orientation of the DA-induced wind anomalies. Although the wind anomalies in 2007 were larger than 2008 in magnitude, the orientation for both years was more favorable for advecting sea ice out of the Arctic into the Nordic seas via the Fram Strait and the Barents Sea. Similarly, wind anomalies in 2007 and 2010 have a similar magnitude and both were larger than those in 2008 and 2009; nevertheless, the orientation of wind anomalies in 2010 was less favorable for advecting more ice out of the Arctic than 2007. The orientation of wind anomalies in 2009 was least favorable, since the primary ice advection was towards the Kara Sea.

Figure 4.6 shows that the simulated sea ice area compares well with the satellite-measurements. The correlation between the observed and modeled time series of ice extent is 0.91 in September, while it is 0.93 for January–September. In particular, the model reproduces good agreement with summer ice minima in 1995, 2002, 2005, 2007, 2008, and 2009, though not for 1999. The 2010 daily summer minimum ice extent (4.8 10^6 km^2) was just higher than in 2007 (4.2 10^6 km^2) and 2008 (4.6 10^6 km^2). This indicates that both the orientation and magnitude of the +DA are key for advecting of summer sea ice out of the Arctic.

The ice-ocean system, particularly the ice-ocean albedo feedback, is investigated in response to the strongest, most persistent (from winter to summer) DA event in 2007. To better understand the linkages between the rapid arctic sea ice retreat and the atmospheric changes during summer (July–September) 2007, it is helpful to examine how the changes are reflected in the NCEP/NCAR surface atmospheric forcing that is used to drive the model. The changes in SLP, surface winds, and surface air temperature (SAT) during summer 2007 are illustrated in Fig. 4.7. It compares the 2007 atmospheric conditions with those averaged over 2000–2006, a period of relatively low summer sea ice extent during the past three decades. Here, the 2000–2006 average is referred to as the "recent average" and the

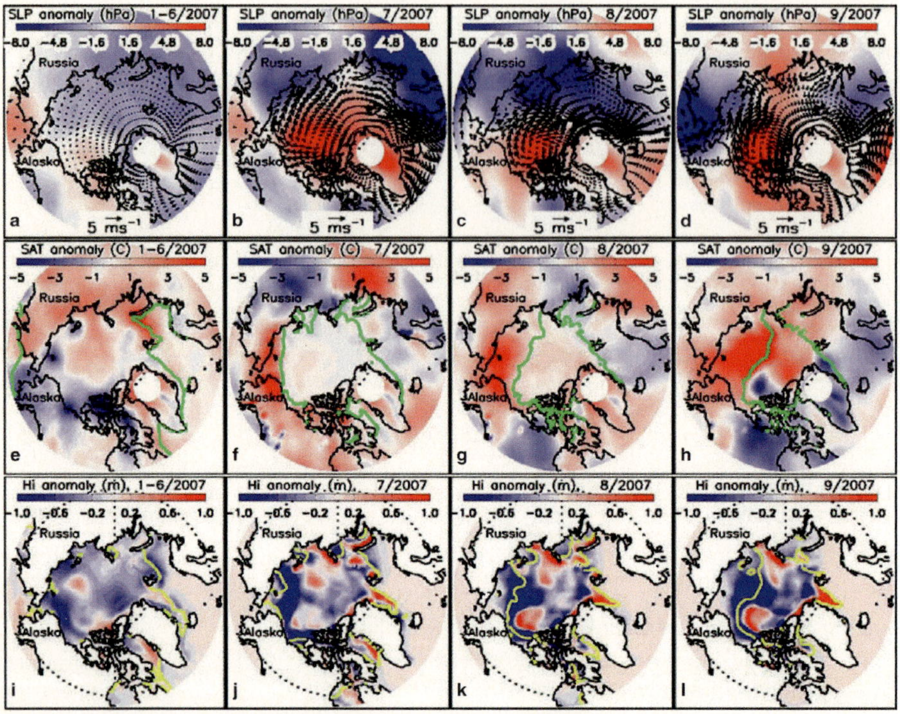

Fig. 4.7 Anomalies of the NCEP/NCAR reanalysis SLP and surface wind (**a–d**) and SAT (**e–h**); anomalies of simulated sea ice thickness (Hi) (**i–l**) In Figs. 4.7, 4.8 and 4.9, an anomaly is defined as the difference between the 2007 value and the 2000–2006 average. One of every 36 wind vectors is plotted in (**a–d**). The *green line* in (**e–h**) represents satellite observed ice edge and *yellow line* in (**i–l**) model simulated ice edge of 2007

difference between a 2007 value and the recent average value is referred to as an "anomaly" (2007 minus the recent average) (Zhang et al. 2008a, b).

As shown in Fig. 4.7a, the Arctic +DA-like SLP and surface wind anomalies were small, but were gradually built up in the first half of 2007. In July, +DA-like SLP was considerably higher in much of the Arctic Basin and lower over a large area in Russia than the recent average (Fig. 4.7b). This is associated with stronger southerly winds in the northern Canada Basin and easterly winds along the East Siberia coast. In August and September, the well-defined +DA (Wang et al. 2009a) was developed: the high SLP anomalous center was mostly confined to the Canada Basin, and a low SLP anomalous center was located in the Barents Sea, producing stronger southerlies in the Pacific sector (Fig. 4.7c–d).

Similar to the SLP and wind anomalies, the arctic SAT anomaly was relatively small in the first half of 2007: SAT was slightly warmer in the central Arctic Basin and slightly cooler in the Chukchi and Beaufort seas compared to the 2000–2006 average (Fig. 4.7e). In July, the SAT anomaly increased significantly in the Chukchi Sea (Fig. 4.7f). In August and September the increase in SAT is most striking: the

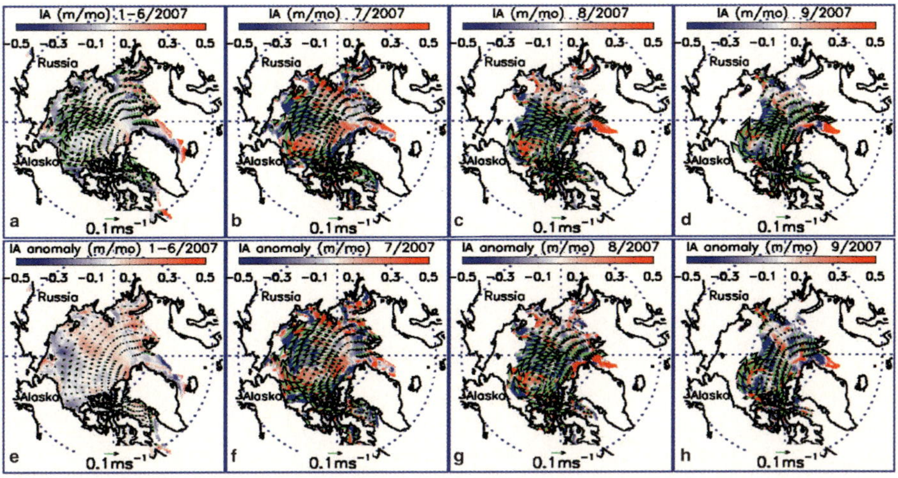

Fig. 4.8 PIOMAS-simulated ice motion (*vectors*) and ice advection (IA) (*color contours*) (**a–d**) and their anomalies (**e–h**). Note the difference between ice motion and ice advection. Ice motion is described by ice velocity and ice advection by ice mass convergence $[-\nabla \cdot (\mathbf{u}h)]$, where u is ice velocity and h is ice thickness. One of every 36 ice velocity vectors is plotted

SAT is up to 5 °C warmer than the recent average over much of the Pacific sector (Fig. 4.7g–h). However, in an area in the Canada Basin, the SAT in September 2007 was lower than the recent average (Fig. 4.7h). This SAT decrease is likely due to the air–ice interactions in that area where ice was thicker than the recent average. Sea ice was also thicker in August in that area (Fig. 4.7k); however, the SAT was not lowered (Fig. 4.7g).

What is the cause of the substantially-reduced ice cover in the Pacific sector during summer 2007? Zhang et al. (2008a, b) found, based on model results, that part (~30 %) of the anomalous reduction in ice extent mainly in the Pacific sector was due to the unusual ice advection, which was caused by the strengthened TDS and the DA-derived meridional wind anomalies. Ice motion in the Arctic Ocean is characterized by an anticyclonic Beaufort gyre and the TDS (Fig. 4.8a–d). In the first half of 2007, the SLP and wind anomalies were small (Fig. 4.7a), so changes in ice motion and changes in ice thickness due to ice advection were small in comparison to the 2000–2006 average (Fig. 4.8e). In July, there were stronger southerly winds in the northern Canada Basin (Fig. 4.7b). The ice motion responded to the winds with a much stronger TDS, so that there were considerable changes in ice thickness due to ice advection almost everywhere (Fig. 4.8b). In a large area of the Pacific sector, the reduction in ice thickness due to ice advection was up to 0.5 m/month more than the recent average (Fig. 4.8f). In August and September, there were even stronger southerly winds in the Pacific sector (Fig. 4.7c–d), which further strengthened the ice motion and TDS (Fig. 4.8g–h). Thus, the Pacific sector continued to lose ice (Kwok 2008). Again, the reduction in ice thickness due to ice advection was up to 0.5 m/month more than usual (Fig. 4.8g–h). In other words, from July to September 2007, the unusual ice motion pattern drove so much ice into the Atlantic sector from

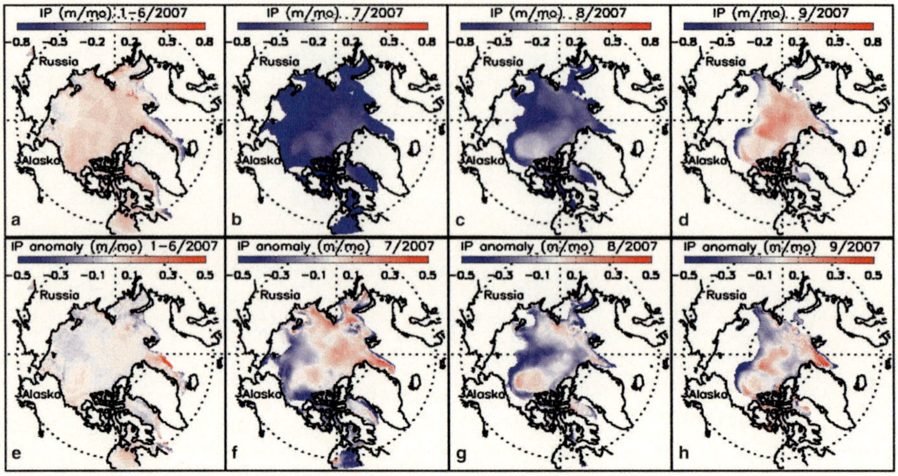

Fig. 4.9 PIOMAS-simulated ice production (IP) (**a–d**) and anomaly (**e–h**). Ice production is the net ice thermodynamic growth or decay due to surface atmospheric cooling/heating and oceanic heat transport, including ice-ocean albedo feedback contributed by the warm Bering water

the Pacific sector that the ice thickness in most of the Pacific sector was reduced by up to 1.5 m more than the recent average, leaving the Pacific sector with an unusually large area of thin ice and open water.

Major (~70 %) anomalous reduction in ice extent in the Pacific sector during summer 2007 was due to reduced ice production or enhanced ice melting (Zhang et al. 2008a), which includes the ice-ocean albedo feedback (Wang et al. 2005b) and ice-cloud feedback (Ikeda et al. 2003). Ice production is the net ice thermodynamic growth or decay due to surface atmospheric cooling/heating and ocean heat transport. In summer, a decrease in ice production is equivalent to an increase in ice melting. Ice production was negative in July and August due to ice melting caused by atmospheric or oceanic heating (Fig. 4.9b–c); it is generally positive in the January–June mean (Fig. 4.9a) and in September (Fig. 4.9d). In the first half of 2007, the ice production anomaly was generally small (Fig. 4.9e). In summer 2007, however, ice production decreased considerably in most of the Pacific sector (Fig. 4.9f–h) where a large reduction in ice thickness due to ice advection also occurred (Fig. 4.8b–d and f–h). This is because that the large reduction in ice thickness due to ice advection is associated with a large area of thin ice and open water, and thin ice and open water tend to allow more surface solar heating owing to the ice-albedo feedback (Wang et al. 2005b), leading to reduced ice production or enhanced ice melting (Zhang et al. 2008a; Lindsay et al. 2009). Therefore, the positive ice/ocean albedo feedback played a key role in the thermodynamic melting during the 2007 summer.

The changes in ice production represent the net effects of all the air–ice–ocean thermodynamic processes, which are also indirectly influenced by the oceanic and atmospheric dynamic forcings that enhance the thermodynamic melting, such as northward oceanic heat transport via the Bering Strait (Mizobata et al. 2010;

Woodgate et al. 2010) and advection of warm air temperature from the south (Overland et al. 2008). Although the reduced ice production or enhanced ice melting in the Pacific sector during summer 2007 was dominated by intensified surface solar heating because of the ice-albedo feedback (Zhang et al. 2008a, b; Wang et al. 2005b), other thermodynamic processes also played a role via both direct and indirect contributions. For example, intensified solar heating also warmed the surface waters in the Pacific sector considerably (Steele et al. 2008; Perovich et al. 2008). The unusually warm surface waters in turn warmed the overlaying atmosphere, elevating summer SAT in the Pacific sector (Fig. 4.7f–h). The stronger southerly winds associated with +DA (Fig. 4.7b–c) may also contributed to increased SAT by bringing warm air from the south to the Arctic (Overland et al. 2008). An increase in SAT led to an increase in surface longwave radiation and turbulent heat fluxes, resulting in additional ice melting (Zhang et al. 2008a; Ikeda et al. 2003). Ocean circulation (advection) and its thermodynamic processes (heat transport via Bering Strait) also played a role in the reduced ice production or enhanced ice melting in the Pacific sector during summer 2007. Therefore, strictly speaking, the 70 % of ice reduction due to thermodynamic processes should also include the indirect contribution by the oceanic and atmospheric dynamic forcings.

In summary, ~30 % of ice loss was due to wind advection (dynamic effect) directly associated with the DA, and ~70 % of ice loss was due to local melting (thermodynamic effect), which is caused by (1) anomalous warm water transport directly caused by the anomalous +DA-derived southerly wind, (2) anomalous high SAT directly advected by the +DA-driven southerly wind, and (3) positive ice-ocean albedo feedback by absorbing anomalous solar radiation. It is clear that the thermodynamic melting also implicitly includes the dynamic effect caused by the DA-derived wind anomaly. So, it was not a simple ice advection that pushed sea ice out of the Arctic to the Nordic Seas. Thus, Fram Strait ice transport must have better correlation to the DA index than the AO index, although the most melting is due to the local (in Pacific Arctic) thermodynamic effect/melting.

Therefore, it must be pointed out that it is impossible to separate the dynamic advection and thermodynamic melting of sea ice. These two processes highly interact and co-exist along with ice/ocean albedo feedback, in particular in the PAR during the +DA phase.

4.5 Bering Strait Heat Transport and the DA

Bering Strait volume and heat transport was continuously monitored since the late 1990s (Woodgate and Aagaard 2005; Woodgate et al. 2005, 2010; Mizobata et al. 2010). There are both seasonal (Hu and Wang 2010) and interannual variations, which are connected to the Pacific-Arctic pressure head (Woodgate et al. 2005; Clement Kinney et al. 2009). However, what mechanisms are responsible to the recent increase of the volume and heat transport since 2004 on the interannual time scale is not well given. In particular, the 2007 heat transport reached a maximum

(Wang et al. 2009a; Woodgate et al. 2010; Mizobata et al. 2010), coinciding with the all-time low in sea ice extent (Wang et al. 2009a). This implies a possible relationship between the volume and heat transport and the DA pattern.

The +DA not only drives sea ice from the western/Pacific Arctic to the eastern/Atlantic Arctic via accelerating the TDS, but also promotes northward flow of warm Pacific water (Woodgate and Aagaard 2005) that contributes above-average heat transport from the Pacific, accelerating the drastic thinning of sea ice (Steele et al. 2004; Shimada et al. 2006; Zhang et al. 2008a). To confirm that the Pacific water heat transport increased in the 2000s, in particular in summer 2007, we calculated the heat transport through the eastern Bering Strait from 2000 to 2010 during the June-October ice free season. Figure 4.10 shows that since 2004, northward heat transport via the Bering Strait has an annual average of 12.14 TW (1 TW = 10^{12} W and 1 W = 1 J s^{-1}) from 2004 to 2007, compared to the annual average of 6.4 TW during 2000–2003, representing a 90 % increase. The heat transport in 2007 (12.36 TW) had a 30.97 % increase compared to the average of 9.27 TW from 2000 to 2007. The heat transport from the Pacific Ocean has two important impacts, both direct and indirect, on sea ice in the western Arctic. The direct impact includes the bottom and lateral melting of sea ice when the warm Pacific water enters the Chukchi Sea, which enhances the melting via instant ice/ocean albedo feedback (see Sect. 4.4). The extreme event was the 2007 summer. The indirect impact involves a time-lag effect: the oceanic heat transport entering in the previous summer may survive winter (Shimada et al. 2006; Hu and Wang 2010) at the subsurface, which enhances the melting in the following spring and summer via ice/ocean albedo feedback, which further amplifies the ice/ocean albedo process, leading to an early melting.

There is a possible relationship between the northward heat transport and the DA. Mizobata et al. (2010) showed that the regression model using the NNW-SSE wind component well reproduced the Bering Strait inflow. Figure 4.11 shows the scatter plot of heat transport against summer +DA indices from 2000 to 2009, since the heat transport data are available after 2000, while the data for 2010 heat transport was not completed (Fig. 4.10). Within the positive territory of summer DA index, the heat transport ramped up with positive DA index, with 2007 being the record high (Woodgate et al. 2010). The reason is that the orientation of the summer DA-induced anomalous wind maximum is almost parallel to the TDS and the Bering

Fig. 4.10 (**a**) Volume transport through the eastern Bering Strait for the period 1999–2010. *Dotted lines* indicate mean volume transport between 1999 and 2003 and between 2004 and 2007. Two-way bars close to x-axis show period from June to October. Numbers above two-way bars show the integrating total volume transports from June to October through the entire eastern channel; (**b**) time series of averaged sea surface temperature (*gray*) and NCEP2-derived 330° wind component (Wnd330ip) of sea surface wind (*black*); and (**c**) estimated heat transport through the eastern channel of Bering Strait from 1999 to 2010. *Black bar* means heat transport through Area 1, while *white bar* indicates heat transport through Area 2, which makes up the entire eastern Bering Strait channel (See Mizobata et al. 2010 for detail)

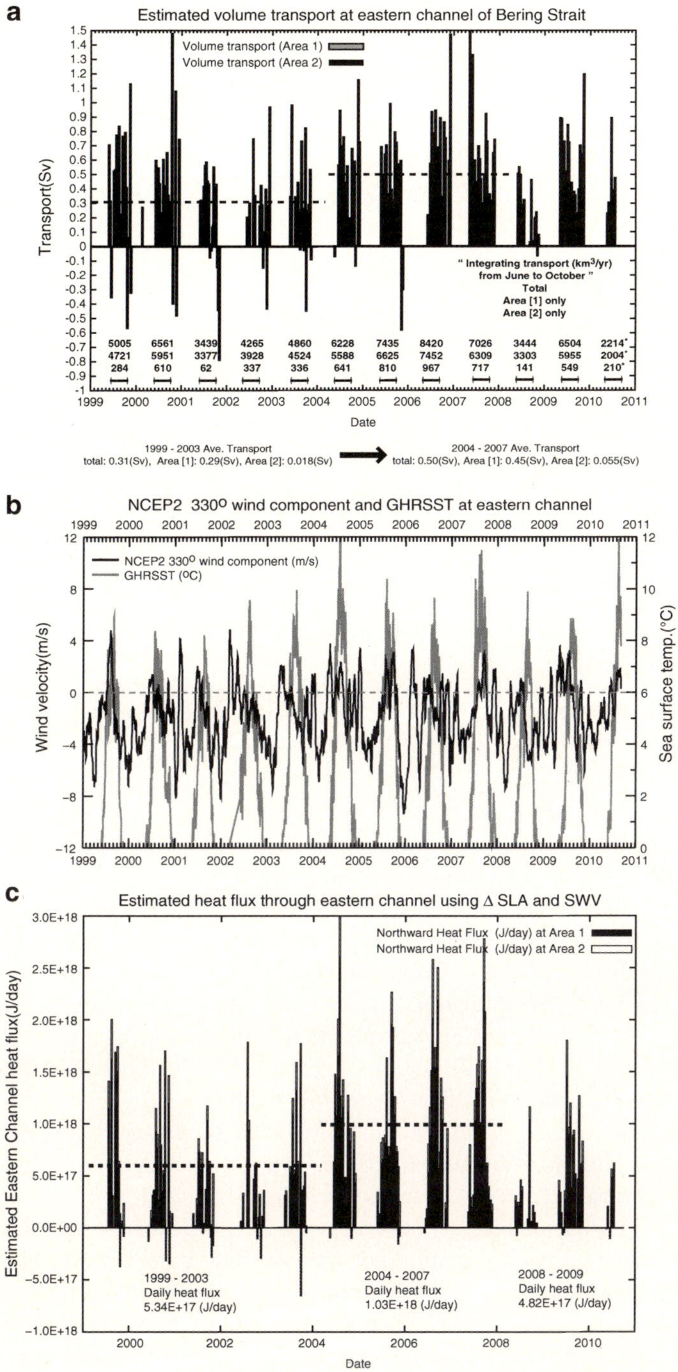
a
Estimated volume transport at eastern channel of Bering Strait
Volume transport (Area 1)
Volume transport (Area 2)
Transport(Sv)
" Integrating transport (km3/yr) from June to October "
Total
Area [1] only
Area [2] only
5005 4721 284
6561 5951 610
3439 3377 62
4265 3928 337
4860 4524 336
6228 5588 641
7435 6625 810
8420 7452 967
7026 6309 717
3444 3303 141
6504 5955 549
2214* 2004* 210*
Date
1999 - 2003 Ave. Transport
total: 0.31(Sv), Area [1]: 0.29(Sv), Area [2]: 0.018(Sv)
2004 - 2007 Ave. Transport
total: 0.50(Sv), Area [1]: 0.45(Sv), Area [2]: 0.055(Sv)
b
NCEP2 330o wind component and GHRSST at eastern channel
NCEP2 330o wind component (m/s)
GHRSST (oC)
Wind velocity(m/s)
Sea surface temp.(°C)
c
Estimated heat flux through eastern channel using Δ SLA and SWV
Northward Heat Flux (J/day) at Area 1
Northward Heat Flux (J/day) at Area 2
Estimated Eastern Channel heat flux(J/day)
1999 - 2003
Daily heat flux
5.34E+17 (J/day)
2004 - 2007
Daily heat flux
1.03E+18 (J/day)
2008 - 2009
Daily heat flux
4.82E+17 (J/day)
Date

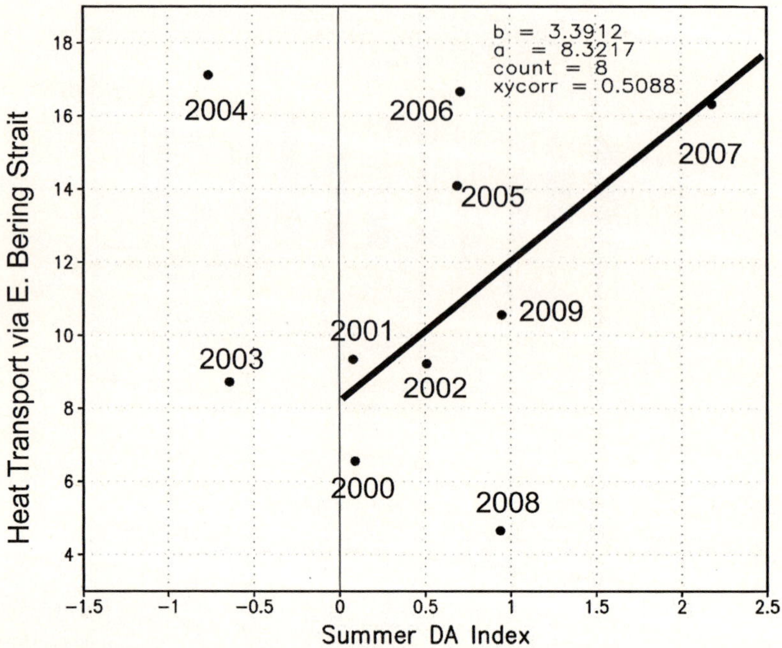

Fig. 4.11 Scatter plot between the summer DA index (x-axis) and northward heat transport via the eastern Bering Strait (y-axis) from 2000 to 2009. Units is in 10^{19} J $year^{-1}$ (June–October). The correlation between the heat transport and summer DA index is 0.51, which is lower than 0.71, the 95 % significance level

Strait, effectively driving Bering warm water (i.e., volume transport and temperature anomalies; Fig. 4.10a, b) northward. In addition, a northward wind anomaly in the Bering Strait could enhance the heat transport to the Arctic Ocean in August 2007 (Fig. 4.10b). By contrast, the heat transport in August 2008 was only 3.52 TW, much smaller than the mean (9.27 TW) of 2000–2007 (Fig. 4.10c). The reduced heat transport in 2008 is likely due to a smaller magnitude of the +DA than that of 2007 (Fig. 4.5b). Another important reason is that over the Bering Strait, a southward local wind anomaly (Fig. 4.5b and 4.10b) can significantly reduce the northward volume and heat transport in 2008. Therefore, the local N-S wind anomalies associated with basin-scale DA pattern should be included as a factor of influencing volume and heat transport (Mizobata et al. 2010). The correlation between the estimated heat transport and summer DA index is 0.51 (Fig. 4.11), although it is not significant at the 95 % significance level (0.71) due to small number of the samples.

Although there is an association between the heat transport and +DA events, it needs further investigation using long-term measurements. Furthermore, in addition to large-scale and local atmospheric forcing, the orientation, bathymetry, and ocean boundary around the Bering Strait may be factors affecting the northward heat transport (Clement Kinney et al. 2009, 2014, this volume), which differs from the atmosphere that does not have a lateral boundary.

4.6 Modeling the Bering Sea Cold Pool Using CIOM

Although summer sea ice in the PAR was experiencing abrupt decline since 1995, the Bering Sea ice showed large interannual variability. This indicates the decoupling of the Arctic summer sea ice that includes multiple-year ice with winter Bering Sea ice (only seasonal ice), due to several factors. First, the Bering Shelf is very shallow, and thus has less memory to the oceanic heat content that varies from year to year; In other words, the year-to-year ice/ocean albedo feedback is negligible in the Bering Shelf. For example, the Bering heat transport, which flows via the Bering Shelf into the Chukchi Sea, caused a rapid decline of sea ice in the Chukchi Sea (Shimada et al. 2006), but had little influence on the winter sea ice in the Bering Shelf. Secondly, the Bering Sea is an open sea, while the Arctic Ocean is a mediterranean sea and its boundary constrains the sea ice extent in the winter. Thus, winter ice thickness (or volume) is a better variable for gauging climate change, although summer ice extent can be used for studies of climate variability (Wang and Ikeda 2001). Thirdly, Bering Sea ice is mainly controlled by atmospheric forcing (Zhang et al. 2010), which includes the competitive impacts of the AO and El Nino and Southern Oscillation (ENSO). By the contrast, Arctic sea ice is influenced both by the warm Atlantic inflow, the leading atmospheric modes such as the AO and the DA, and by the ice/ocean albedo feedback process (Ikeda et al. 2001; Wang et al. 2005b).

Although the Bering Shelf has relatively less memory of oceanic heat content in the previous winter to influence the winter sea ice in the following winter, compared to the Arctic Ocean, it has significant impacts on the local summer water temperature that can have significant impacts on local ecosystems and fisheries on the shelf (Grebmeier et al. 2010). An important feature on the Bering Shelf is the cold water (cold pool) mass on the bottom of the middle shelf (50–100 m isobaths) that persists throughout the summer (Takenouti and Ohtani 1974; Kinder and Schumacher 1981; Wyllie-Echeverria 1995). The cold water was even observed in late September and early October (Kinder and Schumacher 1981). Hu and Wang (2010) investigate the seasonal change of the cold pool using the CIOM, and found that the largest vertical temperature difference, surface minus bottom, is >7 °C in the middle domain, suggesting that the cool pool survives the summer. The cold pool extends from the Gulf of Anadyr in the west with a temperature of <−1.0 °C (Hufford and Husby 1972) to a variable eastern boundary over the southeastern shelf with a temperature of 2 °C (Maeda et al. 1967). The extent and volume of the cold pool vary based on the past winter's meteorological and oceanic (convection) conditions. The minimum annual extent can reach eastward about 170 °W, while the maximum extent can cover Bristol Bay (Fig. 4.12a).

The existence of the cold pool in summer is important not only to the physical environment, but also to marine ecosystems. The cold pool is an ideal habitat to some arctic cold water species such as arctic cod, and acts as a barrier to certain species such as walleye pollock since they prefer water temperature warmer than 2 °C. When the volume and extent of the cold pool change from year to year,

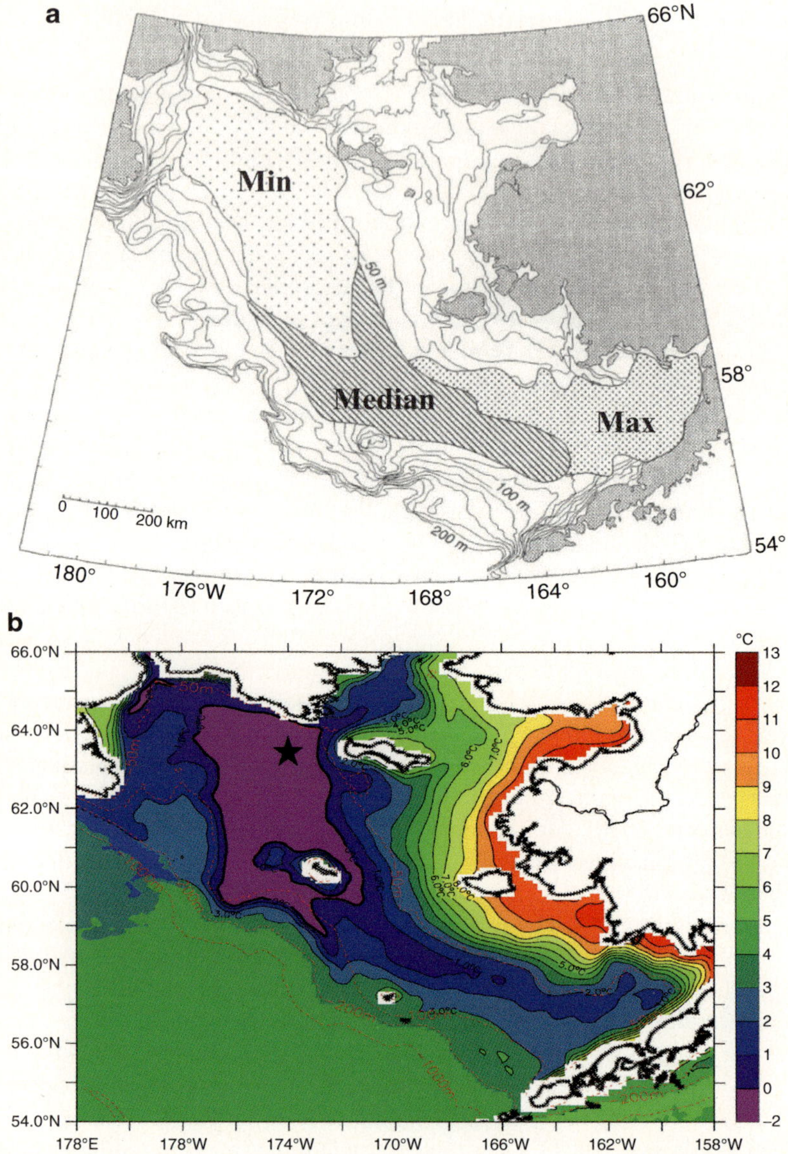

Fig. 4.12 (**a**) The schematic diagram for the cold pool extent based on historical measurements and knowledge. The sparse *dots, strips,* and *dense dots* represent the minimal, median, and maximal extent of the cold pool, respectively, and (**b**) the CIOM-simulated summer bottom water temperature (**b**). The *star* indicates the location of (173.5°W, 63.7°N) used in Fig. 4.13

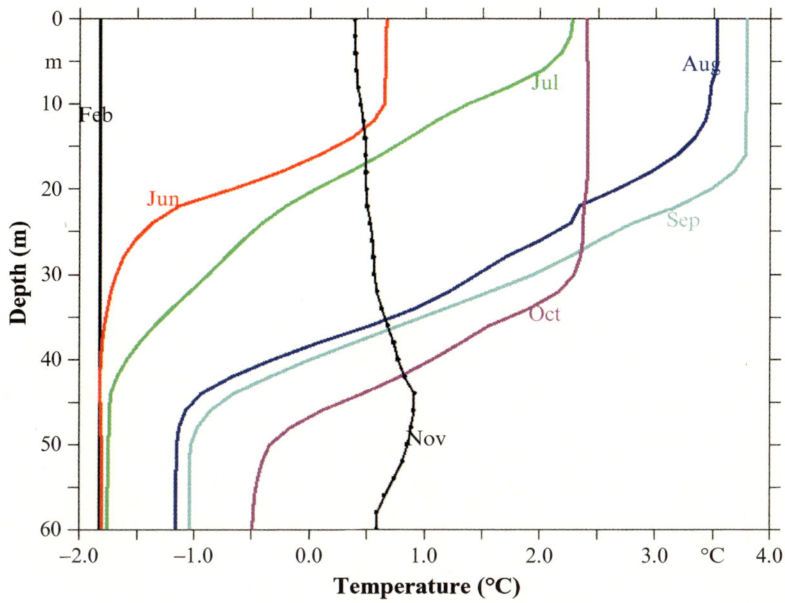

Fig. 4.13 The CIOM-simulated monthly-averaged vertical temperature structure at a northeastern location (see Fig. 4.12b) on the Bering shelf

the population of local species may increase or decrease accordingly to their preference. Thus, the cold pool can affect biomass growth rate and distributions on the Bering Sea shelf.

The summer (August-averaged) cold pool is reproduced by the CIOM (Fig. 4.12b). The cold water lies on the middle shelf between the 50–100 m isobaths, and it may reach the shelfbreak (200-m isobaths) in the western shelf break. The bottom temperature increases gradually from the 50-m isobath to the Alaskan coast. The maximum temperature reaches 11 °C near Norton Sound coast. The basin water at 200 m is basically >3 °C. The minimum bottom temperature in the middle of the southeastern shelf is >0 °C, compared to the northwestern shelf bottom water of <0 °C.

A vertically stable temperature structure inhibits convective heat transfer from surface layer to bottom layer (Hu and Wang 2010). Figure 4.13 shows the progression of monthly-averaged vertical temperature profiles in the cold pool region (see location in Fig. 4.12b). The winter temperature structure is vertically homogeneous (November to March), indicating the surface-to-bottom convection and production of winter shelf water. As solar radiation increases in March, the surface temperature rapidly increases. The surface stratification forms at the upper 10m layer in June, and increases from June to July due to solar heating and sea ice melting, reaching a strongest stratification in July. The SST gradually increases from August to September, reaching the maximum in September. Then, the mixed-layer starts to deepen due to cooling in September. From October to November, the vertical stratification rapidly disappears during strong wind mixing and cooling, becoming vertical homogeneous in December.

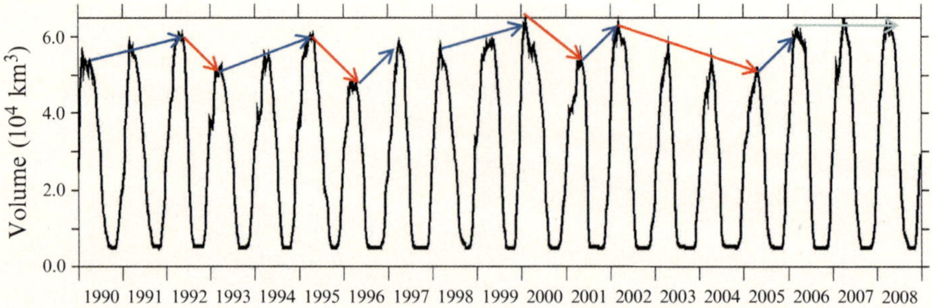

Fig. 4.14 The CIOM-simulated time series of the cold pool volume from 1990 to 2008 (units in 10^4 km^3). The *blue (red) arrows* indicate the increase (decrease) of the cold pool volume, which corresponds to the cooling (warming) winter

It is noted that the bottom cold water, which forms during winter seasons, can survive the summer months (Fig. 4.13). At this specific northern location, the bottom water is as cold as −1 °C, with a well-defined stratification in August and September. Therefore, the cold pool water mass is a seasonal phenomenon on the Bering shelf, which has significant impacts on regional ecosystems and fisheries distribution.

The CIOM, driven with daily atmospheric forcing, was used to calculate instantaneous cold pool volume for years 1990–2008 according to the formula of Hu and Wang (2010)

$$Volume_{\_CP}(t) = \sum_{i,j,k=1}^{N,M,K} \Delta x_{i,j,k} \Delta y_{i,j,k} \Delta z_{i,j,k}, \quad \text{if } \mathrm{T}(\mathrm{x},\mathrm{y},\mathrm{z},\mathrm{t}) \leq 2\ ^\circ\mathrm{C}.$$

Figure 4.14 shows the interannual variability of the cold pool volume, which is an indicator of year-to-year water property change induced by atmospheric forcing (Zhang et al. 2010; Overland et al. 2012). The colder the atmospheric temperature and the stronger the northerly winds, the stronger the vertical convection, leading to a larger cold pool extent and volume. The cold pool had a high extent/volume in 1991, 1992, 1995, 2000, 2002, 2006–2008, and had a low extent/volume in 1993, 1996, 2001, and 2005. The year-to-year changes in volume of the cold pool significantly impact or modify the distribution of marine ecosystems, since temperature-sensitive species are distributed according to their preference of habitat environment.

It is noted that there are two processes involving the year-to-year change in the cold pool volume. The first one is the abrupt change of atmospheric forcing. For instance, an abrupt decrease in the volume in 1992–1993, 1995–1996, and 2000–2001 reflects the abrupt change in atmospheric conditions with warming and weak northerly winds. The abrupt increase in volume occurred in 1996–1997, 2001–2002, and 2005–2006, indicating the sudden cooling and increase in northerly wind anomalies over the Bering Sea. The second process is the water temperature memory effect. During an increased progression in cold pool volume, it often takes 3 years to reach the maximum, such as in 1990–1992, 1993–1995, and 1998–2000, indicating the memory of water heat storage. However, very often it only takes 1 year to detect the abrupt increase in cold pool volume such as in 1996–1997, 2000–2001, and 2005–2006, indicating that a significant change of atmospheric forcing is more important than the heat memory in the Bering Shelf (Overland et al. 2012). The longest decrease in cold pool volume occurred during 2002–2005, followed by the three straight high years from 2006 to 2008, indicating a cold phase in the region, consistent with the cooling of the water temperature in the M2 site in the southeastern Bering shelf (see Fig. 4.5 of Overland et al. 2014, this volume). The persistency of the large cold pool volume during 2006–2008 should have significant impacts on ecosystems and fisheries in the coming years, which should be closely monitored.

4.7 Modeling Landfast Ice in the Beaufort-Chukchi Seas Using CIOM

Landfast ice along the coastline of the PAR plays an important role as a biologically productive habitat and transportation corridor; it also provides important protection to the shoreline and coastal installations (Eicken et al. 2009). However, at the present time it is not clear how the diminishing Arctic summer sea ice and the reduction in multi-year ice extent (Maslanik et al. 2011) have impacted the seasonal cycle and distribution of landfast ice. Thus, while the seasonal and interannual variability of the landfast ice in a diminishing sea ice scenario in the PAR are an important emerging topic; evidence of such variability is somewhat limited. This is largely due to the temporally limited availability of synthetic aperture radar imagery required for accurate assessments of landfast ice extent (Mahoney et al. 2007). Model simulations with coupled ice-ocean models can hence provide insight into longer-term variations on time-scales of decades, although the modeling of landfast ice at these scales is in its infancy (e.g., König Beatty and Holland 2010).

Landfast ice along the Chukchi and Beaufort coast is a seasonal phenomenon with interannual variability (Eicken et al. 2006; Mahoney et al. 2007). It is a great challenge for any coupled ice-ocean model to capture the dynamic and thermodynamic features of landfast ice, since many factors can affect the formation, anchoring, and melting of landfast ice, such as wind forcing, ocean currents, coastal topography and bathymetry, and model resolution. To this end, a 3.8-km resolution CIOM (Wang et al. 2003, 2008; Jin et al. 2008) was used to investigate the seasonal and interannual variability of landfast ice in the Chukchi and Beaufort seas.

There are two approaches to distinguish landfast ice from pack ice in a model. One way is to define landfast ice by an ice velocity criterion that considers ice stationary below a given velocity threshold. In this study, if both the absolute ice velocity is less than 4 cm/s, and the water depths are less than 35 m, then grid cells are designated as landfast ice. The second, prescriptive method stipulates that

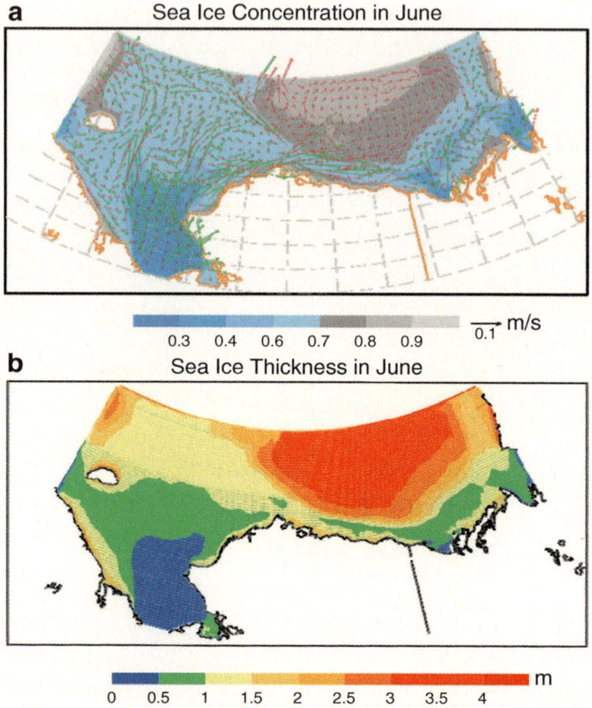

Fig. 4.15 The CIOM-simulated June climatological sea ice concentration (**a**, from 0 to 1) with sea ice velocity vectors superimposed (vector size indicates speed in ms^{-1}), and sea ice thickness (**b**, in meters). Landfast ice remains attached along the Beaufort Sea coast. The *green (orange) arrows* denote the ocean surface current (ice flow)

during the simulation, the wind stress and ice velocity are set to zero shoreward of the 35 m isobaths, roughly corresponding to the extent of landfast ice. This method is widely used in Baltic Sea ice simulations (Haapala et al. 2001; Meier 2002a, b), which may not be capable of representing spatial and interannual or seasonal landfast ice extent.

Figure 4.15 shows the CIOM-simulated climatology of June sea ice concentration (SIC) and thickness using the daily NCEP forcing for the period 1990–2007. The SIC map (Fig. 4.15a) clearly shows what corresponds to simulated landfast ice attaching to the Alaskan Beaufort and Chukchi coast during the melt season as ice of high concentration. During spring, surface melting commences nearshore, but ice concentration first decreases offshore as a result of complete melting and removal of thinner offshore ice in areas of higher open water concentrations that promote absorption of solar heat. This is also reflected in the small magnitude of ice velocity vectors superimposed on ice concentration in Fig. 4.15a; nevertheless, since landfast ice is not modelled explicitly, small residual velocities remain in some areas of effective landfast ice. However, at the same time, the strong contrast between stationary landfast ice and highly mobile ice just offshore from the landfast ice edge appears to be well captured. Even in mid-July, Beaufort landfast ice remains, not melting completely until the end of the month, depending on weather (SAT and wind direction) conditions.

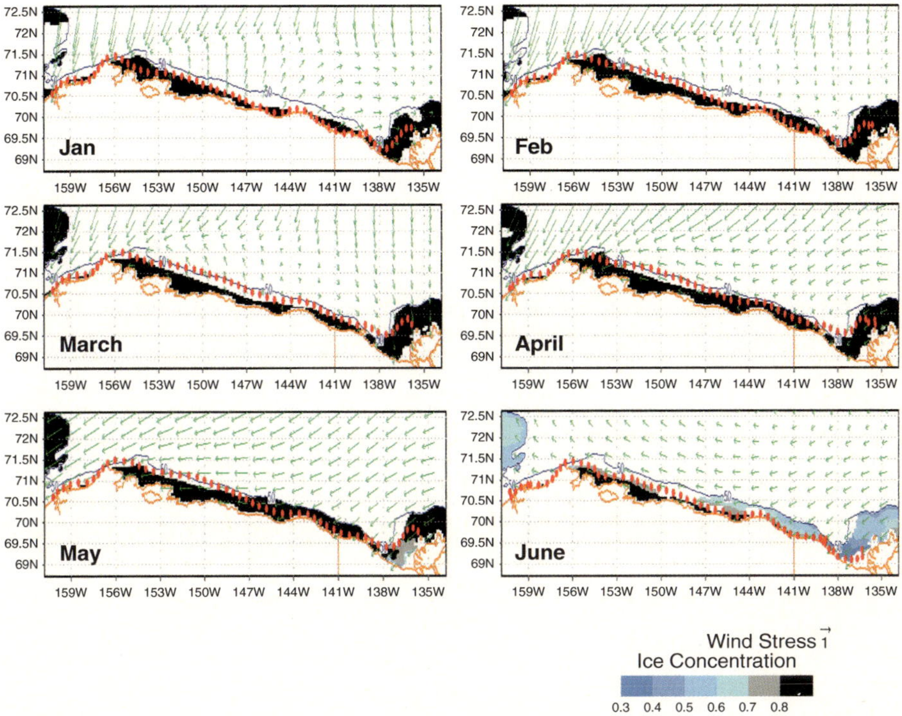

Fig. 4.16 The CIOM-simulated January–June climatological landfast ice extent (from 1990 to 2009, *black shaded*) compared to landfast ice edge locations derived from synthetic aperture radar satellite data (*red dots*) averaged for the period 1996–2004. *Green vectors*: Wind stress in 10^{-5} N m^{-2}

The simulated sea ice thickness map (Fig. 4.15b) in June shows some contrast in landfast ice thickness along the Beaufort and Chukchi coast (~1.5 m) and thinner offshore (<1 m) ice. Since the model is not explicitly simulating processes that contribute to landfast ice stabilization, in particular grounding of pressure ridges (Mahoney et al. 2007), other processes represented in the model can explain the formation and maintenance of landfast ice. These include the following factors based on a series of controlled sensitivity experiments in Wang et al. (2010): (1) an onshore northeast wind due to the Beaufort high pressure system, (2) the eastward ACC and Beaufort Slope Current with its right-turning force due to the Coriolis effect, which advect the warmer Bering water, (3) high resolution topography and bathymetry constraining ice motion in the coastal regions, and (4) sea ice advection. However, these model-inherent factors that help keep landfast ice in place in nature deserve further investigation.

Figure 4.16 shows the climatology (1990–2007) of the simulated landfast ice that was compared to observed landfast ice extents obtained from synthetic aperture radar satellite data for the period 1996–2004 (Eicken et al. 2006; Mahoney et al. 2007). In the model, landfast ice starts to form in autumn due to the Beaufort Gyre and anticyclonic winds induced by the Beaufort High, both of which push sea ice toward the Alaskan Beaufort coast, coupled with the thermal growth of sea ice along the shore (Wang et al. 2010). When sea ice completely covers the entire Arctic from December on, landfast ice is attached to shore, while pack ice offshore still moves with the ocean surface current and wind forcing. During the period of complete ice cover, the radar satellite data indicate quasi-stationary landfast ice with a clearly delineated boundary between pack ice and landfast ice (anchored to the bottom and attached to shore with the velocity almost being zero), while the CIOM-simulated landfast ice still exhibits small movement, since the sea ice produced in CIOM is not resolving the anchoring of grounded pressure ridge keels that stabilize the landfast ice. Thus, more research is required to improve the representation of ice dynamics in coastal regions and landfast ice processes by formulating and including the relevant ice anchoring mechanisms in the model.

The CIOM-simulated landfast ice is generally consistent with landfast ice extent derived from satellite data. The CIOM reproduces the landfast ice boundary in January and February very well. However the model reproduces less landfast ice than the measured boundary in March. During April, the melting season, the CIOM reproduces landfast ice reasonably well in general, but reproduces less ice from 147° to 152°W. In May, CIOM reproduces more landfast ice between 140° and 147°W. In June, the model simulation compares very well with the measurement.

To investigate the interannual variability of the landfast ice, we first constructed the seasonal cycle of landfast ice area based on those grid cells conforming with the ice velocity criterion described above and along the coast with depth of less than 30 m within 160°W–134°W for the period of 1990–2007 (Fig. 4.17a). Three characteristics are apparent: (1) landfast ice exhibits a uniform extent indicative of overall stability of the ice cover from January through April with a maximum in February; (2) the largest variability (i.e., standard deviation) occurs in June and November, during the peak of the melt (decay) and freeze-up (formation) seasons, respectively; and (3) there is no landfast ice in August and September. Figure 4.17b shows the year-to-year variability of the landfast ice. One striking feature is that the landfast ice formed earlier and melted later before 1998, while since then the duration of the landfast ice season has shortened significantly. This is consistent with the increase in Bering Strait heat transport since 2001 (Woodgate et al. 2010), in particular since 2004 during which both temperature and volume transport increased (Fig. 4.10c). In the spring of 2007, landfast ice decayed more rapidly than in other years, since the +DA-derived wind anomaly was directed offshore (i.e. northward) (Wang et al. 2009a), in addition to other forcings such as maximum Bering Strait heat transport (Woodgate et al. 2010; Mizobata et al. 2010), and ice/ocean and cloud albedo feedbacks (Ikeda et al. 2003; Wang et al. 2005b), leading to thinner pack ice. Figure 4.17c shows the time series of landfast ice area anomalies for 1990–2007. It is clear from the figure that from 1990 through 1997, positive anomalies dominated, while since then Beaufort and Chukchi Sea landfast ice extent was characterized by negative anomalies. As evident from Fig. 4.17b, these anomalies are not due to reductions in maximum extent, but rather driven by shifts in the seasonality of the landfast ice. This finding is in line with observations

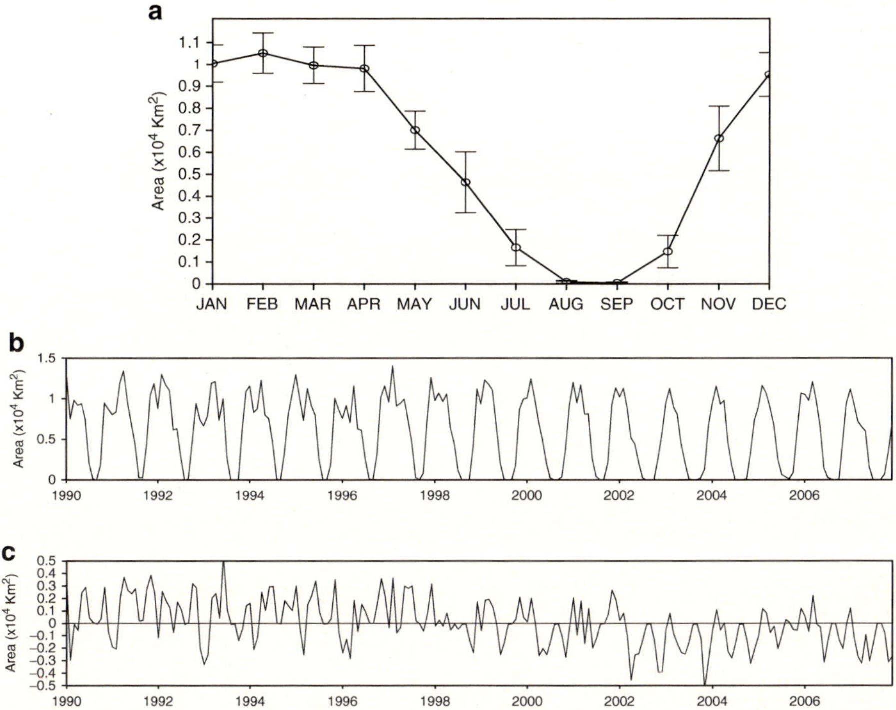

Fig. 4.17 (**a**) Modeled seasonal climatology of landfast ice area with standard (one) deviations (*vertical bars*) for the period of 1990–2007; (**b**) modeled monthly landfast ice area from 1990 to 2007; and (**c**) modeled monthly landfast ice area anomalies from 1990 to 2007. The landfast ice area is calculated with ice concentration being greater than 0.6 within the Beaufort and Chukchi coastal region between 160°W and 134°W

by Mahoney et al. (2007) and points to the importance of decadal-scale variations in landfast ice extent. A more detailed analysis of model simulations over a longer time period, drawing on available remote-sensing and in situ observations, is required to elucidate the role of secular change and regional warming in explaining such changes in landfast ice extent.

4.8 Possible Air-Ice-Sea Feedback Loops in the Western Arctic

In the Pacific Arctic, from the Bering Sea to the Arctic Ocean, there is a regional feedback loop among the atmosphere, sea ice, and ocean (Wang et al. 2005b). The recent increase in Bering Sea heat transport into the Chukchi Sea since 2000 (Woodgate et al. 2010; Mizobata et al. 2010) is closely related to the continuous ice

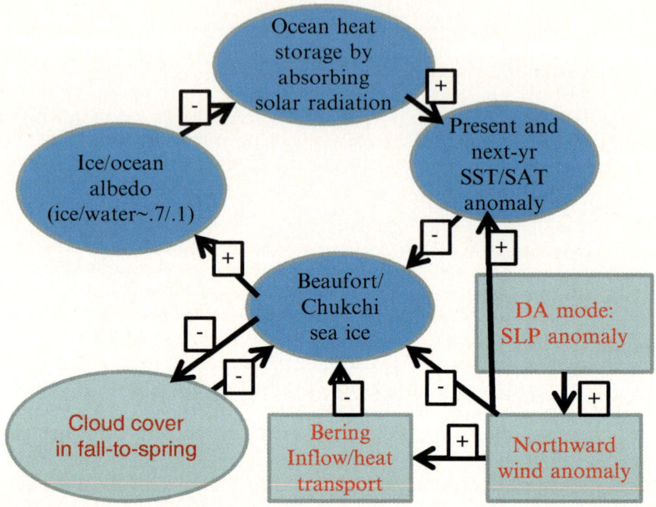

Fig. 4.18 A proposed possible positive air-ice-sea feedback in the western Pacific Arctic Region. +/− indicates positive/negative feedback. Negative feedback between ice cover anomaly and clouds is shown on the *lower left corner*. Less ice cover (or more open water) leads to more clouds cover in fall to spring, which further results in less ice cover due to reduction of long-wave radiation to the upper atmosphere (or space). The long-wave radiation is reflected back to the ice-ocean surface, leading to higher SAT/SST anomalies and less ice cover. The +DA can accelerate and amplify the existing ice/ocean albedo feedback

decline in the Western Pacific (Shimada et al. 2006). Although a well-known positive ice/ocean albedo feedback is in play in general (Wang et al. 2005b), a conceptual connection in the PAR was not established.

Figure 4.18 depicts a possible regional interaction among the atmosphere, sea ice, and the ocean. The +DA-induced wind anomaly enhances the northward transport of the Bering warm water, i.e., heat transport (Mizobata et al. 2010), and also drives sea ice from the Western Arctic to the Eastern Arctic (Wang et al. 2009a; Ikeda 2012) via strengthening of the TDS (Comiso et al. 2008). The oceanic heat transport entering the Chukchi Sea instantly enhances the melting of sea ice, seasonal positive ice/ocean albedo feedback, and increases the local heat storage, as discussed in Sect. 4.4 and by Zhang et al. (2008a, b). Then, the excessive heat will accelerate the melting process during the following winter and spring, and slow down the freezing in the following autumn. In the following summer, more open water absorbs more solar radiation (since the ice and ocean albedos are ~0.7 and 0.1, respectively) and the heat storage increases, which is the well known positive ice/ocean albedo feedback. The heat stored in the upper ocean layer will be present in the following autumn and winter, increasing the SST and SAT, delaying the freezing in the autumn and thinning the sea ice in the winter. The thin ice in winter will lead to early break-up in the following spring, and to more open water in the following summer, absorbing and storing more heat from solar radiation.

In summary, +DA-induced northward wind anomaly has two important impacts on the Western Arctic ice-ocean-ecosystems. The first is a direct, seasonal impact. The northward wind not only drives sea ice away from the Western Arctic to the Eastern Arctic, but also enhances the inflow of the warm Bering water into the Chukchi Sea (Zhang et al. 2008a, b; and see discussion in Sect. 4.4), instantly melting sea ice in the Chukchi Sea and increasing the SST and SAT (Zhang et al. 2008a, b; and see discussion in Sect. 4.4). The seasonal ice/ocean albedo feedback also accelerates ice melting in the Chukchi Sea. The second impact is an indirect, year-to-year cumulative impact. Since the heat transport advected into the Chukchi Sea increases, the excessive heat can be stored in the water column and often can survive the winter. This would lead to the positive ice/ocean albedo feedback in the following year, as depicted in Fig. 4.18 and discussed above.

As suggested by Ikeda et al. (2003), cloud data collected over the Arctic Basin have proved that a cloud increase during the period 1960–1990 made a significant contribution to a reduction in net upward longwave radiation. The importance of clouds applies also to the decadal ice cover variability. The heat transport trend was estimated to be a linear increase in the annual flux to the ice-ocean system at 10 Wm^{-2} in the 30-year period. It is a logical expectation that part of the increase in autumn-to-winter cloud cover is in turn due to increased evaporation from the increased open water area. Thus, an ice-cloud feedback may enhance the ice cover reduction (Fig. 4.18, see the lower left corner). The cloud cover changes over the Arctic induce thermodynamic effects comparable with the albedo reduction caused by the ice reduction. Since Arctic clouds are not convincingly simulated in doubled carbon dioxide experiments, further work on cloud parameterization, and subsequent analysis of model output, are highly recommended.

4.9 Summary

Based on the above investigations of several emerging climate-related ice-ocean processes, the following conclusions can be drawn:

1. Although the ratio of the AO to the DA is about 4:1 in terms of total variance of SLP (i.e., surface wind), the DA-induced wind anomaly is meridional, compared to the zonal wind anomalies induced by the AO. Furthermore, the +DA-induced maximum wind anomaly is nearly parallel to the TDS, thus accelerating the TDS and effectively enhancing ice export. There is seasonal variation of the DA: (1) the summer and autumn DA intensity (or variance, i.e., wind anomaly) is larger than the winter and spring, and (2) the orientation of the DA-induced wind anomaly in spring and summer is more parallel to the TDS than the winter and autumn. Thus, the summer +DA is most effective in advecting sea ice out of the Arctic than the other seasons. The persistency of the +DA from winter, spring, to summer can be used to gauge the ice minimum occurrence under the present thinning ice conditions.

2. The PIOMAS successfully captures almost all of the ice minima since 1995, which compares very well to the satellite measurements. The key mechanisms were identified as DA-driven meridional wind anomaly that enhanced the advection of sea ice out of the PAR, and the enhanced thermodynamic melting via the positive ice/ocean albedo feedback process.
3. The +DA-derived wind anomaly not only drives away sea ice from the Pacific Arctic to the Atlantic Arctic, but also promotes the advection of more warm Bering water volume transport with anomalous heat transport, leading to ice minima in the PAR. In particular, the 2007 summer DA has the most parallel orientation to the TDS and Bering Strait than the summers of 2008, 2009, and 2010, thus leading to high volume and heat transports into the Chukchi Sea. The local N-S wind stress over the Bering Strait, which may not be associated with the DA pattern, is also important (Mizobata et al. 2010).
4. The Pacific Arctic atmosphere-ice-ocean interactions can be summarized by a feedback loop to explain the continuous ice decline and warmer water temperature in the Chukchi Sea in the last decade. The DA plays an important role as an external forcing to the ice-ocean system in the PAR. Once the +DA applies its forcing to the Pacific Arctic, a long-lasting impact on the sea ice reduction was observed. The key is the positive ice-ocean albedo feedback (Wang et al. 2005b; Shimada et al. 2006; Zhang et al. 2008a, b). Therefore, a series of +DA events (Wang et al. 2009a) should apply a series of pulses of positive ice-ocean albedo feedbacks into the PAR ice-ocean system. A possible reversal may be accomplished by –DA events. The ice-cloud feedback is another mechanism for continuous, further reduction of sea ice in the PAR.
5. The Bering shelf cold pool has significant impacts on ecosystems and fisheries distribution. The winter cold water is preserved in the middle shelf throughout the summer due to a stable, stratified temperature structure, and also due to the lack of effective horizontal heat transport and mixing from the deep basin and the inner shelf. Interannual variability of the cold pool extent/volume depends on winter meteorological and oceanic conditions, and winter ocean deep convection.
6. The key aspects of the major dynamic and thermodynamic processes of the landfast ice were captured by the CIOM. The major factors include the shallow, narrow shelf in contrast to the deep shelfbreak, and the deep basin (in terms of heat content), prevailing onshore wind forcing, and the warm ACC and the Slope Current. The largest variability occurs in spring and autumn, indicating that the melting and freezing seasons are the key periods. There are both strong interannual and significant decadal variability in landfast ice. The duration of landfast ice for the period of 1990–1997 was longer than that during 1998–2007. The landfast ice reduced significantly during spring of 2007, possibly due to the impacts of the persistent +DA anomaly from winter to summer of 2007 (Wang et al. 2009a). However, the anchoring process of the landfast ice by grounded pressure ridges is currently not included in the model and needs to be represented for a full evaluation of coastal ice dynamics.

The Arctic Ocean and the PAR in particular have experienced rapid changes not only in sea ice and ocean, but also in ecosystems, in particular in the last decade. This study has discussed several important processes in the region. There is urgent need for a better understanding of these processes or hypotheses using combined long-term measurements and ice-ocean models. The following processes may be of interest:

1. There was a significant change in climate between the early 1990s and last decade. The +AO dominated in the early 1990s (cyclonic anomalous circulation), while the +DA prevailed in the last decade (meridional anomalous circulation), leading to significant ice reduction. These two leading climate modes should be the major forcing to other subsystems, such as sea ice, ocean, and ecosystem. Therefore, to adequately address the Arctic processes, the combined impact of these two modes is necessary.
2. The Bering Strait is a pathway of freshwater (volume transport) and heat to the Arctic Ocean and was more abnormal in the last decades. The connection between the Bering Strait and Fram Strait-Barents Sea opening, along with the Canadian Arctic Archipelagos (CAA) is not well known. A high-resolution coupled ice-ocean model that can resolve CAA is a high priority.

With diminishing summer ice in the Pacific Arctic, a northward migration of ecosystems is an urgent topic to be studied. The changes of food webs, lower trophic level ecosystems, and fisheries can have large impacts on the future planning and management of marine resources.

Acknowledgments We are grateful for supports from NOAA Arctic Program of the NOAA Climate Program Office to J.O. and J.W., NSF grants ARC-0611967, ARC0901987 and ARC0908769 to J.Z., NSF grant ARC-0714078 to Y.Y., H.E., and J.W., and the Ministry of Education, Science, Sports and Culture, Grant-in-Aid for Scientific Research from the Japan Society for the Promotion of Science to K.M. Thanks also go to Cathy Darnell of NOAA GLERL for editing this paper. We also sincerely thank Dr. Stephen Okkonen of UAF for his constructive comments on this paper. This is GLERL Contribution No. 1623, and PMEL Contribution No. 3837.

References

Blumberg A, Mellor G (1987) A description of 3-D coastal ocean circulation model. In: Heaps NS (ed) Coastal and estuarine sciences 4: 3-D coastal ocean models. American Geophysical Union, Washington, DC, pp 1–16

Clement Kinney J, Maslowski W, Okkonen S (2009) On the processes controlling shelf-basin exchange and outer shelf dynamics in the Bering Sea. Deep-Sea Res II 56:1351–1362. doi:10.1016/j.dsr2.2008.10.023

Clement Kinney J et al (2014) Chapter 7: On the flow through Bering Strait: a synthesis of model results and observations. In: Grebmeier JM, Maslowski W (eds) The Pacific Arctic region: ecosystem status and trends in a rapidly changing environment. Springer, Dordrecht, pp 167–198

Comiso J, Parkinson C, Gersten R, Stock L (2008) Accelerated decline in the Arctic sea ice cover. Geophys Res Lett 35:L01703. doi:10.1029/2007GL031972

Deal C, Steiner N, Christian J, Clement Kinney J, Denman K, Elliott S, Gibson G, Jin M, Lavoie D, Lee S, Lee W, Maslowski W, Wang J, Watanabe E (2014) Chapter 12: Progress and challenges in biogeochemical modeling of the Pacific Arctic region. In: Grebmeier JM, Maslowski W (eds) The Pacific Arctic region: ecosystem status and trends in a rapidly changing environment. Springer, Dordrecht, pp 393–446

Eicken H, Shapiro L, Gaylord A, Mahoney A, Cotter P (2006) Mapping and characterization of recurring spring leads and landfast ice in the Beaufort and Chukchi seas. U.S. Department of Interior, Minerals Management Service (MMS), Alaska Outer Continental Shelf Region, Anchorage, 141 pp

Eicken H, Lovecraft AL, Druckenmiller ML (2009) Sea-ice system services: a framework to help identify and meet information needs relevant for Arctic observing networks. Arctic 62(2):119–136

Gascard J et al (2008) Exploring Arctic Transpolar Drift during drastic sea ice retreat. EOS 89(3):21–22

Grebmeier JM, Overland JE, Moore SE et al (2006) A major ecosystem shift in the northern Bering Sea. Science 311(5766):1461–1464. doi:10.1126/science.1121365

Grebmeier J, Moore S, Overland J, Frey K, Gradinger R (2010) Biological response to recent Pacific Arctic sea ice retreats. EOS 91(18):161–162

Haapala J, Meier H, Rinne J (2001) Numerical investigations of future ice conditions in the Baltic Sea. Ambio 30(4–5):237–244

Hibler W (1979) A dynamic and thermodynamic sea ice model. J Phys Oceanogr 9:15959–15969

Hibler W (1980) Modeling a variable thickness sea ice cover. Mon Weather Rev 108:1943–1973

Hu H, Wang J (2010) Modeling effects of tidal and wave mixing on circulation and thermohaline structures in the Bering Sea: process studies. J Geophys Res 115:C01006. doi:10.1029/2008JC005175

Hu H, Wang J, Wang D (2011) A model-data study of the 1999 St. Lawrence Island polynya in the Bering Sea. J Geophys Res 116:C12018. doi:10.1029/2011JC007309

Hufford G, Husby D (1972) Oceanographic survey of the Gulf of Anadyr August 1970, U.S. Coast Guard oceanography report CG-373-52. U.S. Coast Guard Oceanographic Unit, Washington, DC

Ikeda M (2012) Sea-ice cover anomalies in the Arctic Basin associated with the atmospheric variability from multi-decadal trend to intermittent quasi-biennial oscillations. Polar Res 31: 18690. http://dx.doi.org/10.3402/polar.v31i0.18690

Ikeda M, Wang J, Zhao JP (2001) Hypersensitive decadal oscillations in the Arctic/subarctic climate. Geophys Res Lett 28:1275–1278

Ikeda M, Wang J, Makshtas A (2003) Importance of clouds to the decaying trend and decadal variability in the Arctic ice cover. J Meteorol Soc 81:179–189

Jin M, Wang J, Mizobata K, Hu H, Shimada K (2008) Observations and modeling of the ice-ocean conditions in the coastal Chukchi and Beaufort Seas. Acta Oceanol Sin 27(3):79–87

Jin M, Deal C, Lee S, Elliott S, Hunke E, Maltrud M, Jeffery N (2011) Investigation of Arctic sea ice and ocean primary production for the period 1992 to 2007 using a 3-D global ice ecosystem model. Deep-Sea Res II 81–84:28–35. doi:10.1016/j.dsr2.2011.06.003

Kinder TH, Coachman LK (1978) The front overlaying the continental slope of the eastern Bering Sea. J Geophys Res 83:4551–4559

Kinder TH, Schumacher JD (1981) Circulation over the continental shelf of the southeastern Bering Sea. In: Hood DW, Calder JA (eds) The Eastern Bering Sea shelf: oceanography and resource, vol 1. Published by the Office of Marine Pollution Assessment, NOAA and BLM, Distributed by the University of Washington, Seattle, pp 53–76

König Beatty C, Holland DM (2010) Modeling landfast sea ice by adding tensile strength. J Phys Oceanogr 40(1):185–198

Kwok R (2008) Summer sea ice motion from the 18 GHz channel of AMSR-E and the exchange of sea ice between the Pacific and Atlantic sectors. Geophys Res Lett 35:L03504. doi:10.1029/2007GL032692

Kwok R, Cunningham GF, Pang SS (2004) Fram Strait sea ice outflow. J Geophys Res 109:C01009. doi:10.1029/2003JC001785

Lindsay RW, Zhang J, Schweiger AJ, Steele MA, Stern H (2009) Arctic sea ice retreat in 2007 follows thinning trend. J Clim 22:165–176. doi:10.1175/2008JCLI2521

Long Z, Perrie W, Tang CL, Dunlap E, Wang J (2012) Simulated interannual variations of freshwater content and sea surface height in the Beaufort Sea. J Clim 25:1079–1095. doi:10.1175/2011JCI14121.1

Maeda T, Fujii T, Masuda K (1967) On the oceanographic condition and distribution of fish shoals in 1963 (in Japanese). Part 1. Studies on the trawl fishing grounds of the eastern Bering Sea. Bull Jap Soc Sci Fish 33(8):713–720

Mahoney A, Eicken H, Gaylord AG, Shapiro L (2007) Alaska landfast sea ice: links with bathymetry and atmospheric circulation. J Geophys Res 112:C02001. doi:10.1029/2006JC003559

Maslanik J, Drobot S, Fowler C, Emery W, Barry R (2007) On the Arctic climate paradox and the continuing role of atmospheric circulation in affecting sea ice conditions. Geophys Res Lett 34:L03711. doi:10.1029/2006GL028269

Maslanik J, Stroeve J, Fowler C, Emery W (2011) Distribution and trends in Arctic sea ice age through spring 2011. Geophys Res Lett 38:L13502. doi:10.1029/2011GL047735

Meier HEM (2002a) Regional ocean climate simulations with a 3D ice-ocean model for the Baltic Sea. Part 1: Model experiments and results for temperature and salinity. Clim Dyn 19:237–253

Meier HEM (2002b) Regional ocean climate simulations with a 3D ice-ocean model for the Baltic Sea. Part 2: Results for sea ice. Clim Dyn 19:255–266

Mizobata K, Saitoh S (2004) Variability of Bering Sea eddies and primary productivity along the shelf edge during 1998–2000 using satellite multi-sensor remote sensing. J Mar Syst 50:101–111

Mizobata K, Saitoh S, Shiomoto S, Miyamura T, Shiga N, Toratani M, Kajiwara Y, Sasaoka K (2002) Bering Sea cyclonic and anticyclonic eddies observed during summer 2000 and 2001. Prog Oceanogr 55:65–75

Mizobata K, Wang J, Saitoh S (2006) Eddy-induced cross-slope exchange maintaining summer high productivity of the Bering Sea shelf break. J Geophys Res 111:C10017. doi:10.1029/2005JC003335

Mizobata K, Saitoh S, Wang J (2008) Interannual variability of summer biochemical enhancement in relation to the mesoscale eddy at the shelf break in the vicinity of the Pribilof Islands, Bering Sea. Deep Sea Res II 55:1717–1728. doi:10.1016/j.dsr2.2008.03.002

Mizobata K, Shimada K, Woodgate R, Saitoh S, Wang J (2010) Estimation of heat flux through the eastern Bering Strait. J Oceanogr 66(3):405–424. doi:10.1007/s10872-010-0035-7

Nghiem SV, Rigor IG, Perovich DK, Clemente-Colón P, Weatherly JW, Neumann G (2007) Rapid reduction of Arctic perennial sea ice. Geophys Res Lett 34:L19504. doi:10.1029/2007GL031138

Niebauer HJ (1980) Sea ice and temperature variability in the eastern Bering Sea and relation to atmospheric fluctuations. J Geophys Res 85:7507–7515

Overland JE, Stabeno PJ (2004) Is the climate of the Bering Sea warming and affecting the ecosystems? EOS 85(33):309–316

Overland JE, Wang M, Salo S (2008) The recent Arctic warm period. Tellus A 60:589–597

Overland JE, Wang M, Wood KR, Percival DB, Bond NA (2012) Recent Bering Sea warm and cold events in a 95-year context. Deep-Sea Res II 65–70:6–13. doi:10.1016/j.dsr2.2012.02.013

Overland JE, Wang J, Pickart RS, Wang M (2014) Chapter 2: Recent large shifts in the meteorology of the Pacific Arctic region. In: Grebmeier J, Maslowski W (eds) The Pacific Arctic region: ecosystem status and trends in a rapidly changing environment. Springer, Dordrecht, pp 17–30

Pease CH, Salo SA, Overland JE (1983) Drag measurements for first-year sea ice over shallow sea. J Geophys Res 88(C5):2853–2862

Perovich DK, Richter-Menge JA, Jones KF, Light B (2008) Sunlight, water, and ice: extreme Arctic sea ice melt during the summer of 2007. Geophys Res Lett 35:L11501. doi:10.1029/2008GL034007

Pickart RS (2004) Shelfbreak circulation in the Alaskan Beaufort Sea: mean structure and variability. J Geophys Res 109:C04024. doi:10.1029/2003JC001912

PMEL (2000) FY 2000 operating plan. http://www.pmel.noaa.gov/home/accomp/opplan00.html
Proshutinsky AY, Johnson MA (1997) Two circulation regimes of the wind-driven Arctic Ocean. J Geophys Res 102(C6):12493–12514. doi:10.1029/97JC00738
Rigor IG, Wallace JM (2004) Variations in the age of Arctic sea-ice and summer sea-ice extent. Geophys Res Lett 31:L09401. doi:10.1029/2004GL019492
Serreze MC, Holland MM, Stroeve J (2007) Perspectives on the Arctic's shrinking sea ice cover. Science 315:1533. doi:10.1126/science.1139426
Shimada K et al (2006) Pacific Ocean inflow: influence on catastrophic reduction of sea ice cover in the Arctic Ocean. Geophys Res Lett 33:L08605. doi:10.1029/2005GL025624
Springer AM, McRoy CP, Flint MV (1996) The Bering Sea Green Belt: shelf-edge processes and ecosystem productivity. Fish Oceanogr 35:205–223
Stabeno PJ, Reed RK (1994) Lagrangian measurements in the Kamchatka current and Oyashio. J Oceanogr 50:653–662
Stabeno PJ, Schumacher DJ, Ohtani K (1999) The physical oceanography of the Bering Sea. In: Loughlin TR, Ohtani K (eds) Dynamics Bering Sea. North Pacific Marine Science Organization (PICES), Published by University by Alaska Sea Grant, Fairbanks, Alaska, pp 1–28
Steele M, Morison J, Ermold W, Rigor I, Ortmeyer M, Shimada K (2004) Circulation of summer Pacific halocline water in the Arctic Ocean. J Geophys Res 109:C02027. doi:10.1029/2003JC002009
Steele MW, Ermold W, Zhang J (2008) Arctic Ocean surface warming trends over the last 100 years. Geophys Res Lett 35:L0261
Takenouti Y, Ohtani KY (1974) Currents and water masses in the Bering Sea: a review of Japanese work. In: Hood DW, Kelley EJ (eds) Oceanography of the Bering Sea: with emphasis on renewable resources. Institute of Marine Science, University of Alaska, Fairbanks, pp 39–58
Wang J, Ikeda M (1997) Diagnosing ocean unstable baroclinic waves and meanders using quasi-geostrophic equations and Q-vector method. J Phys Oceanogr 27:1157–1172
Wang J, Ikeda M (2000) Arctic oscillation and Arctic Sea-ice oscillation. Geophys Res Lett 27(9):1287–1290
Wang J, Ikeda M (2001) Arctic Sea-ice oscillation: regional and seasonal perspectives. Ann Glaciol 33:481–492
Wang J, van der Baaren A, Mysak LA (1995) A principal component analysis of gridded global sea-level pressure, surface air temperature and sea-ice concentration in the Arctic region, 1953–1993. C^{2}GCR report No. 95–4, McGill University, Montreal, 22 pp+Figures+Fortran source programs. Available from jia.wang@noaa.gov
Wang J, Liu Q, Jin M (2002) A user's guide for a Coupled Ice-Ocean Model (CIOM) in the Pan-Arctic and North Atlantic Oceans. International Arctic Research Center-Frontier Research System for Global Change, Tech Res 02–01, 65 pp
Wang J, Kwok R, Saucier FJ, Hutchings J, Ikeda M, Hibler W III, Haapala J, Coon MD, Meier HEM, Eicken H, Tanaka N, Prentki R, Johnson W (2003) Working towards improved small-scale sea ice and ocean modeling in the Arctic seas. EOS 84(34):329–330
Wang J, Wu B, Tang C, Walsh JE, Ikeda M (2004) Seesaw Structure of subsurface temperature anomalies between the Barents Sea and the Labrador Sea. Geophys Res Lett 31:L19301. doi:10.1029/2004GL019981
Wang J, Liu Q, Jin M, Ikeda M, Saucier FJ (2005a) A coupled ice-ocean model in the pan-Arctic and the northern North Atlantic Ocean: simulation of seasonal cycles. J Oceanogr 61:213–233
Wang J, Ikeda M, Zhang S, Gerdes R (2005b) Linking the northern hemisphere sea ice reduction trend and the quasi-decadal Arctic Sea ice oscillation. Clim Dyn 24:115–130. doi:10.1007/s00382-004-0454-5
Wang J, Mizobata K, Hu H, Jin M, Zhang S, Johnson W, Shimada K (2008) Modeling seasonal variations of ocean and sea ice circulation in the Beaufort and Chukchi seas: a model-data fusion study. Chin J Polar Sci 19(2):168–184
Wang J, Zhang J, Watanabe E, Mizobata K, Ikeda M, Walsh JE, Bai X, Wu B (2009a) Is the dipole anomaly a major driver to record lows in the Arctic sea ice extent? Geophys Res Lett 36:L05706. doi:10.1029/2008GL036706

Wang J, Hu H, Mizobata K, Saitoh S (2009b) Seasonal variations of sea ice and ocean circulation in the Bering Sea: a model-data fusion study. J Geophys Res 114:C02011. doi:10.1029/2008JC004727

Wang J, Mizobata K, Jin M, Hu H (2010) Sea Ice-Ocean-Oilspill Modeling System (SIOMS) for the nearshore Beaufort and Chukchi seas: parameterization and improvement (Phase II), Coastal Marine Institute report and OCS study MMS 2008–021. Coastal Marine Institute, Fairbanks, 88 pp

Wang M, Overland JE (2009) A sea ice free summer Arctic within 30 years? Geophys Res Lett 36:L07502. doi:10.1029/2009GL037820

Watanabe E, Wang J, Sumi T, Hasumi H (2006) Arctic Dipole and its contribution to sea ice exports in the last 20th century. Geophys Res Lett 33:L23703. doi:10.1029/2006GL028112

Woodgate RA, Aagaard K (2005) Revising the Bering Strait freshwater flux into the Arctic Ocean. Geophys Res Lett 32:L02602. doi:10.1029/2004GL021747

Woodgate RA, Weingartner TJ, Aagaard K (2005) A year in the physical oceanography of the Chukchi Sea: moored measurements from autumn 1990–1991. Deep Sea Res II 52:3116–3149

Woodgate RA, Weingartner T, Lindsay R (2010) The 2007 Bering Strait oceanic heat flux and anomalous Arctic sea-ice retreat. Geophys Res Lett 37:L01602. doi:10.1029/2009GL041621

Wu B, Wang J, Zhang R (2004) Effects of intraseasonal variations of the Arctic Oscillation on the Barents Sea. Polar Meteorol Glaciol 18:82–95

Wu B, Wang J, Walsh JE (2006) Dipole anomaly in the winter Arctic atmosphere and its association with Arctic sea ice motion. J Clim 19(2):210–225. doi:10.1175/JCLI3619.1

Wyllie-Echeverria T (1995) Seasonal sea ice, the cold pool and gadid distribution on the Bering Sea Shelf. PhD thesis, University of Alaska, Fairbanks, Alaska, 281 pp

Yao T, Tang CL, Peterson IK (2000) Modeling the seasonal variation of sea ice in the Labrador Sea with a coupled multi-category ice model and the Princeton Ocean Model. J Geophys Res 105:1153–1165

Zhang J, Lindsay RW, Steele M, Schweiger A (2008a) What drove the dramatic retreat of Arctic sea ice during summer 2007? Geophys Res Lett 35. doi:10.1029/2008GL03400

Zhang J, Steele M, Woodgate R (2008b) The role of Pacific water in the dramatic retreat of Arctic sea ice during summer 2007. Chin J Polar Sci 19(2):93–107

Zhang J, Woodgate R, Moritz R (2010) Sea ice response to atmospheric and oceanic forcing in the Bering Sea. J Phys Oceanogr 40:1729–1747. doi:10.1175/2010JPO4323.1

Chapter 5
The Large Scale Ocean Circulation and Physical Processes Controlling Pacific-Arctic Interactions

Wieslaw Maslowski, Jaclyn Clement Kinney, Stephen R. Okkonen, Robert Osinski, Andrew F. Roberts, and William J. Williams

Abstract Understanding oceanic effects on climate in the Pacific-Arctic region requires knowledge of the mean circulation and its variability in the region. This chapter presents an overview of the mean regional circulation patterns, spatial and temporal variability, critical processes and property fluxes from the northern North Pacific into the western Arctic Ocean, with emphasis on their impact on sea ice. First, results from a high-resolution, pan-Arctic ice-ocean model forced with realistic atmospheric data and observations in the Alaskan Stream, as well as exchanges across the Aleutian Island Passes, are discussed. Next, general ocean circulation in the deep Bering Sea, shelf-basin exchange, and flow across the Bering shelf are investigated. Also, flow across the Chukchi Sea, pathways of Pacific summer water and oceanic forcing of sea ice in the Pacific-Arctic region are analyzed. Finally, we hypothesize that the northward advection of Pacific Water together with the excess oceanic heat that has accumulated below the surface mixed layer in the western Arctic Ocean due to diminishing sea ice cover and subsequent increased solar insolation are critical factors affecting sea ice growth in winter and melt the following year.

W. Maslowski (✉) • J. Clement Kinney
Department of Oceanography, Graduate School of Engineering and Applied Sciences, Naval Postgraduate School, Dyer Road, Bldg. SP339B, Monterey, CA 93943, USA
e-mail: maslowsk@nps.edu; jlclemen@nps.edu

S.R. Okkonen
School of Fisheries and Ocean Sciences, University of Alaska, Fairbanks, AK 99775, USA

R. Osinski
Institute of Oceanology, Polish Academy of Sciences, Powstancow Warszawy 55, Sopot 81-712, Poland

A.F. Roberts
Department of Oceanography, Naval Postgraduate School, Dyer Road, Bldg. SP339B, Monterey, CA 93943, USA

W.J. Williams
Fisheries and Oceans Canada, Institute of Ocean Sciences, Sidney, BC, Canada V8L 4B2
e-mail: Bill.Williams@dfo-mpo.gc.ca

J.M. Grebmeier and W. Maslowski (eds.), *The Pacific Arctic Region: Ecosystem Status and Trends in a Rapidly Changing Environment*, DOI 10.1007/978-94-017-8863-2_5,

We argue that process-level understanding and improved model representation of ocean dynamics and ocean-ice-atmosphere interactions in the Pacific-Arctic region are needed to advance knowledge and improve prediction of the accelerated decline of sea ice cover and amplified climate warming in the Arctic.

Keywords Ocean circulation • Alaskan Stream • Bering Sea • Western Arctic • Ice-ocean modeling

5.1 Introduction

This chapter focuses on the physical oceanography of the Pacific-Arctic Region (PAR). For our purposes, PAR is defined to include the Gulf of Alaska, northern North Pacific, and the Bering, Chukchi, Beaufort, and East Siberian seas. Figure 5.1

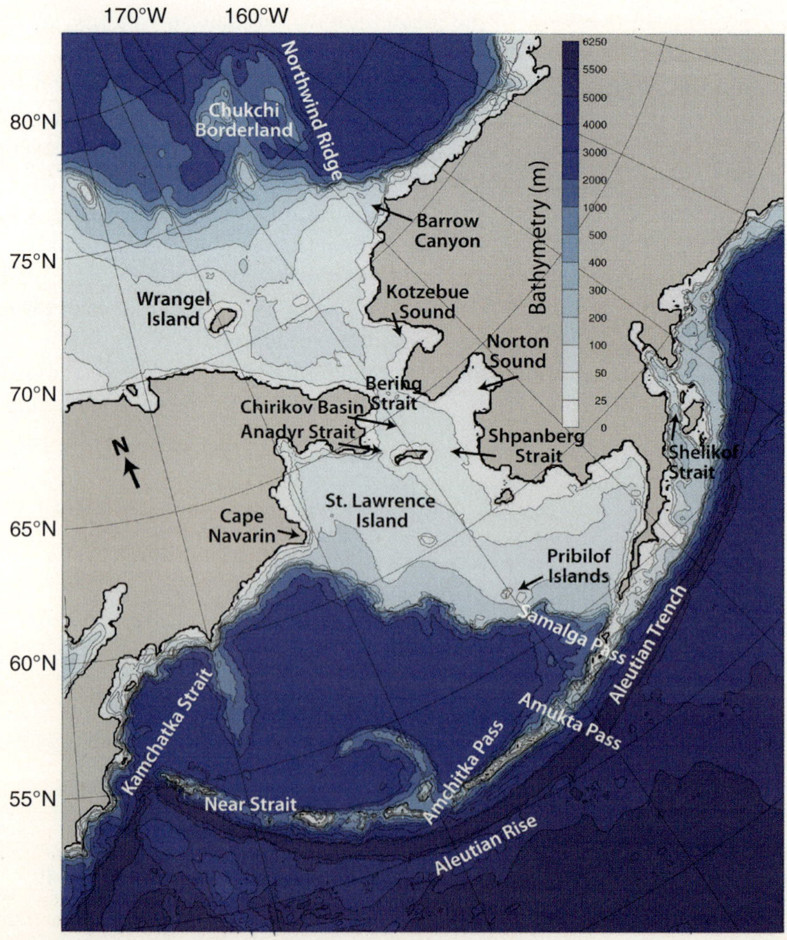

Fig. 5.1 Bathymetry (m) of the Pacific-Arctic Region

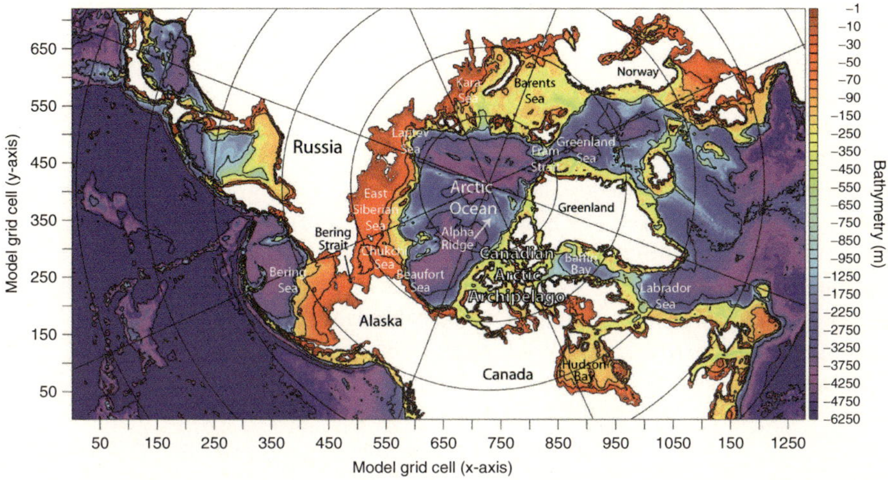

Fig. 5.2 The NAME model domain and bathymetry (m)

shows the bathymetry and geographic place names for the PAR region. To our knowledge, there has been no recent physical oceanography review of this region. Stabeno et al. (1999) reviewed the physical oceanography of the Bering Sea and described observational results on the flow through the Aleutian Island Passes and the general circulation throughout the Bering Sea. They stated a need for more information on: (1) the transport variability and magnitude of the flow through the Aleutian Island Passes, (2) deep Bering Sea circulation, and (3) mechanisms that provide nutrients to the euphotic zone along the Bering Slope. Results presented in this chapter and in Chap. 6 by Williams et al. (2014, this volume) provide more information in these three areas.

Along with observational results from several authors, we will discuss results from the Naval Postgraduate School (NPS) Arctic Modeling Effort (NAME) coupled sea ice-ocean model in this chapter. The model has a horizontal grid spacing of 1/12° (or ~9 km) and 45 vertical depth layers. At this horizontal resolution the model is considered eddy permitting, as mesoscale features down to ~37 km in diameter (four grid points) can be resolved (Maslowski et al. 2008b). In the vertical direction, eight levels are in the upper 50 m and 15 levels in the upper 200 m. The high vertical resolution, especially in the upper water column, allows for more realistic representation of the shallow Arctic and sub-Arctic shelves than a lower vertical resolution typical in many ocean models. The model domain contains the sub-Arctic North Pacific and North Atlantic Oceans, the Arctic Ocean, the Canadian Arctic Archipelago (CAA) and the Nordic Seas (Fig. 5.2). The choice of model domain was in part dictated by the need for realistic prediction of the net northward transport of water mass and properties through Bering Strait and consequently the circulation and variability in volume and heat fluxes entering the Beaufort Sea from the Chukchi Shelf. Additional details of this model are discussed by Clement Kinney et al. (2014, this volume), Maslowski et al. (2004) and Maslowski et al. (2008a, b).

5.2 The Northern North Pacific, Gulf of Alaska, and Alaskan Stream

The Alaskan Stream is the primary circulation feature of the southern Pacific-Arctic Region. It is an intense, narrow (~50–80 km) and deep (>3,000 m) current, which flows westward from the head of the Gulf of Alaska to the westernmost Aleutian Islands (Stabeno and Reed 1992; Reed and Stabeno 1999). Its highly variable transport (e.g. Onishi 2001) reflects the complex air-sea-topographic interactions in the northern North Pacific and strongly influences the exchanges with the Bering Sea (Maslowski et al. 2008a; Clement Kinney and Maslowski 2012), which occur through the straits and passes of the Aleutian-Komandorskii Island Arc.

The transport estimates of the Alaskan Stream based on observations (Thompson 1972; Favorite 1974; Reed 1984; Stabeno and Reed 1992; Reed and Stabeno 1993, 1997) are generally limited by coarse or intermittent temporal and/or spatial sampling. Such limitations hinder investigations of local dynamics, long-term mean total volume and property fluxes and relationships between climate change and transport variability. The deep baroclinic structure and the westward intensification (between 150°W and 180°W; Reed 1984) of this boundary current flowing along the slope of the Aleutian Trench present further challenges for transport estimates based on spatially limited measurements using geostrophic velocity calculations, where an assumption of no motion at depth (e.g. at 1,000 or 1,500 m compared to the total depth of over 6,000 m) is made. For example, Reed (1984) computed a typical transport in the Alaskan Stream (referred to 1,500 dbar) at 12 Sv (1 $Sv = 10^6\ m^3\ s^{-1}$) and at 15–20 Sv when a correction for its baroclinic structure below 1,500 dbar is included. Other commonly reported transports (Royer and Emery 1987; Reed 1984; Reed and Stabeno 1999) are all estimated below 25 Sv, including corrections for the deep baroclinic flow. However, accurate assessment of the total volume transport at any cross section in the Alaskan Stream is still daunting, as it requires knowledge of the absolute velocity structure down to the bottom.

The chain of the Aleutian Islands forms the natural southern boundary of the Bering Sea. The numerous passes between islands play a primary role in determining both circulation and distribution of water properties in the Bering Sea (Stabeno et al. 1999). The variability of the inflow/outflow rate through the southern boundary has a twofold impact on Bering Sea oceanic conditions and ecosystem changes. First, inflow through Aleutian passes is an important source of Bering Sea ecosystem nutrients (Stabeno et al. 1999). Second, strong inflow through the Amchitka, Amukta, and other Aleutian passes may enhance instability of the Aleutian North Slope Current (ANSC) and thus affect Bering Sea eddy formation.

Although the eastern passes are shallow, with depths less than 300 m, the long-term velocity observations in several passes indicate strong and persistent northward flow. For example, Stabeno et al. (2005) reported approximately 4 Sv of mean inflow into the Bering Sea through Amukta Pass during a period of 3 years (2003–2005), which is approximately five times larger than geostrophic estimates (Stabeno et al. 1999). Thus, recent estimates indicating higher exchange rates between the

Bering Sea and Pacific Ocean require extended studies of the robustness of these estimates and, possibly, revising our view on the mechanisms controlling Bering Sea circulation.

Modeling the Alaskan Stream and the inter-basin exchanges between the North Pacific and the Bering Sea has been challenging, as well. Ocean general circulation models (OGCMs) have demonstrated a rather limited ability to realistically represent circulation and inter-basin exchanges of water mass and properties between the North Pacific and the Bering Sea/Arctic Ocean. While global models have commonly lacked the needed horizontal and/or vertical resolution (e.g. Goosse et al. 1997; Xie et al. 2000) regional models of this region have often used closed or fixed (prescribed) boundary conditions in the Bering Sea and/or North Pacific (Overland et al. 1994; Hurlburt et al. 1996; Stabeno and Hermann 1996; Okkonen et al. 2001; Hermann et al. 2002; Ladd and Thompson 2002). Additionally, only few of those models have realistically accounted for the net northward transport (0.8 Sv±0.2 Sv; Clement Kinney et al. 2014, this volume) through Bering Strait (e.g. Goosse et al. 1997).

The NAME model has been developed to address the above limitations and to allow focused studies of the circulation, relevant dynamics and exchanges between the sub-arctic and arctic basins (Maslowski et al. 2004, 2008a; Clement et al. 2005; Clement Kinney et al. 2014, this volume). The pan-Arctic domain of this model, (Fig. 5.2) configured at horizontal resolution of 1/12° and 45 depth levels allows for realistic flow through the many narrow passages in the Aleutian Islands and the northern Bering Sea.

Figure 5.3 shows the 26-year (1979–2004) mean circulation in the upper 100 m based on NAME model results (Clement Kinney and Maslowski 2012), in which the Alaskan Stream is clearly the strongest feature. The Alaskan Stream is the primary source of warm and relatively fresh water for mass and property fluxes through the Aleutian passes and, therefore, has a significant influence on the Bering Sea (Reed 1990; Reed and Stabeno 1999).

NAME model results (not shown) in the upper 100 m of the Alaskan Stream show that maximum speeds close to 90 cm s^{-1} occur in winter (i.e. January, February, March), with a decrease to around 75 cm s^{-1} in summer and autumn, suggesting up to 14 % speed variability due to the seasonal cycle. According to observations, Reed and Stabeno (1997) report 70–125 cm s^{-1} speeds in the upper 100 m across 173.5°W measured during summer 1995. Reed (1984) obtained similar peak speeds (90–110 cm s^{-1}) in the upper ocean across 170°W during winter 1980 and summer 1981. Reed and Taylor (1965) measured velocities of 80 cm s^{-1} around 175°W in June 1959. The above comparison of modeled and observed velocities in the upper 100 m of the Alaskan Stream suggests the model realistically represents this flow.

The modeled 23-year mean westward transport of Alaskan Stream computed over the total water depth ranges from 44.6 Sv to 55.8 Sv, of which 12–18 Sv is below 1,000 m and 7–12 Sv is below 1,500 m. These transports appear rather high compared to the previously reported estimates based on measurements; however some observational evidence exists in support of higher model results. Warren and

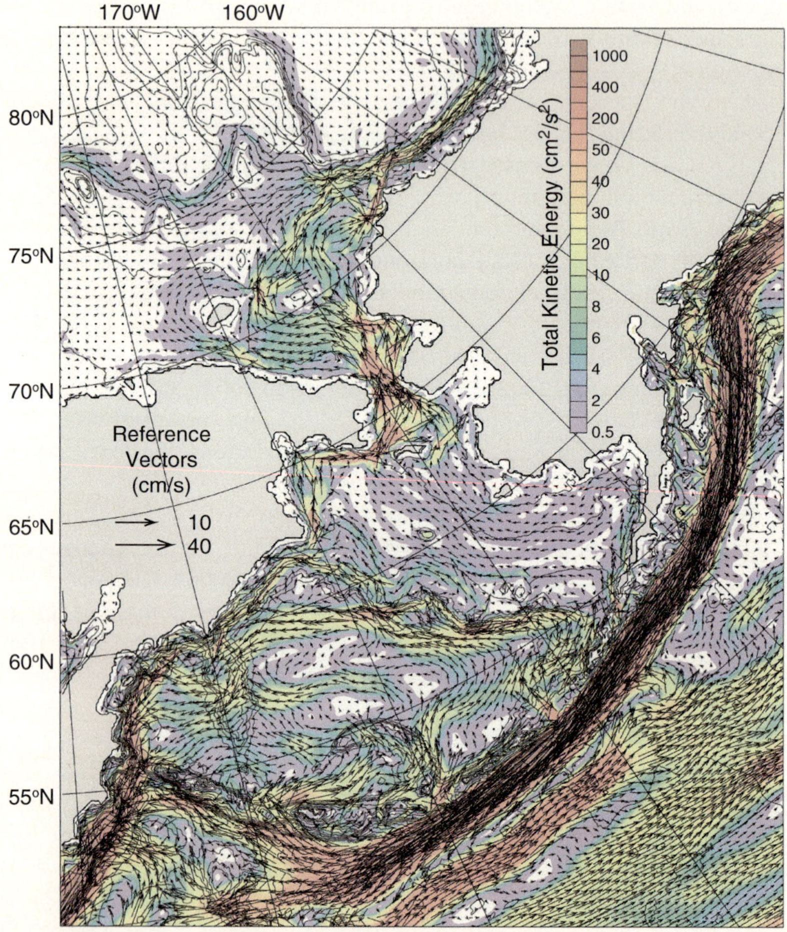

Fig. 5.3 The 26-year mean (1979–2004) upper 100 m modeled circulation (*vectors* shown every 4th grid cell in each direction; cm s^{-1}) and total kinetic energy (*shading*; cm^2 s^{-2}). *Black contour lines* represent bathymetry (m). Please note that the velocity vectors are not scaled in a linear fashion (for figure clarity) and their magnitude can be derived as $|v| = (2 \times TKE)^{1/2}$

Owens (1988) reported 28 Sv based on geostrophic calculation but referred to current meter measurements from a single mooring. Hydrographic measurements by Roden (1995) near 180°W indicate transport of 38 Sv above 6,000 m with 14 Sv below 1,000 dbar and 9 Sv below 1,500 dbar in July 1993. Onishi and Ohtani (1999) computed the Alaskan Stream transport ranging from 21.8 Sv to 41 Sv along 180°W during 1990–1997. Finally, Onishi (2001) estimates the transport referred to 3,000 m ranging from 14.8 Sv to 41.0 Sv along 180°W over 9 years, from 1990 to 1998.

Major disturbances created by southward shifts in the Alaskan Stream and eddies propagating along the Aleutian Island Arc can produce large deviations, which have been explored in multiple studies (e.g. Musgrave et al. 1992; Stabeno et al. 1999; Crawford et al. 2000). In Crawford et al. (2000), 6 years of TOPEX/Poseidon (T/P) data measured the presence of six anticyclonic eddies in the Alaskan Stream. Of the six eddies, some formed near the Alaskan panhandle, while others began propagating just south of Shelikof Strait with an estimated phase speed of ~2.5 km/day. The sea surface height anomalies and eddy diameter approached 72 cm and averaged 160 km, respectively. Because the Alaskan Stream represents the largest source of transport into the Bering Sea, these long-lived eddies (mean lifespan of 1–3 years; Crawford et al. 2000) may significantly alter both the physical and biological structure of the Bering Sea (Okkonen 1992, 1996).

Recent model results (Maslowski et al. 2008a) provided an in-depth examination of the anticyclonic eddies found in the Alaskan Stream and their effect on the strength and location of the stream, as well as their influence on flow through Amukta and Amchitka Passes. For example, sea surface height anomaly (SSHA) and circulation (Fig. 5.4a) indicate the presence of an anticyclonic eddy south of Kodiak Island during January 1993. By December (Fig. 5.4b), the eddy has moved to a position that is just southeast of Amukta Pass. Figure 5.5 shows modeled monthly mean distributions of velocity as the eddy passes by EAS1 during January 1993 and EAS4 during December. In both cases, the velocity core is shifted southward when the eddy is present. The southward shift is strongest at EAS4 by a distance of more than 140 km. Over the course of the entire model simulation (1979–2003), the number of eddies crossing the EAS4 line was calculated (Maslowski et al. 2008a). Twenty "weak" eddies (0.3 m < SSHA < 0.5 m) crossed the line for an average of 0.8 per year and 12 "strong" eddies (SSH > 0.5 m) were identified for an average of 0.5 per year.

Using a combination of these model results, additional model results from Maslowski et al. (2008a), and observational results by Reed and Stabeno (1989, 1999) and Crawford et al. (2000), we conclude that the Alaskan Stream remains relatively intact during the periods of southward shifts. It is also believed that the formation and propagation of eddies along the Alaskan Stream is the main cause of these deviations in the westward flow. Any stationary attempt to monitor the flow of the Alaskan Stream (e.g., moored Acoustic Doppler Current Profiler (ADCP)) would benefit from including instrumentation at least 200 km southward of the mean position of the Alaskan Stream, in order to capture these shifts.

Investigation of eddy effects on exchanges between the Gulf of Alaska and the Bering Sea revealed significant increases in net northward fluxes as an eddy moves along a pass within the Aleutian Islands (Maslowski et al. 2008a). Analysis of modeled time series through the two main passes in the eastern and central Aleutian Archipelago, i.e., Amukta Pass and Amchitka Pass, showed that peaks in the monthly mean anomalies of volume and property transport were often associated with the presence of nearby eddies in the Alaskan Stream (e.g. see Fig. 5.9 in Maslowski et al. 2008a).

Another dynamical aspect of the Aleutian throughflow, which should be taken into account, is tidal mixing and rectification. Tidal currents between the Aleutian Islands are generally strong (100–150 cm/s) and produce a total 70 GW energy flux

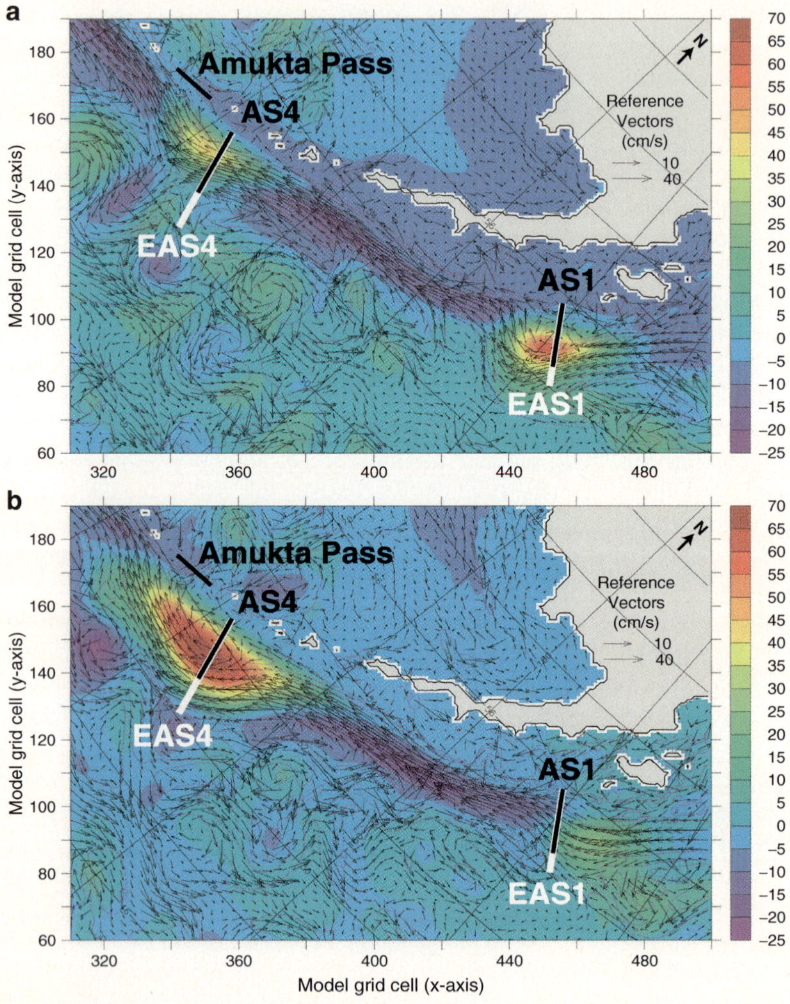

Fig. 5.4 Modeled sea surface height anomaly (m) and 0–477 m circulation during (**a**) Jan. 1993 and (**b**) Dec. 1993. Every 3rd *vector* in each direction shown

into the Bering Sea (Foreman et al. 2006). The most important are the M2 (31 GW), K1 (24.5 GW) and O1 (13 GW) tidal constituents (Foreman et al. 2006). The largest energy conduits into the Bering Sea are via Amukta and Samalga passes (M2) and Amchitka Pass (K1). Modeling studies (e.g. Kowalik 1999) show that tides significantly impact local mixing and generate a northward residual tidal current of 5–15 cm/s between the Aleutian Islands. Recently, Stabeno et al. (2005) hypothesized that tidal rectification may be one of mechanisms responsible for such a high new mean inflow through Amukta Pass.

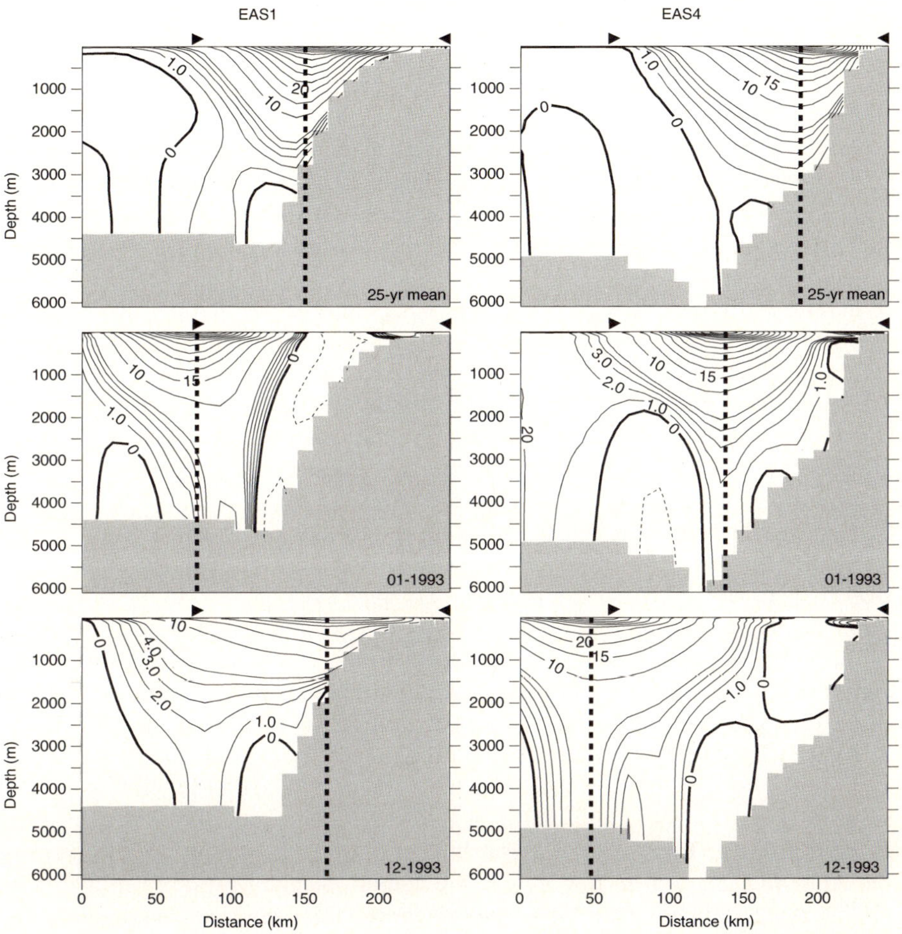

Fig. 5.5 Modeled monthly mean vertical profiles of velocity (both *shading* and *contours*; cm s^{-1}) for cross sections EAS1 (*left*), EAS4 (*right*). Positive velocity is directed westward. Inward-facing *triangles* above each figure indicate the position of the original sections (AS1 on *left*; AS4 on *right*). The *black line* represents the horizontal position of the velocity core

5.3 Western Subarctic Gyre

Focusing on the deep western passes, we know that inflow from the northwestern North Pacific into the Bering Sea occurs primarily through Near Strait, while outflow occurs largely through Kamchatka Pass (Stabeno and Reed 1994). Multi-year continuous observations of the flow through the western Aleutian Island Passes are not available. Available observations include data from satellite-tracked drifters (Stabeno and Reed 1994) and short-term CTD measurements (e.g. Stabeno and Reed 1992; Reed and Stabeno 1993), along with historical datasets (e.g. Arsen'ev 1967; Ohtani 1970; Hughes et al. 1974; Favorite 1974). Observed estimates of the net volume transport through Near Strait are northward and range from ~5 Sv (Reed and Stabeno 1993) to ~10 Sv (historical measurements by Favorite 1974). The long-term mean model result from NAME (5.06 Sv) compares well with the observations from Reed and Stabeno (1993).

Observed estimates of the net volume transport through Kamchatka Strait are southward and range from ~6 to 15 Sv (Ohtani 1970; Verkhunov and Tkachenko 1992; Stabeno and Reed 1992). The long-term mean model result from NAME (8.9 Sv) is within the observed range, however it is much less than recently published results from Panteleev et al. (2006). Panteleev et al. (2006) fulfilled variational analysis of the summer climatological circulation in the major part of the Bering Sea. They estimated the volume transport through Kamchatka Strait to be 24 Sv, which is significantly larger than most of the traditional dynamical method estimates (Ohtani 1970; Verkhunov and Tkachenko 1992; Stabeno and Reed 1992). On the other hand, the Panteleev et al. (2006) transport estimate agrees well with an estimate by Hughes et al. (1974) of 20 Sv, obtained through dynamical calculations referenced to the surface velocities taken from surface drogued floats.

High-resolution and large scale modeling of the Bering Sea basin and the Western Subarctic Gyre has been limited. Previous modeling work by Overland et al. (1994) realized general circulation features (e.g., complex cyclonic flow in the Bering Sea basin) similar to observations. However, they were not able to simulate the meanders in the Kamchatka Current that were identified by Stabeno and Reed (1994), nor were they able to show enough interannual variability in the flow through Near Strait to account for observations by Stabeno and Reed (1992) or Reed et al. (1993). The prescribed boundary condition for the Alaskan Stream and the climatological atmospheric forcing that was used may have prevented Overland et al. (1994) from simulating realistic variability in the flow through Near Strait. While observations before 1994 (e.g., Reed 1984; Stabeno and Reed 1992) showed little eddy energy of the Alaskan Stream near 166°W (i.e. the lateral boundary of the Overland et al. (1994) model domain), more recent observations (e.g. Crawford et al. 2000) are in stark contrast.

Recent modeling work by Clement Kinney and Maslowski (2012) shows how the numerous eddies and meanders of the Bering Sea and within the Alaskan Stream play a critical role in determining the flow through the western Aleutian passes and straits. For example, modeled time series of volume flux through Near Strait over 26 years showed seven time periods when the transport was near-zero or even reversed (Clement Kinney and Maslowski 2012). These anomalies lasted from 3 months to almost 2 years. Based on satellite-tracked drifters released throughout the region, Stabeno and Reed (1992) noted an anomalous lack of inflow through Near Strait that began in summer 1990 and persisted at least through fall 1991. While the model results did not show a lack of inflow at exactly the same time period as Stabeno and Reed (1992), the processes underlying the anomaly appear to be the same. The modeled flow reversals and maxima were related to the presence of eddies and meanders both north and south of the strait. There did not appear to be a consistent circulation regime responsible for inducing a reversal (see Clement Kinney and Maslowski (2012) for further discussion and figures). Instead, flow reversals occurred when mesoscale features were in the proper alignment with the axis of the strait. Therefore, it is believed that short-term observations (months to years) may not be representative of the actual mean flow. This indicates a need for continuous monitoring of the flow through Near Strait, as well as other passes and straits in the Western Aleutian Islands to determine both the mean flow and its variability.

5.4 Bering Sea

The general circulation within the Bering Sea basin is typically described as cyclonic in the long-term mean. However, transport within the gyre can vary by more than 50 % (Stabeno et al. 1999). Causes associated with this variation have been identified as either changes in the Alaskan Stream inflow (Overland et al. 1994) and/or variability of the wind-driven transport within the basin (Bond et al. 1994). Mesoscale eddies within the basin have been identified in observations published by Cokelet et al. (1996) and Stabeno and Reed (1994), among others (e.g., Solomon and Ahlnäs 1978; Kinder et al. 1980; Paluszkiewicz and Niebauer 1984). Instabilities along the Bering Slope and Kamchatka Currents and interactions with canyons and embayments at the landward edge of these currents and inflows through the Aleutian Island Passes may be responsible for eddy generation (Cokelet et al. 1996). Stabeno and Reed (1994) observed several eddies and meanders within the Kamchatka Current and the eastern Bering Sea by utilizing satellite-tracked drifters. Several anticyclonic eddies were observed in the western side of the basin and resulted from the interaction of the Kamchatka Current with topographic features. Stabeno and Reed (1994) suspect that these features are semi-permanent, since they appeared in drifter trajectories for more than 1 year. Stabeno and Reed (1994) observed a large anticyclonic eddy west of Bowers Ridge that had a diameter of approximately 200 km and velocities of 30–40 cm/s, which is similar in size and velocity to those from the NAME model.

Animations of monthly mean NAME model results over the 26-year simulation show frequent and complex eddy activity throughout the Bering Sea basin. For instance, at least 14 mesoscale eddies are present in June 1987 (Clement and Maslowski 2008), as shown in Fig. 5.6. Half of these are anticyclonic and the other half are cyclonic. Diameters of these eddies are 120 km and greater and velocities are up to 40 cm/s. Lifetimes of these eddies are typically a few months. The 9-km horizontal resolution of the model makes it possible to resolve eddies with diameters as small as ~37 km, however, the smallest eddies are likely not resolved, recalling that the Bering Sea has a Rossby radius of deformation of ~12–20 km according to Chelton et al. (1998).

During the 26-year model simulation, meanders and eddies are continuously present throughout the deep Bering Sea basin. This is consistent with results from Cokelet and Stabeno (1997), which show that the background flow in the interior of the Bering Sea is dwarfed by the energetic eddies which populate the region.

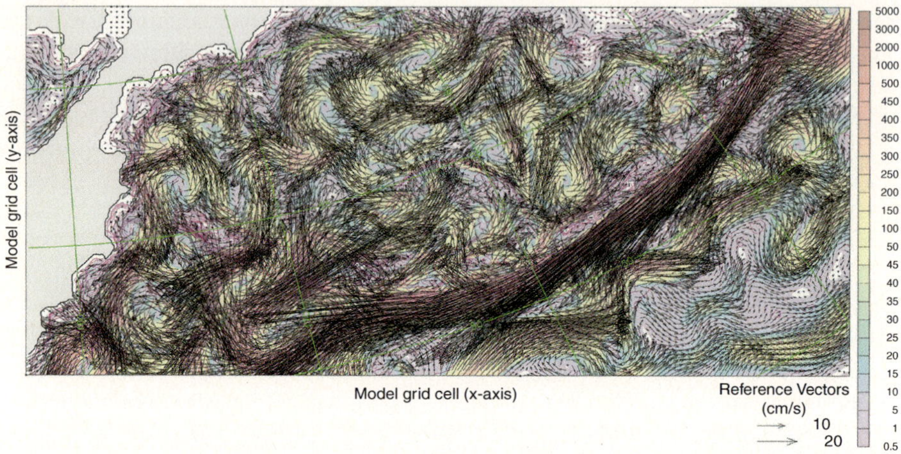

Fig. 5.6 The June 1987 upper 100 m mean modeled circulation (*vectors*) and total kinetic energy (*shading*; $cm^2\ s^{-2}$). *Magenta contour lines* represent bathymetry (m). Every 3rd *vector* in each direction shown

Our results indicate that these eddies are important in redistributing temperature and salinity (Clement Kinney and Maslowski 2012), which may have an effect on biological species in the region by altering environmental conditions.

Numerical model results (Clement Kinney and Maslowski 2012) also show very frequent and complex eddy activity present along the Bering Sea Slope. In fact, the Bering Slope Current (BSC) appears to be more a system of eddies rather than a continuous feature (Okkonen 1993; Clement Kinney et al. 2009). The energetic anticyclonic eddy activity of the BSC is a well-known characteristic of Bering Sea circulation. Analysis of the mooring velocity measurements along the eastern Bergin Sea continental slope between Pribilof and Zhemchug canyons (Schumacher and Reed 1992; Cokelet and Stabeno 1997) showed that mesoscale currents have amplitude >80 cm/s, i.e. 10–15 times higher than the long-term mean BSC velocity. The BSC eddies penetrate to a depth of at least 500 m and transport nutrients from deep regions onto the eastern Bering Sea shelf (Stabeno et al. 1999). Eddy-induced cross-shelf mixing is not well studied, but high-intensity mesoscale variability in the region indicates the importance of this mechanism. Recent climatological studies reveal the linkage between long-term variability of the large-scale dynamic and oceanic eddy activity (Hogg et al. 2005). Currently the mechanisms of this linkage are not clear.

The eddy activity along the Bering Sea slope seems to have an important effect on shelf-basin exchange, especially in canyons such as Zhemchug Canyon (Clement Kinney et al. 2009). A cyclonic eddy located just south of the canyon was shown to greatly increase the on-shelf salt flux during a 3-month period via upwelling of relatively salty water and on-shelf flow. Upwelling of salty and nutrient-rich deep Bering Sea water due to eddies and shelf-basin exchange, in general, could enhance biological production in this region (Clement Kinney et al. 2009). Overall circulation over the

Bering Sea shelf is mainly wind-driven and geostrophic year around, with reversals of both across and along shelf flow and local accelerations due to increased wind stress and bottom friction in response to changing winds between northwesterly and southwesterly (Danielson et al. 2012a, b).

Modeled eddy kinetic energy (EKE) fields show that the northern Bering Sea maintains year-round high energy and mixing, especially in Bering and Anadyr straits (see Figs. 5.16–5.19 of Clement et al. 2005). Notably, these regions of high EKE are found just upstream of highly productive areas in the Bering Sea (e.g., the Chirikov Basin and the region just north of Bering Strait) that have been identified in previous studies such as Grebmeier et al. (1988), Springer and McRoy (1993), and Grebmeier and Dunton (2000). This suggests that high-nutrient Anadyr Water is mixed into the euphotic zone as it flows generally northward and, upon encountering a region of lower EKE, can support water column primary production and the settling of organic matter to the benthos.

5.5 Chukchi Sea

The Arctic receives water from the Pacific through the Bering Strait. The Chukchi Sea connects the Bering Sea to the Arctic Ocean and is, therefore, the Pacific gateway to the Arctic. Three primary water masses enter the Chukchi Sea via Bering Strait, including Alaskan Coastal Water, Bering Shelf Water, and Anadyr Water (Coachman et al. 1975). Alaskan Coastal Water, as its name implies, is found near the coast of Alaska and tends to be the warmest and freshest of the three. Anadyr Water is found in the west, nearer to Siberia and is usually colder and more saline than the others. Bering Shelf water lies in the middle and has more moderate temperature and salinity properties. Analysis of a year long time series of mooring data in 1990–1991 by Woodgate et al. (2005) indicates a fourth branch of water inflow through Bering Strait which exits northwestward through Long Strait. Its content of water originating from the Pacific is less than that contributed by the other three branches, as it is modified by the seasonally occurring Siberian Coastal Current (Weingartner et al. 1999). A number of important bathymetric features, such as Herald Shoal, Hanna Shoal, Barrow Canyon, Herald Canyon, the Chukchi Rise, and the Chukchi Cap influence the circulation on this shallow (mostly <60 m deep) and wide continental shelf.

Previous numerical modeling efforts for the Chukchi and Beaufort seas include Overland and Roach (1987), Winsor and Chapman (2004), and Walsh et al. (2004), with the latter based on results from the same model discussed in this article. In addition, all of the models participating in the Arctic Ocean Model Intercomparison Project (AOMIP; e.g. Proshutinsky et al. 2007; Holloway et al. 2007) as well as a recent study by Watanabe (2011) include the Chukchi Sea. However, many of those models prescribe climatological lateral boundary conditions at or near Bering Strait. As for global climate models, their spatial resolution in ocean models is commonly too coarse to realistically represent exchanges through Bering Strait, and they have to be parameterized (e.g. Goosse et al. 1997). An improvement of the NAME model is the ability to simulate the net northward transport and variability of water mass and properties through Bering Strait (Clement et al. 2005; McClean et al. 2001). Model results from the simulation discussed here have been compared with observations of volume transport, temperature and salinity in Bering Strait. For example, the modeled 1979–2004 mean volume transport of 0.65 Sv compares well with the observed mean of 0.83 Sv estimated for 1990–1994 (Clement et al. 2005; Roach et al. 1995). Please see Clement Kinney et al. (2014, this volume) for a comprehensive discussion of the flow through Bering Strait.

The 26-year (1979–2004) mean circulation from the NAME model in the upper 100 m is shown in Fig. 5.3. Generally northward velocities, as high as 50 cm/s, are found in Bering Strait with three main branches and two coastal flows forming north of the strait. Model results are in agreement with observations at 12 locations within the Chukchi Sea made by Woodgate et al. (2005) that show four northward outflows on the Chukchi Shelf. The main western branch (Herald Valley flow) is found between the 50-m isobaths and flows between Herald Shoal and Wrangel Island. The main eastern branch is directed primarily northward to the east of Herald Shoal and west of the 25-m isobath. This is the "Central Channel" flow identified by Weingartner et al. (2005) based on moored and shipboard data collected in the early 1990s. An offshoot of the western branch flows through Long Strait (between Wrangel Island and Siberia), most of which encircles the island in an anticyclonic fashion and eventually joins the Central Channel flow, however a small portion of the current continues along Siberia to the northwest near the coast. The seasonal, eastward-flowing Siberian Coastal Current (Weingartner et al. 1999) is not clearly showing in the 26-year mean circulation; however, velocity vectors nearest the coast are oriented eastward and snapshots of the ocean circulation (not shown) in early summer (i.e. June–July) have an evident narrow eastward flow. The other seasonal coastal current, modeled with relatively higher skill, is the Alaskan Coastal Current (ACC), which is fresh, warm (with measured temperatures up to 11 °C in August 2007) and narrow (10–40 km; Okkonen et al. 2009; Spall et al. 2008). Still, a more realistic representation of winds and local fresh water discharge from the Siberian and Alaskan coasts as well as higher spatial resolution would further improve model skill in simulating both of these coastal currents (Maslowski and Walczowski 2002; Newton et al. 2008). It is important to note, that observations made by Paquette and Bourke (1974) and Ahlnäs and Garrison (1984) show that the ACC is enhanced by solar-heated waters in Kotzebue and Norton sounds and it impacts the local ice edge position downstream.

Measurements of the oceanic impact on the marginal ice zone and ice edge position in the Chukchi Sea are limited and insufficient to quantify those effects. To address this, a series of cross-sections were established in the NAME model to obtain volume and relative heat flux values from the 26-year long integration (Fig. 5.7). The blue ACC sections were intended to capture the coastal flow, while the CL sections capture the north-south flow across the width of the Chukchi shelf. The CP sections, along with CL4, form a closed polygon surrounding the Chukchi Plateau. Table 5.1 shows the mean volume transport, heat flux and temperature across each

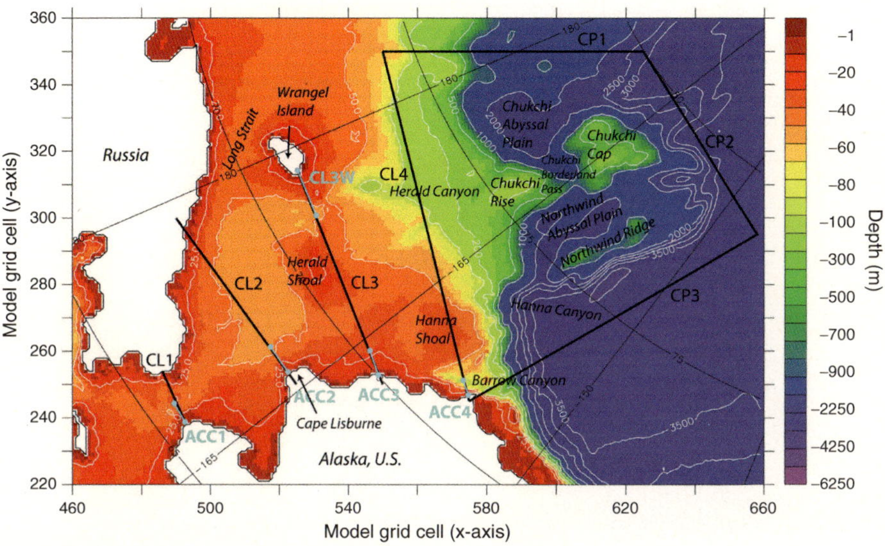

Fig. 5.7 Bathymetry (*shading*; in meters) with the locations of a series of cross-sections (*black* and *blue lines*). Labels corresponding to the cross-sections are either shown in *black* for longer sections or *green* for shorter sections. Place names are shown in *italics*

section. The main purpose of this is to determine relative contributions of heat flux across a given section or depth range relative to the total heat divergence/convergence from/into a specified region or basin and its relative change in time. In particular, the heat fluxes calculated here through a single section are not meant to estimate the change of heat in a closed volume as proposed by Schauer et al. (2008).

The coastal flow through ACC1 and ACC2 makes up almost half of the volume transport through the respective CL sections. More than half of the heat flux through CL1 and CL2 is found in the coastal sections (55 % and 71 %, respectively). The relative increase of heat flux between ACC1 and ACC2 supports the finding of Ahlnäs and Garrison (1984) that the solar-heated waters in the Kotzebue Sound and along the coast are some of the main sources of heat in the Alaskan Coastal Current. However, as one moves farther north, little of the heat or volume flux across CL3 is found in the shorter coastal section ACC3. Instead, the majority of the flow is bathymetrically steered away from the coast and to the west of Herald Shoal (Fig. 5.3). Some mixing and convergence of the Herald Valley flow and the Central Channel flow occurs north of Herald Shoal, where the flow becomes bathymetrically steered eastward as observed by Weingartner et al. (2005). Upon reaching Barrow Canyon, much of the heat (~63 % of the CL4 value) is again found near the coast, within section ACC4 (Fig. 5.8). It is also worth noting that both in the modeled coastal flow as well as across the entire CL4 two to three times more heat is transported below 10 m depth relative to the upper 10 m. We argue that over the Chukchi shelf, heat associated with summer Pacific water is partially lost to sea ice melt and to the

Table 5.1 Twenty-six year mean (1979–2004) and standard deviation (*in parenthesis*) for volume transport, heat flux and mean temperature across selected sections

Section	Volume transport (Sv)	Heat flux (TW)	Mean temperature (°C)
ACC1	0.302	2.315	−0.098
	(0.099)	(3.440)	(2.063)
ACC2	0.285	2.344	−0.450
	(0.151)	(3.735)	(1.688)
ACC3	0.014	0.170	−0.653
	(0.032)	(0.464)	(1.519)
ACC4	0.300	0.736	−1.273
	(0.172)	(1.192)	(0.400)
CL1	0.654	4.172	−0.364
	(0.228)	(5.695)	(1.656)
CL2	0.643	3.288	−0.983
	(0.225)	(4.793)	(0.823)
CL3	0.520	1.416	−1.138
	(0.221)	(1.906)	(0.515)
CL3W	0.196	0.305	−1.345
	(0.067)	(0.255)	(0.226)
CL4	0.678	1.167	−1.092
	(0.216)	(1.271)	(0.101)
CP1	0.526	2.509	−0.902
	(0.177)	(0.759)	(0.058)
CP2	−0.206	−0.447	−1.059
	(0.095)	(0.229)	(0.038)
CP3	−0.927	−2.719	−1.052
	(0.309)	(1.674)	(0.087)

Positive direction is north for all sections except CP1 (East), CP2 (South) and CP3 (West). Salinity-dependent freezing temperature was used as a reference temperature for heat flux calculations. Section locations are shown in Fig. 5.7

atmosphere, while it is gained as a result of the ice-albedo effect, as sea ice retreats, allowing increased absorption of solar insulation (as shown in Fig. 5.8). These processes, combined with the complex flow pattern between CL1 and CL4, provide at least a partial explanation for the relatively low correlation of interannual variability between the heat fluxes through Bering Strait and those entering the Beaufort Sea between Barrow and Barrow Canyon (Shimada et al. 2006). Correlation coefficients for modeled monthly mean heat flux anomalies (after removing the mean annual cycle) between ACC1/CL1 and ACC4/CL4 are 0.24/0.38 (Tables 5.2 and 5.3). The fact that the highest correlations for all sections are at no lag implies that atmospheric forcing, including winds, radiative and sensible heat fluxes, is the dominant factor moving and affecting water mass properties and their variability across the Chukchi Sea. These model results also suggest that the oceanic heat flux from Bering Strait has a direct impact on sea ice melt and advancement of the ice edge northward over the Chukchi shelf, but not on the oceanic heat flux farther downstream over the outer shelf, slope, and into the Beaufort Sea.

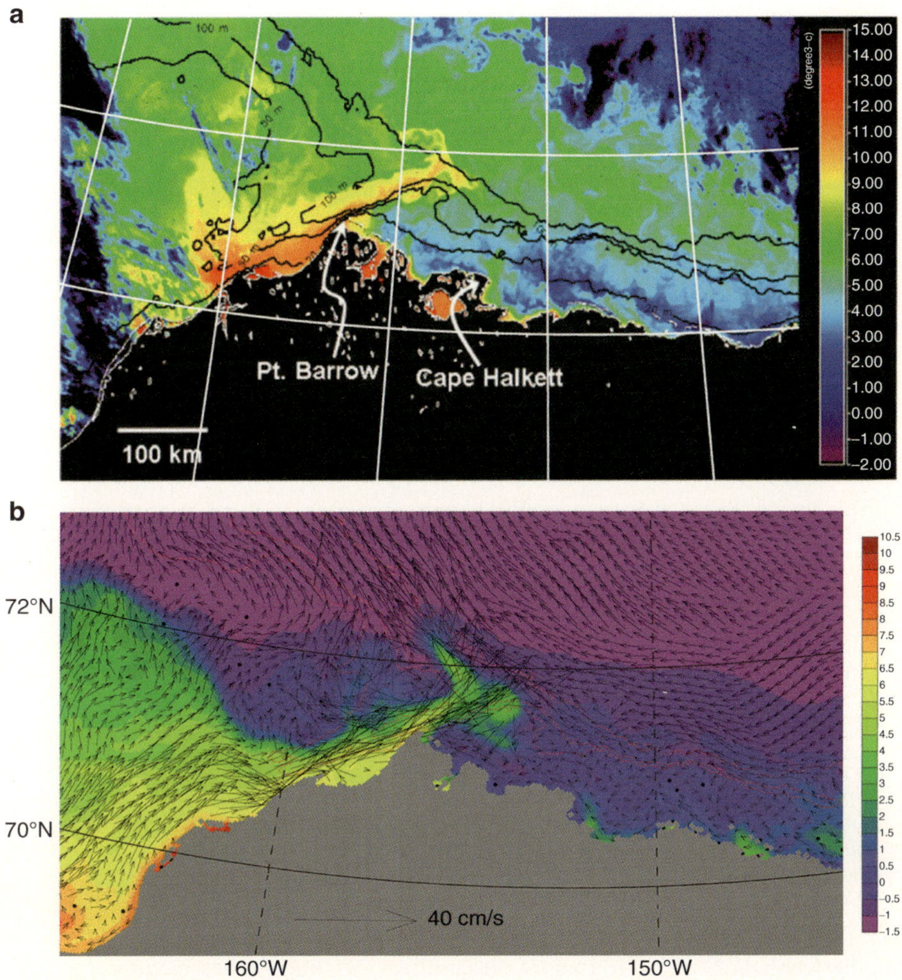

Fig. 5.8 (**a**) Sea surface temperatures (°C) from MODIS for 10 August 2007, 2335UT (Adapted from Okkonen et al. 2009); (**b**) temperature (*shading*; °C) and velocity (*vectors*; cm s^{-1}) at the surface layer (0–5 m) from the NAME model configured at 1/48° (or ~2.3 km) horizontal resolution on 15 August, 1988. The depth contours of 50, 100 and 1,000 m are shown in both panels. Every 10th vector in each direction shown in (**b**)

5.6 Beaufort Sea

Continuing downstream from the Chukchi Shelf, we find a model mean of 0.3 Sv flows northeastward through Barrow Canyon with monthly mean velocities up to 25 cm s^{-1} (Fig. 5.3, Table 5.1). Annually, the warmest waters exiting Barrow Canyon into the Beaufort Sea are typically observed in August–October (Spall et al. 2008; S. Okkonen unpublished data; K. Shimada personal communication). Similarly as

Table 5.2 Correlation matrix for selected ACC sections

	ACC1		ACC2		ACC3	
Section	Absolute	Anomaly	Absolute	Anomaly	Absolute	Anomaly
ACC2	0.69	0.79	*	*	*	*
	0.94	0.90				
ACC3	0.57	0.42	0.55	0.44	*	*
	0.76	0.58	0.75	0.65		
ACC4	0.73	0.61	0.56	0.43	0.83	0.73
	0.59	0.24	0.62	0.33	0.76	0.67

Correlations are provided for the monthly mean values (absolute) and the anomalies (after removing the annual cycle). The *upper number* in each cell is the correlation coefficient for volume transport and the *lower number* represents the correlation coefficient for heat flux. The lag time of 0 months produces the highest correlation in all instances

Table 5.3 Correlation matrix for selected sections

	CL1		CL2		CL3		CL3W	
Section	Absolute	Anomaly	Absolute	Anomaly	Absolute	Anomaly	Absolute	Anomaly
CL2	0.99	0.99	*	*	*	*	*	*
	0.94	0.93						
CL3	0.94	0.90	0.94	0.90	*	*	*	*
	0.75	0.75	0.85	0.83				
CL3W	0.73	0.64	0.73	0.64	0.76	0.68	*	*
	0.26	0.43	0.33	0.45	0.54	0.58		
CL4	0.93	0.89	0.93	0.89	0.95	0.93	0.76	0.68
	0.70	0.38	0.68	0.45	0.72	0.64	0.41	0.58

Correlations are provided for the monthly mean values (absolute) and the anomalies (after removing the annual cycle). The *upper number* in each cell is the correlation coefficient for volume transport and the *lower number* represents the correlation coefficient for heat flux. The lag time of 0 months produces the highest correlation in all instances

in the Chukchi Sea, hydrographic observations off Barrow describe the cold and fresh surface layer due to ice melt with relatively warmer water below (Fig. 5.9). Pacific summer water (PSW) is typically defined by a temperature maximum within salinity range $31 < S < 33$ (Steele et al. 2004; Sumata and Shimada 2007). As this water leaves the Chukchi shelf it converges with water of the boundary current (Rudels et al. 1994; Woodgate et al. 2001; Nikolopoulos et al. 2009) along the shelf-break and flows along the northern coast of Alaska to the east. A portion of PSW enters the basin and moves northward towards and along the eastern flank of the Northwind Ridge (Shimada et al. 2006; Sumata and Shimada 2007) with mean velocities up to 1.6 cm s^{-1} and the highest subsurface temperatures observed in January (Shimada et al. 2001). The importance of this outflow is clearly demonstrated in Fig. 5.8, where observed surface temperatures within the ACC in August 2007 are almost twice as high as the temperatures of ambient water throughout the northern Chukchi and southern Beaufort seas. While ocean surface temperatures increase due to the ice-albedo effect, in absence of organized currents such as the

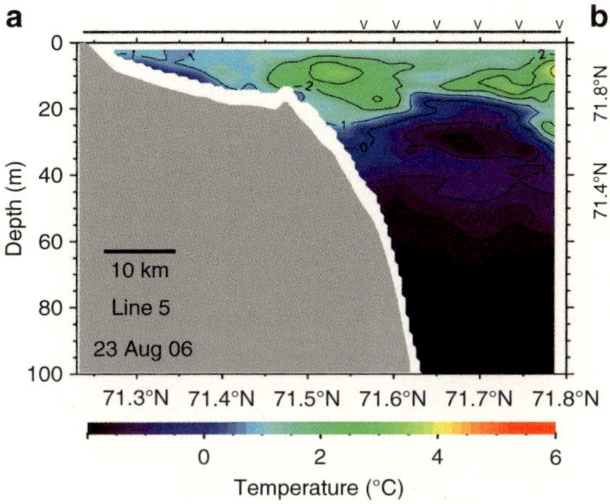

Fig. 5.9 (**a**) Section of observed temperature (°C) along the line from Eielson Lagoon toward the end of Barrow Canyon in the Beaufort Sea on 23 August 2006; (**b**) towed ADCP-measured surface (3–20 m averaged) current vectors along the same section in the Beaufort Sea as in (**a**) (Adapted from Okkonen et al. 2009). The 'v's identify locations of individual CTD casts. The *solid line* identifies the latitudinal extent of sampling by a CTD mounted on a towed undulating vehicle

ACC, their impact on sea ice remains relatively limited. This northward-spreading water, along with associated eddies, is important for introducing relatively warm water into the Canada Basin, which is further discussed below.

The mean ocean circulation in this region is controlled by synoptic, seasonal and interannual atmospheric variability. Okkonen et al. (2009) identified four generalized wind regimes and associated circulation states based on August–September 2005–2007 observations near Barrow, Alaska, combined with numerical simulations for large-scale and long-term context. They found that when winds are weak or from the southwest, PSW intrudes onto the shallow western Beaufort shelf with the majority of PSW flowing eastward along the Beaufort shelf break, while easterly winds favor a more diffuse flow of PSW to the north and northwest, possibly via mesoscale eddies. Moreover, large seasonal variability exists in both water properties exported from the Chukchi Sea and their distribution in the Beaufort Sea (e.g. Münchow and Carmack 1997; Weingartner et al. 1998; Pickart et al. 2005; Spall 2007). In particular, winter-transformed Pacific water is critical to the maintenance of the cold halocline in the Arctic Ocean (Aagaard et al. 1981; Cavalieri and Martin 1994; Chapman and Gawarkiewicz 1995), however its formation and circulation in the western Arctic are not discussed in detail here, as this water mass does not have a direct impact on sea ice.

Recent studies (Spall et al. 2008) estimate that at any time there could be 100–200 eddies in the western Arctic and they could distribute waters originating from the Chukchi shelf (Pickart et al. 2005). Zhang et al. (1999), using an 18-km ice-ocean model, demonstrated that mesoscale ocean eddies play a role in reducing ice concentration and thickness in the western Arctic, in particular along the Chukchi shelf break and the Northwind Ridge. Coincidently, Maslowski et al. (2000) using

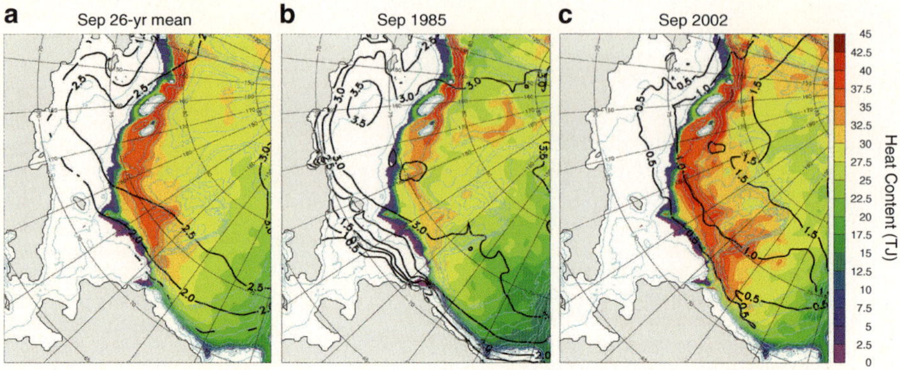

Fig. 5.10 Modeled September monthly mean heat content (TJ; *shading*) at depth 50–120 m and ice thickness contours (m; *black*) for (**a**) 26-year mean, (**b**) 1985 and (**c**) 2002. Bathymetry (m) contours are shown in *blue*

the same model simulated a cyclonic shift in the distribution of Pacific-origin water that accumulated in excess over the Northwind Ridge and along the Chukchi/Beaufort slope between the positive and negative phases of the Arctic Oscillation (AO).

Although the NAME model with horizontal resolution of 9-km does not resolve the smallest eddies (with radius of 10-km or less), larger eddies (with radius of 25 km or more) are commonly simulated in the Canada Basin (Fig. 5.10; Maslowski et al. 2008b). In addition, this limitation is currently being further investigated using the NAME model configured at horizontal resolution of 1/48°. Results from the ongoing model spinup show that at this resolution the model simulates the boundary current around 70 km wide, the ACC between 20 and 30 km wide as well as eddies with radius less than 10 km (Fig. 5.8b). The modeled eddies have variable sizes and origins, however there is a tendency for anticyclonic warm core eddy generation at the mouth of Barrow Canyon and along the Northwind Ridge. The mean core temperature of these eddies at depths of 60–120 m is as much as 1.65 °C above the freezing temperature. The frequency of their occurrence and properties including rotational speeds and temperature are commonly underestimated due to both the spatial resolution limitations and the realism of prescribed atmospheric forcing. However, we believe that the model realistically approximates physical processes that provide sources of heat to the Canada Basin. Based on Figs. 5.8, 5.9 and 5.10, we argue that ocean currents, including mesoscale eddies and topographically driven flow, are important means for distributing warm water and basal melting of the ice pack in the western Arctic.

It is important to note that heat can also be accumulated below the surface mixed layer locally in the Canada Basin, as the sea ice cover continues to retreat farther north and allows increased radiative fluxes into the upper ocean. This is due to the fact that the sunlight can penetrate below the ocean surface. The depth to which the upper ocean can receive heat from the sun depends on the rate at which sunlight is attenuated. It also depends on the wavelength with short wavelengths able to penetrate deeper in the water. Ultimately, in the absence of sea ice, sunlight can often penetrate below the depth of the surface meltwater layer, typically of order 10–15 m thick. Data from summer CTD surveys by icebreaker and from Ice Tethered Profilers (ITP) during 2005–2010 suggest that a near surface temperature maximum (NSTM) can develop in June–July, when solar radiation is at maximum, as a result of trapping heat below the surface mixed layer due to the onset of summer halocline with ice melt (Jackson et al. 2011, 2012). Given the generally insufficient model spatial resolution and limited observations of eddy generation and transport, estimates of eddy-driven shelf-basin heat transport and its redistribution in the western Arctic Ocean are not available at present. We argue that a quantification of this heat source and understanding of its impact on sea ice are needed to explain the regional differences in arctic sea ice decline and to constrain global climate model simulations and predicted scenarios of future arctic climate change.

5.7 Heat/Freshwater Content and Sea Ice

The outflow of warm summer water from the Chukchi and East Siberian shelves plus local absorption of solar radiation, ocean circulation and mixing contribute to the upper ocean (defined above 120 m) heat content in the western Arctic (west of the Lomonosov Ridge). The heat accumulation in the region has significantly increased since the late 1990s according to observations (Shimada et al. 2001, 2006; Jackson et al. 2012) and the NAME model. In Fig. 5.10 we show modeled heat content at depth 50–120 m and ice thickness in the western Arctic for September 26-year mean, September 1985 (for a typical past cold summer) and September 2002 (a warm summer) to determine potential coherence between these two parameters. The 26-year mean modeled ice thickness in September is about 2.0 m between 72 and 73°N in the Beaufort and Chukchi seas and it extends as far south as 71°N in the East Siberian Sea. The ice becomes thicker than 2.5 m roughly 200–300 km farther north. The 26-year mean oceanic heat content is largest (>30 Tera-Joules (TJ)) over the continental slope and it appears to be associated with the cyclonic boundary current flow carrying summer waters from the Chukchi, East Siberian and Laptev shelves and possibly some contribution of Atlantic water (Polyakov et al. 2010). This heat content maximum spreads eastward towards the Barrow Canyon and northward over the Chukchi Rise and Northwind Ridge and, as shown below, roughly coincides with regions in which the changes in mean ice thickness are greatest.

In September of 1985 (i.e. during a cold summer; Fig. 5.10b) ice was present farther south and thicker in the region (0.5–1.0 m thicker or on average 2.5–3.0 m and up to 3.5 m in the East Siberian Sea) compared to the modeled ice thickness climatology. The modeled heat content during that time was respectively lower, with the maximum over the Chukchi Rise and Northwind Ridge around 5 TJ lower compared to modeled September climatological heat content. Note that even during this cold summer, eddy-like mesoscale features with higher heat content can be seen over the Chukchi Cap and close to the Alpha and Northwind (not labeled in any figure) ridges.

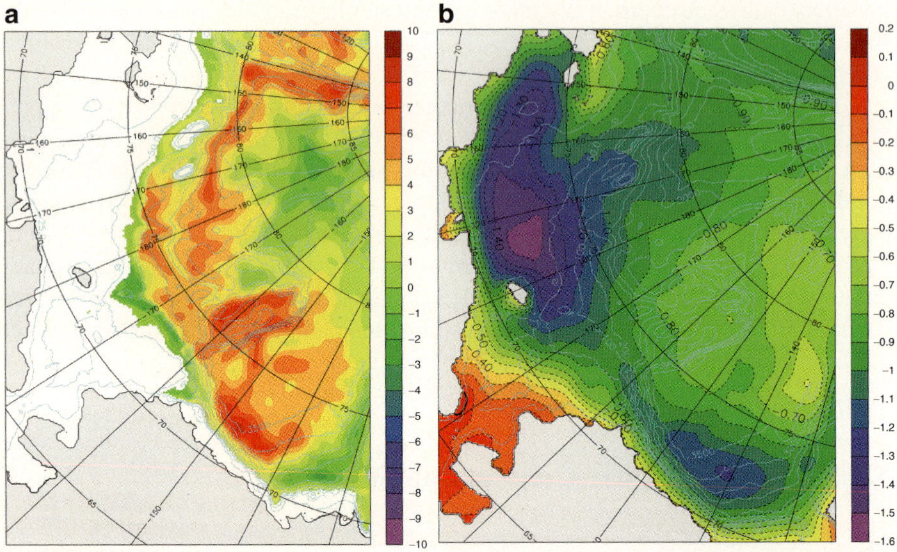

Fig. 5.11 Modeled changes in (**a**) heat content (TJ) at depth 33–120 m and (**b**) sea ice thickness (m) between the mean of 1979–1998 and the mean of 1999–2004

In contrast, 2002 was a relatively warm year with little and thinner sea ice found in the region (Fig. 5.10c). In particular, the thick sea ice over the East Siberian Sea, evident in 1985, has almost disappeared in 2002 leaving open water west of Wrangel Island. Ice thickness between 72 and 73°N is 0.5–1.0 m and only 1.5 m up to 80°N. Coincidently, upper ocean heat content is >30TJ all over that region with a maximum exceeding 40TJ over the continental slope. Recent observations (K. Shimada, personal communication; Jackson et al. 2012) suggest continuing increase of heat content in the western Arctic Ocean, at least through 2010.

A comparison of changes in sea ice thickness and heat content above 120 m between the 20-year mean (1979–1998) and the recent past 5-year mean (1999–2004) is shown in Fig. 5.11. The most striking features in the comparisons of changes in sea ice thickness and heat content above 120 m between the 20-year mean (1979–1998) and the recent past 5-year mean (1999–2004) (Fig. 5.11) are the maximum differences in both sea ice thickness and heat content in the southern Beaufort Sea (140°W–155°W and 71°N–74°N). The reduction in ice thickness of over 1.3 m is co-located with the heat anomaly >15 TJ. In addition, high eddy kinetic energy is simulated in this area both in the 9-km (Maslowski et al. 2008b) and in the 2.3-km (not shown) NAME models. Coincidently, this is also the area where multi-year sea ice has been disappearing and ice cover retreating in summer, despite the increased import of thick and old ice from the region to the north of the Canadian Arctic Archipelago (Stroeve et al. 2011). Finally as discussed below, the maximum freshwater content is co-located in the same region of the Beaufort Sea.

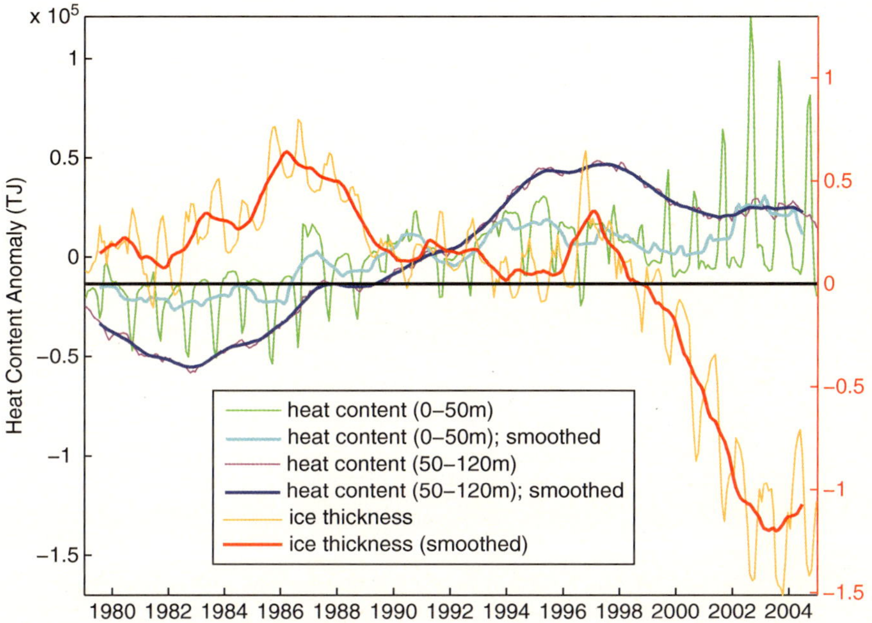

Fig. 5.12 Modeled monthly mean time series of heat content anomaly (TJ) at 0–50 m and 50–120 m and mean ice thickness anomaly (m) in the Western Arctic Ocean (region between 145°E and 135°W and between 70°N and 78°N). Anomaly values equal the annual cycle value subtracted from the monthly mean

The largest sea ice reduction occurring over the East Siberian shelf is associated with a relatively smaller increase in heat content, in part because of its shallow location where long-term heat accumulation is significantly restricted. However, even there increased heat content in the range of 5–10 TJ above the 1979–1998 mean is present on the outer shelf northward of the 100-m isobath and positive heat content anomaly exists southward to the 25-m isobath.

Based on the above results we argue that ocean currents, including mesoscale eddies and topographically driven flow, are important agents for distributing warm water and melting of the overlying ice pack. Next, we analyze time series of heat content in the upper 50 m and at 50–120 m and quantify its negative correlation with sea ice thickness in the Western Arctic Ocean (Fig. 5.12 and Table 5.4). Two depth ranges are defined for these analyses representing the upper ocean (0–50 m) including the surface mixed layer and the pycnocline below and the upper halocline layer (50–120 m). In addition, the surface layer (0–10 m) is considered separately (Table 5.4) as it is most affected by sea ice melt and solar radiation. All correlations between heat content and sea ice thickness shown in Table 5.4 are significant at the 95 % confidence level. For the purpose of this calculation, the Western Arctic Ocean is defined as a region between 70 and 78°N and 145°E–135°W. The region includes the Beaufort, northern Chukchi, and East Siberian seas and accounts for

Table 5.4 Lagged correlation coefficients of heat content anomalies in the top 120 m of ocean with the mean ice thickness anomalies in the western Arctic Ocean

	Monthly	Summer (JAS)
Depth range(m)	Mean (lag)	Mean (lag)
0–10	–0.65 (1)	–0.94 (0)
0–50	–0.56 (1)	–0.91 (0)
0–120	–0.69 (50)	–0.87 (37)
50–120	–0.74 (50)	–0.95 (42)

Anomaly values equal the annual cycle value subtracted from the monthly mean. Positive value (in months) of lag is for heat content anomalies leading ice thickness anomalies. Summer months are July, August, and September (JAS). All correlations shown in this table are significant at the 95 % confidence level

the majority of observed sea ice reduction in the western Arctic over the past decade. The heat content time series (Fig. 5.12) indicate that the heat contents were anomalously low prior to ~1990 but increased since the mid-1980s, which was followed by the loss of excess ice thickness (> 0.5 m) down to more average values through mid-1990s. Heat contents below 10 m (not shown) increased to above normal since the early 1990s and have remained elevated with strong summertime peaks in recent years. It is important to note that especially heat content below 50 m depth has remained anomalously high, including during winter, since the mid-1990s. In contrast, waters in the upper 50 m have been releasing all the excess heat every fall/winter. This implies that the upper halocline layer (roughly 50–120 m) in the western Arctic has been accumulating heat in winter and making it potentially available for melting ice from below during the following spring and summer.

Ice thickness shows an overall decreasing trend during the period 1979–2004; from over 0.5 m above the mean in the second half of the 1980s and in 1996, with a steady decline below the mean since 1998. Ice thickness anomalies over that region reached –1 m in winter and –1.5 m in summer during the recent years. It is possible that a portion of those ice thickness anomalies can be explained by mechanical redistribution (Holloway and Sou 2002). However, the presence of a heat source below the ocean's surface (Shimada et al. 2001) must play a role as well. Recently, Perovich et al. (2008) used autonomous buoys for monitoring thermodynamic mass balance of the ice and estimated that on average bottom ablation in the Beaufort Sea ranged from 0.6 to 1.2 m, and in one case reached ~2 m. They concluded that the positive ice-albedo feedback accelerated the melt of sea ice during the summer of 2007; however they left an open question as to what triggered the increase of open water area at the melt onset.

As already mentioned, there is a strong (at the 95 % confidence level) negative correlation (at positive lag, i.e. heat content anomalies lead ice thickness anomalies) between upper ocean heat content and ice thickness anomalies (Table 5.4). The correlation coefficients are smaller for all monthly means than for summer months only. However, correlations of all monthly means might be hard to interpret due to the fact that heat content anomalies at 0–50 m are reset to zero every winter while ice thickness anomalies are not reset to any values. For this reason we argue that correlation of summer (July–September) anomalies only allows more accurate depiction of relation between the two variables. In this case, much higher correlations are obtained, ranging from −0.87 in the upper 50 m to −0.95 in the 50–120 m layer, with no lag (0 months) for the upper and 37–42 month lag for the lower layer. While no lag in the upper 50 m can reflect simultaneous effects of summer ocean warming and sea ice melting, we note that the lower layer excludes the Chukchi and East Siberian shelves so a positive interannual lag is conceivable. Overall, this relation suggests that a large part of the variance in summertime ice thickness anomalies can potentially be explained by changes in the upper ocean heat content. However, more observations are needed from the evolving marginal ice zone in the Beaufort Sea to confirm this direct relationship between increased subsurface heat content and reduced sea ice thickness. On the modeling side, improvements are necessary toward eddy-resolving models able to accurately represent the magnitude of eddy-driven mass and heat transports and their impact on sea ice.

The overall correlations of heat content and ice thickness anomalies with AO and Pacific Decadal Oscillation (PDO) indices are quite low (below 0.2 for AO and below 0.3 for PDO). It is possible though that the increase of the upper ocean heat content and the associated thinning of sea ice in the western Arctic through the mid-1990s could be related to the combination of high PDO in the 1980s followed by the high AO through the mid-1990s. However, the warming and loss of sea ice during 1996–2004 is much harder to directly connect to the atmospheric forcing (Overland and Wang 2005) and, as such, points to the oceanic role in forcing the recent precipitous decline of sea ice in the western Arctic Ocean.

We finish this review with an analysis of recent freshwater accumulation in the Beaufort Gyre in response to warming climate and increased melt of sea ice. Freshwater content in the Arctic Ocean is usually calculated relative to a salinity of 34.8 for all water above the 34.8 isohaline (the salinity of incoming Atlantic Water through Fram Strait; Proshutinsky et al. 2009), and as such includes the waters of the halocline, formed by influx of Pacific origin water through Bering Strait and large amounts of freshwater from Eurasian and North American rivers. Anticyclonic wind forcing from the Beaufort high pressure system over the Canada Basin then leads to accumulation of the fresh Arctic surface waters via Ekman convergence and transport (Proshutinsky et al. 2009), promoting increased stratification (McLaughlin and Carmack 2010) and enhanced circulation of the anticyclonic Beaufort Gyre. While Ekman convergence is a local phenomenon, the overall accumulation of freshwater in the Beaufort Gyre is thought to be a result of Ekman transport at its periphery (large scale Ekman convergence), thus the freshwater for the Beaufort Gyre can be drawn from elsewhere in the Arctic. However, as discussed earlier, water derived from local ice melt combined with a stronger gyre circulation (e.g. Shimada et al. 2006) might be an increasingly important source of freshwater.

Figure 5.13 shows the summer freshwater content in the Beaufort Gyre from 2003 to 2010. Overall, strong anticyclonic forcing has led to a monotonic increase of the total freshwater content with by far the largest increase occurring in the winter between the 2007 and 2008 surveys when the Beaufort High was exceptionally strong.

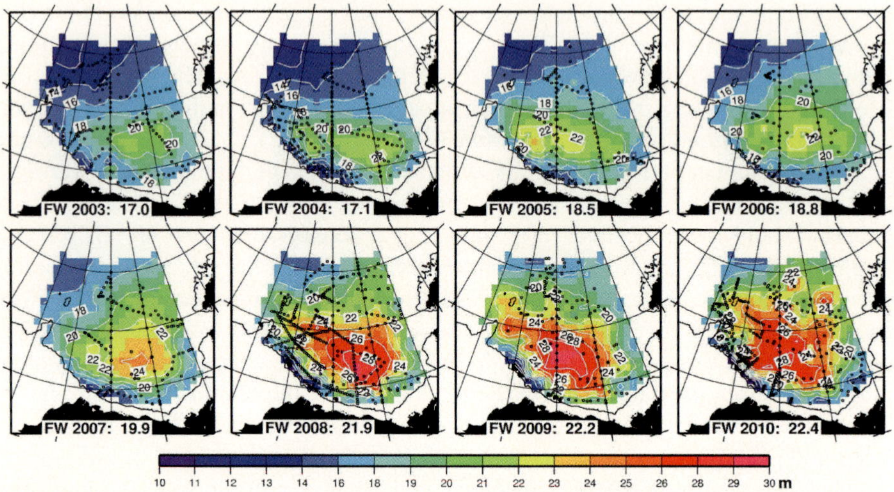

Fig. 5.13 Freshwater content in meters for the Beaufort Gyre region relative to a salinity of 34.8 psu from 2003 to 2010. The numbers at the *bottom* of each plot are the average over the area gridded and show the overall change from year to year. The largest change occurred during the winter between the 2007 and 2008 (Figure courtesy of Rick Krishfield, WHOI.)

The increase in freshwater also occurred during a time period of rapid summer ice retreat and ice melt; large amounts of thick, multi-year ice from the Canadian Arctic Archipelago melted in the southern Canada Basin and extensive summer sea ice retreat occurred in the western Beaufort and over the Northwind Ridge. Additional sources may also include reduced salt or increased freshwater fluxes through Bering Strait.

5.8 Summary

In this chapter, we presented an overview of the model-based mean regional ocean circulation in the Pacific-Arctic region, with emphasis on volume and heat transports into the western Arctic and their impact on sea ice. In particular, we discussed spatial and temporal variability of ocean currents and critical processes controlling northward flow of water. Model analyses and historical observations suggest that mesoscale eddies and meanders are dominant drivers controlling transport and variability of the Alaskan Stream, as well as mass and property fluxes across both the central and eastern Aleutian Island Passes and the western Aleutian Island Passes, between the Western Subarctic Gyre and the deep Bering Sea. Tidal mixing and rectification are also important, especially in shallow passes and on shelves; however their relative role is yet to be determined with eddy- and tide-resolving models. We made a compelling case for multi-year continuous and high spatio-temporal observations in the western passes to validate and constrain modeled estimates of the long-term mean and variability of mass transports. Mesoscale eddies appear to be also important in shelf-basin exchanges, ice-ocean interactions and the expansion of the marginal ice zone (MIZ) in the Bering, Chukchi and Beaufort seas.

In addition, we argue for a number of particular model improvements to advance the realism and skill of simulation in the Pacific-Arctic region, consistent with the pan-Arctic advancements discussed by Maslowski et al. (2012). In particular, we discuss the importance of resolving eddies, tides and air-ice-ocean interactions. Higher horizontal and vertical resolution is required to address some of the present model limitations in representing small scale processes, such as mesoscale eddies, coastal and boundary currents or surface and bottom mixed layers, and to more realistically represent shelf and slope bathymetry, submarine canyons and narrow passes. Spatial resolution sufficient to resolve eddies and mixed layer processes will also likely improve our ability to estimate the on-shelf transport of nutrient-rich, deep Bering Sea water along the Bering Sea slope, which is important for downstream ecosystems throughout the northern Bering, Chukchi, and Beaufort seas. Future ecosystem, biogeochemistry and physical models will also need to be eddy-resolving to simulate small scale processes and quantify their impacts on Arctic-wide budgets of heat, freshwater, nutrients, and carbon.

Based on our model results and limited observations, we found that heat content of the Western Arctic ocean in the top 120 m has been increasing since the late 1990s and appears to be contributing to the rapid sea ice decline there. Our model analysis implies that the advection of warm water from Bering Strait regulates the retreat of sea ice and contributes to the relatively high surface air-sea fluxes over the Chukchi shelf. Earlier ice melt in the Chukchi Sea allows increased absorption of solar radiation and warming of water over the shelf, before it is exported into the Beaufort Sea where it can further contribute to sea ice melt. Based on evidence from observations and from our model, we argue that not all the heat content in the western Arctic Ocean gets removed back to the atmosphere every year before freezing in fall and early winter. We hypothesize that instead the remaining heat above the halocline and below the mixed layer acts to reduce sea ice growth in winter and preconditions an earlier ice melt each year, which further increases heat content in the region, thus resulting in a positive feedback accelerating summer reduction of the sea ice cover. The presence and spreading of a subsurface heat source in the western Arctic Ocean and the large-scale sea ice drift of the Beaufort Gyre help explain why the sea ice has continued to decline most dramatically in that part of the Arctic Ocean, even when large-scale atmospheric conditions were not always most favorable to such a decline (i.e. when the Arctic Oscillation Index was neutral or extremely negative).

Finally, we show that freshwater content of the Western Arctic Ocean increased during the 2000s. The coincidence of the locations in freshwater accumulation and in maximum sea ice thickness anomaly in the Beaufort Sea suggest an increased local freshwater contribution due to ice melt and an intensified circulation within the Beaufort Gyre.

Acknowledgements Funding support for the development and integration of the NAME model as well as analyses of results that contributed to this publication was provided to W. Maslowski, J. Clement Kinney, A. F. Roberts and R. Osinski by multiple grants from the Regional and Global Climate Modeling of the Department of Energy, the Arctic System Science (ARCSS) Program of the National Science Foundation (NSF), and the Office of Naval Research. Computer resources were provided by the Department of Defense High Performance Computer Modernization Program (DOD/HPCMP). The contributions by S. Okkonen were supported by funding from the Cooperative Institute for Alaska Research, NSF-ARCSS, and NASA. Support for W. Williams was provided by Fisheries and Oceans Canada.

References

Aagaard K, Coachman LK, Carmack EC (1981) On the halocline of the Arctic Ocean. Deep-Sea Res 28:529–545

Ahlnäs K, Garrison GR (1984) Satellite and oceanographic observations of the warm coastal current in the Chukchi Sea. Arctic 37:244–254

Arsen'ev VS (1967) Currents and water masses of the Bering Sea, Nauka, Izdatel'stro (translated in 1968, Northwest Fish. Center, Seattle)

Bond NA, Overland JE, Turet P (1994) Spatial and temporal characteristics of the wind forcing of the Bering Sea. J Clim 7:1119–1130

Cavalieri DJ, Martin S (1994) The contributions of Alaskan, Siberian, and Canadian coastal polynyas to the cold halocline layer of the Arctic Ocean. J Geophys Res 99:18343–18362

Chapman DC, Gawarkiewicz G (1995) Offshore transport of dense shelf water in the presence of a submarine canyon. J Geophys Res 100:13373–13387

Chelton DB, deSzoeke RA, Schlax MG, Naggar KE, Siwertz N (1998) Geographical variability of the first baroclinic Rossby radius of deformation. J Phys Oceanogr 28:433–460

Clement JL, Maslowski W, Cooper L, Grebmeier J, Walczowski W (2005) Ocean circulation and exchanges through the northern Bering Sea—1979–2001 model results. Deep-Sea Res II 52:3509–3540

Clement Kinney J, Maslowski W (2008) Results of recent Pacific-Arctic ice-ocean modeling studies at the Naval Postgraduate School. Chin J Polar Sci 19(2):230–236

Clement Kinney J, Maslowski W (2012) On the oceanic communication between the Western Subarctic Gyre and the deep Bering Sea. Deep-Sea Res I 66: 11–25 http://dx.doi.org/10.1016/j.dsr.2012.04.001

Clement Kinney J, Maslowski W, Okkonen S (2009) On the processes controlling shelf-basin exchange and outer shelf dynamics in the Bering Sea. Deep-Sea Res II 56:1351–1362. doi:10.1016/j.dsr2.2008.10.023

Clement Kinney J, Maslowski W, Aksenov Y, de Cuevas B, Jakacki J, Nguyen A, Osinski R, Steele M, Woodgate RA, Zhang J (2014) Chapter 7: On the flow through Bering Strait: a synthesis of model results and observations. In: Grebmeier JM, Maslowski W (eds) The Pacific Arctic region: ecosystem status and trends in a rapidly changing environment. Springer, Dordrecht, pp 167–198

Coachman LK, Aagaard K, Tripp RB (1975) Bering Strait: the regional physical oceanography. University of Washington Press, Seattle

Cokelet ED, Stabeno PJ (1997) Mooring observations of the thermal structure, density stratification and currents in the southeast Bering Sea basin. J Geophys Res 102(C10):22947–22964

Cokelet ED, Schall ML, Dougherty DM (1996) ADCP-referenced geostrophic circulation in the Bering Sea Basin. J Phys Oceanogr 26:1113–1128

Crawford WR, Cherniawski JY, Forman MGG (2000) Multi-year meanders and eddies in the Alaskan Stream as observed by TOPEX/Poseidon altimeter. Geophys Res Lett 27(7):1025–1028

Danielson S, Weingartner T, Aagaard K, Zhang J, Woodgate R (2012a) Circulation on the central Bering Sea shelf, July 2008 to July 2010. J Geophys Res. doi:10.1029/2012JC008303

Danielson S, Hedstrom K, Aagaard K, Weingartner T, Curchitser E (2012b) Wind-induced reorganization of the Bering shelf circulation. Geophys Res Lett. doi:10.1029/2012GL051231

Favorite F (1974) Flow into the Bering Sea through Aleutian Island passes. In: Hood DW, Kelley EJ (eds) Oceanography of the Bering Sea. University of Alaska, Fairbanks

Foreman MGG, Cummins PF, Cherniawsky JY, Stabeno PJ (2006) Tidal energy in the Bering Sea. J Mar Res 64(6):797–818. doi:10.1357/002224006779698341

Goosse H, Campin JM, Fichefet T, Deleersnijder E (1997) Sensitivity of a global ice–ocean model to the Bering Strait throughflow. Clim Dyn 13(5):349–358

Grebmeier JM, Dunton KH (2000) Benthic processes in the northern Bering/Chukchi seas: status and global change. In: Huntington HP (ed) Impacts of changes in sea-ice and other environmental parameters in the Arctic. Report of the Marine Mammal Commission Workshop, 15–17 Feb 2000, Girdwood, Alaska

Grebmeier JM, McRoy CP, Feder HM (1988) Pelagic-benthic coupling on the shelf of the northern Bering and Chukchi seas. I. Food supply source and benthic biomass. Mar Ecol Prog Ser 48:57–67

Hermann AJ, Stabeno PJ, Haidvogel DB, Musgrave DL (2002) A regional tidal/subtidal circulation model of the southeastern Bering Sea: development, sensitivity analyses and hindcasting. Deep-Sea Res II 49:5945–5967

Hogg AM, Killworth PD, Blundell JR, Dewar WK (2005) Mechanisms of decadal variability of the wind-driven ocean circulation. J Phys Oceanogr 35:512–531

Holloway G, Sou T (2002) Has Arctic Sea ice rapidly thinned? J Clim 15:1691–1701

Holloway G et al (2007) Water properties and circulation in Arctic Ocean models. J Geophys Res 112:C04S03. doi:10.1029/2006JC003642

Hughes FW, Coachman LK, Aagaard K (1974) Circulation, transport and water exchange in the western Bering Sea. In: Hood DW, Kelley EJ (eds) Oceanography of the Bering Sea. University of Alaska, Fairbanks

Hurlburt H, Wallcraft A, Schmitz W, Hogan P, Metzger E (1996) Dynamics of the Kuroshio/Oyashio current system using eddy-resolving models of the North Pacific Ocean. J Geophys Res 101(C1):941–976. doi:10.1029/95JC01674

Jackson JM, Allen SE, McLaughlin FA, Woodgate RA, Carmack EC (2011) Changes to the near-surface waters in the Canada Basin, Arctic Ocean from 1993–2009: a basin in transition. J Geophys Res 116:C10008. doi:10.1029/2011JC007069

Jackson JM, Williams WJ, Carmack EC (2012) Winter sea-ice melt in the Canada Basin, Arctic Ocean. Geophys Res Lett 39:L03603. doi:10.1029/2011GL050219

Kinder TH, Schumacher JD, Hansen DV (1980) Observations of a baroclinic eddy: an example of mesoscale variability in the Bering Sea. J Phys Oceanogr 10:1228–1245

Kowalik Z (1999) Bering Sea tides. In: Loughlin TR, Ohtani K (eds) Dynamics of the Bering Sea. University of Alaska Sea Grant, Fairbanks

Ladd C, Thompson L (2002) Decadal variability of North Pacific central mode water. J Phys Oceanogr 32:2870–2881

Maslowski W, Walczowski W (2002) Circulation of the Baltic Sea and its connection to the Pan-Arctic region – a large scale and high-resolution modeling approach. Boreal Environ Res 7(4):319–325

Maslowski W, Newton B, Schlosser P, Semtner AJ, Martinson DG (2000) Modeling recent climate variability in the Arctic Ocean. Geophys Res Lett 27(22):3743–3746

Maslowski W, Marble D, Walczowski W, Schauer U, Clement JL, Semtner AJ (2004) On climatological mass, heat, and salt transports through the Barents Sea and Fram Strait from a pan-Arctic coupled ice-ocean model simulation. J Geophys Res 109:C03032. doi:10.1029/2001JC001039

Maslowski W, Roman R, Clement Kinney J (2008a) Effects of mesoscale eddies on the flow of the Alaskan Stream. J Geophys Res 113:C07036. doi:10.1029/2007JC004341

Maslowski W, Clement Kinney J, Marble DC, Jakacki J (2008b) Towards eddy-resolving models of the Arctic Ocean. In: Hecht MW, Hasumi H (eds) Ocean modeling in an eddying regime, vol 177, Geophysics monograph series. American Geophysical Union, Washington, DC

Maslowski W, Clement Kinney J, Higgins Roberts AF (2012) The future of Arctic Sea ice. Annu Rev Earth Planet Sci 40:625–654
McClean JL, Maslowski W, Maltrud M (2001) Towards a coupled environmental prediction system. In: Alexandrov VN, Dongorra JJ, Juliano BA, Renner RS, Tan CJK (eds) Computational science – ICCS 2001, Lecture notes in computer science 2073. Springer, Berlin
McLaughlin FA, Carmack EC (2010) Deepening of the nutricline and the chlorophyll maximum in the Canada Basin interior, 2003–2009. Geophys Res Lett 37:L24602. doi:10.1029/2010GL045459
Münchow A, Carmack EC (1997) Synoptic flow and density observations near an Arctic shelf break. J Phys Oceanogr 27:1402–1419
Musgrave DL, Weingartner TJ, Royer TC (1992) Circulation and hydrography in the northwestern Gulf of Alaska. Deep-Sea Res 39:1499–1519
Newton R, Schlosser P, Martinson DG, Maslowski W (2008) Freshwater distribution in the Arctic Ocean: simulation with a high-resolution model and model-data comparison. J Geophys Res 113:C05024. doi:10.1029/2007JC004111
Nikolopoulos A, Pickart RS, Fratantoni PS, Shimada K, Torres DJ, Jones EP (2009) The western Arctic boundary current at 152°W: structure, variability, and transport. Deep-Sea Res II 56:1164–1181
Ohtani K (1970) Relative transport in the Alaskan Stream in winter. J Oceanogr Soc Jpn 26:271–282
Okkonen SR (1992) The shedding of an anticyclonic eddy from the Alaskan Stream as observed by the GEOSAT altimeter. Geophys Res Lett 19(24):2397–2400
Okkonen SR (1993) Observations of topographic planetary waves in the Bering Slope Current using GEOSAT altimeter. J Geophys Res 98(12):22603–22613
Okkonen SR (1996) The influence of an Alaskan Stream eddy on flow through Amchitka Pass. J Geophys Res 101(C4):8839–8851
Okkonen SR, Jacobs GA, Metzger EJ, Hurlburt HE, Shriver JF (2001) Mesoscale variability in the boundary currents of the Alaska Gyre. Cont Shelf Res 21:1219–1236
Okkonen SR, Ashjian CJ, Campbell RG, Maslowski W, Clement-Kinney JL, Potter R (2009) Intrusion of warm Bering/Chukchi waters onto the shelf in the western Beaufort Sea. J Geophys Res 114:C00A11. doi:10.1029/2008JC004870
Onishi H (2001) Spatial and temporal variability in a vertical section across the Alaskan Stream and Subarctic Current. J Oceanogr 57:79–91
Onishi H, Ohtani K (1999) On seasonal and year to year variation in flow of the Alaskan Stream in the central North Pacific. J Oceanogr 55:597–608
Overland JE, Roach AT (1987) Northward flow in the Bering and Chukchi seas. J Geophys Res 92:7097–7105
Overland JE, Wang M (2005) The Arctic climate paradox: the recent decrease of the Arctic Oscillation. Geophys Res Lett 32(6):L06701. doi:10.1029/2004GL021752
Overland JE, Spillane MC, Hurlburt HE, Wallcraft AJ (1994) A numerical study of the circulation of the Bering Sea basin and exchange with the North Pacific Ocean. J Phys Oceanogr 24:736–758
Paluszkiewicz T, Niebauer HJ (1984) Satellite observations of circulation in the eastern Bering Sea. J Geophys Res 89:3663–3678
Panteleev GG, Stabeno P, Luchin VA, Nechaev DA, Ikeda M (2006) Summer transport estimates of the Kamchatka Current derived as a variational inverse of hydrophysical and surface drifter data. Geophys Res Lett 33:L09609. doi:10.1029/2005GL024974
Paquette R, Bourke R (1974) Observations on the coastal current of arctic Alaska. J Mar Res 32:195–207
Perovich DK, Richter-Menge JA, Jones KF, Light B (2008) Sunlight, water, and ice: extreme Arctic sea ice melt during the summer of 2007. Geophys Res Lett 35:L11501. doi:10.1029/2008GL034007
Pickart RS, Weingartner TJ, Zimmermann S, Torres DJ, Pratt LJ (2005) Flow of winter-transformed water into the Western Arctic. Deep-Sea Res II 52:3175–3198
Polyakov IV et al (2010) Arctic Ocean warming contributes to reduced polar ice cap. J Phys Oceanogr 40:2743–2756
Proshutinsky A, Ashik I, Häkkinen S, Hunke E, Krishfield R, Maltrud M, Maslowski W, Zhang J (2007) Sea level variability in the Arctic Ocean from AOMIP models. J Geophys Res 112:C04S08. doi:10.1029/2006JC003916

Proshutinsky A, Krishfield R, Timmermans M-L, Toole J, Carmack E, McLaughlin F, Williams WJ, Zimmermann S, Itoh M, Shimada K (2009) Beaufort Gyre freshwater reservoir: state and variability from observations. J Geophys Res 114: C00A10. http://dx.doi.org/10.1029/2008JC005104

Reed RK (1984) Flow of the Alaskan Stream and its variations. Deep-Sea Res 31(4):369–386

Reed RK (1990) A year-long observation of water exchange between the North Pacific and the Bering Sea. Limnol Oceanogr 35:1604–1609

Reed RK, Stabeno PJ (1989) Circulation and property distributions in the Central Bering Sea, Spring 1988, NOAA technical report, ERL 439-PMEL 39, NTIS PB90–155847, 13 pp

Reed RK, Stabeno PJ (1993) The recent return of the Alaskan Stream to Near Strait. J Mar Res 51:515–527

Reed RK, Stabeno PJ (1997) Long-term measurements of flow near the Aleutian Islands. J Mar Res 55:565–575

Reed RK, Stabeno PJ (1999) The Aleutian North Slope current. In: Loughlin TR, Ohtani K (eds) Dynamics of the Bering Sea: a summary of physical, chemical, and biological characteristics, and a synopsis of research on the Bering Sea, North Pacific Marine Science Organization (PICES), University of Alaska Sea Grant, AK-SG-99-03

Reed RK, Taylor NE (1965) Some measurements of the Alaska Stream with parachute drogues. Deep-Sea Res 12:777–784

Reed RK, Khen GV, Stabeno PJ, Verkhunov AV (1993) Water properties and flow over the deep Bering Sea basin, summer 1991. Deep-Sea Res 40:2325–2334

Roach AT, Aagaard K, Pease CH, Salo SA, Weingartner T, Pavlov V, Kulakov M (1995) Direct measurements of transport and water properties through the Bering Strait. J Geophys Res 100:18443–18457

Roden GI (1995) Aleutian Basin of the Bering Sea: thermohaline, oxygen, nutrient, and current structure in July 1993. J Geophys Res 100:13539–13554

Royer TC, Emery WJ (1987) Circulation in the Gulf of Alaska, in 1981. Deep-Sea Res 34:1361–1377

Rudels B, Jones EP, Anderson LG, Kattner G (1994) On the intermediate depth waters of the Arctic Ocean. In: Johannesen OM, Muench RD, Overland JE (eds) The polar oceans and their role in shaping the global environment: the Nansen centennial volume. American Geophysical Union, Washington, DC

Schauer U, Beszczynska-Möller A, Walczowski W, Fahrbach E, Piechura J, Hansen E et al (2008) Variation of measured heat flow through the Fram Strait between 1997 and 2006. In: Dickson RR (ed) Arctic-subarctic ocean fluxes: defining the role of the northern seas in climate. Springer Science+Business Media B.V., Dordrecht

Schumacher JD, Reed RK (1992) Characteristics of currents over the continental slope of the eastern Bering Sea. J Geophys Res 97:9423–9433

Shimada K, Carmack E, Hatakeyama K, Takizawa T (2001) Varieties of shallow temperature maximum waters in the Western Canadian Basin of the Arctic Ocean. Geophys Res Lett 28:3441–3444

Shimada K, Kamoshida T, Itoh M, Nishino S, Carmack E, McLaughlin F, Zimmermann S, Proshutinsky A (2006) Pacific Ocean inflow: influence on catastrophic reduction of sea ice cover in the Arctic Ocean. Geophys Res Lett 33:L08605. doi:10.1029/2005GL025624

Solomon H, Ahlnäs K (1978) Eddies in the Kamchatka Current. Deep-Sea Res 25:403–410

Spall MA (2007) Circulation and water mass transformation in a model of the Chukchi Sea. J Geophys Res 112:C05025. doi:10.1029/2005JC002264

Spall MA, Pickart RS, Fratantoni P, Plueddemann A (2008) Western Arctic shelfbreak eddies: formation and transport. J Phys Oceanogr 38:1644–1668

Springer AM, McRoy CP (1993) The paradox of pelagic food webs in the northern Bering Sea. III. Patterns of primary production. Cont Shelf Res 13:575–599

Stabeno PJ, Hermann AJ (1996) An eddy resolving model of circulation on the western Gulf of Alaska shelf. II. Comparison of results to oceanic observations. J Geophys Res 101:1151–1161

Stabeno PJ, Reed RK (1992) A major circulation anomaly in the western Bering Sea. Geophys Res Lett 19:1671–1674

Stabeno PJ, Reed RK (1994) Circulation in the Bering Sea Basin observed by satellite-tracked drifters: 1986–1993. J Phys Oceanogr 24:848–854

Stabeno PJ, Schumacher JD, Ohtani K (1999) The physical oceanography of the Bering Sea. In: Loughlin TR, Ohtani K (eds) Dynamics of the Bering Sea: a summary of physical, chemical, and biological characteristics, and a synopsis of research on the Bering Sea, North Pacific Marine Science Organization (PICES), Univ. of Alaska Sea Grant, AK-SG-99-03

Stabeno PJ, Kachel DG, Kachel NB, Sullivan ME (2005) Observations from moorings in the Aleutian Passes: temperature, salinity and transport. Fish Oceanogr 14(1):39–54. doi:10.1111/j.1365-2419.2005.00362.x

Steele M, Morison J, Ermold W, Rigor I, Ortmeyer M, Shimada K (2004) Circulation of summer Pacific halocline water in the Arctic Ocean. J Geophys Res 109:C02027. doi:10.1029/2003JC002009

Stroeve JC, Maslanik J, Serreze MC, Rigor I, Meier W, Fowler C (2011) Sea ice response to an extreme negative phase of the Arctic Oscillation during winter 2009/2010. Geophys Res Lett 38:02502. doi:10.1029/2010GL045662

Sumata H, Shimada K (2007) Northward transport of pacific summer water along the Northwind Ridge in the Western Arctic Ocean. J Oceanogr 63:363–378

Thompson RE (1972) On the Alaskan Stream. J Phys Oceanogr 2:363–371

Verkhunov AV, Tkachenko YY (1992) Recent observations of variability in the western Bering Sea current system. J Geophys Res 97:14369–14376

Walsh JJ, Dieterle DA, Maslowski W, Whitledge TE (2004) Decadal shifts in biophysical forcing of marine food webs in the Arctic: numerical consequences. J Geophys Res 109:C05031. doi:10.1029/2003JC001945

Warren BA, Owens WB (1988) Deep currents in the central Subarctic Pacific Ocean. J Phys Oceanogr 18:529–551

Watanabe E (2011) Beaufort shelf break eddies and shelf-basin exchange of Pacific summer water in the western Arctic Ocean detected by satellite and modeling analyses. J Geophys Res 116:C08034. doi:10.1029/2010JC006259

Weingartner TJ, Cavalieri DJ, Aagaard K, Sasaki Y (1998) Circulation, dense water formation and outflow on the northeast Chukchi Sea shelf. J Geophys Res 103:7647–7662

Weingartner TJ, Danielson S, Sasaki Y, Pavlov V, Kulakov M (1999) The Siberian Coastal Current: a wind- and buoyancy-forced Arctic coastal current. J Geophys Res 104:29697–29713. doi:10.1029/1999JC900161

Weingartner T, Aagaard K, Woodgate R, Danielson S, Sasaki Y, Cavalieri D (2005) Circulation on the North Central Chukchi Sea Shelf. Deep-Sea Res II 52(24–26):3150–3174

Williams W, Shroyer E, Clement Kinney J, Itoh M, Maslowski W (2014) Chapter 6: Shelf-break exchange in the Bering, Chukchi and Beaufort seas. In: Grebmeier JM, Maslowski W (eds) The Pacific Arctic region: ecosystem status and trends in a rapidly changing environment. Springer, Dordrecht, pp 133–166

Winsor P, Chapman DC (2004) Pathways of Pacific water across the Chukchi Sea: a numerical model study. J Geophys Res 109:C03002. doi:10.1029/2003JC001962

Woodgate RA, Aagaard K, Muench RD, Gunn J, Björk G, Rudels B, Roach AT, Schauer U (2001) The Arctic Ocean Boundary Current along the Eurasian slope and the adjacent Lomonosov Ridge: water mass properties, transports and transformations from moored instruments. Deep-Sea Res I 48(8):1757–1792

Woodgate RA, Aagaard K, Weingartner TJ (2005) A year in the physical oceanography of the Chukchi Sea: moored measurements from autumn 1990–1991. Deep-Sea Res II 52:3116–3149. doi:10.1016/j.dsr2.2005.10.016

Xie S-P, Kunitani T, Kubokawa A, Nonaka M, Hosoda S (2000) Interdecadal thermocline variability in the North Pacific for 1958–97: a GCM simulation. J Phys Oceanogr 30:2798–2813

Zhang Y, Maslowski W, Semtner A (1999) Impact of mesoscale ocean currents on sea ice in high-resolution Arctic ice and ocean simulations. J Geophys Res 104(C8):18409–18429. doi:10.1029/1999JC900158

Chapter 6
Shelf-Break Exchange in the Bering, Chukchi and Beaufort Seas

William J. Williams, Emily Shroyer, Jaclyn Clement Kinney, Motoyo Itoh, and Wieslaw Maslowski

Abstract The Bering and Chukchi/Beaufort shelf-breaks form the beginning and end of the dramatic sea-level and wind-forced flow of Pacific Ocean water across the Bering and Chukchi continental shelves between the Pacific and Arctic Oceans. Recent model results suggest that the on-shelf flow in the Bering is distributed along the shelf-break, wind-dependant, focused by Zhemchug and Bering canyons and modified by shelf-break eddies. Similarly, the off-shelf flow in the Chukchi/Beaufort is mediated by canyons, shelf-break jets, eddies, and wind forcing. In the Chukchi, flow is channeled through Barrow and Herald canyons to the shelf-break, where across-slope flow in Ekman boundary layers and instabilities result in the exchange of water and properties across the slope. In addition, dense shelf-water, created from brine rejection during ice formation in coastal polynyas, has been observed to flow downslope through Barrow Canyon. Along the Beaufort shelf, the shelf-break current is unstable, shedding eddies that populate the deep Beaufort basin. Upwelling favorable winds in summer have been observed to modify the structure of the shelf-break current and drive exchange across the Chukchi and Beaufort slopes. Most of

W.J. Williams (✉)
Fisheries and Oceans Canada, Institute of Ocean Sciences, Sidney, BC, Canada V8L 4B2
e-mail: Bill.Williams@dfo-mpo.gc.ca

E. Shroyer
College of Oceanic and Atmospheric Sciences, Oregon State University,
104 COAS Administration Building, Corvallis, OR 97331-5503, USA
e-mail: eshroyer@coas.oregonstate.edu

J. Clement Kinney • W. Maslowski
Department of Oceanography, Graduate School of Engineering and Applied Sciences,
Naval Postgraduate School, Monterey, CA 93943, USA
e-mail: jlclemen@nps.edu; maslowsk@nps.edu

M. Itoh
Japan Agency for Marine-Earth Science and Technology, Yokosuka-city,
Kanagawa 237-0061, Japan
e-mail: motoyo@jamstec.go.jp

J.M. Grebmeier and W. Maslowski (eds.), *The Pacific Arctic Region: Ecosystem Status and Trends in a Rapidly Changing Environment*, DOI 10.1007/978-94-017-8863-2_6,

our understanding of shelf-break flow in Bering and Chukchi Seas is based on numerical model results and broad-scale observations. There is thus a need for more detailed shelf-break observational programs.

Keywords Shelf-break • Canyon • Bering Sea • Chukchi Sea • Beaufort Sea • Polynya

6.1 Introduction

The shelf-break at the edge of a continental shelf is a transition region between the shallow shelf and the much deeper slope and basin. In comparison to basins, shelves are directly influenced by river inflow and their shallowness strongly amplifies the response to tidal and wind-forcing. The rapid increase in bottom slope at the shelf-break, from the relatively flat continental shelf to the much steeper slope, forms a barrier to rotationally dominated (i.e. geostrophically balanced) flows, which are strongly constrained to flow along, rather than across isobaths and changes the interaction of stratified flow with the bathymetry. Since the shelf-break delineates all these basic differences, fronts, eddies and jets contribute to significant variability at the shelf-break. In addition, the shelf-break is incised by a wide range of undersea canyons formed during glacial periods of low sea level by river runoff and glaciation. The interaction of shelf-break flows and eddies with these canyons results enhanced shelf-break exchange of mass, heat, salt, and other tracers.

Shelf-breaks can be characterized by their depth, width and topographic variations such as undersea headlands and canyons, where the width is the transition from the bottom slope of the shelf to that of the slope. The shelf-break of the Bering, bounded by the Aleutian Islands in the southeast and Siberia in the northwest, is ~150–200 m deep and contains a number of large cross shelf canyons of which Zhemchug Canyon dominates the central shelf and Pribilof and Bering canyons influences the southeastern shelf (Fig. 6.1a). The Chukchi shelf-break, between Herald Canyon in the west and Barrow Canyon in the east, is ~75 m deep and varies from ~20 km wide in the east to being very broad in the west (Fig. 6.1b). The Alaskan Beaufort Shelf, east of the Chukchi Sea, sits between Barrow Canyon and Mackenzie Trough and its shelf-break is only 60 m deep and very sharp – the transition from shelf to slope is only ~5 km wide (Fig. 6.1c).

The Bering and Chukchi/Beaufort shelf-breaks are also unique in that they are the beginning and end of the flow of Pacific-origin water from the deep Bering Sea through Bering Strait to the Canadian Basin of the Arctic Ocean. This sea-level and wind-forced flow across the broad continental shelves of the Bering and Chukchi forms an advective highway on which there is high biological production fed from the nutrients within the Pacific-origin water. To balance the northward flow through Bering Strait, there is on-shelf flow across the Bering shelf-break in addition to the along-shelf flow through the Unimak Pass in the Aleutian Island Chain. The Bering Strait through-flow crosses the Chukchi Sea, becoming off-shelf flow across the Chukchi-Beaufort shelf-break, as the Pacific water spills off the shelf and into the surface layers and halocline of the Canadian Basin. The on-shelf flow in the Bering

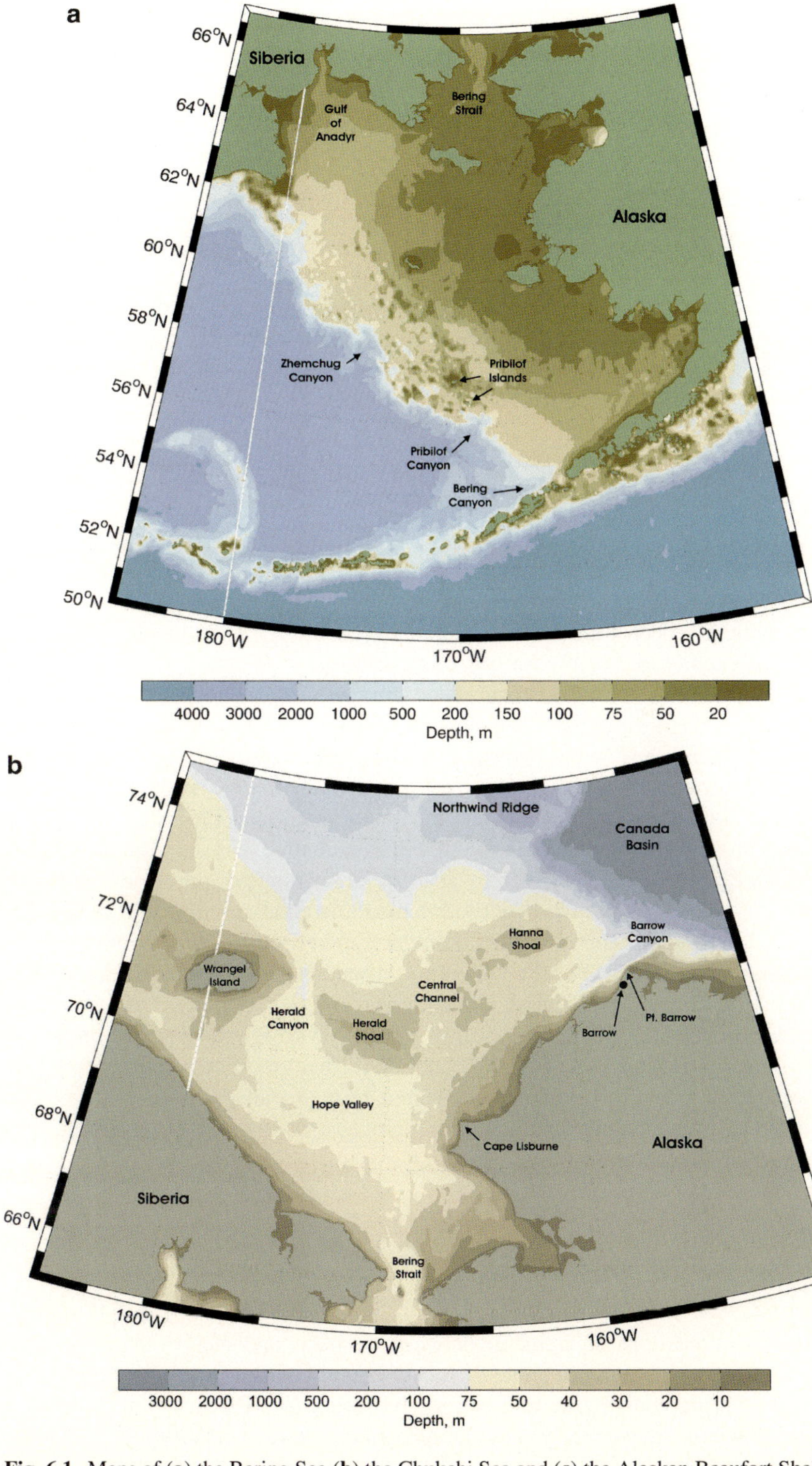

Fig. 6.1 Maps of (**a**) the Bering Sea (**b**) the Chukchi Sea and (**c**) the Alaskan Beaufort Shelf. The bathymetry changes color to highlight the shelf-break and canyons. In (**c**) the *black dots* at the mouth of Barrow Canyon show the location of the mooring data used in Figs. 6.12 and 6.13

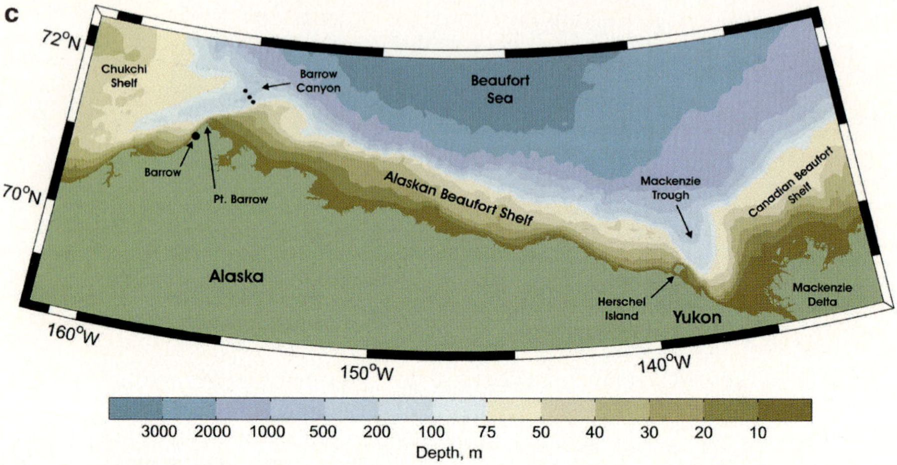

Fig. 6.1 (continued)

Sea has typically been thought of as occurring in the northwest near the Gulf of Anadyr, but see Fig. 6.2 below for modeling results.

This chapter reviews our historical and current understanding of on- and off-shelf flows at the Bering and Chukchi/Beaufort shelf-breaks, including contributions from eddies and wind-driven upwelling and downwelling, and, for the Chukchi/Beaufort the production of dense shelf water in coastal polynyas and its subsequent flow across the shelf-break. Particular attention is paid to the role that the many canyons play in enhancing shelf-break exchange in this region. In general, shelf-break exchange can be driven by an even larger range of processes, including the above as well as tidal dynamics (both barotropic and baroclinic), frontal meanders, filaments, and intrusions, bottom boundary layer friction, and coastally trapped waves (Huthnance 1995). For the generation of eddies by instabilities, the interaction of eddies with the shelf-break and canyons, and the interaction of along-shelf flow with canyons the constraint that geostrophic flow follows isobaths is broken by the non-linear terms in the momentum equation further modified by stratification. In addition spatial variability in wind stress and Ekman transport (for example that introduced by the coastline) leads to convergence/divergence on the shelf and consequent onshelf/offshelf flow.

6.2 The Bering Shelf-Break

The earliest suggestions of exchange between the basin and shelf in the Bering Sea can be inferred from charts of generalized surface circulation schemes proposed by a number of researchers in the mid-twentieth century (see Hughes et al. 1974 for a review). Because these early circulation charts were typically derived from

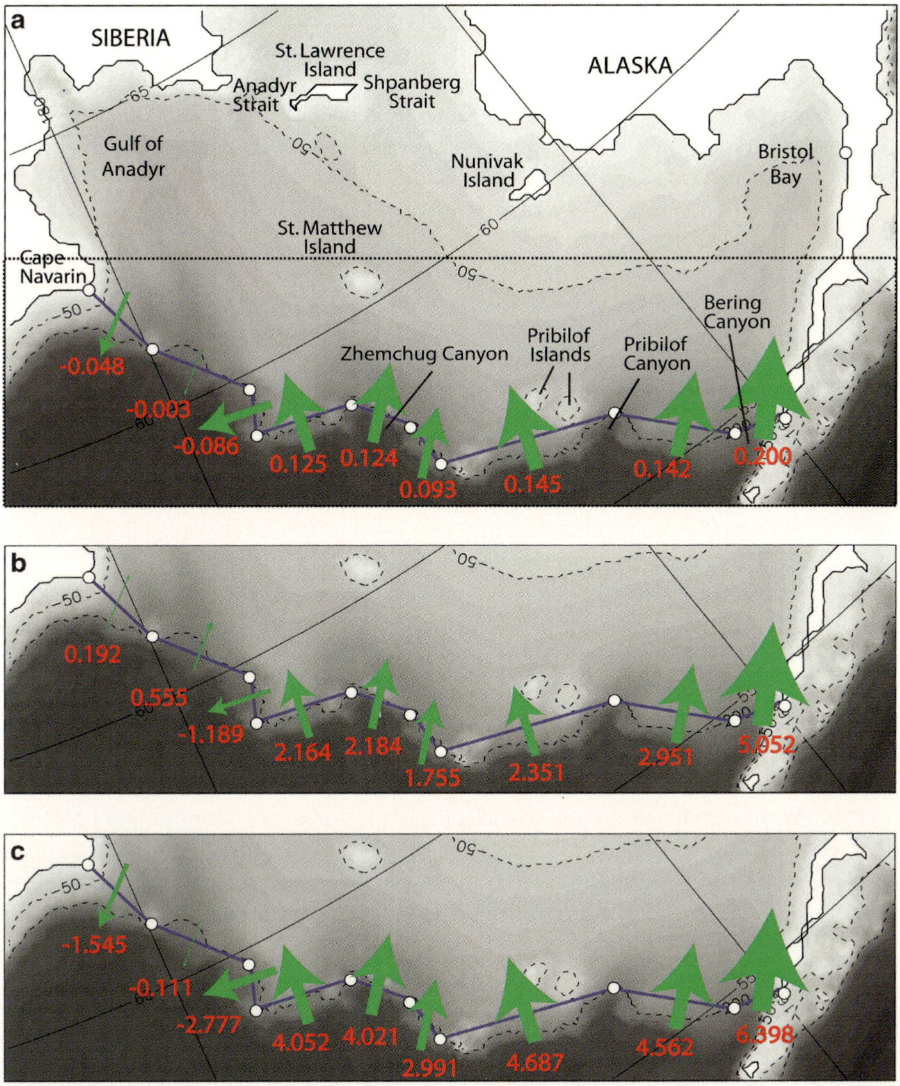

Fig. 6.2 Twenty six year (1979–2004) mean (**a**) volume transport (Sv), (**b**) heat flux (TW; relative to −1.9 °C) and (**c**) salt flux (10^6 kg/s) across various sections along the Bering shelf-break using output from the NAME numerical model. *Arrows* indicate net direction (positive is North or West) and are scaled relative to the largest value in each Figure. The *shading* indicates depth (m). The 50 m and 200 m isobaths are shown as *dotted lines* (Figure adapted from Fig. 3 of Clement Kinney et al. (2009) with permission from Elsevier)

compilations of disparate, temporally- and spatially-limited data sets, the depicted exchange pathways between basin and shelf were rudimentary at best.

A significant improvement to our understanding of basin-shelf exchange in the Bering Sea occurred with the description of the Bering Slope Current (BSC) system (Kinder et al. 1975). A transport budget derived from continuity constraints on geostrophic transport calculations for the BSC implied the existence of large exchanges between basin and shelf, exceeding 1 Sv, that were proposed to preferentially occur in association with the large submarine canyons incising the continental slope. Additionally, the authors proposed that the BSC could be characterized as a system of mesoscale eddies that exhibits seasonal modulation. Both these aspects of the BSC, the eddies and the canyons, have been found to be important to shelf-basin exchange.

Kinder and Coachman (1978) estimated cross-shelf exchange through consideration of the persistent haline front that overlies the 1,000-km long continental slope in the central Bering Sea. Maintenance of this front was found to require a $\sim 5 \times 10^4$ kg s^{-1} shoreward, tidally-driven diffusion of salt. Using the best estimates of Bering Strait transport, mean salinity, and freshwater runoff available at the time, they also computed a salt and freshwater balance for the entire Bering Shelf from which an on-shelf advective salt flux of 4.9×10^7 kg s^{-1} was estimated. This is approximately 1,000 times the estimated flux required to maintain the shelf-break haline front within the central Bering. They inferred that most of this advective salt flux occurs near the Siberian coast, where the haline front is poorly defined, and then passes directly through the Gulf of Anadyr to Bering Strait. This supported a previous paper by Coachman et al. (1975), which was the first to show hydrography that clearly indicated flow around the Gulf of Anadyr, implying shelf-basin exchange near the Siberian coast.

Although not explicitly identified as such, a broad-scale indicator of on-shelf (heat) transport in the Bering Sea was inferred from Pease (1980). She first identified that southwestward ice advection driven by northerly winds coupled with ice melt at the thermodynamic limit of the southern ice-edge are the principal processes that influence sea ice production and extent in the Bering Sea. She further suggested that inter-annual changes in sea ice extent were dependent upon the interplay of these processes and thus dependent on the heat flux to the Bering shelf from the deep basin.

Although Kinder et al. (1980) reported on a satellite-tracked drifter that circumscribed an eddy and then moved onto the shelf, Paluszkiewicz and Niebauer (1984), using satellite imagery of sea surface temperature, and Karl and Carlson (1987), using distributions of suspended sediments, were the first to implicate both (1) mesoscale eddies as agents promoting exchange between basin and shelf in the Bering Sea and (2) submarine canyons as locations of enhanced eddy-induced basin-shelf exchange. Paluszkiewicz and Niebauer (1984) provide rough estimates for the temporal (2–6 months) and spatial (order 100 km) scales of the BSC eddy field using satellite imagery and suggest the variability in the eddy field was associated with changes in flow through the passes between the eastern Aleutian Islands. Okkonen (2001) later used 4 years of TOPEX satellite altimeter data to refine these estimates and show that the most energetic activity in the BSC eddy field typically occurs in the spring and summer months and is characterized by along-slope wavelengths of ~200 km and wave periods of ~4 months.

The first direct evidence for shelf-slope exchange in the Bering Sea was acquired by moored instruments deployed at slope and mid-slope locations in Pribilof Canyon, Zhemchug Canyon, and midway between these canyon sites (Schumacher and Reed 1992). Whereas computed momentum and heat fluxes indicated both on-shelf and off-shelf flows, salt fluxes were preferentially on-shelf. Schumacher and Stabeno (1994) re-examined these data in conjunction with satellite-tracked drifter trajectories from the southeast Bering Sea and concluded that the current reversals and coincident shelf-slope exchanges of water properties were due to the passage of small (<25 km) eddies at periodicities of 45–60 days. Subsequent satellite-tracked drifter studies also indicated an association between eddies and on-shelf flow (Stabeno and van Meurs 1999; Stabeno et al. 1999).

Mizobata and Saitoh (2004) used 3 years (1998–2000) of satellite measurements of sea surface height anomalies, ocean color and sea surface temperature to show that surface chlorophyll concentrations and primary production along the outer shelf and above the continental slope in the central Bering Sea were enhanced when the variability of the BSC eddy field was elevated. They attributed elevated eddy field variability to increased transport in the BSC fed by increased transports of the Alaskan Stream and Aleutian North Slope Current. Okkonen et al. (2004) used satellite images of ocean color to illustrate eddy-induced on-shelf intrusions of low chlorophyll basin water and off-shelf entrainment of high chlorophyll shelf water.

These observational findings were complemented by the modeling efforts of Mizobata et al. (2006), who used a 5-km resolution numerical model of the southeastern Bering Sea to study eddy-induced exchange between basin and shelf. Results showed that subsurface, on-shelf nutrient flux was greater when eddies were adjacent to the shelf-break than when absent from the shelf-break region. Model results also showed that surface waters from the outer shelf were transported ~100 km into the basin by shelf-break eddies.

Models have also been used to constrain the total mean volume and property transport budgets across the outer Bering Sea shelf (defined between the 50-m and 200-m isobaths), Clement et al. (2009) analyzed output from the Naval Postgraduate School Arctic Modeling Effort ice-ocean model (NAME; see e.g. Maslowski et al. 2004, 2014, this volume). This model has ~9 km horizontal resolution and 45 levels in the vertical. The horizontal resolution permits eddies as small as 36 km. The model was first spun-up for 21 years using ECMWF winds and then the final run used realistic daily-averaged ECMWF forcing beginning in 1979 and continuing through 2004. A series of cross-shelf sections along the shelf break of the long-term (1979–2004) mean volume, heat, and salt transports from NAME are shown in Fig. 6.2. Figure 6.2a shows the 26-year mean value of volume transport, in Sv ($1\ Sv = 10^6\ m^3/s$), across each section with an arrow indicating the mean direction of flow. Arrows are scaled as a percentage of the largest cross-sectional value shown in the figure (e.g. a volume transport of 0.2 Sv would have an arrow twice as wide as a section with a volume transport of 0.1 Sv). Along the central and eastern part of the slope, we can see that the net volume transport is positive (on-shelf); however the western part of the slope shows relatively small negative (off-shelf) values.

Table 6.1 Twenty-six-year (1979–2004) mean on-shelf, off-shelf, and net volume, heat and salt transport across the sections shown in Fig. 6.2, which approximate the 200 m isobath

Parameter	On-shelf	Off-shelf	Net
Volume transport (Sv)	0.829	−0.137	0.692
Heat transport (TW)	17.204	−1.189	16.015
Salt transport (10^6 kg/s)	26.711	−4.433	22.278

The Bering Slope Current tends to separate from the shelf-break west of Zhemchug Canyon, thereby allowing off-shelf transports for sections further west (Clement Kinney et al. 2009). The mean total on-shelf, off-shelf and net volume transports are 0.829 Sv (S.D. = 0.728 Sv), −0.137 Sv (S.D. = 0.384 Sv), and 0.692 Sv (S.D. = 0.421 Sv), respectively (Table 6.1). The net on-shelf volume transport is the mean flow though Bering Strait in the model, and comparable to the observed Bering Strait through-flow of ~0.8 Sv (Woodgate et al. 2006). However, Aagaard et al. (2006), in a volume and salinity balance for the Bering Shelf, found on-shelf volume transport to be roughly equally split between the flow of the Alaskan Coastal Current through Unimak Pass (the first opening in the Aleutian Island chain) and on-shelf flow across the shelf-break, highlighting the need for observations of shelf-break exchange in the Bering and of the flow through Unimak Pass and subsequent gaps in the Aleutian Island chain. Note that these modeling results are in contrast to the historical notions of the Bering Strait through-flow crossing the Bering shelf-break in the Gulf of Anadyr (e.g. Coachman et al. 1975) which warrants further investigation, such as Danielson et al. (2012).

The long-term mean heat transports, calculated using a reference temperature of −1.9 °C, are shown in Fig. 6.2b. The largest on-shelf oceanic heat flux is through Zhemchug Canyon (a total of 6.1 TW through the three sections that encompass Zhemchug Canyon), with the second largest on-shelf heat flux through Bering Canyon (5.1 TW). The mean total on-shelf, off-shelf and net heat transports are 82.2432 TW (S.D. = 30.718 TW), 56.542 TW (S.D. = 19.894 TW) and 25.701 TW (S.D. = 17.447 TW), respectively. The mean salt transports through the sections along the 200-m isobath (Fig. 6.2c) follow the mean volume transports, due to the strong dependence of salt transport on volume transport. The mean total on-shelf, off-shelf, and net volume, heat and salt transports are given in Table 6.1.

Danielson et al. (2012) break down the long-term mean view by considering the response of the Bering Shelf to no-wind or along-shelf wind that is either northwesterly (upwelling favorable) or southeasterly (downwelling favorable). They illustrate the effect of wind-direction using a one layer idealized barotropic model and linearized bottom friction (Fig. 2 in Danielson et al. 2012). Under no-wind conditions, the model shows the Bering Strait source waters to cross the shelf-break in the Gulf of Anadyr, similar to a study by Kinder et al. (1986) that suggested the notion of topographic beta effect focusing the flow through the Gulf of Anadyr. However, during southeasterly winds, the through-flow is spread out over the entire shelf-break, reminiscent of the long-term mean presented by Clement et al. (2009). Northwesterly wind generated off-shelf flow over most of the shelf-break that was compensated by additional on-shelf flow, with the Bering Strait through-flow, at the northwestern end of the shelf. These results from their idealized barotropic model were also confirmed with realistic three-dimensional modeling.

Although this idealized model is instructive, the Bering shelf-break is much more complex with variation on a variety of time and space scales, e.g. those related to eddies and cross-shelf exchange in canyons. To illustrate the potential effects of the interaction of eddies and canyons on shelf-basin exchange in the Bering Sea, Figs. 6.3 and 6.4 show three cases from the NAME model (July 1987, November 1993, and May 2002) when heat and salt flux anomalies across the Zhemchug Canyon were at local maxima (see Fig. 7 of Clement Kinney et al. 2009 for time series of heat and salt flux anomalies) caused by modeled cyclonic eddies located just south of the canyon. Figure 6.3 shows that the sea surface height anomalies decrease to −20 cm (or less) in the center of these modeled cyclones. When the cyclones are present, salinities increase both near the bottom and near the surface by up to 0.4 psu and on-shelf flow occurs with speeds between 8 cm/s (July 1987) and over 14 cm/s (May 2002).

Clement Kinney et al. (2009) calculated the on shelf nutrient fluxes associated with a cyclonic eddy located over the slope at Zhemchug Canyon beginning in November 1993 (Fig. 6.3b). First the vertically integrated on-shelf salt flux, via both upwelling of relatively salty water and on-shelf flow, was calculated and then this was converted to a flux of silica using a relation between salinity and silica concentration from data in Cooper et al. (1997). It was estimated that a total of 8,350 kg of silica could be advected onto the Bering Sea shelf through Zhemchug Canyon over 3 months The transport associated with the upwelled water only (deeper than 100 m) was 2,426 kg of silica so that 70 % of the transport was in the upper water column.

Stabeno et al. (1999) suggest that high productivity in the waters surrounding the Pribilof Islands results as a consequence of flow on the outer shelf passing through the narrow space between the northern flank of Pribilof Canyon and the Pribilof Islands. Here, the flow on the outer shelf accelerates and entrains nutrient-rich slope water from Pribilof Canyon. This increase in the flow can also be seen in model results from Clement Kinney et al. (2009; see Fig. 6.2), as elevated total kinetic energy in the area between the northern flank of Pribilof Canyon and the Pribilof Islands.

It can be concluded that upwelling of nutrient-rich, deep Bering Sea water due to eddies and shelf-basin exchange enhances biological production in this region.

6.3 The Chukchi and Beaufort Shelf-Break

6.3.1 Shelf-Basin Connections

Pacific Water is a source of heat, freshwater, nutrients, and carbon for the Arctic Ocean. After passing through Bering Strait, Pacific Water must transit across the shallow, broad Chukchi before reaching the shelf-break and the edge of the deep Arctic Ocean basin. Physical mechanisms regulating shelf-break exchange control

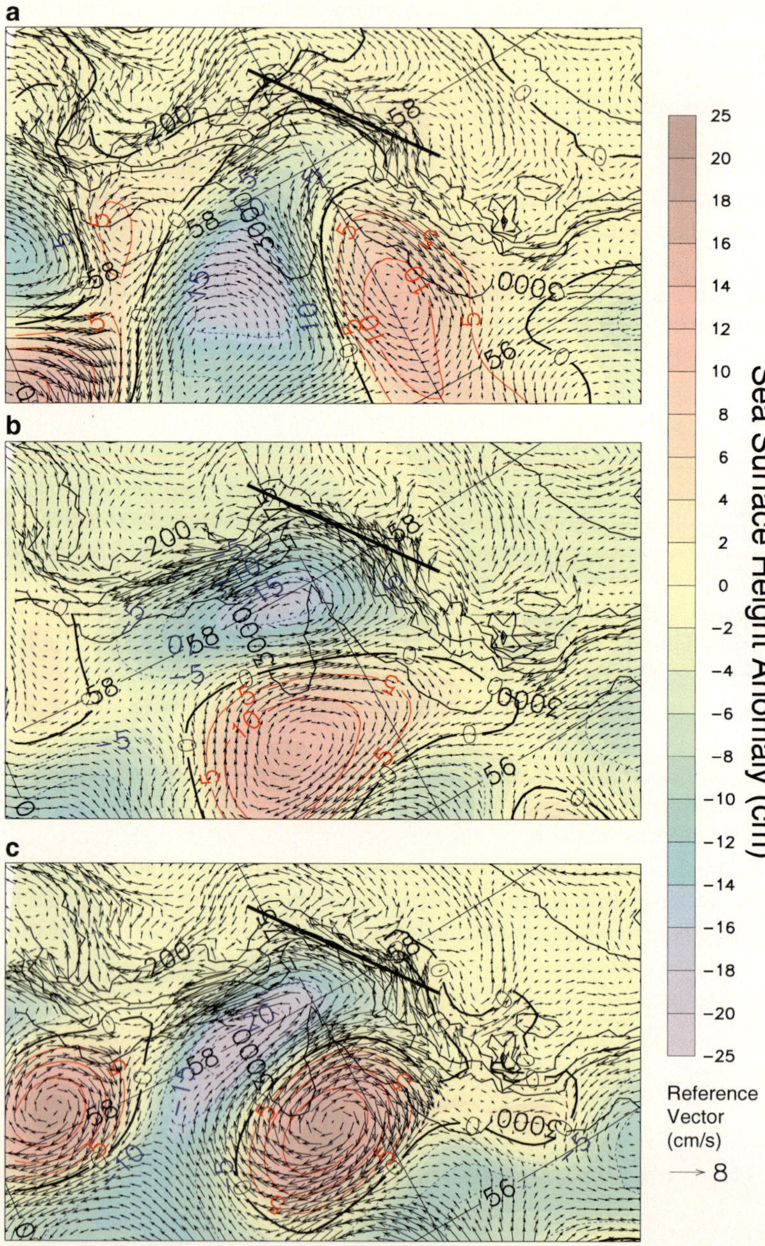

Fig. 6.3 Sea surface height anomaly (cm; *shading*) and velocity vectors (cm/s; *arrows*) averaged over the entire water column during (**a**) July 1987, (**b**) November 1993, and (**c**) May 2002 using output from the NAME numerical model. *Red* and *blue contour lines* indicate sea surface height anomalies. The location is the Bering shelf-break encompassing Zhemchug Canyon (see Fig. 6.1a). The cross-section across Zhemchug Canyon that is displayed in Fig. 6.4 is shown as a solid *black line* (Figure adapted from Fig. 8 of Clement Kinney et al. (2009) with permission from Elsevier)

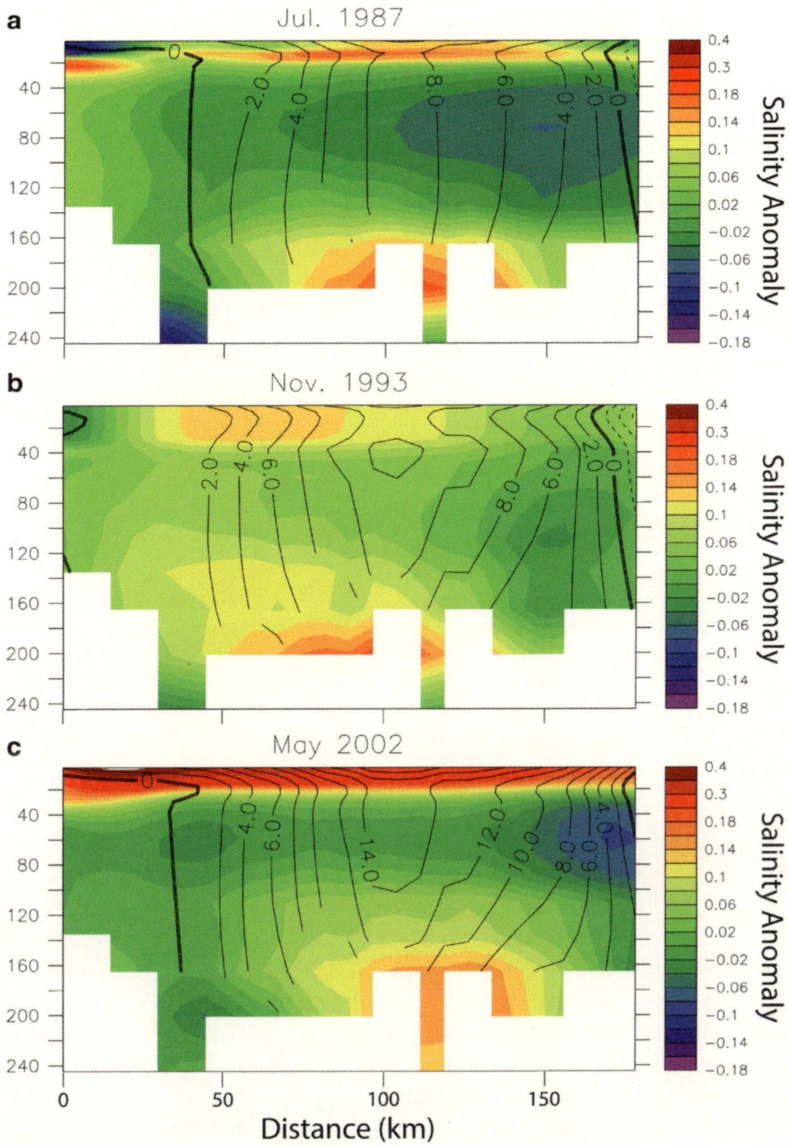

Fig. 6.4 Vertical section of salinity anomaly (*shading*) and velocity (*contours*) across Zhemchug Canyon during (**a**) July 1987, (**b**) November 1993, and (**c**) May 2002 using output from the NAME numerical model. The location of the cross-section is shown in Fig. 6.3 (Figure adapted from Fig. 9 of Clement Kinney et al. (2009) with permission from Elsevier)

how, when, and where Pacific Water enters the deep Arctic basin (see Fig. 5.3 in Chap. 5 of this volume for the mean ocean circulation pattern). Although many questions exist regarding the processes, we know that exchange between the shelf and basin occurs since Pacific Water dominates the halocline of the Canadian Basin. In the Beaufort Gyre region, for instance, Pacific Summer Water makes up a warm

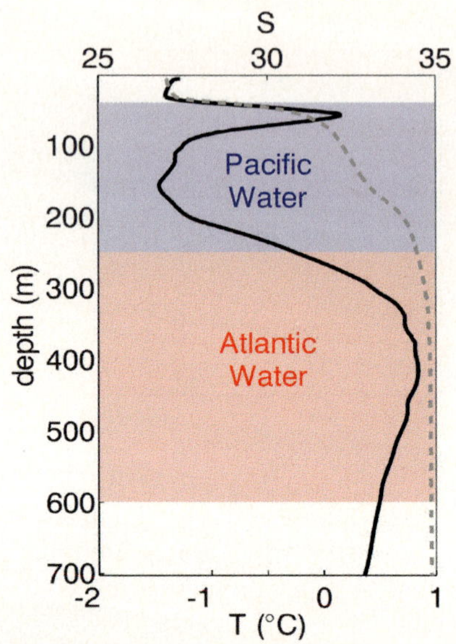

Fig. 6.5 Temperature (*solid black*) and salinity (*dashed grey*) from an ice-tethered profiler within the Canada Basin. Data show month-long average beginning May 15, 2007. The signature of warm, summer Pacific water is evident in the upper halocline. Cool, winter Pacific water contributes to water found deeper within the halocline. The Ice-Tethered Profiler data were collected and made available by the Ice-Tethered Profiler Program (Toole et al. 2011; Krishfield et al. 2008) based at the Woods Hole Oceanographic Institution (http://www.whoi.edu/itp)

layer that is regularly observed at roughly 50 m depth (e.g. Toole et al. 2010), while colder and saltier Pacific Winter Water contributes to deeper halocline layers (e.g. Aagaard et al. 1981). An example profile of temperature and salinity data acquired by an ice-tethered profiler within the Canada Basin is shown in Fig. 6.5.

Due to connections between the shelf and deep basin, variability in the circulation and conditions over the shelf and shelf-break may affect the entire Arctic. For example, Woodgate et al. (2010) suggest that the 2007 record minima in sea-ice extent may have been related to an unusually large flux of heat through the Bering Strait. The consequent lateral and basal melting of sea ice over the Chukchi shelf may have amplified the effect of local solar heating by initiating an ice-albedo feedback. Satellite imagery of ice melt and sea-surface temperature have provided striking visual records of the pathways of warm Pacific Water through the Chukchi Sea (e.g. Martin and Drucker 1997; Woodgate et al. 2010) and out into the deep Beaufort (Shimada et al. 2006; Pickart and Stossmeister 2008).

Pacific water inflow is steered by topography through the Chukchi Sea (Winsor and Chapman 2004; Weingartner et al. 2005; Woodgate et al. 2005a; Maslowski et al. 2014, this volume); and, upon reaching the shelf-break, it is generally thought that flow turns east establishing a shelf-break current along the upper Chukchi and Beaufort slopes (e.g. Mathis et al. 2007; Spall 2007; Nikolopoulos et al. 2009;

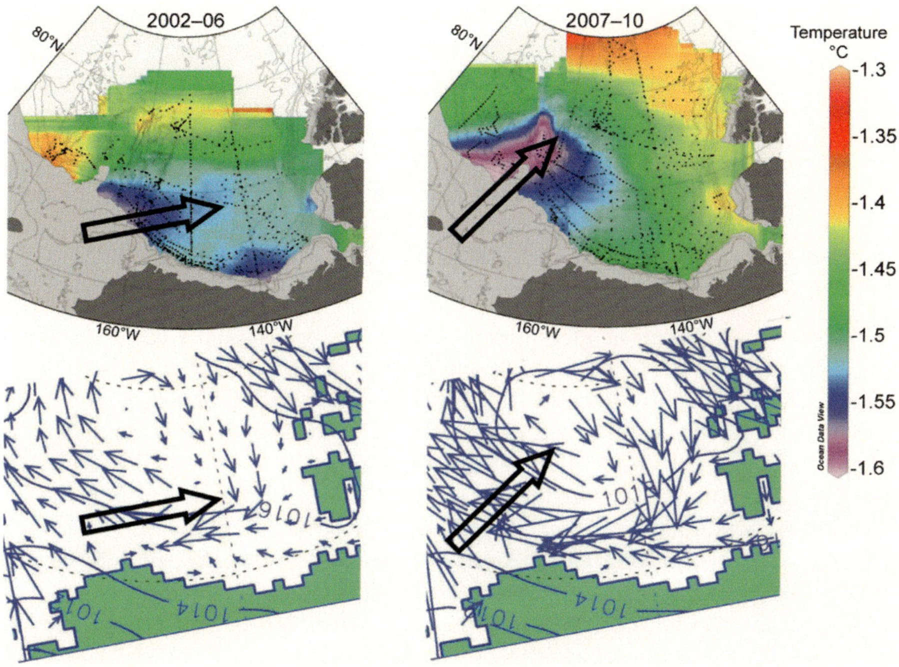

Fig. 6.6 Changes in the temperature minimum of the Pacific Winter Water in the Canadian Basin between 2002–2006 and 2007–2010 and their relationship to changes in ice motion. The *top panels* show potential temperature on the 33.1 psu isohaline to identify the core of the Pacific Winter Water and the *large arrows* suggest the direction of movement of shelf water from the Chukchi shelf-break. The *bottom panels* show sea level atmospheric pressure (hPa) and simulated wind-driven component of ice drift (Figure 5.9 from Proshutinsky et al. 2011, reproduced with permission of the American Meteorological Society)

Pickart et al. 2010). The Pacific water that is integrated into the Chukchi/Beaufort shelf-break current system enters the deep basin via eddies, filaments, or wind-driven exchange. Data from 2002 to 2004 show the annual mean volume transport within the shelf-break current at 152°W over the Alaskan Beaufort Shelf represents only 20 % of the mean transport through the Bering Strait (Nikolopoulos et al. 2009), suggesting that much of the Pacific Water enters the deep basin by crossing the Chukchi shelf-break, before it reaches the Alaskan Beaufort Shelf. Analysis of broad-scale hydrographic data from 2002 to 2010 shows the core of the Pacific Winter Water (33.1 psu) is coldest adjacent to the shelf-break of the Chukchi Sea, suggesting the greatest shelf-break exchange is in this region and thus supporting the results from 152°W. Splitting the hydrographic data into two time periods (2002–2006 and 2007–2010) shows a westward shift in the location of shelf-break exchange and the path of Pacific Winter Water in the Canadian Basin that correlates with an increase in westward mean wind and ice surface-stress (Fig. 6.6, Proshutinsky et al. 2011). The Barrow Canyon event shown in Fig. 6.9, and further discussed below, offers an example of off shelf flow in the Chukchi, yet, the dynamics that regulate this type of breakout event are not well understood. We choose to focus this summary on those mechanisms, dynamic instabilities of the shelf-break current and wind-driven exchange, which have received more attention to-date.

6.3.2 Instabilities of the Shelf-Break Jet

The shelf-break current as measured at 152°W on the Alaskan Beaufort Shelf is relatively narrow (10–15 km) and, in the annual mean, is a bottom-intensified jet with speeds of roughly 15 cm/s (Nikolopoulos et al. 2009). The structure of the shelf-break current in this location varies greatly depending on both season and wind conditions (Nikolopoulos et al. 2009; von Appen and Pickart 2011). In winter and spring, a bottom-intensified jet carries cold Pacific Winter Water. In late summer and early fall, the shelf-break current transports warm, Pacific Summer Water often in the form of a surface-intensified jet. Strong easterlies can cause the current to completely reverse, and, although reversals may happen at anytime of year, they appear to be most common in early fall when the ocean is ice-free and easterly winds are strong (Pickart et al. 2009).

In the absence of wind, von Appen and Pickart (2011) further classify the summer configuration at 152°W into two distinct cases – a surface intensified jet carrying Alaskan Coastal Water and a bottom-intensified jet carrying weakly stratified Chukchi Summer Water. In the mean, the Alaskan Coastal Water configuration exceeded speeds of 30 cm/s near the surface. This configuration carried stratified water with maximum (mean) temperatures greater than 5 °C and salinity less than 31 psu. In contrast, the Chukchi Summer Water configuration consisted of a subsurface jet (mean velocities near 25 cm/s) with temperatures near 0 °C and salinities around 31.5 psu. These authors also summarized the mean winter condition in the absence of wind forcing (Fig. 6.7), although also a sub-surface jet, the winter-configurations transports cooler (roughly −1.5 °C) and saltier (33 psu) water in comparison to the Chukchi Summer Water configuration.

Subsurface eddies containing shelf-origin water populate the Canada Basin (e.g. Manley and Hunkins 1985; Muench et al. 2000; Timmermans et al. 2008). The majority of eddies are anticyclonic and subsurface with cold Pacific Winter Water found in their core (Krishfield et al. 2002); however, observations have also documented surface-intensified warm anticyclones composed of Alaskan Coastal Water (Pickart and Stossmeister 2008). While previous studies have suggested that Barrow Canyon may provide a localized source of eddy generation (e.g. D'Asaro 1988; Watanabe and Hasumi 2009), other studies have also shown that the shelf-break current is generally unstable without the influence of complex topography (Spall et al. 2008; von Appen and Pickart 2011).

Modeling work by Spall et al. (2008) indicates that the winter-time shelf-break jet is susceptible to baroclinic instabilities. Consistent with the numerous observations

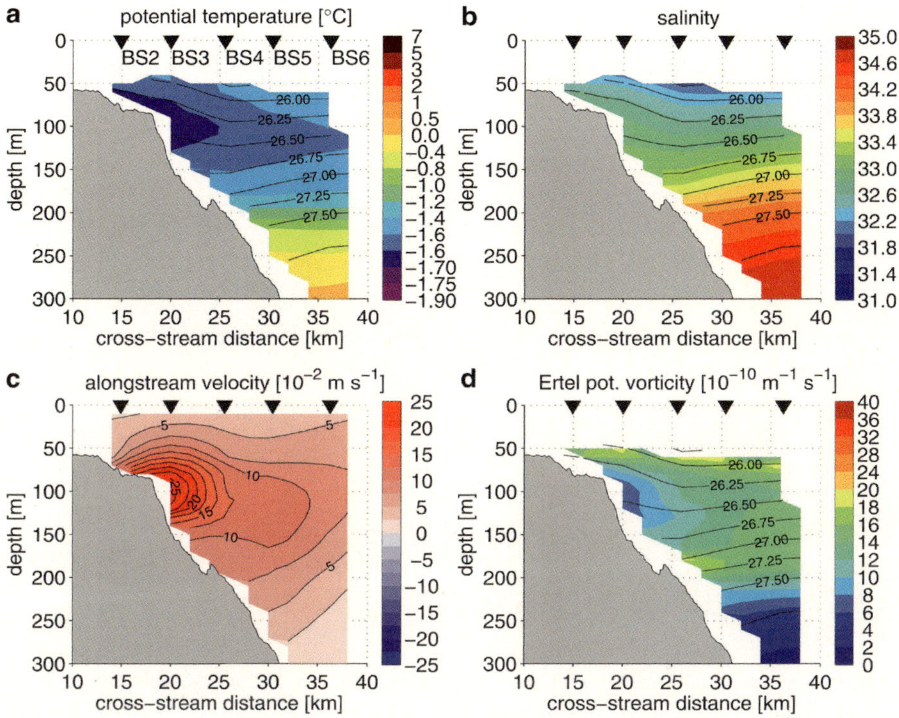

Fig. 6.7 (**a**) Temperature, (**b**) salinity, (**c**) along-shelf velocity, and (**d**) across-shelf velocity for the Pacific Winter Water configuration of the Beaufort shelf-break current (Reproduction of Fig. 10 from von Appen and Pickart 2011, with permission of the American Meteorological Society)

of eddies containing Pacific Winter Water, the modeled system sheds eddies readily. Interestingly, idealized inclusion of the Beaufort Gyre circulation further enhanced eddy production in the model, raising the possibility that the integrated wind response and circulation within the gyre feeds back into shelf-break current dynamics. Observations have also suggested that the summer-time configurations are baroclinically (both the Alaskan Coastal Water and Chukchi Summer Water configuration) and barotropically (only the Alaskan Coastal Water configuration) unstable (von Appen and Pickart 2011). Using energetics, von Appen and Pickart argue that the summer-time configurations decay over an e-folding scale of a few hundred kilometers along shelf, so that Pacific Water integrated into the shelf-break current is routed into the deep basin at some point along the Beaufort Slope. In contrast, the winter-time configuration decays slowly over a scale of ~1,400 km, raising the possibility that some Pacific Water may exit the Arctic Ocean directly via the Canadian Arctic Archipelago rather than being routed into the deep Arctic. Eddy generation throughout the year by these instabilities is likely responsible for a significant part of the mass-flux of Pacific Water from the shelf to the Canada Basin.

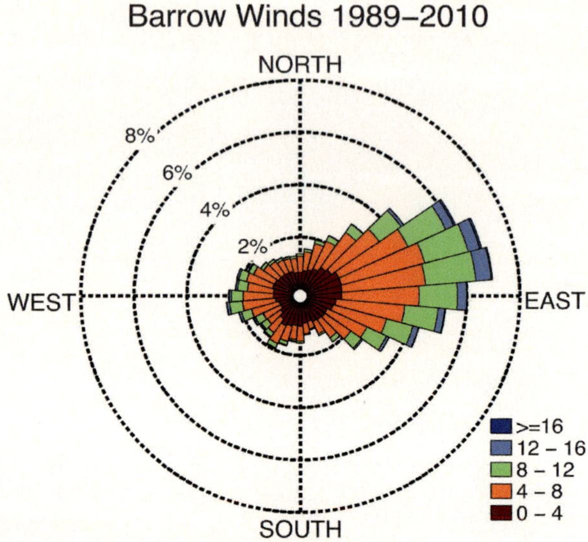

Fig. 6.8 Polar histogram showing hourly wind direction and speed from 1989 to 2010. Histogram bars point in the direction the winds originate. Data are measured 10 m above the surface at the NOAA/ESRL/GMD Pt. Barrow Observatory (http://www.cmdl.noaa.gov/info/ftpdata.html)

6.3.3 Wind-Driven Exchange

As with many places in the coastal ocean, wind-driven surface-stress is thought to play an important role in the total across-slope transport along the Chukchi and Beaufort Seas. Wind speed and direction from 1989 to 2010 at Point Barrow, Alaska are summarized in Fig. 6.8. Winds typically consist of an easterly component, which over the Chukchi/Beaufort shelf corresponds to upwelling favorable conditions. Upwelling circulation drives surface waters offshore and brings up cold, typically nutrient-rich water from depth. Upwelling of deep Atlantic Water has been regularly observed across the Chukchi and Beaufort shelves (e.g. Aagaard et al. 1981; Pickart 2004; Nikolopoulos et al. 2009); and, enhanced exchange has been indicated in canyons that cut across the slope (e.g. Carmack and Kulikov 1998; Williams et al. 2006; Mountain et al. 1976; Munchow and Carmack 1997; Weingartner et al. 1998).

Aleutian Low storm systems centered over the eastern end of the Alaskan Peninsula and Aleutian Island chain appear to be responsible for the strongest upwelling on the Beaufort Shelf (Pickart et al. 2009). Although these storms typically occur throughout the fall-winter, upwelling tends to be strongest before the onset of heavy pack ice (Pickart et al. 2009). Through the course of the year, the Chukchi and Beaufort shelves display among the largest and most variable Ekman transports within the Arctic (Yang 2006). Consistent with in situ studies, maximum upwelling occurs in November. Considering the seasonality of the winds and shelf-break current system, wind-driven exchange of Pacific Water properties (e.g. freshwater and heat) will also vary. For example, forcing of the shelf-break current in the Alaskan Coastal Water configuration by upwelling winds may contribute to significant heat transport but negligible nutrient transport. A similar wind event applied to the Winter Water configuration would result in different fluxes of heat, nutrients, and freshwater.

Pickart et al. (2011, 2013a) examined the response of the coastal current to a strong northeasterly wind event measured in November 2002. Surface-stress was enhanced during this storm due to loose ice (50–70 % concentration) responding to the wind and generating ice-ocean surface-stress higher than wind-ocean surface-stress (Pickart et al. 2013a) The westward surface-stresses were sufficient to reverse the Beaufort shelf-break jet, generating a surface intensified westward jet, and upwell Atlantic Water onto the shelf. The barotropic alongshore pressure gradient eventually 'spins-down' the reversed jet to leave behind a baroclinic structure in which upwelled Atlantic Water is integrated into a deep, eastward flowing jet that persists after the relaxation of the wind. A regional model used to interpret these observations suggests that spatial variability in ice cover, and consequently wind-stress, caused the creation of an anticyclonic gyre in the northern Chukchi. The end result being that, although upwelling and westward flow was prevalent across the Beaufort shelf and within Barrow Canyon during the storm, flow remained eastward along the Chukchi shelf. These results highlight the large temporal and spatial variability anticipated for wind forcing in this region.

Further examination of Ekman fluxes during the November 2002 storm by Pickart et al. (2013a) shows that the offshore fluxes of heat and freshwater, and onshore fluxes of nitrate are significant. For example, only 4–5 such storms can provide the nitrate needed for the estimated annual primary production of the Alaskan Beaufort Shelf. Pickart et al. (2013b) continues the analysis to show that there are 9–10 upwelling events per year that have winds greater than 10 m/s and lasting longer than 4 days. The upward nitrate flux associated with these storms during the open water season can drive primary production, and the magnitude of this storm-driven primary production is similar to that during less windy times.

6.4 Undersea Canyons of the Chukchi and Beaufort Shelves

Undersea canyons, which cut across the continental shelf, have long been noted as regions that impact not only coastal circulation but also biological productivity and sediment transport. The complex topography of canyons introduces along-shelf variation to the physical system which leads to enhanced across-shelf exchange. The circulation within and around canyons is influenced by a variety of processes, including internal wave activity and mixing, modification of coastally-trapped waves, and topographic steering of coastal currents. In turn, these physical processes influence the ecosystem by controlling sediment transport and suspension, and creating an environment that is often favorable for biological productivity. Not surprisingly, previous studies from the Chukchi and Beaufort suggest that canyons have a profound effect on the circulation and ecosystem in the Pacific Arctic.

Here, we examine the circulation in and around three canyons, two of which are found on the Chukchi Shelf, Herald and Barrow Canyon, and one of which cuts across the Canadian Beaufort Shelf, Mackenzie Trough. All three of these canyons cut across the shelf-break isobaths, yet all three differ from one another in essential ways. A primary difference between the Chukchi canyons as compared to Mackenzie Trough is due to the physical set-up, in particular the incident angle of the primary flow. Herald and Barrow Canyons extend off the wide, shallow Chukchi Sea; and, because of northward inflow at the Bering Strait, significant along-canyon flow is anticipated and observed in these two cases even in the absence of wind. For Herald Canyon, the primary topography is on the shelf and above the isobath of the shelf-break. Thus Herald Canyon is considered to channel part of the Bering Strait through-flow and deliver it to the shelf-break and so modify the distribution of Pacific Origin water at the edge of the Chukchi Shelf. For Barrow Canyon, most of the topography is below the shelf break and the constriction in the isobaths (steep topography) against Point Barrow leads to down-canyon flow. In contrast to both of the Chukchi canyons, the axis of Mackenzie Trough runs perpendicular to the Beaufort shelfbreak current. Along-canyon transport in Mackenzie Trough, and shelf-break exchange, is found to be strongly influenced by wind-driven along-shelf flow and along-shelf dynamics (e.g. meanders in the Beaufort Slope Current or eddies).

6.4.1 Herald Canyon

Herald Canyon is located on the western Chukchi shelf along the eastern side of Wrangel Island. Its axis is primarily aligned north-south and the mouth of Herald Canyon opens at roughly the 150-m isobath on the slope separating the Chukchi Sea from the Chukchi Abyssal Plain. In comparison to the eastern Chukchi, observational work has been limited in Herald Canyon; however, the general character of the current structure and water mass properties in the canyon has been documented by a handful of studies.

Herald Canyon creates a conduit for Pacific Water from the Bering Strait toward the shelf-break of the Chukchi (Winsor and Chapman 2004; Woodgate et al. 2005a; Spall 2007) rather than directly influencing shelf-break exchange. The Pacific Water tends to occupy the eastern flank of the canyon (e.g. Woodgate et al. 2005a) and, upon exiting the canyon, is thought to turn eastward following the shelf-break (e.g. Mathis 2007; Spall 2007; Maslowski et al. 2014, this volume). In winter, the canyon is occupied by Bering Winter Water that has cooled slightly en-route through the Chukchi (Woodgate et al. 2005a). In summer, relatively warm Pacific-origin water and remnant winter water, which is believed to be formed via cooling and brine rejection within the Chukchi Sea (Kirillova et al. 2001; Weingartner et al. 2005), co-exist in Herald Canyon. These water masses are divided by sharp temperature gradients (e.g. Pickart et al. 2010).

Dynamics of flow through Herald Canyon have received little attention, perhaps due to the dearth in observations. Pickart et al. (2010) analyze a set of synoptic surveys progressing northward through the Canyon, which show the winter layer thickened and switched from the western to the eastern flank of the canyon, consistent with theories of hydraulic control in straits. The observations suggest that hydraulic control may be important in setting both the northward transport of winter water and the mixing, which can be significant, between summer and winter. Due to limited observations, whether and how Herald Canyon impacts shelf-break exchange is not well-known. Numerical modeling efforts (e.g., Chapman and Gawarkiewicz 1995; Gawarkiewicz and Chapman 1995; Chapman 2000) suggest that dense water created on the shelf and routed through canyons may be carried across the shelf break as deep gravity currents and episodic eddies, although such flows have yet to be observed in Herald Canyon. Similarly, the role of wind-driven circulation in this region is not fully understood.

6.4.2 Barrow Canyon

Barrow Canyon is located at the northeast corner of the Chukchi Sea at the juncture between the Chukchi Shelf and the Alaskan Beaufort Shelf where the shelf narrows dramatically. Due to a nearly right degree turn in the Alaskan coastline, the axis of the canyon is located parallel to the Alaskan coast on the Chukchi shelf, but perpendicular to the Beaufort shelf. The canyon stretches roughly 200 km from the shallow Chukchi (40 m deep) out across the Beaufort slope. Its cross-canyon scale is ~50 km; and, depths exceed 250 m. The complexities of the topography and coastline combined with wind forcing create a dynamic physical environment, which despite numerous observations is not fully understood.

Pacific Water, flowing northward along the coastline of Alaska, exits the Chukchi through Barrow Canyon. Similar to Herald Canyon, Barrow Canyon primarily transports Bering Winter Water in winter; however, its source is from the eastern Bering, which is relatively nutrient poor compared to western Bering water. In summer, water mass properties in Barrow Canyon vary, with contributions from warm, buoyant Alaskan Coastal Water, cold Bering Winter Water, fresh and cold melt water, and relatively warm, salty Arctic halocline water upwelled from the Beaufort (Mountain et al. 1976; Munchow and Carmack 1997; Weingartner et al. 1998). In addition, water from Herald Canyon can be re-routed through the head of Barrow Canyon after traveling eastward along the Chukchi shelf-break and around Hanna Shoal situated to the west of Barrow Canyon (e.g. Spall 2007).

Circulation in Barrow Canyon depends upon the large-scale pressure gradient, wind forcing, coastally-trapped waves, hydraulic control (of winter water), and (in summer) buoyancy forcing. Short-term variability (days-weeks) has been linked to wind forcing (Weingartner et al. 1998; Woodgate et al. 2005a) and shelf wave activity (Aagaard and Roach 1990). As Barrow Canyon is at a sharp change in shelf width, the scattering of shelf waves into shorter modes will likely also affect the complex and variable flow at the mouth of the canyon (Wilkin and Chapman 1987). In contrast, long-term variability (months) appears to be related to the large-scale pressure gradient between the Pacific and Arctic (Mountain et al. 1976). As with Herald Canyon, hydraulics and mixing may be a factor in the flow of dense winter water down Barrow Canyon (Pickart et al. 2005).

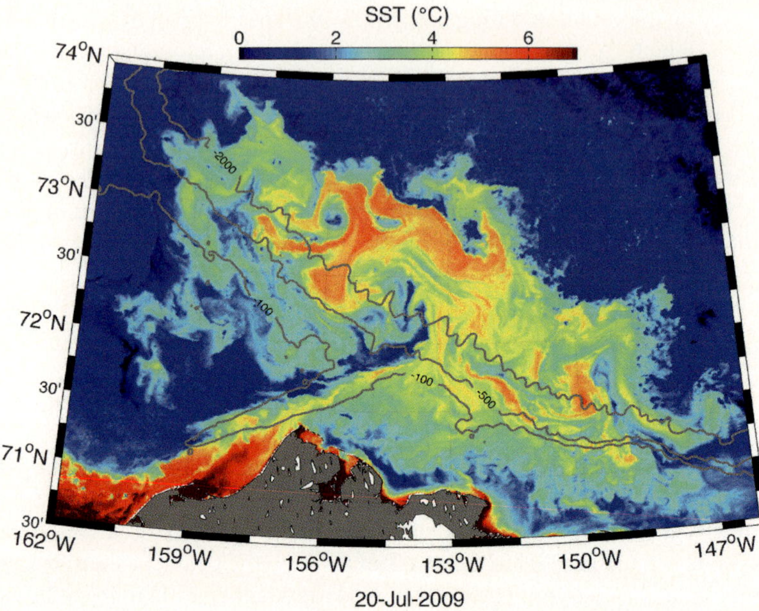

Fig. 6.9 Sea surface temperature (Modis Aqua) from 20 July 2009 in the vicinity of Barrow Canyon. The 100-, 500-, and 2,000-m isobaths are shown in *grey*

The momentum balance within Barrow Canyon is complicated. A secondary circulation in the across-shelf momentum equation results in a wedge shaped pycnocline (with a sharper, deeper interface on the shoreward side of the canyon) formed between the relatively warm, fresh surface water and cold salty winter water at depth (Signorini et al. 1997). This signature has also been seen in other observations (Pickart et al. 2005). Using a numerical model, Signorini et al. (1997) quantify terms in the along-shelf momentum balance, which is influenced strongly by the propagation of coastally-trapped waves initiated through temporal variability in the boundary conditions. The resultant interaction of a variable barotropic flow with canyon topography yields an up-canyon flow of deep water. Thus, both wind-driven local upwelling (e.g. Weingartner et al. 1998) and internal canyon dynamics (potentially affected by upstream changes in wind forcing) may result in up-canyon flow of Arctic halocline water.

The fate of the coastal current upon exiting the canyon depends upon wind conditions (Okkonen et al. 2009) and current stability. Sea surface temperature imagery shows instabilities that appear to emanate from the mouth of the canyon (Fig. 6.9). These dynamical processes are likely of importance to the shelf-basin flux of heat, fresh water, carbon, and nutrients, thus impacting the ecosystem, hydrography, and ice cover in the deep Arctic. A combination of mechanisms have been suggested as candidates for eddy formation in the vicinity of Barrow Canyon, including lateral boundary separation at Point Barrow (D'Asaro 1988), barotropic instability (Watanabe 2011), and outflow of relatively low potential vorticity water from the

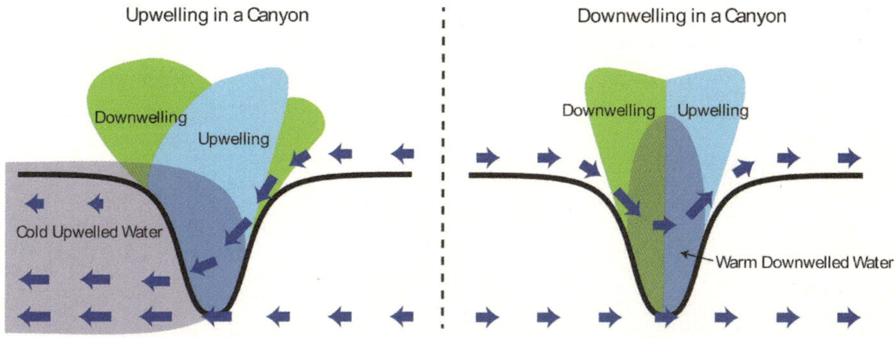

Fig. 6.10 Schematics showing canyon circulation during upwelling and downwelling winds after modeling results by Klinck (1996). Shown are approximate regions of upwelling (*blue*), downwelling (*green*), and cool upwelled water (*transparent purple*). *Blue arrows* are representative of horizontal currents

canyon (Shimada et al. 2001; Watanabe 2011). In addition, modeling efforts have shown that dense water formed in coastal polynyas is advected down Barrow Canyon and carried across the slope in eddies that form near the mouth of Barrow Canyon (Nguyen et al. 2012), in agreement with the idealized models of (e.g., Chapman 2000; Gawarkiewicz and Chapman 1995). The range of processes believed relevant in exchange near Barrow Canyon is indicative of several factors including strong seasonality in hydrographic and current properties, as well as complex topography and forcing (i.e., the large-scale flow of Pacific water, shelf wave activity, and local wind forcing).

6.4.3 Mackenzie Trough

Mackenzie Trough is located offshore of the Mackenzie River delta in between the Alaskan Beaufort Shelf and Mackenzie Shelf. This undersea canyon is roughly 200 km long, ~60-km wide, and ~400-m deep at its mouth. Circulation within the canyon is strongly influenced by wind forcing, and observations (Williams et al. 2006) suggest that the dynamics are in agreement with that expected for other canyons of the same dynamic scale as Mackenzie Trough for which the width is 2–3 times the internal Rossby radius (see Hyun 2004).

Upwelling of deep water in the canyon is correlated with northeasterly winds and is amplified in the canyon compared to the surrounding shelf (Carmack and Kulikov 1998). Upwelling winds result in an along-canyon pressure gradient at the canyon rim with a region of relatively high pressure near the mouth compared to that at the canyon head; across-isobath, onshore flow results from the along-canyon pressure gradient (for details see Klinck 1996). A schematic for upwelling wind-induced circulation following Klinck (1996) is shown in Fig. 6.10a, although not apparent in this schematic the strongest upwelling occurs near the head of the canyon. Also

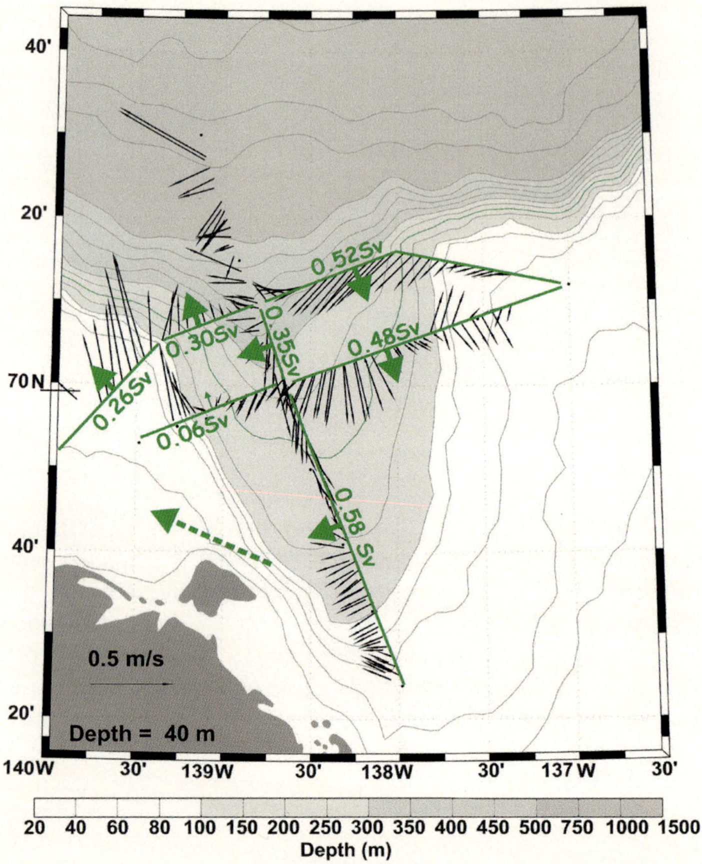

Fig. 6.11 Current data from Mackenzie Trough acquired during a strong upwelling wind event in September 2002 (see Williams et al. 2006 for further description). Data from 40 m deep are shown. Although flow fields are complex, the general tendency is up-canyon along the down-shelf (eastward) side and a return back along the canyon on the up-shelf (westward) side. Transports across each section (*green arrows*) show that there is much more inflow than outflow suggesting shelfbreak upwelling immediately north of Herschel Island (*dotted green arrow*) (Reproduced from Fig. 5 of Williams and Carmack 2012)

shown are velocity data from Mackenzie Trough (Fig. 6.11, reproduced from Fig. 9 in Williams et al. 2006). There is strong onshore flow on the down-shelf side of the canyon (reminiscent of the idealized modeling by Klinck 1996) but much weaker return flow on the up-shelf side of the canyon. Integration of the data to calculate transports shows that the flow crosses the head of the canyon and suggests it upwells onto the Alaskan Beaufort Shelf immediately north of Herschel Island. Data shown in Mathis et al. (2012) provide an example of this occurring.

An asymmetry exists in the comparison of upwelling and downwelling wind events, such that cross-shelf exchange within a canyon is much weaker in the case of downwelling (Klinck 1996). Downwelling winds result in regions of upwelling and downwelling that tend to be symmetric about the canyon's axis and of small amplitude (Fig. 6.10b); in such cases, flow is primarily steered along-isobaths rather than across-isobaths (Klinck 1996). This asymmetry combined with the observation that under the influence of ice cover upwelling-favorable surface stresses are preferred (since downwelling favorable wind-forced ice motion in this region tends to be blocked by internal ice stress) results in a net flux of deep, nutrient-rich water on to the shelf (Williams et al. 2006). Observations from Kugmallit Valley, one of the many shallow valleys near Mackenzie Trough, suggest that enhanced cross-shelf transport may also be associated with less-pronounced topographic features that need not extend all the way to the shelf edge (Williams et al. 2008). These results suggest the possibility of many localized regions of enhanced exchange contributing to the overall physical setting and marine ecosystem on the Chukchi and Beaufort shelves.

6.5 Polynya-Formed Dense Shelf Water

Wind-driven polynyas occur when offshore wind pushes ice away from the coast or from the land-fast ice edge forming areas of recurrent open water in winter. The large heat flux to the atmosphere from the water leads to large amounts of ice formation and, as the brine is rejected from the forming sea ice, results in the creation of cold, salty dense shelf water in the polynya (for a review of polynya types and their forcing see Williams et al. 2007). The dense shelf water has a tendency to flow downhill and ultimately cross the shelf-break (Melling and Lewis 1982; Melling and Moore 1995). However, many factors – rotational control, bottom boundary layer processes, along-shelf flows, instabilities leading to eddies, and mixing with other water masses – all affect how and where this water crosses the shelf-break and dilute its properties (e.g. Gawarkiewicz 2000). We include a description of polynyas in Chukchi Sea as these are relatively well studied in terms of their effect on the ocean and they modify the waters that flow into Barrow Canyon.

One of the largest Arctic polynyas occurs in the Chukchi Sea along the Alaskan coast between Cape Lisburne and Point Barrow, forced by offshore winds during winter (e.g. Cavalieri and Martin 1994; Martin et al. 2004). This polynya lies along the pathway of Pacific Water inflow from the Bering Strait that goes via Barrow Canyon into the Canada Basin. In general, heat loss over a coastal polynya is one or two orders of magnitude larger than that over thick ice (Maykut 1978), thus coastal polynyas are very active areas of ice formation. Brine rejection from this rapid ice formation in the polynya increases the salinity of Pacific Winter Water flowing through Barrow Canyon. The maximum salinity observed is up to 34.5 psu and is potentially the densest water that forms in the western Arctic Ocean (e.g. Weingartner et al. 1998, 2005) thus we call it *hyper-saline* Pacific Winter Water. Transport of winter water through Barrow Canyon into the basin is estimated to be 0.1–0.3 Sv from mooring and shipboard data (Woodgate et al. 2005a; Pickart et al. 2005; Itoh et al. 2012).

The formation of hyper-saline Pacific Winter Water is dependent on the occurrence of the Barrow – Cape Lisburne polynya. Thus, accurate observation of the polynya's area is essential, as the salt flux to the Pacific Winter Water is due to ice formation in the polynya. Martin et al. (2004) and Tamura and Ohshima (2011) examined an algorithm that estimates the thin ice thickness within a polynya using Special Sensor Microwave/Imager (SSM/I) data, and, using that, estimated sea ice production and salt flux within the coastal polynyas via a heat budget. Tamura and Ohshima (2011) estimated annual accumulated sea ice production is 6–14 m/year within the Barrow – Cape Lisburne polynya for 1992–2007, far larger than typical annual sea ice production in the Chukchi Sea of ~2 m/year. Sea-ice production in the Barrow-Cape Lisburne polynya can vary from year to year by as much as a factor of 3–4 (Martin et al. 2004; Tamura and Ohshima 2011)

Winsor and Chapman (2002) examined dense water formation for the coastal polynyas on the Chukchi shelf using an ocean model for the winters of the 1978–1998 period. They showed that the accumulated ice growth is 9–15 m/year over the polynya area and the maximum salinity anomaly is typically 1–1.5 psu. They proposed that its interannual variability is high and that most of the observed variability can be explained by the varying wind field. Winsor and Chapman (2002) also suggested that variation in the initial salinity of the Chukchi shelf, which is determined by Bering Strait through-flow, is equally important to the salinization within the polynya and thus in determining the interannual variability of dense water formation. Martin et al. (2004) compared the ice volume, derived using the SSM/I algorithm, with that obtained by Winsor and Chapman (2002) based on Pease's (1987) analytical model, and showed a significant difference, even in relative variabilities of sea ice production between the two estimates. Martin et al.'s (2004) results suggested that interannual variability of Barrow – Cape Lisburne polynya cannot be explained by simple analytical polynya model developed by Pease (1987), implying that Pease's assumption of free-drift of ice does not apply adequately in this polynya.

Kawaguchi et al. (2011) examined winter water formation for the winters of 1992–2006 using an ocean model forced by wind and surface salinity flux derived from SSM/I thin-ice thickness estimate. Their experiment shows that the Barrow Cape Lisburne polynya preferentially opens under strong northeasterly wind conditions. Additionally, under the northeasterly wind conditions, northward through-flow across Barrow Canyon tends to be reduced and results in salinity build up and ultimately in greater enhancement of salinity in water carried into the Canada Basin. Such wind conditions are established when the Beaufort high pressure system is intensified and Aleutian low pressure system is located west of its mean position. Kawaguchi et al. (2011) also pointed out that the ice production in the Kotzebue Sound coastal polynya is comparable to that in the Barrow – Cape Lisburne polynya, and it is also crucially important for salinity enhancement of winter water transported through Barrow Canyon. Together with moored salinity in the Bering Strait from Woodgate et al. (2005b), Kawaguchi et al.'s (2011) model results show that the salinity across the Barrow and Herald Canyons is 32.9 ± 0.8 psu and 32.7 ± 0.3 psu, respectively. Through-flow via Herald Valley is estimated to have smaller and more intermittent salinity enhancement than that via Barrow Canyon. This is expected since winter water carried via Herald Valley is less likely to have been modified by coastal polynyas.

Hyper-saline Pacific Winter Water formation within the Barrow – Cape Lisburne polynya and its interannual variability have been mostly examined by heat budget analysis and modeling, with limited oceanographic observational studies. From oceanographic observations during spring in 1975, Garrison and Becker (1976) found that highly saline water (up to 34 psu) is capable of ventilating the lower halocline layer (see Fig. 6.5 for a profile through the halocline), and suggested that it flows into the Beaufort Sea via Barrow Canyon. Weingartner et al. (1998, 2005) investigated circulation and water modification processes in the Chukchi Sea using mooring and shipboard data between 1992 and 1995. Weingartner et al. (2005) found that winter water salinity values on the Hanna Shoal ranged widely: from 31.3 to 34.5 psu in the winter of 1993–1994, and from 31.5 to 32.7 in the winter of 1992–1993. They suggested that reduced fall and winter ice cover and a large winter polynya area in 1993–1994 result in enhancement of hyper-saline winter water production. Weingartner et al. (2005) also pointed out that hyper-saline winter water derived from the Barrow – Cape Lisburne polynya exits the Chukchi Shelf into the basin not only via the Barrow Canyon but also via the Central Channel.

Itoh et al. (2012) examined the interannual variability of Pacific Winter Water both upstream in the northeastern Chukchi Sea, using moored observations, and downstream in the Canada Basin, using hydrographic data for 2000–2006. They proposed that the formation of hyper-saline Pacific Winter Water depends jointly on the salinity of incoming water at the Bering Strait gateway and the on occurrence of the Barrow – Cape Lisburne polynya. A large transport of cold and saline winter water via Barrow Canyon was observed in the winters of 2000/2001 and 2001/2002. In the former winter, enhanced ice formation in the polynya contributed to the increased salinity of Pacific Winter Water entering the basin, whereas in the latter winter, the high salinity of water entering the Bering Strait gateway was essential in delivering saline Pacific Winter Water to the basin. Itoh et al. (2012) also found that warm water within the polynya constrained ice formation in the winter of 2003/2004, resulting in reduced transport of hyper-saline winter water despite the fact that the coastal polynya was frequently open. Figure 6.12 shows that subsurface water at $S = 31.5$ psu, corresponding to the core of Pacific Summer Water, was above freezing point even in mid winter, possibly due to up-canyon flow from the basin onto to the Chukchi Shelf.

The Barrow Cape Lisburne polynya is believed to be a wind-driven polynya, and such polynyas are very active regions of ice formation when maintained by offshore wind. However, accumulated upper ocean heat in the Arctic Ocean due to global warming could change this situation. Woodgate et al. (2006) showed that heat flux through the Bering Strait increased from 2002 to 2004 due to the warming of Pacific Summer Water. Steele et al. (2008) also reported increased heat in the upper layer of the Chukchi Sea after 2000. In addition, less extensive sea ice has been observed in the Arctic Ocean during summers in recent years (Serreze et al. 2007) and the absence of ice cover for a longer season in the Chukchi Sea will promote surface warming. Figure 6.13 shows that surface temperature through Barrow Canyon has

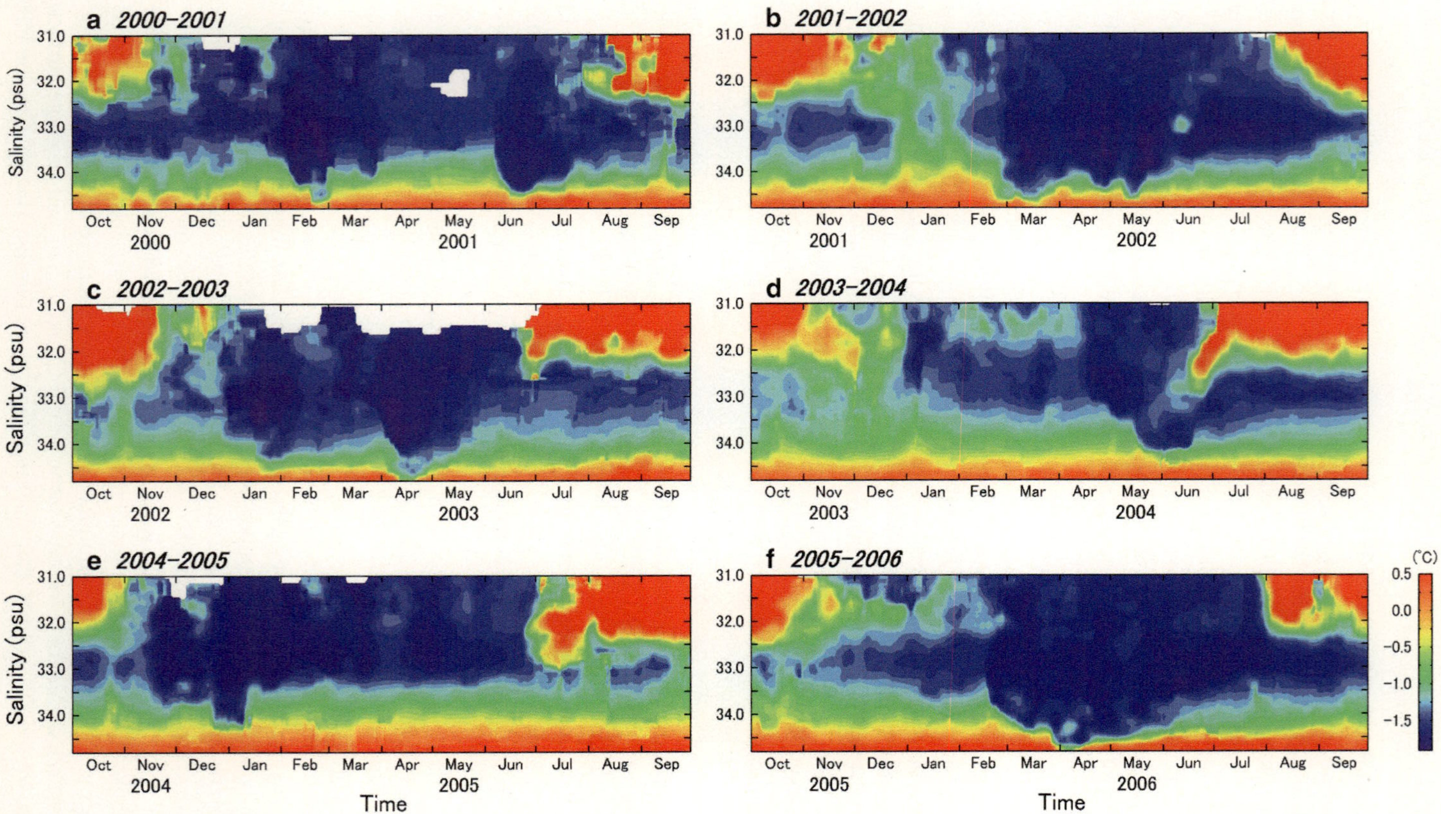

Fig. 6.12 Time series of temperature plotted as a function of salinity at the mouth of Barrow Canyon obtained by combining data from moorings BCE, BCC, BCW that spanned the mouth of the canyon (see Fig. 6.1c for locations) and contained instruments from ~ 30 m deep to close to the sea-floor. Panel (**a**) 2000–2001, (**b**) 2001–2002, (**c**) 2002–2003, (**d**) 2003–2004, (**e**) 2004–2005, and (**f**) 2005–2006 (Adapted from Fig. 13 of Itoh et al. (2012) with permission from Springer)

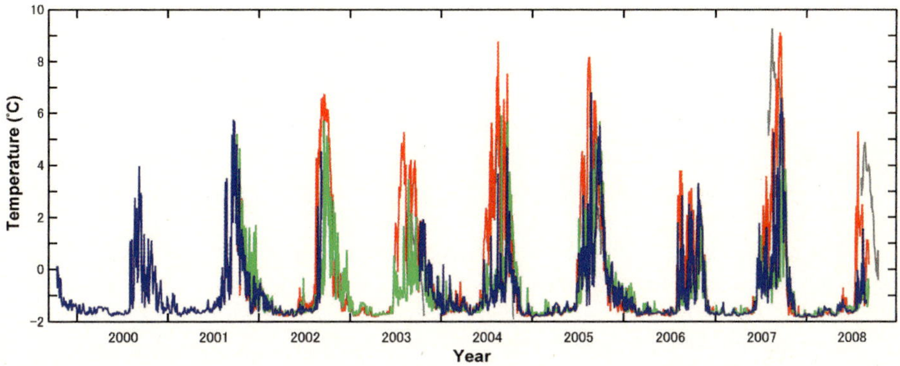

Fig. 6.13 Time series of upper ocean temperature (40 m) from the mouth of Barrow Canyon obtained from moorings BCE, BCC and BCW (see Fig. 6.1c for locations). *Red, blue and green curves* indicate data from the eastern (BCE), central (BCC) and western (BCW) canyon. *Gray curves* indicate SST data derived from microwave satellite AMSRE

increased since early 2000s. These results, showing warming, suggest an increase in the time in autumn that heat loss merely cools the water column, rather than leading to the formation of sea ice. Further, the water column within the polynya was warm enough to constrain ice formation even in mid-winter of 2003/2004, as suggested by Itoh et al. (2012). Although this warm water constrained ice formation in mid-winter in this polynya, it must be a local, advective phenomenon as ice formation does occur in the Bering and Chukchi Seas in winter following rapid heat loss in autumn. In addition, Wang et al. (2012) predict that there will always be ice formation on the Bering Shelf in winter under IPCC climate warming scenarios.

6.6 Summary

6.6.1 Bering Shelf-Break

Much of our detailed understanding of the Bering shelf-break comes from numerical models. These results need support from complementary observational data, in particular to assess the relative importance of wind-driven exchange, large slope eddies, canyon dynamics, and internal tides that propagate across the basin from the Aleutians. Such observational data could be achieved by deployment of high- and medium-resolution mooring arrays across the shelf-break and upper slope. Recent high-resolution observations by BEST-BSIERP increased our understanding of processes on the shelf and we suggest similar effort needs to be focused on the shelf-break, at a range of temporal and spatial scales. The evident success of the high-resolution array deployed across the Alaskan Beaufort shelf-break (Pickart 2004) could be emulated here.

6.6.2 *Chukchi/Beaufort Shelf-Break*

Transport via current instabilities and wind-driven exchange are important components of shelf-break exchange at the Chukchi/Beaufort shelf-break. However, many questions remain regarding how Pacific Water is carried into the deep basin. The role of time-varying winds and instabilities in the presence of a complex coastline (e.g. Point Barrow) and topography (canyons) needs further study. Long-term, high-frequency observations detailing circulation in canyons, which are thought to be important avenues for enhanced exchange, would further add to our understanding. In addition, since the data suggest that much of the off shelf flow of Pacific Water, crosses the Chukchi shelf-break, further research on the mechanisms of shelf-break exchange there would greatly enhance our understanding of how Pacific-origin water moves into the Canadian Basin. The Chukchi shelf-break is much more rounded than the Alaskan Beaufort shelf-break and possibly behaves substantially differently because of this. Considering the impact that cross-shelf exchange has on the deep basin hydrography and ice cover, addressing these unknowns is key to understanding the current state of the Arctic ecosystem and how this system might evolve under climate change.

Observational work in the Bering, Chukchi and Beaufort Seas has additional logistical complexity as these seas border Russia, Canada and the USA. Our growing comprehensive understanding of the Pacific-Arctic Region can only be nurtured though continued international collaboration that is able to transcend cultural and bureaucratic limitations.

Acknowledgements Support for W. Williams was provided by Fisheries and Oceans Canada. E. Shroyer was supported as a WHOI Postdoctoral Scholar through the WHOI Ocean and Climate Change Institute. Funding support for contributions by J. Clement Kinney and W. Maslowski was provided by multiple grants from the Climate Change Prediction Program of the Department of Energy, the Arctic System Science (ARCSS) Program of the National Science Foundation (NSF), and the Office of Naval Research. Funding for M. Itoh was provided by Japan Agency for Marine-Earth Science and Technology. We thank Robert Pickart and Albert Plueddemann and Seth Danielson for insightful discussions regarding circulation in the Chukchi-Beaufort region. We also thank the editors of this book and two anonymous reviewers for insightful comments, which improved an earlier version of this manuscript.

References

Aagaard K, Roach AT (1990) Arctic ocean-shelf exchange: measurements in Barrow Canyon. J Geophys Res 95:18163–18175

Aagaard K, Coachman LK, Carmack EC (1981) On the halocline of the Arctic Ocean. Deep-Sea Res 28:529–545

Aagaard K, Weingartner TJ, Danielson SL, Woodgate RA, Johnson GC, Whitledge T (2006) Some controls on flow and salinity in Bering Strait. Geophys Res Lett 33:L19602. doi:10.1029/2006GL026612

Carmack EC, Kulikov EA (1998) Wind-forced upwelling and internal Kelvin wave generation in Mackenzie Canyon, Beaufort Sea. J Geophys Res 103:18447–18458

Cavalieri D, Martin S (1994) The contribution of Alaskan, Siberian, and Canadian coastal polynyas to the cold halocline layer of the Arctic Ocean. J Geophys Res 99:18343–18362
Chapman DC (2000) The influence of an alongshelf current on the formation and offshore transport of dense water from a coastal polynya. J Geophys Res 105(C10):24007–24019. doi:10.1029/2000JC000296
Chapman DC, Gawarkiewicz G (1995) Offshore transport of dense shelf water in the presence of a submarine canyon. J Geophys Res 100(C7):13373–13387. doi:10.1029/95JC00890
Clement Kinney J, Maslowski W, Okkonen S (2009) On the processes controlling shelf–basin exchange and outer shelf dynamics in the Bering Sea. Deep Sea Res Part II Top Stud Oceanogr 56(17):1351–1362. doi:10.1016/j.dsr2.2008.10.023
Coachman LK, Aagaard K, Tripp RB (1975) Bering Strait: the regional physical oceanography. University of Washington Press, Seattle, p 172
Cooper LW, Whitledge TE, Grebmeier JM, Weingartner TJ (1997) The nutrient, salinity, and stable oxygen isotope composition of Bering and Chukchi Seas waters in and near the Bering Strait. J Geophys Res 102:12563–12573
D'Asaro EA (1988) Generation of submesoscale vortices: a new mechanism. J Geophys Res 93:6685–6693
Danielson S, Hedstrom K, Aagaard K, Weingartner T, Churchitser E (2012) Wind-induced reorganization of the Bering shelf circulation. Geophys Res Lett 39:L08601. doi:10.1029/2012GL051231
Garrison GR, Becker P (1976) The Barrow Submarine Canyon: a drain for the Chukchi Sea. J Geophys Res 81:4445–4453. doi:10.1029/JC081i024p04445
Gawarkiewicz G (2000) Effects of ambient stratification and shelf break topography on offshore transport of dense water on continental shelves. J Geophys Res 105(C2):3307–3324
Gawarkiewicz G, Chapman DC (1995) A numerical study of dense water formation and transport on a shallow, sloping continental shelf. J Geophys Res 100(C3):4489–4507. doi:10.1029/94JC01742
Hughes FW, Coachman LK, Aagaard K (1974) Circulation, transport, and water exchange in the western Bering Sea. In: Hood DW, Kelley EJ (eds) Oceanography of the Bering Sea, with emphasis on renewable resources, Publ. 2. Institute of Marine Science, University of Alaska Fairbanks, Fairbanks
Huthnance JM (1995) Circulation, exchange and water masses at the ocean margin: the role of physical processes at the shelf edge. Prog Oceanogr 35:353–431
Hyun KH (2004) The effect of submarine canyon width and stratification on coastal circulation and across shelf exchange. PhD thesis, Old Dominion University
Itoh M, Shimada K, Kamoshida T, McLaughlin F, Carmack E, Nishino S (2012) Interannual variability of Pacific Winter Water inflow through Barrow Canyon from 2000 to 2006. J Oceanogr 68:575–592. doi:10.1007/s10872-012-0120-1
Karl HA, Carlson PR (1987) Surface current patterns suggested by suspended sediment distribution over the outer continental margin Bering Sea. Mar Geol 74:301–308
Kawaguchi Y, Tamura T, Nishino S, Kikuchi T, Itoh M, Mitsudera F (2011) Numerical study of winter water formation estimation in the Chukchi Sea: roles and impacts of coastal polynya. J Geophys Res. doi:10.1029/2010JC006606
Kinder TH, Coachman LK (1978) The front overlying the slope in the eastern Bering Sea. J Geophys Res 83:4551–4559
Kinder TH, Coachman LK, Galt JA (1975) The Bering slope current system. J Phys Oceanogr 5:231–244
Kinder TH, Schumacher JD, Hansen DV (1980) Observation of a baroclinic eddy: an example of mesoscale variability in the Bering Sea. J Phys Oceanogr 101:1228–1245
Kinder TH, Chapman DC, Whitehead JA (1986) Westward intensification of the mean circulation on the Bering Sea Shelf. J Phys Oceanogr 16:1217–1229. doi:10.1175/15200485
Kirillova EP, Stepanov OV, Weingartner TJ (2001) Distribution and variability of nutrients in the northwestern part of the Chukchi Sea. In: Proceedings of the Arctic Regional Centre 3

Klinck JM (1996) Circulation near submarine canyons: a modeling study. J Geophys Res 101:1211–1223

Krishfield RA, Plueddemann AJ, Honjo S (2002) Eddies in the Arctic Ocean from IOEB ADCP data. Woods Hole Oceanographic Institution technical report WHOI-2002-09

Krishfield R, Toole J, Proshutinsky A, Timmermans M-L (2008) Automated ice-tethered profilers for seawater observations under pack ice in all seasons. J Atmos Oceanic Tech 25:2091–2095

Manley TO, Hunkins K (1985) Mesoscale eddies of the Arctic Ocean. J Geophys Res 90(C3):4911–4930

Martin S, Drucker R (1997) The effect of possible Taylor columns on the summer ice retreat in the Chukchi Sea. J Geophys Res 102:10473–10482

Martin S, Drucker R, Kwok R, Holt B (2004) Estimation of the thin ice thickness and heat flux for the Chukchi Sea Alaskan coast polynya from Special Sensor Microwave/Imager data, 1990–2001. J Geophys Res 109:C10012. doi:10.1029/2004JC002428

Maslowski W, Marble D, Walczowski W, Schauer U, Clement JL, Semtner AJ (2004) On climatological mass, heat, and salt transports through the Barents Sea and Fram Strait from a pan-Arctic coupled ice-ocean model simulation. J Geophys Res 109:C03032. doi:10.1029/2001JC001039

Maslowski W, Clement Kinney J, Okkonen SR, Osinski R, Roberts AF, Williams W (2014) Chapter 5: The large scale ocean circulation and physical processes controlling Pacific-Arctic interactions. In: Grebmeier JM, Maslowski W (eds) The Pacific Arctic region: ecosystem status and trends in a rapidly changing environment. Springer, Dordrecht, pp 101–132

Mathis JT, Pickart RS, Hansell DA, Kadko D, Bates NR (2007) Eddy transport of organic carbon and nutrients from the Chukchi Shelf: impact on the upper halocline of the western Arctic Ocean. J Geophys Res 112, C05011. doi:10.1029/2006JC003899

Mathis JT, Pickart RS, Byrne RH, McNeil CL, Moore GWK, Juranek LW, Liu X, Ma J, Easley RA, Elliot MM, Cross JN, Reisdorph SC, Bahr F, Morison J, Lichendorf T, Feely RA (2012) Storm-induced upwelling of high pCO_2 waters onto the continental shelf of the western Arctic Ocean and implications for carbonate mineral saturation states. Geophys Res Lett 39:L07606. doi:10.1029/2012GL051574

Maykut G (1978) Energy exchange over young sea ice in the central Arctic. J Geophys Res 83:3646–3658

Melling H, Lewis EL (1982) Shelf drainage flows in the Beaufort Sea and their effect on the Arctic Ocean pycnocline. Deep Sea Res Part A 29:967–985

Melling H, Moore RM (1995) Modification of halocline source waters during freezing on the Beaufort Sea shelf: evidence from oxygen isotopes and dissolved nutrients. Cont Shelf Res 15:89–113

Mizobata K, Saitoh SI (2004) Variability of Bering Sea eddies and primary productivity along the shelf edge during 1998–2000 using satellite multi-sensor remote sensing. J Mar Syst 50:101–111

Mizobata K, Wang J, Saitoh SI (2006) Eddy-induced cross-slope exchange maintaining summer high productivity of the Bering Sea shelf break. J Geophys Res 111:C10017

Mountain DG, Coachman LK, Aagaard K (1976) On the flow through Barrow Canyon. J Phys Oceanogr 6:461–470

Muench R, Gunn J, Whitledge T, Schlosser P, Smethie W Jr (2000) An Arctic Ocean cold core eddy. J Geophys Res 105:23997–24006. doi:10.1029/2000JC000212

Munchow A, Carmack EC (1997) Synoptic flow and density observations near an Arctic shelf break. J Phys Oceanogr 27:1402–1419

Nguyen AT, Kwok R, Menemenlis D (2012) Source and pathway of the western Arctic upper halocline in a data-constrained coupled ocean and sea ice model. J Phys Oceanogr 42:802–823

Nikolopoulos A, Pickart RS, Fratantoni PS, Shimada K, Torres DJ, Jones EP (2009) The western Arctic boundary current at 152°W: structure, variability, and transport. Deep-Sea Res Part II 56:1164–1181. doi:10.1016/j.dsr2.2008.10.014

Okkonen SR (2001) Altimeter observations of the Bering Slope Current eddy field. J Geophys Res 106:2465–2476

Okkonen SR, Schmidt GM, Cokelet ED, Stabeno PJ (2004) Satellite and hydrographic observations of the Bering Sea 'Green Belt'. Deep-Sea Res Part II 51:1033–1051

Okkonen SR, Ashjian CJ, Campbell RG, Maslowski W, Clement Kinney JL, Potter R (2009) Intrusion of warm Bering/Chukchi waters onto the shelf in the western Beaufort Sea. J Geophys Res. doi:10.1029/2008JC004870

Paluszkiewicz T, Niebauer HJ (1984) Satellite observations of circulation in the eastern Bering Sea. J Geophys Res 89:3663–3678

Pease CH (1980) Eastern Bering Sea ice processes. Mon Weather Rev 108:2015–2023

Pease CH (1987) The size of wind-driven coastal polynyas. J Geophys Res 92:7049–7059. doi:10.1029/JC092iC07p07049

Pickart RS (2004) Shelf-break circulation in the Alaskan Beaufort Sea: mean structure and variability. J Geophys Res 109:C04024. doi:10.1029/2003JC001912

Pickart RS, Stossmeister G (2008) Outflow of Pacific water from the Chukchi Sea to the Arctic Ocean. Chin J Polar Sci 19:135–148

Pickart RS, Weingartner TJ, Pratt LJ, Zimmermann S, Torres DJ (2005) Flow of winter-transformed Pacific water into the Western Arctic. Deep-Sea Res Part II 52:3175–3198

Pickart RS, Moore GWK, Torres DJ, Fratantoni PS, Goldsmith RA, Yang J (2009) Upwelling on the continental slope of the Alaskan Beaufort Sea: storms, ice, and oceanographic response. J Geophys Res 114:C00A13. doi:10.1029/2008JC005009

Pickart RS, Pratt LJ, Torres DJ, Whitledge TE, Proshutinsky AY, Aagaard K, Agnew TA, Moore GWK, Dail HJ (2010) Evolution and dynamics of the flow through Herald Canyon in the western Chukchi Sea. Deep-Sea Res Part II 57:5–26. doi:10.1016/j.dsr2.2009.08.002

Pickart R, Spall M, Moore G, Weingartner T, Woodgate R, Aagaard K, Shimada K (2011) Upwelling in the Alaskan Beaufort Sea: atmospheric forcing and local versus non-local response. Prog Oceanogr 88:78–100. doi:10.1016/j.pocean.2010.11.005

Pickart RS, Spall MA, Mathis JT (2013a) Dynamics of upwelling in the Alaskan Beaufort Sea and associated shelf-basin fluxes. Deep Sea Res Part I: Oceanogr Res Pap 76:35–51. doi:10.1016/j.dsr.2013.01.007, ISSN 0967–0637

Pickart RS, Schulze LM, Moore GWK, Charette MA, Arrigo KR, van Dijken G, Danielson SL (2013b) Long-term trends of upwelling and impacts on primary productivity in the Alaskan Beaufort Sea. Deep-Sea Res 79:106–121

Proshutinsky A, Timmermans M-L, Ashik I, Beszczynska-Moeller A, Carmack E, Frolov I, Itoh M, Kikuchi T, Krishfield R, McLaughlin F, Rabe B, Schauer U, Shimada K, Sokolov V, Steele M, Toole J, Williams W, Woodgate R, Yamamoto-Kawai M, Zimmermann S (2011) The Arctic. In: Blunden J, Arndt DS, Baringer MO (eds) State of the climate in 2010. Bulletin of the American Meteorological Society 92(6), S143–S160

Schumacher JD, Reed RK (1992) Characteristics of currents over the continental slope of the eastern Bering Sea. J Geophys Res 97:9423–9433

Schumacher JD, Stabeno PJ (1994) Ubiquitous eddies of the eastern Bering Sea and their coincidence with concentrations of larval Pollock. Fish Ocean 3:182–190

Serreze M, Holland M, Stroeve J (2007) Perspective on the Arctic's shrinking sea ice cover. Science 315:1533–1536

Shimada K, Carmack EC, Hatakeyama K, Takizawa T (2001) Varieties of shallow temperature maximum waters in the western Canadian Basin of the Arctic Ocean. Geophys Res Lett 28:3441–3444

Shimada K, Kamoshida T, Itoh M, Nishino S, Carmack E, McLaughlin FA, Zimmermann S, Proshutinsky A (2006) Pacific Ocean inflow: influence on catastrophic reduction of sea ice cover in the Arctic Ocean. Geophys Res Lett 33:L08605. doi:10.1029/2005GL025624

Signorini SR, Munchow A, Haidvogel D (1997) Flow dynamics of a wide Arctic canyon. J Geophys Res 102:18661–18680

Spall MA (2007) Circulation and water mass transformation in a model of the Chukchi Sea. J Geophys Res 112:C05025. doi:10.1029/2005JC003364

Spall MA, Pickart RS, Fratantoni P, Plueddemann A (2008) Western Arctic Shelf-break eddies: formation and transport. J Phys Oceanogr 38:1644–1668. doi:10.1175/2007JPO3829.1

Stabeno PJ, van Meurs P (1999) Evidence of episodic onshelf flow in the southeast Bering Sea. J Geophys Res 104:29715–29720

Stabeno PJ, Schumacher JD, Salo SA, Hunt GL Jr, Flint M (1999) Physical environment around the Pribilof Islands. In: Loughlin TR, Ohtani K (eds) Dynamics of the Bering Sea, AK-SG-99-03. University of Alaska Sea Grant, Fairbanks

Steele M, Ermold W, Zhang J (2008) Arctic Ocean surface warming trends over the past 100 years. Geophys Res Lett. doi:10.1029/2007GL031561

Tamura T, Ohshima KI (2011) Mapping of sea ice production in the Arctic coastal polynyas. J Geophys Res. doi:10.1029/2010JC006586

Timmermans M, Toole J, Proshutinsky A, Krishfield R, Plueddemann A (2008) Eddies in the Canada Basin, Arctic Ocean, observed from ice-tethered profilers. J Phys Oceanogr 38:133–145

Toole JM, Timmermans ML, Perovich DK, Krishfield RA, Proshutinsky A, Richter-Menge JA (2010) Influences of the ocean surface mixed layer and thermohaline stratification on Arctic Sea ice in the central Canada Basin. J Geophys Res 115:C10018. doi:10.1029/2009JC005660

Toole JM, Krishfield RA, Timmermans M-L, Proshutinsky A (2011) The Ice-Tethered Profiler: Argo of the Arctic. Oceanography 24(3):126–135

von Appen W-J, Pickart RS (2011) Two configurations of the Western Arctic shelf-break current in summer. J Phys Oceanogr 42:329–351. doi:10.1175/JPO-D-11-026.1

Wang M, Overland JE, Stabeno P (2012) Future climate of the Bering and Chukchi Seas projected by global climate models. Deep-Sea Res Part II 65–70:46–57

Watanabe E (2011) Beaufort shelf break eddies and shelfbasin exchange of Pacific summer water in the western Arctic Ocean detected by satellite and modeling analyses. J Geophys Res 116:C08034. doi:10.1029/2010JC006259

Watanabe E, Hasumi H (2009) Pacific water transport in the Western Arctic Ocean simulated by an eddy-resolving coupled sea ice–ocean model. J Phys Ocean 39:2194–2211. doi:http://dx.doi.org/10.1175/2009JPO4010.1

Weingartner TJ, Cavalieri DJ, Aagaard K, Sasaki Y (1998) Circulation, dense water formation, and outflow on the northeast Chukchi shelf. J Geophys Res 103:7647–7662. doi:10.1029/98JC00374

Weingartner TJ, Aagaard K, Woodgate R, Danielson S, Sasaki Y, Cavalieri DJ (2005) Circulation on the north central Chukchi Sea shelf. Deep-Sea Res Part II 52:3150–3174

Wilkin JL, Chapman DC (1987) Scattering of continental shelf waves at a discontinuity in shelf width. J Phys Oceanogr 17:713–724

Williams WJ, Carmack EC (2012) Physical oceanography. In: Burn CR (ed) Herschel Island Qikiqtaruk – a natural and cultural history of Yukon's Arctic island. Wildlife Management Advisory Council (North Slope)

Williams WJ, Carmack EC, Shimada K, Melling H, Aagaard K, Macdonald RW, Ingram RG (2006) Joint effects of wind and ice motion in forcing upwelling in Mackenzie Trough, Beaufort Sea. Cont Shelf Res 26:2352–2366. doi:10.1016/j.csr.2006.06.012

Williams WJ, Carmack EC, Ingram RG (2007) Physical oceanography of polynyas. In: Smith WO Jr, Barber DG (eds) Polynyas: windows to the world, vol 74, Elsevier oceanography series. Elsevier, Amsterdam

Williams WJ, Melling H, Carmack EC, Ingram RG (2008) Kugmallit Valley as a conduit for cross-shelf exchange on the Mackenzie Shelf in the Beaufort Sea. J Geophys Res 113:C02007. doi:10.1029/2006JC003591

Winsor P, Chapman DC (2002) Distribution and interannual variability of dense water production from coastal polynyas on the Chukchi shelf. J Geophys Res. doi:10.1029/2004JC000984

Winsor P, Chapman DC (2004) Pathways of Pacific water across the Chukchi Sea: a numerical model study. J Geophys Res 109:C03002. doi:10.1029/2003JC001962

Woodgate RA, Aagaard K, Weingartner TJ (2005a) A year in the physical oceanography of the Chukchi Sea: moored measurements from autumn 1990–1991. Deep-Sea Res Part II 52:3116–3149

Woodgate R, Aagaard K, Weingartner T (2005b) Monthly temperature, salinity, and transport of the Bering Strait flow. Geophys Res Lett 32:L04601. doi:10.1029/2004GL021880

Woodgate R, Aagaard K, Weingartner T (2006) Interannual changes in the Bering Strait fluxes in volume, heat and freshwater between 1991 and 2004. Geophys Res Lett 33:L15609. doi:10.1029/2006GL026931

Woodgate RA, Weingartner T, Lindsay R (2010) The 2007 Bering Strait oceanic heat flux and anomalous Arctic sea-ice retreat. Geophys Res Lett 37:L01602. doi:http://dx.doi.org/10.1029/2009GL041621

Yang J (2006) The seasonal variability of the Arctic ocean Ekman transport and its role in the mixed layer heat and salt fluxes. J Climate 19:5366–5387

Chapter 7
On the Flow Through Bering Strait: A Synthesis of Model Results and Observations

Jaclyn Clement Kinney, Wieslaw Maslowski, Yevgeny Aksenov, Beverly de Cuevas, Jaromir Jakacki, An Nguyen, Robert Osinski, Michael Steele, Rebecca A. Woodgate, and Jinlun Zhang

Abstract Bering Strait is the only ocean connection between the Pacific and the Arctic. The flow through this narrow and shallow strait links the Pacific and Arctic oceans and impacts oceanic conditions downstream in the Chukchi Sea and the Western Arctic. We present a model synthesis of exchanges through Bering Strait at monthly to decadal time scales, including results from coupled ice-ocean models and observations. Significant quantities of heat and freshwater are delivered annually into the southern Chukchi Sea via Bering Strait. We quantify seasonal signals, along with interannual variability, over the course of 26 years of multiple model integrations. Volume transport and property fluxes are evaluated among several high-resolution model runs and compared with available moored observations. High-resolution models represent the bathymetry better, and may have a more realistic representation of the flow through the strait, although in terms of fluxes and mean properties, this is not always the case. We conclude that, (i) while some of the models used for Arctic studies achieve the correct order of magnitude for fluxes of volume, heat and freshwater, and have significant correlations with observational results, there is still a need for improvement and (ii) higher spatial

J. Clement Kinney (✉) • W. Maslowski
Department of Oceanography, Graduate School of Engineering and Applied Sciences, Naval Postgraduate School, Dyer Road, Bldg. SP339B, Monterey, CA 93943, USA
e-mail: jlclemen@nps.edu; maslowsk@nps.edu

Y. Aksenov • B. de Cuevas
Marine Systems Modelling, National Oceanography Centre, Southampton, UK

J. Jakacki • R. Osinski
Institute of Oceanology, Polish Academy of Sciences, Powstancow Warszawy 55, Sopot 81-712, Poland

A. Nguyen
Program in Atmospheres, Oceans, and Climate, Massachusetts Institute of Technology, Cambridge, MA, USA

M. Steele • R.A. Woodgate • J. Zhang
Applied Physics Laboratory, University of Washington, Seattle, WA, USA

J.M. Grebmeier and W. Maslowski (eds.), *The Pacific Arctic Region: Ecosystem Status and Trends in a Rapidly Changing Environment*, DOI 10.1007/978-94-017-8863-2_7,

resolution is needed to resolve features such as the Alaska Coastal Current (ACC). At the same time, additional measurements with better spatial coverage are needed to minimize uncertainties in observed estimates and to constrain models.

Keywords Bering Strait • Ocean modeling • Pacific water • Numerical modeling

7.1 Introduction

The Pacific Arctic Region spans the sub-Arctic Bering Sea northward through the Chukchi and Beaufort seas and the Arctic Ocean. The Bering Strait, a narrow passageway, connects the wide and shallow shelves of the Bering and Chukchi seas and is the only Pacific connection to the Arctic Ocean. The narrow (~85 km wide) and shallow (~50 m deep) strait provides low-salinity and high-nutrient Pacific Water to the Chukchi Sea and the Arctic Ocean. Many global and regional models face challenges with resolving oceanic exchanges across this narrow and shallow strait, mainly due to the requirement of high spatial resolution and the associated high computational cost to resolve it. In fact, many coarse-resolution models either have a closed Bering Strait or use a prescribed boundary condition. However, Goosse et al. (1997) demonstrated that there is a significant improvement in modeled ocean dynamics in a coarse resolution ($3° \times 3°$) model with an opened Bering Strait. They also found that opening Bering Strait produced a more realistically positioned sea ice edge in the Bering Sea, because warm water was allowed to advect further north onto the Bering-Chukchi shelf. Arctic freshwater budgets were also improved, with increased freshwater storage in the Greenland and Norwegian Seas.

Scientific access across Bering Strait has been restricted due to the political boundary between the United States and Russia. The Russian-US Convention line, dividing the Exclusive Economic Zones (EEZs) of the two countries, lies between two islands near the center of the strait: Ratmanova Island (part of Russia, also called Big Diomede in the U.S.) and Little Diomede Island (part of the U.S.). While U.S. research has maintained moorings in the Bering Strait almost continuously since 1990, only for limited portions of that time has U.S. access been granted to the western side of the strait.

The first goal of this work is to compare state-of-the-art output on the Bering Strait throughflow from several regional and global Arctic-focused models. We will analyze the volume and property fluxes over a long time series (up to 26 years depending on available results from individual models). In addition to interannual changes, we will also examine seasonal cycles in these parameters. The second related goal of this work is to compare model results with the available observational data. These data are from moored instruments placed near-bottom in three point locations in the vicinity of the strait (Fig. 7.1a). Both observations and models have their own limitations in Bering Strait. Numerical models are limited by relatively coarse resolution in the strait, errors in forcing and omitted processes

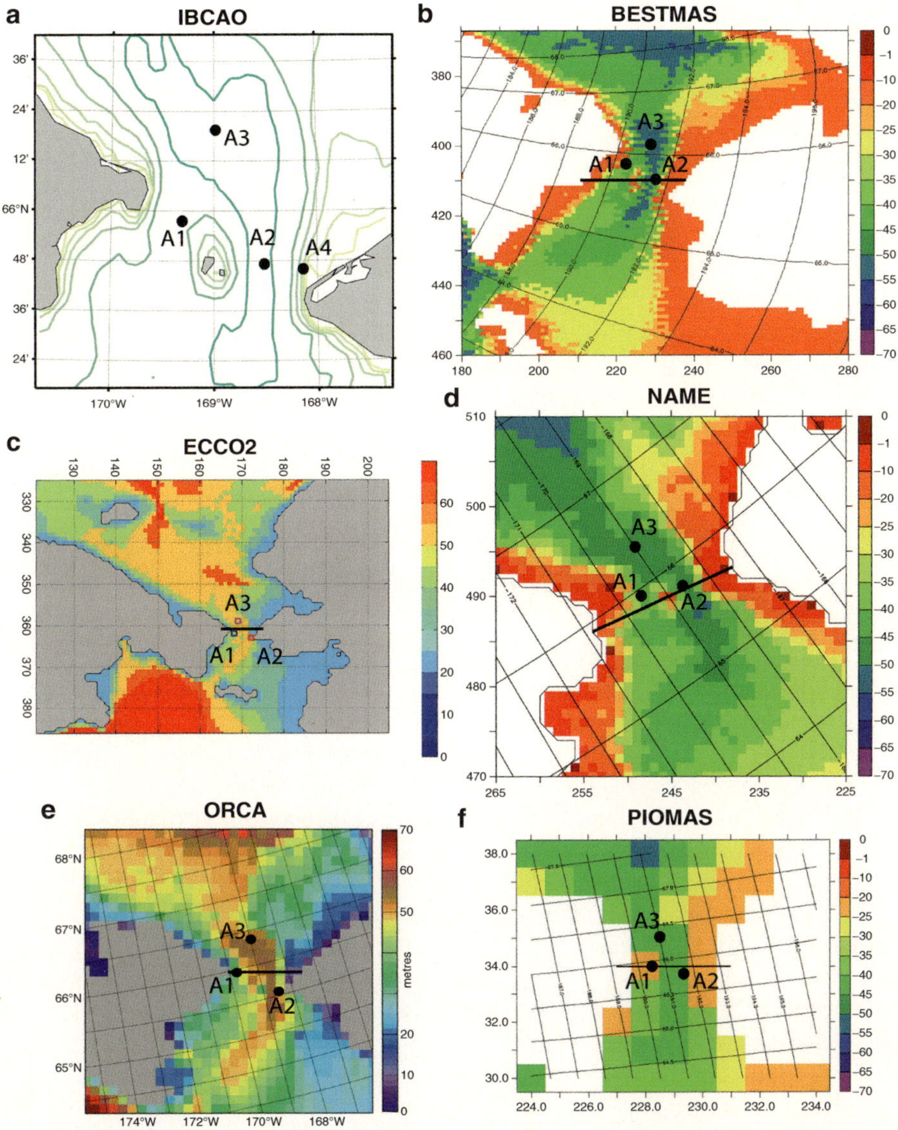

Fig. 7.1 Bathymetry (m) in the vicinity of the Bering Strait (**a**). Depth contours are every 10 m from the International Bathymetric Chart of the Arctic Ocean (Jakobsson et al. 2000). Model bathymetry (m) from (**b**) BESTMAS, (**c**) ECCO2, (**d**) NAME, (**e**) ORCA, and (**f**) PIOMAS. The approximate locations of the moored observations are indicated with *black circles*. The cross-sections across Bering Strait are shown as *black lines* in each model bathymetry figure

(e.g., tides), whereas observational results are limited by spatial coverage across the strait, and lack of upper layer measurements.

7.2 Model Descriptions

In this section we describe five global and regional sea ice-ocean coupled models employed to investigate Bering Strait inflow (Table 7.1). The models used in the study have various design features, including resolution, atmospheric forcing, restoring terms, coefficients, and parameterizations. Details of these features for each model are discussed below and shown in Tables 7.1 and 7.2. The goal here is to present results on the flow through Bering Strait from a variety of models and assess differences among them and observed data.

7.2.1 Bering Ecosystem Study Ice-Ocean Modeling and Assimilation System (BESTMAS)

BESTMAS (Zhang et al. 2010) is based on the coupled Parallel Ocean and sea Ice Model (POIM) of Zhang and Rothrock (2003). The sea ice model is the multicategory thickness and enthalpy distribution (TED) sea ice model (Zhang and Rothrock 2001; Hibler 1980). It employs a teardrop viscous-plastic rheology (Zhang and Rothrock 2005), a mechanical redistribution function for ice ridging (Thorndike et al. 1975; Hibler 1980), and a LSR (line successive relaxation) dynamics model to solve the ice momentum equation (Zhang and Hibler 1997). The TED ice model also includes a snow thickness distribution model following Flato and Hibler (1995). The ocean model is based on the Parallel Ocean Program (POP) developed at Los Alamos National Laboratory (Smith et al. 1992; Dukowicz and Smith 1994). Given that tidal energy accounts for 60–90 % of the total horizontal kinetic energy over the southeastern shelf region of the Bering Sea (Kinder and Schumacher 1981), tidal forcing arising from the eight primary constituents (M2, S2, N2, K2, K1, O1, P1, and Q1) (Gill 1982) is incorporated into the POP ocean model. The tidal forcing

Table 7.1 Basic information on the five models used in this study

Model	Global/regional	Atmospheric forcing	Resolution in Bering Strait	Data assimilation?
BESTMAS	Regional	NCEP/NCAR reanalysis	~4 km	No
ECCO2	Regional	Japanese 25-year reanalysis	~23 km	No
NAME	Regional	ECMWF reanalysis	~9 km	No
ORCA	Global	DRAKKAR Forcing Set (DFS 3.1) reanalysis	~13 km	No
PIOMAS	Regional	NCEP/NCAR reanalysis	~40 km	No

Table 7.2 Cross-sectional area across Bering Strait for the models and observations and friction coefficients for the models

Model/observed	Area (km^2)	Bottom friction coefficient	Lateral boundary condition	Lateral friction coefficient	Surface friction coefficient: Ice-Ocean	Surface friction coefficient: Air-Ocean
BESTMAS	3.24	Quadratic bottom drag: 1.225×10^{-3}	No-slip	Variable momentum harmonic horizontal mixing depending on variable grid size	5.5×10^{-3}	1.0×10^{-3}
ECCO2	4.50	Quadratic bottom drag: 2.1×10^{-3}	Free-slip	Modified Leith [Fox-Kemper and Menemenlis 2008]	5.4×10^{-3}	Large and Pond (1981, 1982)
NAME	2.37	Quadratic bottom drag: 1.225×10^{-3}	No-slip	Momentum biharmonic horizontal mixing: -1.25×10^{18}	5.5×10^{-3}	0.6×10^{-3}
Observed	2.60	N/A	N/A	N/A	N/A	N/A
ORCA	4.17	Quadratic bottom drag: 1.0×10^{-3}	Free-slip	Bi-harmonic (−1.5e+11 m4/s)	Quadratic, 5.0×10^{-3}	CORE bulk formulae, Large and Yeager (2004)
PIOMAS	2.38	Quadratic bottom drag: 1.225×10^{-3}	No-slip	Variable momentum harmonic horizontal mixing depending on variable grid size	5.5×10^{-3}	1.0×10^{-3}

consists of a tide generating potential with corrections due to both the earth tide and self-attraction and loading following Marchuk and Kagan (1989). The model domain of BESTMAS covers the northern hemisphere north of 39°N. The BESTMAS finite-difference grid is based on a generalized orthogonal curvilinear coordinate system with a horizontal dimension of 600 × 300 grid points. The "north pole" of the model grid is placed in Alaska. Thus, BESTMAS has its highest horizontal resolution along the Alaskan coast and in the Bering, Chukchi, and Beaufort seas, with an average of about 7 km for the whole Bering Sea and 10 km for the combined Chukchi and Beaufort seas. There are 26 grid cells across Bering Strait (Fig. 7.1b), which allows a good connection between the Bering Sea and the Arctic Ocean. The TED sea ice model has eight categories each for ice thickness, ice enthalpy, and snow depth. The centers of the 8 ice thickness categories are 0, 0.38, 1.30, 3.07, 5.97, 10.24, 16.02, and 23.41 m. The POP ocean model has 30 vertical levels of varying thicknesses to resolve surface layers and bottom topography. The first 13 levels are in the upper 100 m and the upper six levels are each 5 m thick. The model bathymetry is obtained by merging the IBCAO (International Bathymetric Chart of the Arctic Ocean) dataset and the ETOPO5 (Earth Topography Five Minute Gridded Elevation Data Set) dataset (see Holland 2000). BESTMAS is forced by daily NCEP/NCAR reanalysis (Kalnay et al. 1996) surface forcing fields. Model forcing also includes river runoff of freshwater in the Bering and Arctic seas. For the Bering Sea, monthly climatological runoffs of the Anadyr, Yukon, and Kuskokwim rivers are used (Zhang et al. 2010). For the Arctic Ocean, monthly climatological runoffs of the Pechora, Ob, Yenisei, Olenek, Yana, Indigirka, Kolyma, Mackenzie, Dvina, Lena, Khatanga, Taimyra, and Piasina rivers are from the Alfred Wegener Institute (Prange and Lohmann 2004). Although BESTMAS has a large model domain that includes the Arctic and the North Pacific, realistic lateral open boundary conditions are still necessary to create the right water masses and fluxes. The POP ocean model has been further modified to incorporate open boundary conditions so that BESTMAS is able to be one-way nested to a lower resolution but global POIM (Zhang 2005). Monthly mean open boundary conditions of ocean temperature, salinity, and sea surface height from the global POIM are imposed at the southern boundaries along 39°N. No data were assimilated in BESTMAS.

7.2.2 Estimating the Circulation and Climate of the Ocean, Phase II (ECCO2)

The ECCO2 regional Arctic Ocean solution uses a configuration of the Massachusetts Institute of Technology general circulation model (MITgcm; Marshall et al. 1997; Losch et al. 2010; Nguyen et al. 2011). The domain boundaries are at ~55° North in both the Atlantic and Pacific sectors. These boundaries coincide with grid cells in a global, cubed-sphere configuration of the MITgcm (Menemenlis et al. 2005).

The grid covering the Arctic domain is locally orthogonal with horizontal grid spacing of approximately 18 km. There are 50 vertical levels ranging in thickness from 10 m near the surface to approximately 450 m at a maximum model depth of 6,150 m. The model employs the rescaled vertical coordinate "z*" of Adcroft and Campin (2004) and the partial-cell formulation of Adcroft et al. (1997), which permits accurate representation of the bathymetry. Bathymetry is from the S2004 (W. Smith, 2010, personal communication) blend of the Smith and Sandwell (1997) and the General Bathymetric Charts of the Oceans (GEBCO) one arc-minute bathymetric grid. The non-linear equation of state of Jackett and McDougall (1995) is used. Vertical mixing follows Large et al. (1994). A 7th-order monotonicity-preserving advection scheme of Daru and Tenaud (2004) is employed and there is no explicit horizontal diffusivity. Horizontal viscosity follows Leith (1996) but is modified to sense the divergent flow (Fox-Kemper and Menemenlis 2008).

The ocean model is coupled to the MITgcm sea ice model described in Losch et al. (2010). Ice mechanics follow a viscous-plastic rheology and the ice momentum equations are solved numerically using the line-successive-over-relaxation (LSOR) solver of Zhang and Hibler (1997). Ice thermodynamics use a zero-heat-capacity formulation and seven thickness categories, equally distributed between zero to twice the mean ice thickness in each grid cell. Ice dynamics use a 2-category thickness with one for open water and one for ice. Salt rejected during ice formation is treated using a sub-grid-scale salt-plume parameterization described in Nguyen et al. (2009). The model includes prognostic variables for snow thickness and for sea ice salinity.

Initial and lateral boundary conditions come from the globally optimized ECCO2 solution (Menemenlis et al. 2008). Surface atmospheric forcing fields are from the Japanese 25-year reanalysis (JRA25; Onogi et al. 2007). Monthly mean river runoff is based on the Arctic Runoff Data Base (ARDB) as prepared by P. Winsor (2007, personal communication). No restoring is used.

Ocean and sea ice parameters, such as mixing and drag coefficients and albedos, were optimized regionally based on observations (Nguyen et al. 2011). The model results presented here are from a 1992 to 2008 forward model run using the optimized parameters and do not assimilate any data. The model bathymetry in the vicinity of Bering Strait and the location of the Bering Strait cross-section are shown in Fig. 7.1c. The mean horizontal grid spacing of the model across Bering Strait is 23 km.

7.2.3 *Naval Postgraduate School Arctic Modeling Effort (NAME)*

The NAME coupled sea-ice–ocean model (Maslowski et al. 2004) has a horizontal grid spacing of 1/12° (or ~9 km). In the vertical direction, there are 45 vertical depth layers ranging from 5 m near the surface to 300 m at depth, with eight levels in the upper 50 m. The high vertical resolution, especially in the upper water column, allows for more realistic representation of the shallow Arctic and sub-Arctic shelves. In addition, the horizontal grid permits calculation of flow through the narrow straits of the northern Bering Sea (Clement et al. 2005). The model domain is configured in a rotated spherical coordinate system to minimize changes in grid cell area. It contains the sub-Arctic North Pacific (including the Sea of Japan and the Sea of Okhotsk) and North Atlantic Oceans, the Arctic Ocean, the Canadian Arctic Archipelago (CAA) and the Nordic Seas (see Fig. 7.1a of Maslowski et al. 2004 for model domain). The region of interest, the Bering Sea, is therefore far away from the artificially closed lateral boundaries in the North Pacific at 30°N, greatly reducing any potential effects of boundary conditions. In an effort to balance the net flow of Pacific Ocean water into the Arctic Ocean, a U-shaped 500 m deep, 162 km (18 grid point) wide channel was created through North America connecting the Atlantic Ocean to the Pacific Ocean. A westward wind forcing of 1.75 dyne cm^{-2} is prescribed along the channel (see Maslowski et al. 2004 for further details). Flow through the Bering Strait and the channel is not prescribed. There are 15 grid cells across Bering Strait in this model (Fig. 7.1d). Model bathymetry is derived from two sources: ETOPO5 at 5 km resolution for the region south of 64°N and International Bathymetric Chart of the Arctic Ocean (IBCAO; Jakobsson et al. 2000) at 2.5 km resolution for the region north of 64°N.

The ocean model was initialized with climatological, 3-dimensional temperature and salinity fields (PHC; Steele et al. 2001) and integrated for 48 years in a spinup mode. During the spinup, daily averaged annual climatological atmospheric forcing derived from 1979 to 1993 reanalysis from the European Centre for Medium-Range Weather Forecasts (ECMWF) was used for 27 years. Next an additional run was performed using repeated 1979 ECMWF annual cycle for 6 years and then 1979–1981 interannual fields for the last 15 years of the 48-year spinup. This approach is especially important in establishing realistic ocean circulation representative of the time period at the beginning of the actual interannual integration. This final run with realistic daily averaged ECMWF interannual forcing starts in 1979 and continues through 2004. Results from this integration (26 years) are used for the analyses in this chapter. Daily climatological runoff from the Yukon River (and all other major Arctic rivers) is included in the model as a virtual freshwater flux at the river mouth. However, in the Gulf of Alaska the freshwater flux from runoff (Royer 1981) is introduced by restoring the surface ocean level (of 5 m) to climatological (Polar Science Center Hydrographic Climatology; PHC) monthly mean temperature and salinity values over a monthly time scale (as a correction term to the explicitly calculated fluxes between the ocean and underlying atmosphere or sea-ice). Additional details on the model including sea-ice and river runoff have been provided elsewhere (Maslowski et al. 2004).

7.2.4 Nucleus for European Modelling of the Ocean (NEMO) with ORCA Configuration

The ORCA025-N102 model configuration of the National Oceanography Centre Southampton is an "eddy-permitting" z-level global coupled sea ice-ocean model. ORCA025-N102 was developed within the Nucleus for European Modelling of the

Ocean (NEMO) framework for ocean climate research and operational oceanography (http://www.nemo-ocean.eu/; Madec 2008) as part of the DRAKKAR configurations (DRAKKAR Group 2007) and is largely based on the ORCA025-G70 configuration (e.g., Lique et al. 2009). ORCA025-N102 includes the ocean circulation model OPA9 (Madec et al. 1998) coupled to the Louvain-la-Neuve Ice Model sea ice model LIM2 (Fichefet and Morales Maqueda 1997). The ocean model is configured on a tri-polar Arakawa C-grid (Arakawa 1966) with the model poles at the geographical South Pole, in Siberia and in the Canadian Arctic Archipelago (CAA). The horizontal resolution is approximately 28 km at the equator, increasing to 6–12 km in zonal and ~3 km in meridional directions in the Arctic Ocean. The model resolves large eddies (~30–50 km), while "permitting" most of smaller eddies. ORCA025-N102 has a higher vertical resolution than the ORCA025-G70 configuration, utilizing 64 vertical levels with thicknesses ranging from approximately 6 m near the surface to 204 m at 6,000 m. The high vertical resolution in the upper ocean (8 levels in the upper 50 m and 13 levels in the upper 100 m) greatly improves the model representation of the shallow Arctic continental shelves, Bering and Chukchi Seas. There are eight model cells across Bering Strait (Fig. 7.1e). The fine model resolution in the both, horizontal and vertical, together with high resolution model bathymetry adapted from ETOPO2 and partial steps in the model bottom topography accurately approximates the steep seabed relief near the Arctic shelves, resulting in the more realistic along-shelf flow (e.g., Barnier et al. 2006; Penduff et al. 2007). The LIM2 sea ice model uses the Viscous-Plastic (VP) ice rheology (Hibler 1979) and the 3-layer Semtner (1976) thermodynamics updated with sub-grid scale sea ice thickness distribution (Fichefet and Morales Maqueda 1997) and sea ice thickness-dependent albedo (Payne 1972). To obtain more distinct sea ice edges, the model employs the positive-definite, second moments conserving advection scheme by Prather (1986). The sea ice model is coupled to the ocean model every five oceanic time steps through a non-linear quadratic drag law (Timmermann et al. 2005).

For the 1958–2001 simulations used in the present study, the ORCA025 model was driven by the DRAKKAR Forcing Set (DFS 3.1) atmospheric reanalysis (Brodeau et al. 2010). The reanalysis combines monthly precipitation, daily downward shortwave and longwave radiation from the CORE forcing data set (Large and Yeager 2004) and 6-hourly 10 m wind, 2 m air humidity and 2 m air temperature from ERA40 reanalysis. The turbulent exchanges between atmosphere and ocean and atmosphere and sea ice are computed during model integration using the bulk formulae from Large and Yeager (2004). Climatological monthly continental runoff (Dai and Trenberth 2002) is included as an additional freshwater source, applied along the coastline. Initial conditions for temperature and salinity are derived from a monthly climatology that merges the Levitus et al. (1998) World Ocean Atlas climatology with the PHC2.1 database (Steele et al. 2001) in high latitudes. To avoid salinity drift, the sea surface salinity is restored toward the monthly mean climatological values on the timescale of 180 days for the open ocean and 12 days under sea-ice.

7.2.5 *Pan-Arctic Ice-Ocean Modeling and Assimilation System (PIOMAS)*

PIOMAS is a variant of BESTMAS (see description above) with a coarser horizontal resolution (~40 km) and smaller model domain (north of 49°N; Zhang et al. 2008). However, it has 12 categories each for ice thickness, enthalpy, and snow depth (Zhang et al. 2008). The centers of the 12 ice thickness categories are 0, 0.26, 0.71, 1.46, 2.61, 4.23, 6.39, 9.10, 12.39, 16.24, 20.62, and 25.49 m. The model bathymetry in the vicinity of Bering Strait and the location of the Bering Strait cross-section are shown in Fig. 7.1f.

7.3 Bering Strait Observational Mooring Data

Year-round moorings have been deployed in the strait almost continuously since 1990 (see Woodgate et al. 2006, 2010; and http://psc.apl.washington.edu/BeringStrait.html), generally at 2–4 locations, as shown in Fig. 7.1a. Site A1 is in the western channel of the strait and thus in the Russian EEZ. Access was only granted to this site in the early 1990s (data available from 1990 to 1991; 1992 to 1993, 1993 to 1994) and since 2004. Site A2 is in the eastern portion of the strait (U.S. waters). A third site, A3, was established in 1990 at a site just north of the strait (and in the US EEZ), hypothesized to provide a useful average of the flow through both of the channels (Woodgate et al. 2005a, b, 2006, 2007). For some years (1992–1993, 1993–1994, 1994–1995) the A3 mooring was deployed ~120 nm further north, but these data are not considered here. Observations from A2 and A3 are available since autumn 1990, except for a few missing months, and for the deployment year autumn 1996–1997 when no moorings were deployed in the strait. A fourth mooring site A4, was established near the U.S. coast in 2001 to measure the Alaskan Coastal Current (Woodgate and Aagaard 2005; see discussion below). A high-resolution array was deployed in the strait starting in 2007; for more details see http://psc.apl.washington.edu/BeringStrait.html.

Since the region is ice-covered in winter, all mooring instrumentation has traditionally been kept near-bottom to avoid damage by ice keels. The moorings provide measurements of temperature, salinity and velocity approximately 10 m above bottom. High correlation (0.95; Woodgate et al. 2005b) in velocity is found between all sites in the strait region (Woodgate et al. 2005b) suggesting that extrapolation of velocity between mooring sites is reasonable. All available ADCP data (some moorings, and ship-based ADCP sections from the eastern channel) and newer mooring data, show strong coherence in the vertical (see e.g., Roach et al. 1995, where the first EOF at a central channel site explains 90 % of the variance), with some surface intensification of the flows, especially within the Alaskan Coastal Current. Thus, assuming the near-bottom flow correlates well with the total volume transport also seems reasonable (see Woodgate et al. 2005b for a discussion).

In terms of water properties, the near-bottom data do not capture the upper layer, which in the summer/autumn period of the year is likely 10–20 m thick, about 1–2 °C warmer and about 1 psu fresher than the lower layer (Woodgate and Aagaard 2005; Woodgate et al. 2010).

The flow through the Bering Strait is generally believed to be driven by some far field forcing (often described as the pressure head forcing) modulated by local wind effects (see Woodgate et al. 2005b for discussion and historic references). Woodgate et al. (2005b) suggest this large-scale forcing likely explains the high velocity correlation between sites. On the Alaskan Coast on the edge of the eastern channel there is seasonally a strong surface-intensified current. This is the Alaskan Coastal Current, which is present from midsummer until about the end of the year (Paquette and Bourke 1974; Ahlnäs and Garrison 1984; Woodgate and Aagaard 2005), and in summer CTD sections it is present as a ~10 km wide, 40 m deep warm, fresh current (Woodgate and Aagaard 2005). Much less is known about the Siberian Coastal Current (SCC), which is present sometimes on the Russian coast (Weingartner et al. 1999). Observations from the western side of Bering Strait indicate that the SCC can, at times, flow southward here under strong northerly winds. These events tend to occur during autumn and winter and appear to be short-lived (1–10 days; Weingartner et al. 1999). The SCC transport is estimated to be small (~0.1 Sv; Weingartner et al. 1999).

7.4 Results

Model representations of the geographical width across Bering Strait range from 90 to 160 km (Fig. 7.2), in part due to the choice made of the representative section in the model. ORCA and PIOMAS have widths most similar to reality (~85 km), while BESTMAS, NAME, and ECCO2 are wider than reality. The various horizontal resolutions from the five models and the different bathymetry schemes make the results appear disparate upon first glance. In fact the cross-sectional area of the strait varies from 2.4 to 4.5 km^2 for the models (Table 7.2). However, a closer look suggests agreement that horizontal shear is frequently present in the model results and that the highest speeds tend to be in the eastern channel. It is likely that this is at least in part an artifact due to the way the model sections cross the bathymetry, with the ends of the sections being either north or south of the strait proper. Certainly, observational results (e.g., Woodgate et al. 2005b) show no significant differences in the near-bottom velocity between the two channels away from the ACC. Vertical shear is present in some model results, particularly the NAME, ECCO2, and ORCA models. In NAME, the velocity tends to increase from surface to bottom, in contrast to other models. It appears that the velocity maxima are located deeper in the channels where frictional effects are less, as compared to the surface and nearby the coasts.

We also present northward velocity, temperature, and salinity sections for the summer period (Jul.–Sep.) from the five models (Figs. 7.3, 7.4, and 7.5). Strong vertical mixing is expected during the winter period within the northern Bering Sea

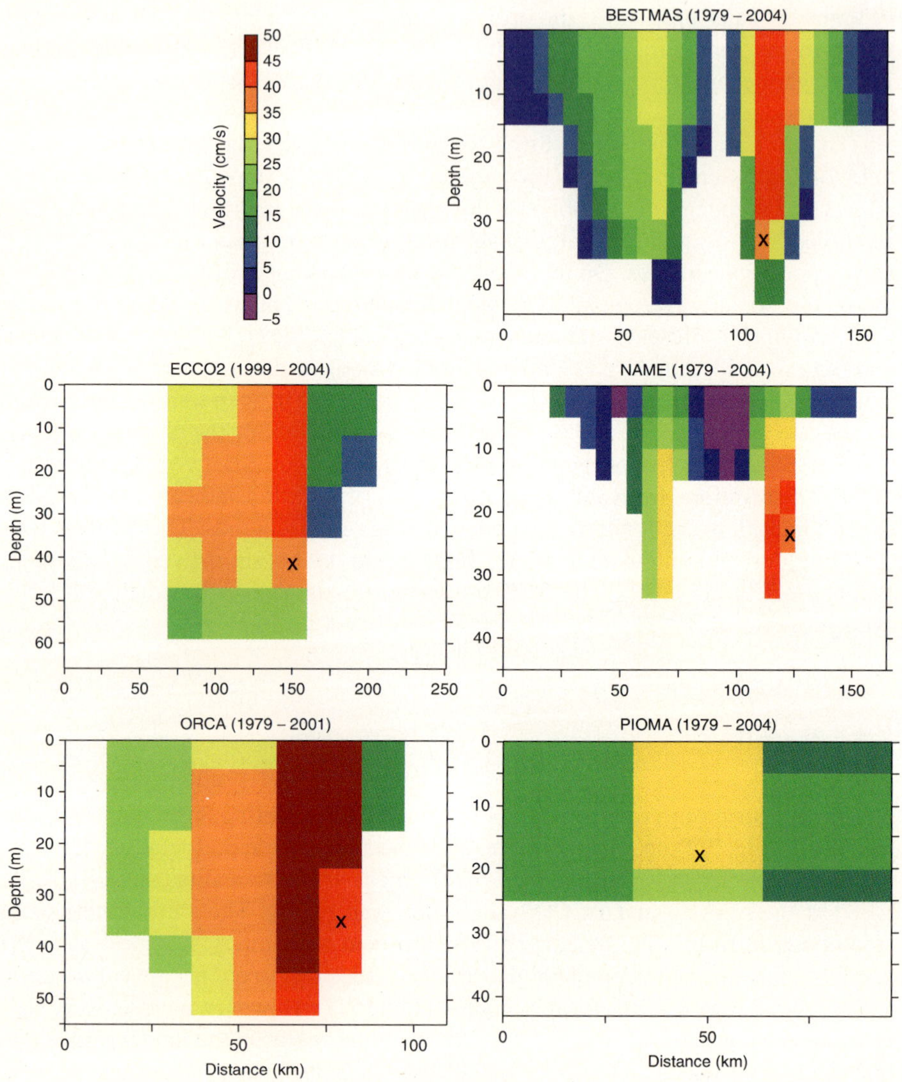

Fig. 7.2 Vertical section of the long-term mean northward velocity (cm/s) across Bering Strait from all models. Positive velocity is northward. A *black X marks* the approximate location of the A2 mooring within each model domain

(Clement et al. 2004; Woodgate and Aagaard 2005). Therefore, we present the mean summer results for comparison. The mean summer velocity sections from the models show slightly higher speeds than the long-term annual mean, especially in the upper water column (Fig. 7.3). There tends to be less vertical shear in the mean summer sections, as compared to the long-term mean sections (Fig. 7.2). Temperature sections (Fig. 7.4) indicate higher values in the upper water column near the Alaskan

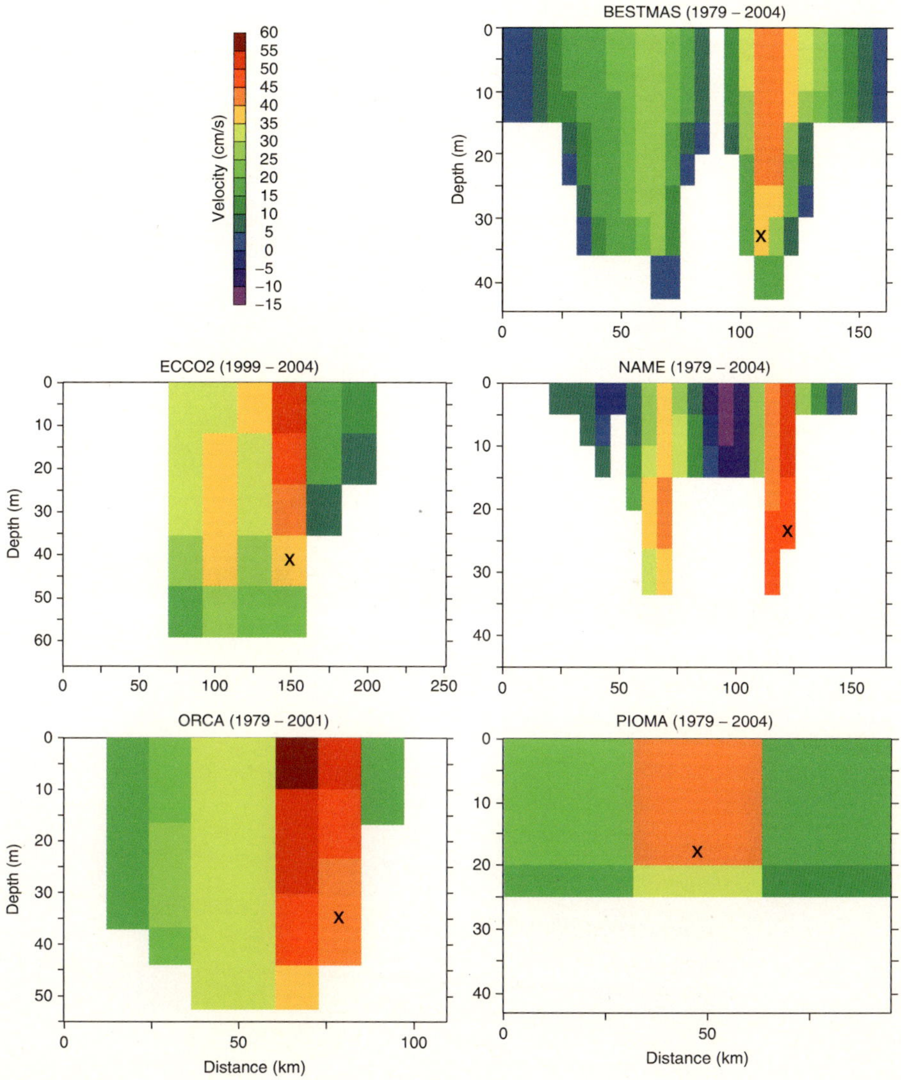

Fig. 7.3 Vertical section of the long-term summer (Jul.–Sep.) mean northward velocity (cm/s) across Bering Strait from all models. Positive velocity is northward. A *black X marks* the approximate location of the A2 mooring within each model domain

coast. BESTMAS and NAME have temperature values up to ~10 °C here. Similarly, for salinity (Fig. 7.5) an east–west gradient is present with the lower values found on the eastern side. Particularly, ECCO2, BESTMAS, and NAME show salinities less than 30 psu in this location. Multiple summertime CTD sections of temperature and salinity (http://psc.apl.washington.edu/BeringStrait.html) indicate elevated temperature and decreased salinity nearby the U.S. coast due to the presence of the

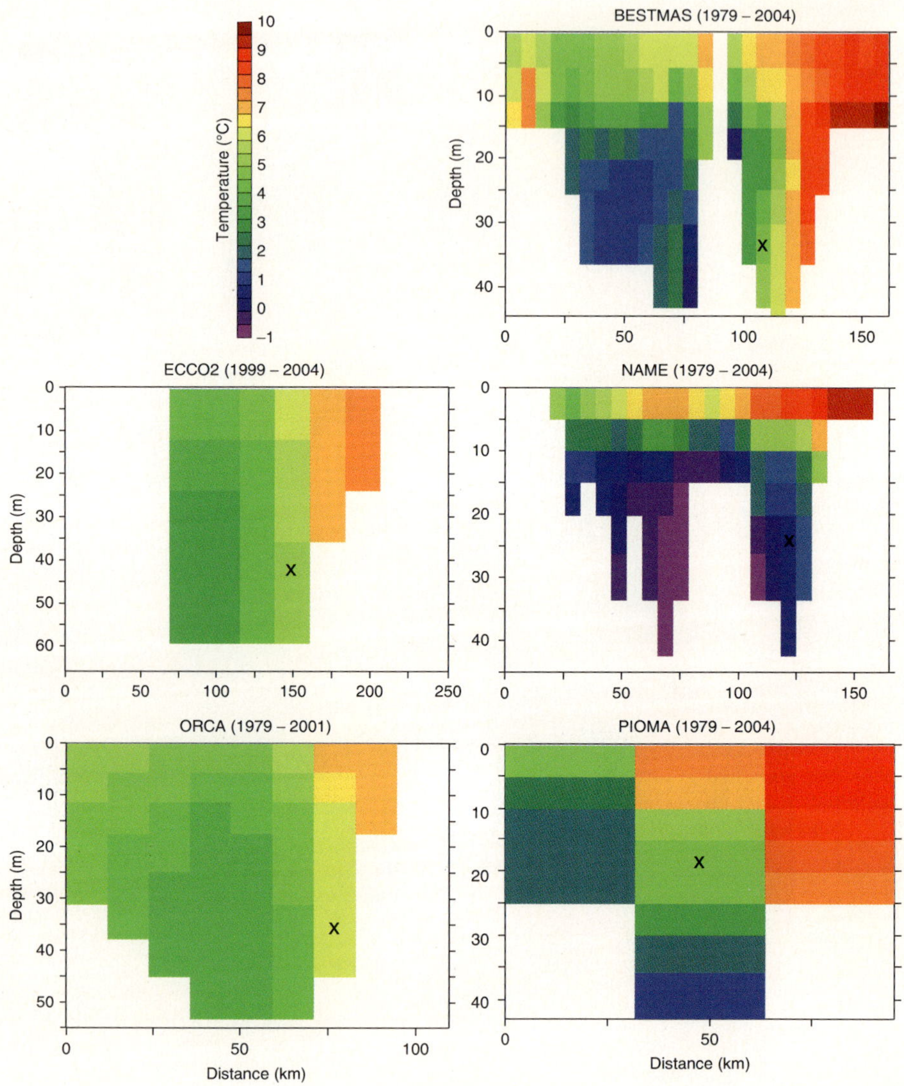

Fig. 7.4 Vertical section of the long-term summer (Jul.–Sep.) mean temperature (°C) across Bering Strait from all models. A *black X marks* the approximate location of the A2 mooring within each model domain

Alaska Coastal Current (ACC). The width of this current is on the order of ~10 km and, therefore, it is not properly resolved by the models due to spatial resolution limitations. However, model results do show the proper east–west gradients in temperature and salinity, as expected from observations (see e.g., Coachman et al. 1975).

To compare with long-term moored observations, we present monthly mean northward near-bottom velocity at sites A2 and A3 for models (color) and data

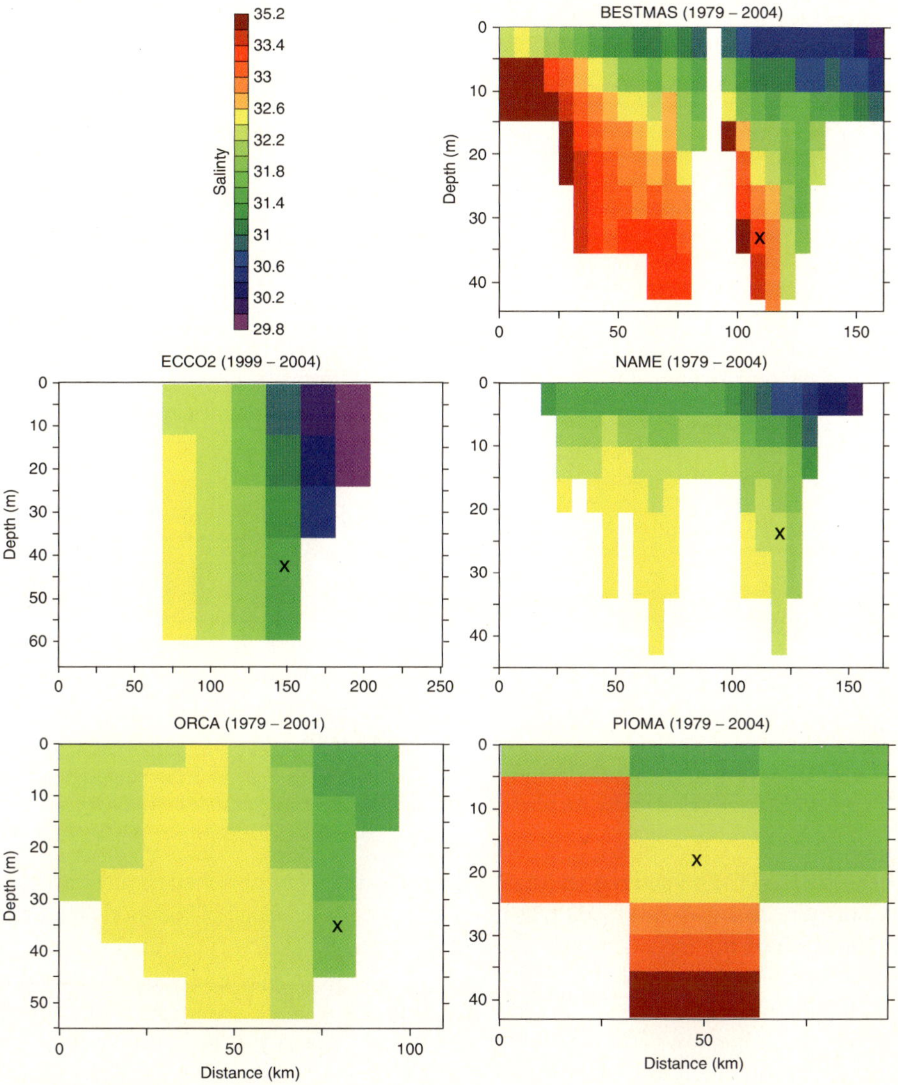

Fig. 7.5 Vertical section of the long-term summer (Jul.–Sep.) mean salinity (psu) across Bering Strait from all models. A *black X marks* the approximate location of the A2 mooring within each model domain

(black) for 1979–2004 in Fig. 7.6. For A2 (eastern channel, Fig. 7.1a), model velocities range from ~5 cm/s southward to over 80 cm/s northward. Predominantly, the flow is northward with the mean northward velocity ranging from 28.6 (+/−1.0) to 40.1 (+/−1.9) cm/s among models, over the time period when observations are available (Table 7.3). The range is 29.5 (+/−0.49) to 43.2 (+/−0.88) cm/s over the larger 1979–2004 time period (Table 7.4.) Two of the lower resolution models (ORCA and ECCO2) have the highest velocities, while the higher resolution

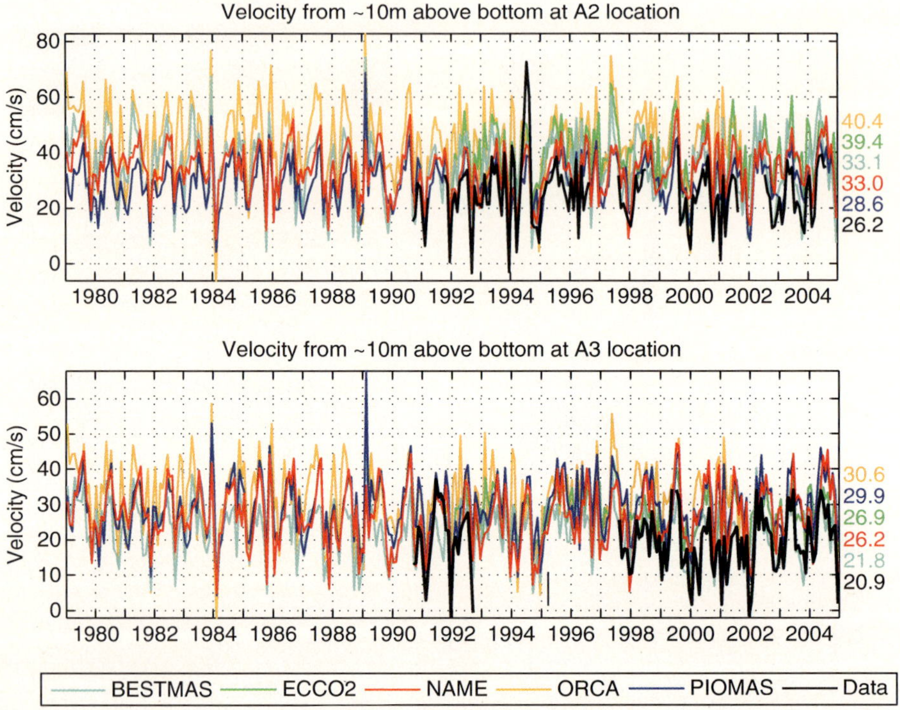

Fig. 7.6 Monthly mean velocity at ~10 above the bottom from the A2 mooring location (*upper*) and A3 mooring location (*lower*). Model results are shown in *color* and the observations are shown in *black*. Mean values for the time period when data are available are shown on the far right

models (BESTMAS and NAME) have lower velocities. The observed mean northward velocity is 26.2 (+/−2.8) cm/s, which matches the lower range of the modeled mean values. All of the models show a significant (at the 99 % level) correlation with the observed velocities at this location. The correlation coefficients range from 0.67 to 0.78 for the monthly means at A2 (Table 7.5).

Figure 7.3b shows the near-bottom northward velocity at the A3 location. The model spread of velocities is slightly narrower for A3, with the BESTMAS model having the lowest mean velocity (21.8 +/− 0.8 cm/s) and ORCA having the highest mean velocity (30.6 +/− 1.3 cm/s) over the same time period as observations. The observed mean northward velocity is 20.9 +/− 2.3 cm/s. The correlations between the models and the data at A3 are significant (at the 99 % level), with correlation coefficients ranging between 0.70 and 0.82 (Table 7.5).

It is important to recognize that a comparison between point measurements and model results is difficult. In the data, velocity is measured at a single point, while in models it is a grid-cell mean, which may range from a few to tens of kilometers in the horizontal and several meters in the vertical. In addition, the discrepancy

Table 7.3 Mean velocity, volume transport, near-bottom temperature, and near-bottom salinity from the models and from observations, for the time period when observations are available, as shown in Figs. 7.3 and 7.4

Model/data	Mean velocity ~10 m above bottom at the A2 location (cm/s)	Mean velocity ~10 m above bottom at the A3 location (cm/s)	Mean volume transport (Sv)	Mean temperature ~10 m above bottom at the A2 location (°C)	Mean temperature ~10 m above bottom at the A3 location (°C)	Mean salinity ~10 m above bottom at the A2 location (psu)	Mean salinity ~10 m above bottom at the A3 location (psu)
BESTMAS	**33.1** (1.3)	**21.8** (0.8)	**0.69** (0.03)	**0.79** (0.25)	**0.65** (0.24)	**33.16** (0.06)	**33.20** (0.06)
Data	**26.2** (2.8)	**20.9** (2.3)	**0.80** (0.20)[a]	**0.27** (0.3)	**−0.11** (0.2)	**32.26** (0.08)	**32.49** (0.06)
ECCO2	**39.4** (1.2)	**26.9** (0.7)	**1.06** (0.03)	**1.10** (0.27)	**0.62** (0.23)	**31.72** (0.06)	**32.29** (0.04)
NAME	**33.0** (1.0)	**26.2** (0.7)	**0.67** (0.03)	**−0.96** (0.09)	**−1.26** (0.05)	**32.45** (0.04)	**32.61** (0.04)
ORCA	**40.4** (1.9)	**30.6** (1.3)	**1.29** (0.06)	**0.96** (0.25)	**0.79** (0.24)	**32.30** (0.05)	**32.47** (0.03)
PIOMAS	**28.6** (1.0)	**29.9** (0.9)	**0.81** (0.03)	**0.34** (0.21)	**0.60** (0.23)	**33.09** (0.05)	**32.80** (0.05)

However, the values from ECCO2 are for 1992–2004 only and the values from ORCA are for 1990–2001 only. Error estimates are shown in parenthesis. All model errors are calculated as the standard error of the mean (sample standard deviation divided by the square root of the sample size)

[a]The uncertainty for the data estimate is ~25 % (Woodgate et al. 2005a, b)

Table 7.4 Long-term mean velocity, volume transport, near-bottom temperature, and near-bottom salinity from the models, for the time periods shown in Figs. 7.3 and 7.4

Model/data	Mean velocity ~10 m above bottom at the A2 location (cm/s)	Mean velocity ~10 m above bottom at the A3 location (cm/s)	Mean volume transport (Sv)	Mean temperature ~10 m above bottom at the A2 location (°C)	Mean temperature ~10 m above bottom at the A3 location (°C)	Mean salinity ~10 m above bottom at the A2 location (psu)	Mean salinity ~10 m above bottom at the A3 location (psu)
BESTMAS	**34.0** (0.69)	**22.8** (0.45)	**0.72** (0.02)	**0.76** (0.17)	**0.71** (0.16)	**33.18** (0.05)	**33.17** (0.04)
ECCO2	**39.9** (0.78)	**27.4** (0.52)	**1.07** (0.02)	**1.08** (0.25)	**0.63** (0.20)	**31.72** (0.05)	**32.25** (0.03)
NAME	**34.1** (0.52)	**27.0** (0.65)	**0.65** (0.01)	**−1.02** (0.06)	**−1.27** (0.03)	**32.46** (0.02)	**32.58** (0.02)
ORCA	**43.2** (0.88)	**31.6** (0.65)	**1.33** (0.03)	**0.89** (0.17)	**0.69** (0.15)	**32.33** (0.02)	**32.47** (0.02)
PIOMAS	**29.5** (0.49)	**29.2** (0.49)	**0.79** (0.02)	**0.26** (0.14)	**0.53** (0.15)	**33.13** (0.03)	**32.78** (0.03)

Error estimates are shown in parenthesis

Table 7.5 Correlation coefficients between models and the observations of velocity, temperature, and salinity at A2 and A3 locations

	Velocity		Temperature		Salinity	
Model	A2	A3	A2	A3	A2	A3
BESTMAS	0.78	0.71	0.78	0.70	0.67	0.53
ECCO2	0.67	0.72	0.77	0.76	0.60	0.48
NAME	0.69	0.75	0.73	0.86	0.70	0.57
ORCA	0.68	0.70	0.79	0.76	0.60	0.39
PIOMAS	0.70	0.82	0.88	0.79	0.66	0.59

All correlations are significant at the 95 % level

Table 7.6 Depth information (m) for the models and the observations at the A2 and A3 mooring locations

Location	Model/data	Water column depth (m)	Mid-depth of model grid cell or depth of observation ~10 m above bottom (m)
A2	Data	53	44
	BESTMAS	51	39.5
	ECCO2	50	35
	NAME	53	37.7
	ORCA	57.9	35.5
	PIOMAS	43	33
A3	Data	56	47
	BESTMAS	51	39.5
	ECCO2	50	35
	NAME	53	37.7
	ORCA	57.9	35.5
	PIOMAS	43	33

between the real and model bathymetry introduces a difference in bathymetric gradients, displacing model currents from their "true" geographical positions. Choice of model section location is also very important, for example, there will be obvious discrepancies between an observational section taken across the narrowest point of the strait and a model section crossing shallow regions to the north or south of the strait. Table 7.6 and Fig. 7.2 illustrate these points. Table 7.6 shows the depth at the moorings A2 and A3 and model depth at the co-located virtual moorings; the difference between real and model bathymetry is clear. Moreover, in the models, velocity can vary significantly between the adjacent model grid cells (Fig. 7.2), although this is not seen in observations outside the ACC. Thus the results of a model-observations comparison would depend upon the exact geographical position of model virtual moorings. Finally, the stochastic nature of the oceanic turbulence cannot be simulated by the models used in this study. Therefore, it is likely more informative to evaluate model results using integrated fluxes, as discussed below.

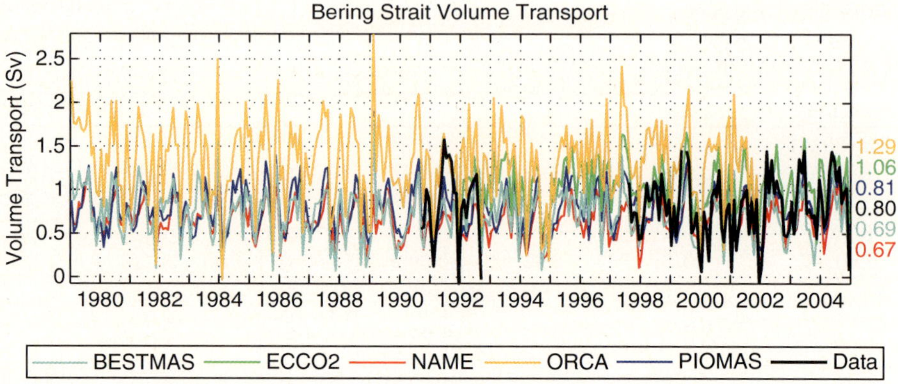

Fig. 7.7 Monthly mean volume transport from the models and observations. The observations are based on the near-bottom velocity at the A3 mooring location multiplied by a cross-sectional area of 4.25 km^2, as per Woodgate et al. (2010). Mean values for the time period when data are available are shown on the far right

Monthly mean Bering Strait volume transport from the models and observations is shown in Fig. 7.7. The observations are based on the near-bottom velocity at the A3 mooring location multiplied by a cross-sectional area of 4.25 km^2, as per Woodgate et al. (2010). Model means range from 0.67 (+/–0.03) to 1.29 (+/–0.06) Sv (Table 7.3) over the time period when observations are available. The volume transport is highest for the ORCA and ECCO2 models and is lowest for the PIOMAS, BESTMAS and NAME models. The observed estimate of the long-term mean (1991–2004) volume transport through Bering Strait is 0.8 +/– 0.2 Sv (Woodgate et al. 2005a). This estimate is based on observations at the A3 mooring location, although numbers do not differ significantly if using observations from the other mooring sites.

Near-bottom monthly mean temperatures at the A2 and A3 mooring locations are shown in Fig. 7.8. (Temperature at A1 is not shown because there are too few data available at this time.) Temperatures tend to be warmer at the southern A2 location, with model means ranging between –0.96 (+/–0.09) and 1.1 (+/–0.27) °C. The mean observed near-bottom temperature for the same location is 0.27 (+/–0.3). ORCA, ECCO2, and BESTMAS models tend to overestimate the temperature by 0.5–0.8 °C in the mean, while NAME underestimates the temperature by 1.2 °C in the mean. We speculate that the colder temperatures for the NAME model may be related to excessive ice production, especially in polynya regions of the northern Bering Sea. Surprisingly, the PIOMAS temperatures are closest to the observed, despite the fact that it is the lowest resolution model in this study and only has 3 grid points across the strait (Fig. 7.2). Temperatures at the A3 location are, again, underestimated in the NAME model and overestimated in ORCA, ECCO2, BESTMAS and also in PIOMAS. While the magnitude of the model-data differences may be up to ~1 °C in the mean, the models' results are significantly correlated (at the 99 % confidence level) with

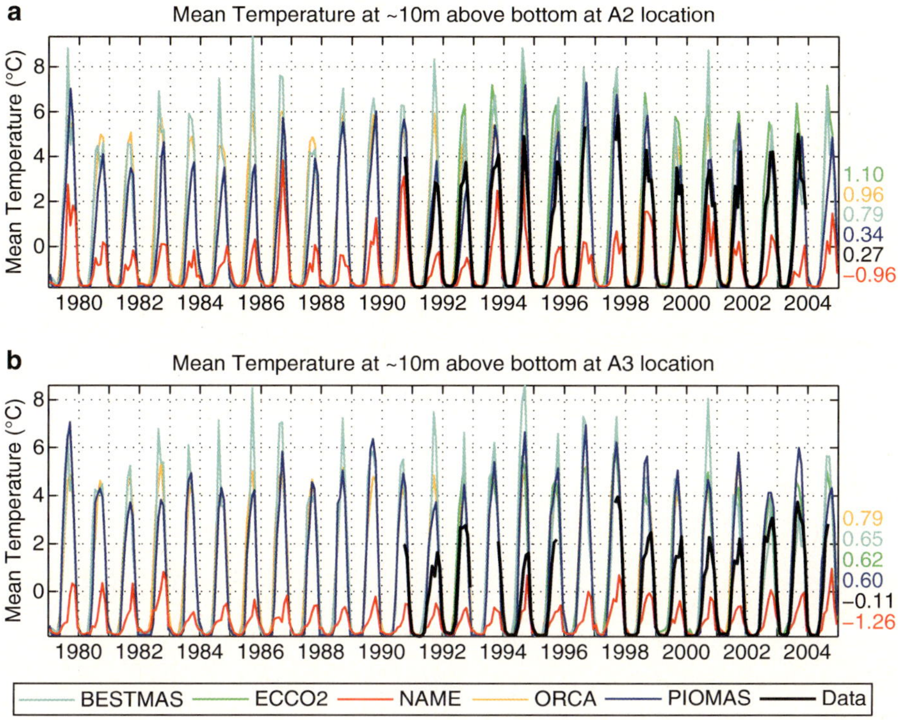

Fig. 7.8 Monthly mean near-bottom temperature (°C) at the (**a**) A2 and (**b**) A3 mooring locations. Model results are shown in *various colors* and observations are shown in *black*. Mean values for the time period when data are available are shown on the far right

the observations. The correlation coefficients range between 0.73 and 0.88 at A2 and between 0.70 and 0.86 at A3 (Table 7.5). There is no trend, either observed or modeled, in the time series shown here. There is, however, a strong seasonal cycle present, which enhances the correlations. The seasonal cycle has been identified by many authors (e.g. Fedorova and Yankinam 1964; Coachman et al. 1975; Roach et al. 1995 and references therein) and was later quantified into a modern climatology by Woodgate et al. (2005a). This seasonal cycle will be discussed below.

A similar analysis was performed for salinity at the A2 and A3 mooring locations (Fig. 7.9). The mean modeled salinity ranges between 31.7 (+/−0.06) and 33.2 (+/−0.06) psu at A2 and between 32.2 (+/−0.04) and 33.2 (+/−0.06) psu at A3. The mean observed salinities are 32.3 (+/−0.08) at A2 and 32.5 (+/−0.06) at A3. The BESTMAS and PIOMAS models tend to overestimate the salinity, by up to 0.9 psu above the observed mean value, whereas the NAME, ECCO2, and ORCA models have values close to the observed. All of the models' results are significantly correlated (at the 99 % confidence level) with the observations of salinity at A2 and

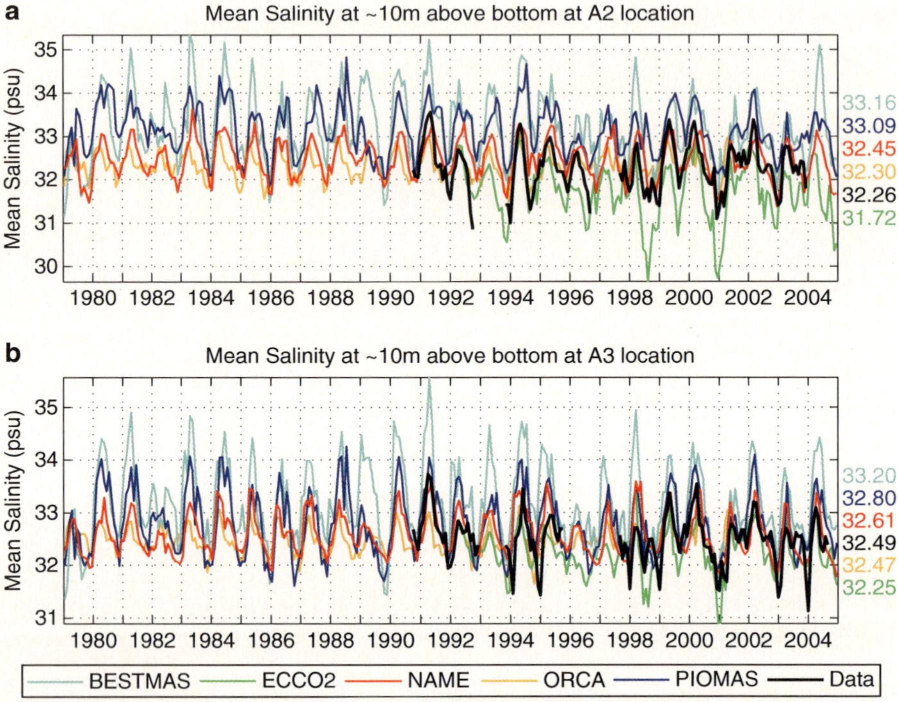

Fig. 7.9 Monthly mean near-bottom salinity at the (**a**) A2 and (**b**) A3 mooring locations. Model results are shown in *various colors* and observations are shown in *black*. Mean values for the time period when data are available are shown on the far right

A3. The correlation coefficients range between 0.60 and 0.70 at A2 and between 0.39 and 0.59 at A3 (Table 7.5). The correlations are not as high for salinity as they are for temperature, especially at the A3 location. Again, a seasonal cycle of salinity is apparent in the time series (also see Woodgate et al. 2005a), however it is not as strong as the seasonal cycle of temperature.

Annual mean volume transport from models and observations is shown in Fig. 7.10a. Observed volume transport (not including the ACC) ranges from 0.6 to 1 Sv (+/−0.2 Sv, Woodgate et al. 2006), which is most similar to the estimates from the BESTMAS, NAME, and PIOMAS models. The ACC likely adds around 0.1 Sv to the estimates (Woodgate and Aagaard 2005), thus the true flux is likely slightly higher than shown in Fig. 7.10a, and closer to the ECCO2 values.

Heat fluxes through Bering Strait and through the Chukchi shelf appear to influence the distribution and thickness of sea ice (Coachman et al. 1975; Shimada et al. 2006; Woodgate et al. 2010). Previously published observations of heat flux (e.g., Woodgate et al. 2010) use a reference temperature of −1.9 °C. Therefore, for the model calculations, we used the same value for a reference temperature. However, we note that the PIOMAS and BESTMAS models use −1.8 °C as the freezing temperature for an ease in conserving heat in the models. Oceanic heat flux through

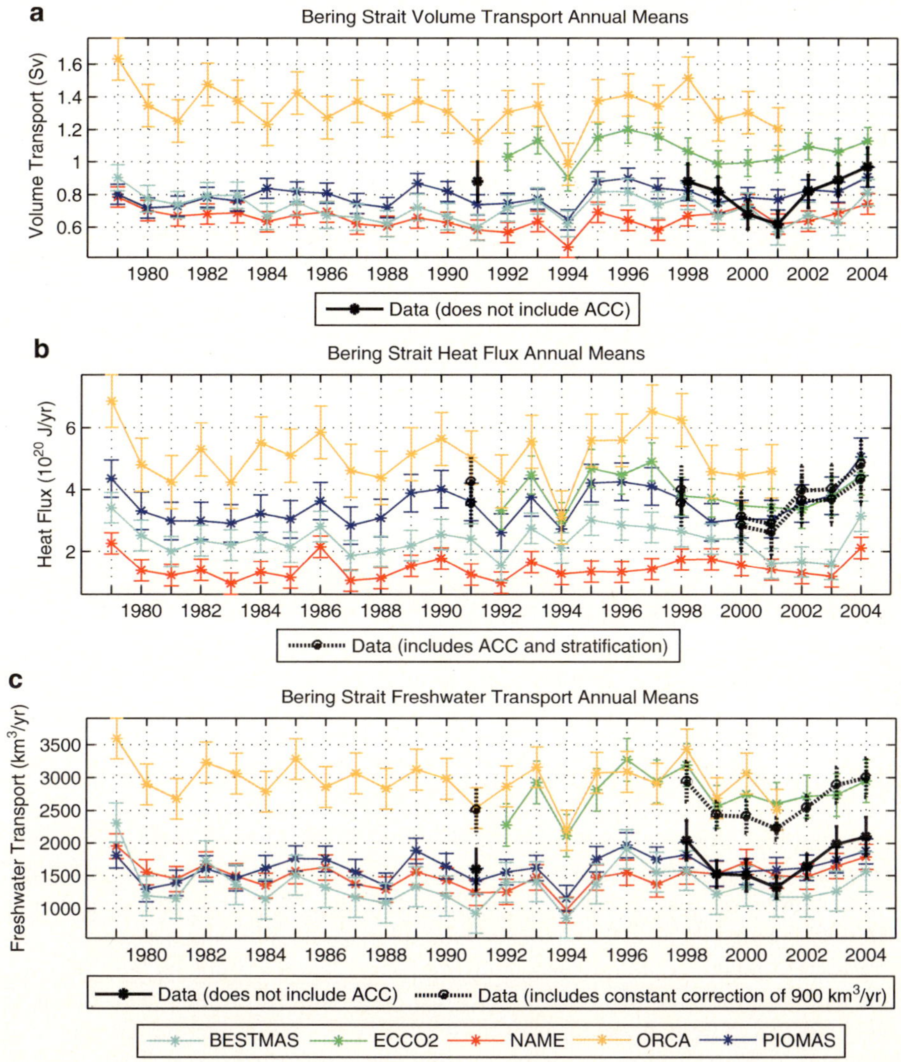

Fig. 7.10 Annual mean (**a**) volume transport, (**b**) heat, and (**c**) freshwater fluxes. Heat flux is referenced to −1.9 °C for the models, in order to compare with cited observations in the text. Freshwater is referenced to 34.8 psu. Observed volume transport values (**a**) do not include the ACC and stratification, which likely add ~0.1 Sv (see Woodgate et al. 2006). The observed heat flux values (**b**) include an estimate for the ACC using SST for a 10 m surface layer (*lower bound*) and a 20 m surface layer (*upper bound*). Observed heat flux values are described further in Woodgate et al. (2010). The *dashed black line* (**c**) represents the observed freshwater flux with an estimated ACC and stratification correction of an additional 900 km^3/year (Woodgate et al. 2006)

Bering Strait in the models was calculated as the vertical and horizontal integral of: the heat (heat capacity multiplied by the difference between the temperature and the reference temperature) multiplied by velocity normal to the cross-section on a monthly mean time scale.

The annual mean oceanic heat flux time series for the models and observations (as per Woodgate et al. 2010) are shown in Fig. 7.10b. In the models, peaks in the heat flux occurred during several years (e.g., 1979, 1986, 1993, and 1997) and consistently showed up in results from all five models. However, data coverage is not sufficient to confirm these peaks in the real world. A peak in 2004 is noted in observations (see Woodgate et al. 2010) and is apparent in all of the models, except ORCA, which does not have results for that time period. ECCO2 is also able to simulate a recent increase in heat flux in 2007 (not shown), similar to the observations (Woodgate et al. 2010).

The long-term model mean heat flux ranged between 1.5×10^{20} J/year and 5.1×10^{20} J/year. This is, admittedly, a wide range of values. ORCA and ECCO2 have much higher values than BESTMAS and NAME. Observations of the annual heat flux based on near-bottom measurements, a correction for the ACC, and SST from satellite data were published in Woodgate et al. (2010). The observed range of heat flux estimates is ~$2.8–4.5 \times 10^{20}$ J/year with estimated uncertainty of 0.8×10^{20} J/year, based on years 1991, 1998, 2000–2006. However, the 2007 heat flux was estimated at $5–6 \times 10^{20}$ J/year.

Freshwater flux from the Bering Sea into the Chukchi Sea is an important factor affecting stratification and the maintenance of the Arctic Ocean halocline (e.g., Aagaard et al. 1985a). As discussed in Aagaard et al. (2006), the salinity field in Bering Strait is influenced by a number of processes primarily within the Bering Sea, including inflow from the Gulf of Alaska, on-shelf transport from the deep basin, precipitation minus evaporation, river runoff, and formation/degradation of sea ice. The combined net effect of these processes determine, in large part, the downstream salinity (and to a lesser extent freshwater fluxes) found in the strait. For the calculation of freshwater fluxes, a reference salinity of 34.8 psu was used because this value is considered to be the mean salinity of the Arctic Ocean and has been used in most other Arctic studies (based on original work by Aagaard and Carmack 1989). Integrated annual mean oceanic freshwater fluxes were calculated on a monthly mean timescale (see Eq. 1 in Melling 2000) from each of the models and are shown in Fig. 7.10c. An observationally-based lower bound of annual mean freshwater fluxes is also shown, however these values do not include the ACC or stratification and thus likely underestimate the freshwater flux by about 800–1,000 km^3/year (Woodgate et al. 2006). With this correction, the observed freshwater annual means are similar to results from the ECCO2 and ORCA models, with the other models appearing to underestimate the total freshwater flux. No long-term trend is apparent in either the heat or freshwater flux for this time period, however a gradual increase in freshwater during the early 2000s has occurred in the model results, ending with a peak in freshwater flux in 2004, similar to observations (also see Woodgate et al. 2006, 2010).

It is important to note that both the models and the data have limitations with respect to calculations of heat and freshwater fluxes. The models used here are too coarse to represent the narrow (~10 km) ACC, which is estimated to carry 25 % of the freshwater flux and 20 % of the heat flux (Woodgate et al. 2006) through the

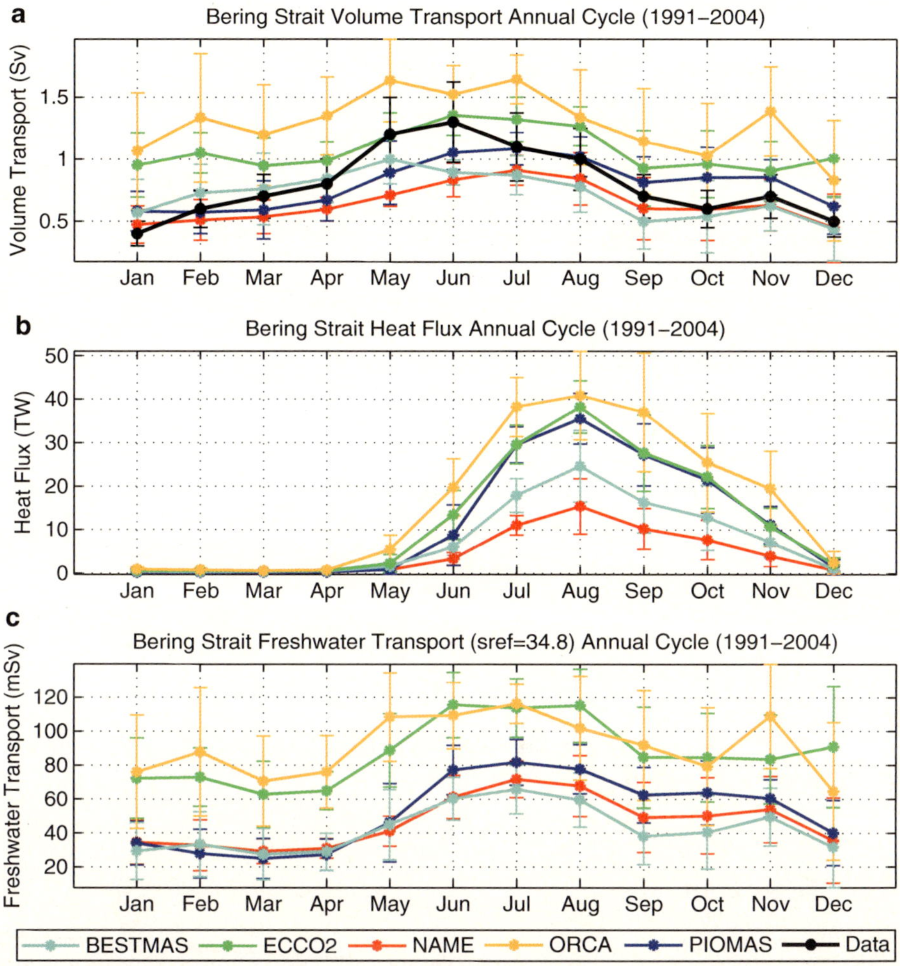

Fig. 7.11 Seasonal cycles of (**a**) volume transport, (**b**) heat flux, and (**c**) freshwater transport. The seasonal cycles are averaged over 1991–2004, except for ORCA (1991–2001) and ECCO2 (1992–2004). Heat flux is referenced to –1.9 °C for the models, in order to compare with cited observations in the text. The freshwater transport is referenced to 34.8 psu

strait. The historic near-bottom data used here does not measure the ACC, which is a surface/coastal feature. Thus, on-going research is using extra moorings, hydrographic data and upper water column sensors to estimate stratification (see e.g., http://psc.apl.washington.edu/BeringStrait.html).

Arctic shelf seas have a strong seasonal cycle of temperature and salinity; some areas may also exhibit strong seasonal changes in the oceanic circulation. The Bering Strait region is no different in this respect. Observations have shown stronger northward flows in summer (e.g., Coachman and Aagaard 1988; Roach et al. 1995; Woodgate et al. 2005a, b). According to the model results, volume transport peaks in summer (May–July) and is lowest in winter (December–March; Fig. 7.11). This agrees reasonably with observational results (peaking in May/Jun, minimum in Dec–Feb; although variability is high; Woodgate et al. 2005a). In general, the data have a larger seasonal cycle, with a range of 0.4–1.3 Sv (errors order 25 %; Woodgate et al. 2005a, b). PIOMAS, BESTMAS, ECCO2, and NAME models have similar seasonal cycles to the data, however they are not as strong.

As shown by Woodgate et al. (2010), the heat flux seasonal cycle is also very strong. Observational results (see Fig. 3 of Woodgate et al. 2010) suggest strong interannual variability in the timing of the summer peak, although the computation presented there does not include the seasonality of the ACC. In the models (Fig. 7.11), heat flux peaks in summer and is near-zero in winter. However, the models do not agree on the magnitude of the summertime peak, which ranges between 15 (+/–6.4) to over 40 (+/–14) TW. The heat flux is near zero for December–April (when water temperatures are around freezing). The models with the highest resolutions (BESTMAS and NAME) show lower peaks in the summertime heat flux [15 (+/–6.4) and 22.5 (+/–7.9) TW], while the lower resolution models have higher heat fluxes.

Seasonal cycles of freshwater flux through Bering Strait are similar for PIOMAS, BESTMAS, and NAME, with peaks in the summer (June–August) and lowest in winter (December–April). Again, interannual variability makes these peaks less certain. The freshwater flux maxima for these models are between 65 (+/–14.3) and 80 (+/–13.4) mSv in July. Seasonal cycles for ECCO2 and ORCA have somewhat similar shapes, however they transport more freshwater (up to 115 (+/–11.7) mSv in summer and more than 60 (+/–40) mSv in winter for ORCA) to the north.

7.5 Discussion

Model volume transports ranged from 0.67 (+/–0.03) to 1.29 (+/–0.06) Sv in the mean, compared to observational estimates of 0.8 +/–0.2 Sv; the observations may still underestimate the ACC contribution. Thus, most of the models are in agreement with the observational estimate to within errors. ORCA and ECCO2 showed the highest volume transports, while NAME and BESTMAS showed the lowest transports. Oddly, higher resolution models seem to give lower transport estimates; we do not fully understand why this is. Note that of the models, ORCA and ECCO2 also have the largest cross sectional area of the strait. The cross-sections in each model were chosen to approximate the locations of moored observations as closely as possible (Fig. 7.1), however different cross-sectional areas would arise from choosing a slightly different position of the section. The models are using both lateral and vertical friction parameterization to represent the flow next to the boundary (bottom/surface or lateral; see Table 7.2). Some uncertainty of model estimates of the volume transport throughout the strait might be related to the estimation of the frictional layers, subject to the parameterization used. Penduff et al. (2007) have demonstrated that enstrophy-conserving momentum advection schemes produce a spurious numerical sidewall friction, leading to a weaker topographic alignment of the mean flow and weaker barotropic transports. They reported a ~10 % reduction in Bering Strait transports in simulations with the enstrophy conserving advection, compared to the runs with the energy-enstrophy conserving scheme, characterized by low numerical friction. The effect of spurious friction on transports is similar to the one of explicit lateral friction. Both spurious sidewall and explicit non-slip lateral friction could explain lower transports in BESTMAS and NAME models compared to ECCO2 and ORCA, as the last two models feature free-slip lateral boundary conditions. Besides, ORCA utilizes energy-enstrophy conserving advection, which may result in higher transport than in ECCO2 (Tables 7.3 and 7.4). However, this cannot explain a higher transport in the PIOMAS model compared to BESTMAS, since these two models share the same configuration, except for the resolution and different number of sea ice categories (12 and 8 respectively). Thus, another possibility is that different transports reflect different large-scale forcings of the flow.

Panteleev et al. (2010) applied an inverse model (with ten grid points across the strait) to reconstruct the flow using available data for 1990–1991 and recently calculated the transport through Bering Strait as 0.57 Sv (no stated uncertainty). The data used to reconstruct the circulation were from 12 moorings that were deployed in the Bering Strait and Chukchi Sea from September 1990 to October 1991(Woodgate et al. 2005b). This estimate from Panteleev et al. (2010) tends to agree with the estimates from the BESTMAS and NAME models. In fact, the mean volume transports during the same time period (Sept. 1990–Oct. 1991) were 0.62 (+/–0.03) and 0.59 (+/–0.03) Sv for the BESTMAS and NAME models, respectively.

The model sections presented here show significant vertical and horizontal velocity shear across the strait. This is in contrast to observational results, which show strong coherence of flow and agreement of speeds in the centers of the two channels of the strait, and stronger flow in the ACC. The only currently published sections of observed velocity in the strait are those of Coachman et al. (1975), but as the authors themselves point out, these sections are subject to time aliasing being taken over a period of days. Mooring data shows that the cross-strait velocity variability found outside the ACC on those sections can be explained by temporal variability of the flow.

It seems likely that the variability found in the models is due to edge effects and/or the poor resolution of the real world bathymetry and the exact choice of model section. The lesson to be learned here is that a coarse resolution model cannot be used to study the details of features at the same resolution as the model (e.g. the ACC width of ~10 km). It must also be remembered however, that the observational transports presented here are based on an assumption of homogeneity of flow at all locations in the strait. This is being tested currently by an increased mooring effort in the strait region. Preliminary results suggest this assumption to be reasonably sound outside the ACC region, but more analysis remains to be done.

It seems inevitable that the seasonally-intensified ACC volume transport is not accounted for in models due to spatial resolution limitations. In order to resolve the ACC, models with higher spatial resolution will need to be employed, while maintaining large model domains to obtain proper water mass transformations and circulations. At the same time, the estimates presented here from observations also lack continuous measurements in the surface layers and near the coast. Although estimates of the contributions from the ACC and stratification have been made by Woodgate et al. (2006, 2010), interannual quantification of the seasonal contribution by the ACC to the overall Bering Strait transport is yet to be computed from either observations or models. The freshwater flux, which has a significant influence on the density structure of the Arctic Ocean (e.g., Aagaard et al. 1985b), would also be better measured if more salinity information could be obtained in the upper layers and especially nearby the coast (see e.g., Woodgate and Aagaard 2005 for discussion). Similarly, for heat flux, it is crucial to get information on the upper layers where maybe 1/3rd of the heat is advected [see Woodgate et al. (2010) who used satellite-derived sea surface temperatures to estimate the contribution from the upper layers]. An international effort is currently underway with eight moorings placed in the Bering Strait region. New information from these moorings will be important for better understanding details of the flow through the strait.

7.6 Summary

While it is encouraging that, in many of the larger-scale models, fluxes of volume, heat and salt are of the right order of magnitude and in interannual terms show correlated variations with observations, there are still significant discrepancies. These have to be considered when using model results to look at the role of Pacific waters in the Arctic. We also see a need for model results with higher spatial resolution in the strait region. The ACC is only in the order of 10 km in width and thus not resolved by global or regional Arctic models with resolutions of 4–40 km (Table 7.1). The implementation of higher-resolution (2 km or less) regional models should improve estimates of the volume, heat and freshwater fluxes in the strait, if the issues of large-scale boundary conditions for such a model can be solved. The challenge is to be able to capture small-scale features, such as the Alaska Coastal Current and mesoscale eddies in the strait itself and its immediate vicinity. The modeling community is working toward that goal to properly represent such features.

Acknowledgments We thank the Department of Energy Earth System Modeling program (J. C. K and W. M.), National Science Foundation Office of Polar Programs (J. C. K, W. M., M. S., and J. Z.), and the Office of Naval Research (J. C. K and W. M.) for support of this research. We also thank the Arctic Ocean Model Intercomparison Project (J. C. K. and W. M.) for travel support and for conference opportunities to present and discuss this research. Support for this work was provided (in part) by NSF grants ARC-0632154, ARC-0855748, and the NOAA-RUSALCA program (R. W.). The mooring data used in this study was collected under funding from ONR, NSF, MMS, AOOS and NOAA-RUSALCA (R. W). At the National Oceanography Centre Southampton (Y.A. and B. d C.) the study was supported by the UK Natural Environment Research Council as a contribution to the Marine Centres' Strategic Research Programme Oceans 2025. The NOCS-ORCA simulations were performed as part of the DRAKKAR collaboration (Barnier et al. 2006). NOCS also acknowledges the use of UK National High Performance Computing Resource. We also thank the editors of this book and two anonymous reviewers for insightful comments, which improved an earlier version of this manuscript.

References

Aagaard K, Carmack EC (1989) The role of sea-ice and other fresh water in the Arctic circulation. J Geophys Res 94:14485–14498

Aagaard K, Roach AT, Schumacher JD (1985a) On the wind-driven variability of the flow through Bering Strait. J Geophys Res 90:7213–7221

Aagaard K, Swift JH, Carmack EC (1985b) Thermohaline circulation in the arctic Mediterranean seas. J Geophys Res 90:4833–4846

Aagaard K, Weingartner TJ, Danielson SL, Woodgate RA, Johnson GC, Whitledge TE (2006) Some controls on flow and salinity in Bering Strait. Geophys Res Lett 33:L19602. doi:10.1029/2006GL026612

Adcroft A, Campin JM (2004) Rescaled height coordinates for accurate representation of free-surface flows in ocean circulation models. Ocean Model 7(3–4):269–284

Adcroft A, Hill C, Marshall J (1997) The representation of topography by shaved cells in a height coordinate model. Mon Weather Rev 125(9):2293–2315

Ahlnäs K, Garrison GR (1984) Satellite and oceanographic observations of the warm coastal current in the Chukchi Sea. Arctic 37:244–254

Arakawa A (1966) Computational design of long-term numerical integration of the equations of fluid motion. J Comput Phys 1:119–143

Barnier B, Madec G, Penduff T, Molines JM, Treguier AM, Le Sommer J, Beckmann A, Biastoch A, Böning C, Dengg J, Derval C, Durand E, Gulev S, Remy E, Talandier C, Theetten S, Maltrud M, McClean J, de Cuevas BA (2006) Impact of partial steps and momentum advection schemes in a global ocean circulation model at eddy permitting resolution. Ocean Dyn 56:543–567. doi:10.1007/s10236-006-0082-1

Brodeau L, Barnier B, Treguier AM, Penduff T, Gulev S (2010) An ERA40-based atmospheric forcing for global ocean circulation models. Ocean Model 31(3–4):88–104. doi:10.1016/j.ocemod.2009.10.005

Clement JL, Cooper LW, Grebmeier JM (2004) Late winter water column and sea ice conditions in the northern Bering Sea. J Geophys Res 109(C3):C03022. doi:10.1029/2003JC002047

Clement JL, Maslowski W, Cooper L, Grebmeier J, Walczowski W (2005) Ocean circulation and exchanges through the northern Bering Sea – 1979–2001 model results. Deep Sea Res II 52:3509–3540. doi:10.1016/j.dsr2.2005.09.010

Coachman LK, Aagaard K (1988) Transports through Bering Strait: annual and interannual variability. J Geophys Res 93:15535–15539

Coachman LK, Aagaard K, Tripp RB (1975) Bering Strait: the regional physical oceanography. University of Washington Press, Seattle

Dai A, Trenberth KE (2002) Estimates of freshwater discharge from continents: latitudinal and seasonal variations. J Hydrometeorol 3:660–687

Daru V, Tenaud C (2004) High order one-step monotonicity-preserving schemes for unsteady compressible flow calculations. J Comput Phys 193(2):563–594, doi: http://dx.doi.org/10.1016/j.jcp.2003.08.023

DRAKKAR Group (2007) Eddy-permitting ocean circulation hindcasts of past decades. CLIVAR Exchanges No 42:12(3):8–10

Dukowicz JK, Smith RD (1994) Implicit free-surface method for the Bryan-Cox-Semtner ocean model. J Geophys Res 99:7791–8014

Fedorova AP, Yankinam AS (1964) The passage of Pacific Ocean water through the Bering Strait into the Chukchi Sea. Deep-Sea Res 11:427–434

Fichefet T, Morales Maqueda MA (1997) Sensitivity of a global sea ice model to the treatment of ice thermodynamics and dynamics. J Geophys Res 102(C6):12609–12646

Flato GM, Hibler WD III (1995) Ridging and strength in modeling the thickness distribution of Arctic sea ice. J Geophys Res 100:18611–18626

Fox-Kemper B, Menemenlis D (2008) Can large eddy simulation techniques improve mesoscale rich ocean models? In: Hecht M, Hasumi H (eds) Ocean modeling in an eddying regime. AGU, Washington, DC

Gill A (1982) Atmosphere–ocean dynamics. Academic Press Inc., Burlington

Goosse H, Campin JM, Fichefet T, Deleersnijder E (1997) Sensitivity of a global ice–ocean model to the Bering Strait throughflow. Clim Dyn 13(5):349–358

Hibler WD III (1979) A dynamic thermodynamic sea ice model. J Phys Oceanogr 9(4):815–846

Hibler WD III (1980) Modeling a variable thickness sea ice cover. Mon Weather Rev 108: 1943–1973

Holland DM (2000) Merged IBCAO/ETOPO5 global topographic data product. National Geophysical Data Center (NGDC), Boulder, CO. http://www.ngdc.noaa.gov/mgg/bathymetry/arctic/ibcaorelatedsites.html. Cited 25 Dec 2000

Jackett DR, McDougall TJ (1995) Minimal adjustment of hydrographic profiles to achieve static stability. J Atmos Oceanic Tech 12(2):381–389

Jakobsson M, Cherkis N, Woodward J, Macnab R, Coakley B (2000) New grid of Arctic bathymetry aids scientists and mapmakers. Eos Trans AGU 81(9):89

Kalnay E et al (1996) The NCEP/NCAR 40-year reanalysis project. Bull Am Meteorol Soc 77:437–471

Kinder TH, Schumacher JD (1981) Circulation over the continental shelf of the Southeastern Bering Sea. In: Hood DW, Calder JA (eds) The eastern Bering Sea shelf: oceanography and resources. University of Washington Press, Seattle

Large WG, Pond S (1981) Open ocean momentum flux measurements in moderate to strong winds. J Phys Ocean 11(3):324–336

Large WG, Pond S (1982) Sensible and latent-heat flux measurements over the ocean. J Phys Ocean 12(5):464–482

Large WG, Yeager SG (2004) Diurnal to decadal global forcing for ocean and sea-ice models: the data sets and flux climatologies. Technical report TN-460+STR, NCAR, 105 pp

Large WG, McWilliams JC, Doney S (1994) Oceanic vertical mixing: a review and a model with a nonlocal boundary layer parameterization. Rev Geophys 32(4):363–403

Leith CE (1996) Stochastic models of chaotic systems. Physica D 98:481–491

Levitus S, Boyer TP, Conkright ME, O'Brian T, Antonov J, Stephens C, Stathopolos L, Johnson D, Gelfeld R (1998) World ocean database 1998, NOAA atlas NESDID 18. US Government Printing Office, Washington, DC

Lique C, Treguier AM, Scheinert M, Penduff T (2009) A model-based study of ice and freshwater transport variability along both sides of Greenland. Clim Dyn 33:685–705. doi:10.1007/s00382-008-0510-7

Losch M, Menemenlis D, Heimbach P, Campin JM, Hill C (2010) On the formulation of sea-ice models. Part 1: effects of different solver implementations and parameterizations. Ocean Model 33:129–144

Madec G (2008) NEMO reference manual, ocean dynamic component: NEMO-OPA. Rep. 27, Note du pôle de modélisation, Institut Pierre Simmon Laplace (IPSL), France. ISSN No. 1288–1619

Madec G, Delecluse P, Imbard M, Levy C (1998) OPA 8.1 ocean general circulation model reference manual. IPSL technical report. Tech Rep. 11, Institut Pierre-Simon Laplace, 91 pp

Marchuk GI, Kagan BA (1989) Dynamics of ocean tides. Kluwer Academic Publishers, Heidelberg

Marshall J, Adcroft A, Hill C, Perelman L, Heisey C (1997) A finite-volume, incompressible Navier–Stokes model for studies of the ocean on parallel computers. J Geophys Res 102(C3):5753–5766

Maslowski W, Marble D, Walczowski W, Schauer U, Clement JL, Semtner AJ (2004) On climatological mass, heat, and salt transports through the Barents Sea and Fram Strait from a pan-Arctic coupled ice-ocean model simulation. J Geophys Res 109:C03032. doi:10.1029/2001JC001039

Melling H (2000) Exchanges of fresh-water through the shallow straits of the North American Arctic. In: Lewis EL et al (eds) The fresh-water budget of the Arctic Ocean. Proceedings of a NATO Advanced Research Workshop, Tallinn Estonia, 27 April–1 May 1998, Kluwer Academic Publishers, Dordrecht, The Netherlands

Menemenlis D, Hill C, Adcroft A, Campin J, Cheng B, Ciotti B, Fukumori I, Heimbach P, Henze C, Koehl A, Lee T, Stammer D, Taft J, Zhang J (2005) NASA supercomputer improves prospects for ocean climate research. Eos Trans AGU 86(9):89, 95–96

Menemenlis D, Campin J, Heimbach P, Hill C, Lee T, Nguyen A, Schodlok M, Zhang H (2008) ECCO2: high resolution global ocean and sea ice data synthesis. Mercat Ocean Q Newsl 31:13–21

Nguyen AT, Menemenlis D, Kwok R (2009) Improved modeling of the Arctic halocline with a subgrid-scale brine rejection parameterization. J Geophys Res 114:C11014. doi:10.1029/2008JC005121

Nguyen AT, Menemenlis D, Kwok R (2011) Arctic ice-ocean simulation with optimized model parameters: approach and assessment. J Geophys Res 116:C04025. doi:10.1029/2010JC006573

Onogi K, Tsutsui J, Koide H, Sakamoto M, Kobayashi S, Hatsushika H, Matsumoto T, Yamazaki N, Kamahori H, Takahashi K, Kadokura S, Wada K, Kato K, Oyama R, Ose NMT, Taira R (2007) The jra-25 reanalysis. J Meteorol Soc Jpn 85(3):369–432

Panteleev GG, Nechaev D, Proshutinsky AY, Woodgate R, Zhang J (2010) Reconstruction and analysis of the Chukchi Sea circulation in 1990–1991. J Geophys Res 115:C08023. doi:10.1029/2009JC005453

Paquette RG, Bourke RH (1974) Observations on the coastal current of arctic Alaska. J Mar Res 32:195–207

Payne RE (1972) Albedo at the sea surface. J Atmos Sci 29:959–970

Penduff T, Le Sommer J, Barnier B, Treguier AM, Molines JM, Madec G (2007) Influence of numerical schemes on current-topography interactions in 1/4° global ocean simulations. Ocean Sci 3:509–524

Prange M, Lohmann G (2004) Variable freshwater input to the Arctic Ocean during the Holocene: implications for large-scale ocean-sea ice dynamics as simulated by a circulation model. In: Fischer H et al (eds) The KIHZ project: towards a synthesis of Holocene proxy data and climate models. Springer, New York

Prather MC (1986) Numerical advection by conservation of second-order moments. J Geophys Res 91:6671–6681

Roach AT, Aagaard K, Pease CH, Salo SA, Weingartner T, Pavlov V, Kulakov M (1995) Direct measurements of transport and water properties through the Bering Strait. J Geophys Res 100:18443–18457

Royer TC (1981) Baroclinic transport in the Gulf of Alaska, part II. A freshwater driven coast current. J Mar Res 39:251–266

Semtner AJ (1976) A model for the thermodynamic growth of sea ice in numerical investigation of climate. J Phys Oceanogr 6:376–389

Shimada K, Kamoshida T, Itoh M, Nishino S, Carmack E, McLaughlin F, Zimmermann S, Proshutinsky A (2006) Pacific Ocean inflow: influence on catastrophic reduction of sea ice cover in the Arctic Ocean. Geophys Res Lett 33:L08605. doi:10.1029/2005GL025624

Smith WHF, Sandwell DT (1997) Global sea floor topography from satellite altimetry and ship depth soundings. Science 277(5334):1956–1962

Smith RD, Dukowicz JK, Malone RC (1992) Parallel ocean general circulation modeling. Physica D 60:38–61

Steele M, Morley R, Ermold W (2001) PHC: a global ocean hydrography with a high quality Arctic Ocean. J Clim 14(9):2079–2087

Thorndike AS, Rothrock DA, Maykut GA, Colony R (1975) The thickness distribution of sea ice. J Geophys Res 80:4501–4513

Timmermann R, Goose H, Madec G, Fichefet T, Ethe C, Duliere V (2005) On the representation of high latitude processes in the ORCA-LIM global coupled sea ice-ocean model. Ocean Model 8:175–201

Weingartner TJ, Danielson S, Sasaki Y, Pavlov V, Kulakov M (1999) The Siberian Coastal Current: a wind- and buoyancy-forced Arctic coastal current. J Geophys Res 104:29697–29713. doi:10.1029/1999JC900161

Woodgate RA, Aagaard K (2005) Revising the Bering Strait freshwater flux into the Arctic Ocean. Geophys Res Lett 32:L02602. doi:10.1029/2004GL021747

Woodgate RA, Aagaard K, Weingartner TJ (2005a) Monthly temperature, salinity, and transport variability of the Bering Strait throughflow. Geophys Res Lett 32:L04601. doi:10.1029/2004GL021880

Woodgate RA, Aagaard K, Weingartner TJ (2005b) A year in the physical oceanography of the Chukchi Sea: moored measurements from autumn 1990–1991. Deep Sea Res II 52:3116–3149. doi:10.1016/j.dsr2.2005.10.016

Woodgate RA, Aagaard K, Weingartner TJ (2006) Interannual changes in the Bering Strait fluxes of volume, heat and freshwater between 1991 and 2004. Geophys Res Lett 33:L15609. doi:10.1029/2006GL026931

Woodgate RA, Aagaard K, Weingartner TJ (2007) First steps in calibrating the Bering Strait throughflow: preliminary study of how measurements at a proposed climate site (A3) compare to measurements within the two channels of the strait (A1 and A2). 20 pp, University of Washington, Seattle, WA

Woodgate RA, Weingartner TJ, Lindsay RW (2010) The 2007 Bering Strait oceanic heat flux and anomalous Arctic sea-ice retreat. Geophys Res Lett 37:L01602. doi:10.1029/2009GL041621

Zhang J (2005) Warming of the arctic ice-ocean system is faster than the global average since the 1960s. Geophys Res Lett 32:L19602. doi:10.1029/2005GL024216

Zhang J, Hibler WD (1997) On an efficient numerical method for modeling sea ice dynamics. J Geophys Res 102:8691–8702

Zhang J, Rothrock DA (2001) A thickness and enthalpy distribution sea-ice model. J Phys Oceanogr 31:2986–3001

Zhang J, Rothrock DA (2003) Modeling global sea ice with a thickness and enthalpy distribution model in generalized curvilinear coordinates. Mon Weather Rev 131:681–697

Zhang J, Rothrock DA (2005) The effect of sea-ice rheology in numerical investigations of climate. J Geophys Res 110:C08014. doi:10.1029/2004JC002599

Zhang J, Steele M, Lindsay RW, Schweiger A, Morison J (2008) Ensemble one-year predictions of arctic sea ice for the spring and summer of 2008. Geophys Res Lett 35:L08502. doi:10.1029/2008GL033244

Zhang J, Woodgate R, Moritz R (2010) Sea ice response to atmospheric and oceanic forcing in the Bering Sea. J Phys Oceanogr 40:1729–1747. doi:10.1175/2010JPO4323.1

Chapter 8
Carbon Fluxes Across Boundaries in the Pacific Arctic Region in a Changing Environment

Wei-Jun Cai, Nicholas R. Bates, Laodong Guo, Leif G. Anderson, Jeremy T. Mathis, Rik Wanninkhof, Dennis A. Hansell, Liqi Chen, and Igor P. Semiletov

Abstract While the inflow of dissolved inorganic carbon (DIC) from the Pacific Ocean is relatively well quantified, the intermittent input from the East Siberian Sea (ESS) is not. The export flux to the Atlantic Ocean has unknown uncertainty due to a paucity of DIC data from the Canadian Archipelago. Within the region, the Chukchi Sea is the dominant site for atmospheric carbon dioxide (CO_2) uptake, while the Beaufort Sea and the Canadian Archipelago take-up much less CO_2 with latter potentially a weak source of CO_2 during certain times of the year. Additionally, the ESS shelf is a net source of CO_2. Summertime CO_2 uptake capacity in the deep

W.-J. Cai (✉)
School of Marine Science and Policy, The University of Delaware, Newark, DE 19716, USA
e-mail: wai@udel.edu

N.R. Bates
Bermuda Institute of Ocean Sciences (BIOS), Ferry Reach, Bermuda

L. Guo
School of Freshwater Sciences, University of Wisconsin-Milwaukee, Milwaukee, WI, USA

L.G. Anderson
Department of Chemistry, University of Gothenburg, Gothenburg, Sweden

J.T. Mathis
Pacific Marine Environmental Laboratory, National Oceanic and Atmospheric Administration, 7600 Sand Point Way, Seattle, WA, USA

School of Fisheries and Ocean Sciences, University of Alaska Fairbanks, Fairbanks, AK, USA
e-mail: jeremy.mathis@noaa.gov

R. Wanninkhof
Atlantic Oceanographic and Meteorological Laboratory, National Oceanic and Atmospheric Administration, Miami, FL, USA

D.A. Hansell
Rosenstiel School of Marine and Atmospheric Science, University of Miami, Miami, FL, USA

J.M. Grebmeier and W. Maslowski (eds.), *The Pacific Arctic Region: Ecosystem Status and Trends in a Rapidly Changing Environment*, DOI 10.1007/978-94-017-8863-2_8,

Canada Basin has increased greatly recently as sea-ice retreat progresses rapidly. The region appears to export more DIC than it receives by a small amount, suggesting that it is probably weakly net heterotrophic. In addition to labile organic carbon (OC) produced in the productive marginal seas, some riverine and coastal erosion-derived OC likely is also recycled. As warming progresses, the Arctic Ocean may produce and export more DIC. Whether this change will turn the Arctic Ocean into a weaker CO_2 sink or even a CO_2 source for the atmosphere is uncertain and dependent on multiple factors that control the rate of surface water CO_2 increase versus the rate of the atmospheric CO_2 increase.

Keywords Arctic Ocean • Carbon cycle • Air-sea CO_2 flux • Riverine POC • DOC and DIC fluxes

8.1 Introduction

Accumulation of greenhouse gases such as carbon dioxide (CO_2) in the atmosphere, global climate and environmental change, and the sustainability of the Earth's biosphere are of great scientific and societal concern. Improved understanding of the global carbon cycle and potential responses of the climate system requires better knowledge about key physical, chemical and biological processes that control global biogeochemical cycles, an assessment of the complex interactions and feedbacks, and a better understanding of the vulnerabilities that impact carbon pools in the Earth system. These needs are particularly true for the Arctic region, which is widely viewed as one of the most sensitive of the Earth's physical and biological systems (Perovich et al. 2007; Serreze and Francis 2006; ACIA 2004). On land, these changes include reorganization of the hydrologic cycle (e.g., reductions of permafrost and glacial ice), expansion of temperate ecosystems into subpolar/polar zones, and mobilization of terrestrial carbon pools (McGuire et al. 2009). In the Arctic Ocean, changes include atmospheric and ocean warming, sea-ice loss (Serreze and Francis 2006) and increased phytoplankton primary production and biomass of Arctic surface waters (Arrigo et al. 2008). As a framework for supporting carbon cycle studies, "*A U.S. Carbon Cycle Science Plan (1999)*" (Sarmiento and Wofsy 1999) proposed two long-term research goals for North America: (1) understanding the northern hemisphere land CO_2 sink, and (2) understanding the ocean carbon sink. These goals will continue to be emphasized for the next decade (Michalak et al. 2011). As part of this framework, determining atmosphere-land-ocean interactions, fluxes and exchanges in the high latitude areas of North America

L. Chen
Third Institute of Oceanography, The State Oceanic Administration of China, Xiamen, China

I.P. Semiletov
Pacific Oceanological Institute, Far Eastern Branch of the Russian, Academy of Sciences, Vladivostok, Russia and International Arctic Research Center, University of Alaska Fairbanks, Fairbanks, AK, USA

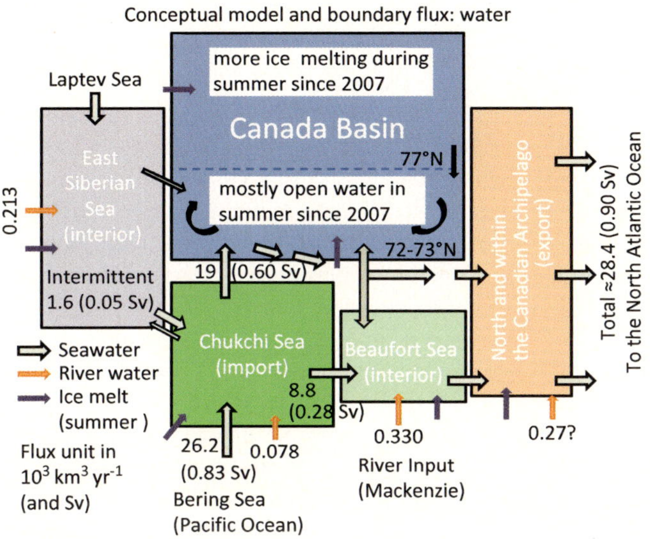

Fig. 8.1 Water fluxes within the Pacific Arctic Region. Bering Strait water inflow is set to 0.83 Sv as in Anderson et al. (1998). 1 Sv = 1×10^6 m^3 s^{-1}. The *larger hollow arrows* represent seawater fluxes. The *orange color arrows* represent river water fluxes. The *purple arrows* represent water input from summer ice melt. The ice-melt water fluxes do not indicate flux from one box into the other. The Chukchi Sea is classified as an import shelf as it receives water and nutrients from the Pacific Ocean, the East Siberian Sea and the Beaufort Sea are classified as interior seas, and the Canadian Archipelago as an export sea. A negative value in Table 8.4 (air to sea flux) is converted to a positive one here to show a gain for the ocean

(i.e., the Arctic Ocean and its margins) is critical for understanding high-latitude carbon cycling, assessment of vulnerable carbon pools (McGuire et al. 2009), and decision-support activities.

In this chapter, we use the geographic terms the Pacific Arctic Region (PAR) and the western Arctic Ocean interchangeably even though they are not strictly equal (as defined in Fig. 8.1). Our review and analyses focus on assessing carbon exchanges across subregions within the ocean domain of the PAR and on carbon fluxes across the land-ocean and air-sea boundaries. This assessment includes organic carbon (OC) and inorganic carbon (e.g., CO_2 gas and total dissolved inorganic carbon, DIC). Carbon cycling in the PAR is highly dynamic and seasonally driven by extreme variability in physical conditions. A conceptual model depicting both boundary processes (this chapter) and interior processes (Mathis et al. 2014, this volume) has been developed (Fig. 8.1, Mathis et al. 2014, this volume). Water column physical and biogeochemical processes are reviewed in detail in the Mathis et al. 2014 chapter as well. A brief description of critical processes is given here.

The inflow of nutrient-rich Pacific water through Bering Strait into the Chukchi Sea (Codispoti et al. 2005), coupled with abundant light and the seasonal retreat and melting of sea ice, supports a brief but intense period of marine phytoplankton photosynthesis and growth (Cota et al. 1996; Hill and Cota 2005; Springer and

McRoy 1993). This biological production of OC is much stronger in the Chukchi Sea than in other Arctic Ocean shelves where nutrients concentrations are lower. The seasonally high biological production in the Chukchi Sea leads to sea surface CO_2 depletion and subsequent atmospheric CO_2 uptake, which must influence CO_2 flux in downstream subsystems (i.e. the Beaufort Sea and deep Canada Basin).

Atmospheric CO_2 is also fixed into OC through terrestrial plant photosynthesis. Terrestrial primary productivity will likely increase as the Arctic warms and more productive communities of plants establish themselves in regions that have traditionally been dominated by tundra. Increasing river runoff (Peterson et al. 2002; Wu et al. 2005), which is likely to occur as a result of warming, will subsequently enhance transport of terrestrial OC in river water (e.g., estuaries and wetlands of the Arctic) to the coastal ocean. Once in the ocean, OC may be metabolized back to CO_2 (Anderson et al. 2009; Hansell et al. 2004) and either released back to the atmosphere or exported into the deep ocean (Smith and Hollibaugh 1993; Mackenzie et al. 2004). Within the ocean margins of the Arctic, net fixation of CO_2 into phytoplankton biomass occurs, with a small portion of the ocean derived particulate organic carbon (with addition of terrestrial OC) exported to the deep ocean and an even smaller fraction buried in sediments (Moran et al. 2005; Lepore et al. 2007). Thus CO_2 flux across the atmosphere-ocean interface is a critical variable in global carbon cycle models (Liu et al. 2010; Takahashi et al. 2009), which in the Arctic is complicated by the seasonal presence of sea-ice that can act as a barrier to the exchange of gases.

In pre-industrial times, the global ocean and in particular the coastal oceans released CO_2 as a result of the respiration and remineralization of terrestrial OC supplied through rivers (Smith and Hollibaugh 1993). In the Anthropocene, the atmosphere-ocean exchange of CO_2 is reversed with the global ocean acts as a net sink for atmosphere CO_2 due to the increase in the atmospheric CO_2 concentration (Takahashi et al. 2009). However, the situation in the ocean margins is more complex. In general, the tropical and nearshore coastal seas as well as areas more influenced by rivers are more likely to be net sources of CO_2 while temperate and polar coastal seas and those regions more distant to delivery of terrestrial OC are more likely to be net sinks of CO_2 (Ducklow and McCallister 2004; Mackenzie et al. 2004; Borges et al. 2005; Chen and Borges 2009; Cai et al. 2006; Cai 2011). However, an argument can be made that increasing terrestrial inputs and discharge of anthropogenic materials associated with climate and land use changes, as well as changes in the terrestrial hydrological cycle, would tend to shift the ocean margins toward being sources of CO_2 as most terrestrial OC is respired in adjacent coastal zones. This argument is particularly relevant in the Arctic Ocean margins and probably even its interior, as riverine organic and inorganic carbon fluxes are likely to increase due to increased river runoff, permafrost thawing, and coastal erosion under current conditions and future climate change scenarios (Guo and Macdonald 2006; McGuire et al. 2009). For example, it was shown that coastal erosion and rivers bring to the East Siberian Arctic shelf an integrated signal of terrestrial organic matter released from thawing permafrost (Gustafsson et al. 2011; Pipko et al. 2011; Semiletov et al. 2007, 2011; van Dongen et al. 2008;

Vonk et al. 2010). This climate change-driven scenario of increased terrestrial OC export may have particularly important and perhaps non-linear consequences on ocean biogeochemistry, given that the Arctic Ocean has only 1 % of global ocean volume but receives about 10 % global freshwater input and associated suspended and dissolved materials.

8.2 Geographic and Water Mass Features

8.2.1 Geographic Definition and Description

The PAR discussed here includes the Canada Basin in the north and, to the south, from west to east, the East Siberian, Chukchi and Beaufort Seas, and shelf waters of the Canadian Archipelago (Fig. 8.1). In addition, as a gateway to the Arctic from the Pacific, the northern Bering Sea and Bering Strait are considered in this synthesis. The Arctic Ocean (excluding the Nordic Seas and Baffin Bay) covers ~2.6 % of the surface area of the global ocean, but occupies a volume of <1 %. The shallow continental shelves make up ~50 % of the total area of the Arctic Ocean (Table 8.1). In contrast, the continental shelves of the global ocean comprise only ~7 % of the total ocean area (Walsh 1988; Cai 2011). Given the physical setting of the Arctic Ocean, fluxes at land-ocean, atmosphere-ocean, and ocean-ocean boundaries have great influence on the carbon cycle.

The East Siberian Sea (ESS) has the shallowest shelf in the PAR, with a mean depth of only 58 m and is highly influenced by river inputs from the Lena, Kolyma and Indigirka Rivers (Semiletov et al. 2005, 2011). The Chukchi Sea shelf, with a mean depth of ~80 m, is the broadest coastal sea in the entire region and is the area most impacted by inflow of Pacific Ocean water. The Beaufort Sea shelf is narrow with a mean depth of 125 m, while the mean shelf depth further increases towards the Canadian Archipelago (Table 8.1), with both shelves highly influenced by Mackenzie River input (Yamamoto-Kawai et al. 2009b; Shadwick et al. 2011). The adjacent deep Canada Basin receives outflows from the shelf through submarine features like Barrow Canyon and Herald Valley (Codispoti et al. 2005; Woodgate et al. 2005b) as well as the northern flank of the ESS (Anderson et al. 2010), facilitating the export of OC to the deep Arctic Ocean, while shelf-slope topography facilitates upwelling of dissolved nutrients and DIC, and generation of mesoscale eddies that promote shelf-basin exchanges of water and materials (Mathis et al. 2007).

8.2.2 Water-Mass Characterizations

There are several water masses in the slope and basin areas of the Pacific Arctic Region (PAR) (Anderson et al. 1990; McLaughlin et al. 2004; Woodgate et al. 2005b) (Table 8.1): (1) the Surface or Polar Mixed Layer (SML, 0–40 m) with

Table 8.1 Basic information and carbon inventory in the Pacific Arctic Region

Basin and sea	Area (10^3 km^2)	Depth (m)	Volume (10^3 m^3)	Residence time (year)	Concentration (μmol/kg) DIC	DOC	Inventory (Tmol) DIC	DOC	Inventory (PgC) DIC	DOC
Canada Basin	2,797	2,700	6,900	2–30*						
SML		0–80	228	<10	2,050	100	467.4	22.8	5.61	0.27
UH		80–220	398	<10	2,210	75	879.6	29.9	10.55	0.36
LH		220–300	228	>10	2,155	75	491.3	17.1	5.90	0.21
AL		300–1,000	1,992		2,159	53	4,300.7	105.6	51.61	1.27
DBW		>1,000	4,054		2,050	55	8,310.7	223.0	99.73	2.68
Chukchi	620	80	50	0.2–1.2	2,050	150	102.5	7.5	1.23	0.09
East Siberian	987	58	57	1.5–5.5	2,050	200	116.9	11.4	1.40	0.14
Beaufort	178	125	22	0.5–1	2,050	100	45.1	2.2	0.54	0.03
Archipelago										
(north of)	240	290	69.6	~0.5	2,145	100	149.3	7.0	1.79	0.08
(within)	1,250	250	312.5	~12	2,145	100	670.3	31.3	8.04	0.38
Total shelf	3,275		511.1				1,084.1	59.3	13.01	0.71
Total PAR	6,072		7,411.1				15,533.8	457.6	186.41	5.49
% of shelf	53.9									
%	56.7		59.7							
Eurasian Basin	1,704.25		4,556		2,120	100	9,658.7	455.6	115.90	5.47
Lapteve	498		24		2,120	100	50.9	2.4	0.61	0.03
Kara	926		121		2,120	100	256.5	12.1	3.08	0.15
Barents	1,512		302		2,120	100	640.2	30.2	7.68	0.36
Total Arctic	10,712		12,414				26,140	957.9	313.7	11.5

Modified from McGuire et al. (2009) and Macdonald et al. (2010) and references therein

Most DIC concentrations were from L. Anderson in Macdonald et al. (2010). During summer 2008 average DIC values in the upper 75-m water however are: 2,040 μmol kg^{-1} (Bering Strait), 2,062 μmol kg^{-1} (Chukchi Sea), 1,997 μmol kg^{-1} (Beaufort Sea), and 2,012 μmol kg^{-1} (Canada Basin) (W-J. Cai unpublished). DOC data are from Hansell and coworkers (see Fig. 8.4 in the Mathis et al. 2014, this volume)

SML the Surface Mixing Layer, *UHL* Upper Halocline Layer, *LHL* Lower Halocline Layer

$S<31$, DIC = ~1,750–1,900 μmol kg^{-1} (Bates et al. 2005) and water residence time ~10 years; (2) the Upper Halocline Layer (UHL, ~40–145 m) with $S\sim33.1$, DIC = ~2,160–2,190 μmol kg^{-1}, [NO_3^-] = ~14 μmol L^{-1}, [PO_4] = ~1.8 μmol L^{-1}, and water residence time ~10 years; (3) the Lower Halocline Layer (LHL, ~150–220 m) with $S\sim34.3$, DIC = ~2,170–2,190 μmol kg^{-1}, [NO_3^-] = ~12 μmol L^{-1}, [PO_4] = ~0.8 μmol L^{-1}, and residence time ~15 years; (4) Atlantic Water (AW, >250–900 m) with $S\sim34.8$–34.9, DIC = ~2,140–2,160 μmol kg^{-1}, [NO_3^-] = ~14 μmol L^{-1}, [PO_4^{3-}] = ~1.0 μmol L^{-1}, and residence time ~30 years; and (5) the Arctic Ocean Deep Water (>900 m deep). The SML and UHL waters have a Pacific Ocean origin, with some fluvial contributions, while the SML also includes contributions from local sea ice melt. The LHL water is mostly of Atlantic Ocean and eastern Arctic origin but with some influence of Pacific Ocean water by western shelf plumes.

8.3 Pacific Ocean Inflow

Water masses entering and forming the upper layers of the western Arctic Ocean come from the Bering Sea, which holds nutrient-rich waters derived from the North Pacific Ocean that shoal onto the shelf toward Bering Strait. Waters transiting through Bering Strait during summer include the relatively colder, saltier, and more nutrient-rich and CO_2-poor water of the Anadyr Current (AC) in the west, Bering Shelf water (BSW) in the middle, and warmer, fresher, and relatively high CO_2 Alaskan Coastal Current (ACC) waters in the east. The inflow is higher in summer and lower in winter with an annual average of ~0.83 Sv (Fig. 8.1) (Roach et al. 1995; Woodgate et al. 2005a; Anderson et al. 1998). It is likely that the inflow has increased in recent time giving the fact that the yearly ice-free period at Bering Strait has extended seasonally (Woodgate et al. 2010).

With an average DIC value of 1997 ± 70 μmol kg^{-1} in the northern Bering Sea from the 2010 RUSALCA (Russian-American Long-term Census of the Arctic) and the 2003–2004 SBI (Shelf-Basin Interactions) data (Bates et al. 2005), we estimate a DIC flux of 52.3×10^{12} mol year^{-1} (or 627 ± 79 Tg C year^{-1}) entering the Arctic Ocean from the Pacific Ocean (N.R. Bates, unpublished; Table 8.2; Bates et al. 2011). Our estimate is slightly lower than an earlier estimate of 56.7×10^{12} mol year^{-1} based on fewer data from limited areas (Anderson et al. 1998; Chen et al. 2003). Uncertainty attached to this Bering Strait DIC flux is primarily associated with uncertainty in water flow (±0.1 Sv or 12.5 %), which leads to an uncertainty of ±78 Tg C year^{-1} in DIC flux. Variability in DIC content (3.3 %, Table 8.2) of the three water masses (i.e., AC, BSW and ACC) flowing through Bering Strait leads to an uncertainty of ±21 Tg C year^{-1} in DIC flux. Thus an overall uncertainty of 81 Tg C year^{-1} is derived (i.e., $\mathrm{SQRT}(78^2+21^2)=81$). This DIC flux accounts for approximately 50 % of the total DIC inventory in the Chukchi Sea, requiring that the flushing time of water and the physical residence time of DIC in the Chukchi Sea be not longer than 6 months. In addition, there is an intermittent inflow of

Table 8.2 Bering Strait and East Siberian Shelf water and DIC inflow

Bering Strait inorganic carbon inflow						
	% of flow	DIC range (μmol/kg)	Mean DIC (μmol/kg)	Water flux (m^3/year)	DIC flux (Tmol C/year)	DIC flux (Tg C/year)
Anadyr water	50	1,980–2,040	2,010	1.309E+13	26.3	316
Bering Sea water	30	1,900–1,980	1,940	7.852E+12	15.2	183
Alaskan Coastal	20	2,020–2,080	2,050	5.235E+12	10.7	129
Mean			1,997		**52.3**	**627**
Total flow	0.83	Sv				
Uncertainty	0.1	Sv	or 12.5 %			78
DIC error estimate			Error			
Anadyr Water			60	1.309E+13	0.79	9.4
Bering Sea water			80	7.852E+12	0.63	7.5
Alaskan Coastal			60	5.235E+12	0.31	3.8
						21 or 3.3 %
East Siberian Sea inflow						
	Water flow	DIC range	Mean DIC	Water flux	DIC flux	DIC flux
	0.05 Sv	1,600–1,750	1,700	1.577E+12	2.68	32.2
Mean					**2.7**	**32**
Total flow	0.1	Sv				
Net flow	0.05	Sv				
Total flux to the Chucki Sea					**55**	**659**

0–0.1 Sv to the Chukchi Sea from the ESS (Woodgate et al. 2005b). ESS inflow into the Chukchi Sea through Long Strait has a DIC content of ~1,700 μmoles kg^{-1} (RUSALCA data), equivalent to a transport of ~32 ± 16 Tg C $year^{-1}$ if a net water flux of 0.05 Sv is assumed (Table 8.2). Here, the large uncertainty is associated with both the water flux (±50 %) and DIC values (±20 %). Taken together, the total DIC entering the Chukchi Sea is around 55.0×10^{12} mol $year^{-1}$ (or 659 Tg C $year^{-1}$) with an estimated error of 13 %. As we assume a balanced water budget in the region, the uncertainty in the water inflow and outflow is the same and it is systematic, and thus it does not affect our net DIC mass balance analysis. Therefore, we may view that only the DIC concentration uncertainty contributes to the DIC budget uncertainty, i.e., at the level of only 3.3 % or ±21 Tg C $year^{-1}$.

Early studies of the DOC content of waters passing Bering Strait suggested a large range of values (34–134 μmol L^{-1}; see Wheeler et al. (1996) and Walsh et al. (1997)). However, more recent measurements at Bering Strait (RUSALCA data) indicate that the DOC of AC and BSW waters ranged from ~80 to 90 μM while the DOC content of the ACC was higher (~90–110 μM). Assuming an average DOC content of 90 μM, we estimate a DOC flux of 2.36×10^{12} mol year^{1} (or 28.3 ± 3.3 Tg C year^{1}). At Long Strait, DOC content varied from ~90 to 140 μM, with higher DOC contents in the nearshore ESS current (see distributions of surface DOC in (Letscher et al. 2011)). Assuming an average DOC content of 100 μM, we estimate a DOC flux of 0.16×10^{12} mol year^{1} (or 1.9 ± 0.4 Tg C year^{1}). Thus the DOC flux entering the Chukchi Sea is around 2.52×10^{12} mol year^{-1} (or 30.2 ± 3.3 Tg C year^{-1}). We would like to caution our readers that these DOC estimates are based on late summertime measurements, and they mostly reflect surface water enriched in DOC. The flux is probably higher during the spring freshet and lower during the winter when the ACC is not operative. Further measurements, in particular during other seasons, are needed to confirm and improve our conclusions.

On the broad and shallow continental shelf of the Chukchi Sea, the prevailing view is that a major part (we assume a total flux of 0.6 Sv here) of the Pacific inflow water travels northward to exit the shelf into the Canada Basin through four major outflows: Herald Valley, Long Strait, Central Channel, and Barrow Canyon (~0.1–0.3 Sv each), and a smaller part (we assume a total flux of 0.28 Sv) flows eastward along the shelf break of the Beaufort Sea towards the Canadian Archipelago (Woodgate et al. 2005a, b; Overland and Roach 1987; Codispoti et al. 2005; Pickart 2004) (Fig. 8.1). A large part of the northward flow is thought to join the eastward flow of the shelf break current (Weingartner et al. 1998; Pickart et al. 2005). Despite the overall complexity of the circulation, most of the inflow of water and DIC from the Pacific and the ESS is eventually balanced by the southward flow through the Canadian Archipelago (Codispoti et al. 2009; McGuire et al. 2009). For the purpose of balancing the water budget, we adopted the sum of all input water fluxes, 0.90 Sv, as the total outflow water flux. Finally, a small part of the Bering Strait inflow goes to the ESS, which is evidenced from the chemical signals in the eastern part of the ESS (Anderson et al. 2009, 2011; Semiletov et al. 2005; Codispoti and Richards 1963). However, the net flux is not known and more than likely the water rejoins the eastward flow towards the North Atlantic. This transport is one reason for assigning a smaller value of 0.05 Sv for the ESS current to the Chukchi Sea rather than the suggested maximum flux of 0.1 Sv.

8.4 Fluxes Across the Arctic Land-Sea Interface

As stated earlier, the Arctic Ocean receives a disproportionately large amount of global freshwater runoff (Aagaard and Carmack 1989) and therefore river inputs of terrigenous OC and DIC are important to the biogeochemical cycling of carbon in the region. Arctic terrestrial ecosystems have been considered sinks for atmospheric CO_2, resulting in the storage of large quantities of soil-based organic matter in high latitude regions (Ping et al. 2011; Waelbroeck et al. 1997; Oechel et al. 2000). Increased terrestrial OC inputs to the Arctic Ocean through permafrost thawing, accelerated coastal erosion, and increasing river runoff could dramatically change

Table 8.3 Carbon fluxes from rivers to the North American side of the Pacific Arctic Region

River	Basin area (10^3 km^2)	Discharge (km^3 year^{-1})	Concentration (mg C L^{-1}) DOC	POC	DIC	Flux (Tg C year^{-1}) DOC	POC	DIC
Mackenzie	1,787	330	5.2	7.2	14.7	1.72	2.3	4.84
Yukon	839	205	5.14	3.8	19.5	1.7	1.16	4.41
Colville	57	16	7.7	6.2	–	0.12	0.099	–
Sagavanirktok	15	6.5	4.1	2.4	22.1	0.027	0.016	0.14
Kuparuk	8	1.2	11.4	1.6	–	0.014	0.0019	–
Others	726	37	–	–	–	0.055	0.24	–
Total	16,369	3,531	–	–	–	3.63	3.87	9.39

Yukon River data are from Guo et al. (2012) and fluxes are average values of recent studies listed in Table 8.5 of Guo et al. (2012, and references therein). DOC and POC data from Colville, Sagavaniktov and Kuparuk Rivers are from Rember and Trefry (2004) using average of high flow and low flow samples. Other data are taken from McGuire et al. (2009). Tg = 10^{12} g

the turnover and transport rates of tundra carbon, and alter biogeochemical cycles and the marine ecosystem as a whole. Indeed, warming and environmental changes in the Arctic regions have resulted in the remobilization of soil OC and subsequent terrestrial OC inputs into the Arctic Ocean through increased river runoff and accelerated degradation of permafrost and peatlands (Frey and Smith 2005; Jorgenson and Brown 2005; Guo et al. 2004, 2007; Jorgenson et al. 2006; Savelieva et al. 2000; Peterson et al. 2002; Costard et al. 2007). Thus, quantitative measurements of riverine export fluxes are needed to determine the response of the Arctic river basins to a changing climate (Finlay et al. 2006; McGuire et al. 2009; Semiletov et al. 2011; Guo et al. 2012). However, DOC concentrations and discharge in Arctic rivers are highly variable between seasons, making quantitative estimation of carbon fluxes difficult for Arctic rivers (Finlay et al. 2006; Guo and Macdonald 2006). As pointed out by Guo et al. (2012), flux estimates without spring freshet sampling would underestimate organic species but overestimate inorganic species, whereas flux estimates without winter sampling overestimates organic species but underestimates inorganic species.

On the North American side of the PAR margins, inputs of terrestrial DOC and particulate organic carbon (POC) via river discharge are mainly derived from the Mackenzie River to the Beaufort Sea and the Yukon River to the Bering Sea. Many recent studies have estimated river fluxes of DOC and DIC from the Yukon River (Guo and Macdonald 2006; Raymond et al. 2007; Striegl et al. 2005, 2007; Spencer et al. 2009; Guo et al. 2012). Based on data compiled in Guo et al. (2012), average DOC and POC flux from the Yukon River was 1.7 and 0.85 Tg C year^{-1}, respectively, while DIC flux was 4.6 Tg C year^{-1} (Table 8.3). Riverine TOC (=DOC+POC) flux from the Mackenzie River is ~4.1 Tg C year^{-1} while DIC flux is ~4.8 Tg C year^{-1}. Fluxes from other northern Alaska rivers remain poorly quantified, but may contribute 3–5 % of fluxes from large rivers in the North America (Rember and Trefry 2004; McGuire et al. 2009).

Riverine fluxes of DOC and POC to the ESS have recently been estimated based on more intensive sampling (Lobbes et al. 2000; Finlay et al. 2006). The OC export flux from the Kolyma River was estimated to be 0.46 Tg C $year^{-1}$ for DOC and 0.31 Tg C $year^{-1}$ for POC (Lobbes et al. 2000). With intensive sampling during snowmelt period, the DOC flux from the Kolyma River was significantly higher, at 0.96 Tg C $year^{-1}$ (Finlay et al. 2006). Combining fluxes from both the Kolyma and Indigirka Rivers to the ESS, the TOC export flux is 1.1–1.7 Tg C $year^{-1}$, significantly lower than the TOC flux from either the Mackenzie or Yukon Rivers (Table 8.3). DIC data for the East Siberian rivers are scarce and some are from total alkalinity (TA) data (Cooper et al. 2008) assuming DIC equals TA in river waters.

Export of terrestrial organic matter exerts major influences in the carbon cycle on the ESS shelf. There is a significant amount of terrestrial OC stored within sediments, especially in the inner shelf most strongly influenced by coastal erosion between Dmitry Laptev Strait and the Kolyma River mouth, where the OC is mostly of terrestrial origin (80–100 %, (Semiletov et al. 2005)). However, only 50 % of the OC in the more offshore sediments underneath Pacific-origin water is of terrestrial origin (Semiletov et al. 2011; Vonk et al. 2010).

It is a challenge to precisely quantify annual river fluxes of DIC, TA, POC and DOC due to the very high seasonal variability, although the recently completed PARTNERS project greatly improved the data abundance and quality (Cooper et al. 2008). However, it is still difficult to quantify the percentage of riverine POC and DOC that is labile and will subsequently be respired to CO_2 once in the ocean margins. Although a major part, if not most, of the terrestrial organic carbon is expected to be respired in estuarine and shelf areas, the ultimate fate of this material is not well known (de Haas et al. 2002; Hansell et al. 2004). Riverine DIC plays a significant role in the DIC mass balance in the Arctic Ocean (~20 %, Anderson et al. 1990), as will the mineralized fraction of terrigenous DOC (Hansell et al. 2004; Letscher et al. 2011; Anderson et al. 2009). Taken together, the total Arctic riverine C flux is ~77 Tg C $year^{-1}$, similar to the direct CO_2 flux from the atmosphere to the sea (Bates and Mathis 2009). From our synthesis the total riverine C flux in the PAR is ~20 Tg C $year^{-1}$, which is only somewhat lower than the direct oceanic CO_2 uptake flux from the atmosphere (~40 Tg C $year^{-1}$, see next section). This river carbon input may increase rapidly as the Arctic warms and permafrost thaws. Thus, it is important to delineate this component of land-ocean carbon input.

In addition to Arctic fluvial fluxes into the Arctic Ocean, recent studies have shown the importance of lateral fluxes across the soil-water interface through Arctic coastal erosion (Jorgenson and Brown 2005; Ping et al. 2011). Total OC fluxes derived from coastal erosion have been estimated at 6–7 Tg C $year^{-1}$ (Stein and Macdonald 2004; McGuire et al. 2009), although large uncertainties exist. Jorgenson and Brown (2005) first estimated that eroding shorelines of the Beaufort Sea contribute ~0.18 Tg C $year^{-1}$. Based on estimated fluxes and scaled up to pan-Arctic regions (McGuire et al. 2009), coastal erosion contributes ~15 % of the total terrigenous OC flux to the Arctic Ocean (41 Tg C $year^{-1}$). This lateral export from pan-Arctic erosion is on the same order as riverine fluxes of POC, which have been estimated at 6–7 Tg C $year^{-1}$ (McGuire et al. 2009). Similar to coastal erosion, the river POC fluxes are mostly derived from the erosion of river banks and permafrost, based on evidence of radiocarbon composition of riverine POC and estuarine sediments (Guo and Macdonald 2006; Guo et al. 2007; Goñi et al. 2005). In addition, while river export is restricted in the estuarine region, materials exported from coastal erosion are dispersed along the entire Arctic Ocean coastline, allowing extensive biogeochemical cycling after erosion.

Lateral export fluxes of OC from coastal erosion have significant implications for the Arctic Ocean carbon cycle and marine ecosystem changes. Although only ~1–2 % of the total soil organic carbon could be released in dissolved form during soil leaching experiments (Dou et al. 2010), inputs of soil organic matter through both coastal erosion and river export and the subsequent degradation of both dissolved and particulate organic matter could significantly alter the water and environmental quality, shifting the Arctic coastal ecosystem to net heterotrophy (i.e., respiring more OC than it produces). Unfortunately, the fate of organic matter exported from coastal erosion and rivers is poorly known (Holmes et al. 2002). Further studies are needed to examine the lability and biogeochemical cycling of soil OC in the Arctic Ocean.

8.5 CO_2 Flux Across the Air-Sea Boundary

8.5.1 *Sea Surface* pCO_2 *Distribution*

A recent regional snapshot: Low seawater partial pressure of CO_2 (pCO_2) was observed in the Chukchi Sea during the 2008 China National Arctic Research Expedition (CHINARE), which includes the newly ice-free basin areas (Fig. 8.2). During this survey, the lowest seawater pCO_2 (120–250 μatm) was in marginal seas. However, in the ice-free region of the Canada Basin, there was a large area of relatively high seawater pCO_2 (320–365 μatm), though still below the atmospheric level (375 μatm). Further north (≥77°N), where melting of ice was less extensive, pCO_2 was <280 μatm. Surface pCO_2 also was low (<320 μatm) in areas west of 170°W where ice-cover was relatively heavy, temperature was low, and salinity was high.

In early studies of the Chukchi Sea, it was observed that seawater pCO_2 was much lower than that in the atmosphere during the sea-ice free period (i.e., the Arctic Ocean Section (AOS) expedition in 1994, (Jutterström and Anderson 2010). Subsequent studies have similarly observed low seawater pCO_2 on the Chukchi Sea shelf during summertime (Bates 2006; Murata and Takizawa 2003; Pipko et al. 2002; Bates et al. 2005; Chen and Gao 2007; Fransson et al. 2009; Andreev et al. 2010). Seasonal changes in surface pCO_2 on the Chukchi Sea shelf have been largely attributed to cooling of surface waters during the northward transit of waters across the Chukchi Sea shelf (Bates 2006; Murata and Takizawa 2003) and high rates of summertime phytoplankton primary production that decreases

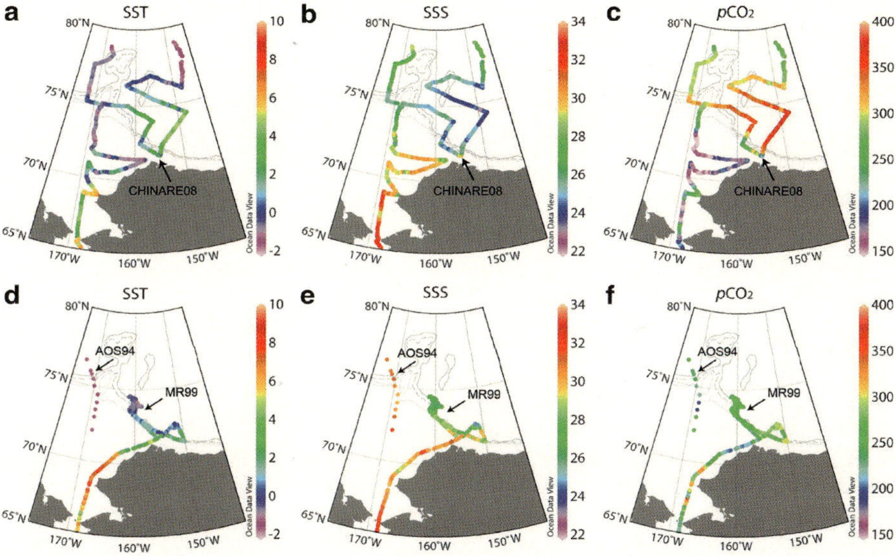

Fig. 8.2 Sea surface temperature (SST) (°C), sea surface salinity (SSS), and pCO_2 (μatm) distributions during the CHINARE08 survey and during AOS94 (Jutterström and Anderson 2010) and MR99 (Murata and Takizawa 2003) (Previously published in Cai et al. 2010. Use with permission)

seawater DIC and pCO_2 (Bates 2006). These processes act to produce a dynamic shelf-to-basin carbon pump (Anderson et al. 2010; Bates 2006). The seasonal rebound of seawater pCO_2 and DIC during wintertime likely results from a continued uptake of CO_2 through gas exchange during sea-ice formation and brine rejection (Anderson et al. 2004; Omar et al. 2005), continued transport of Pacific Ocean waters into the Chukchi Sea through Bering Strait and vertical entrainment by mixing with CO_2-enriched subsurface waters. These processes are discussed in details in the Mathis et al. (2014, this volume).

On the ESS shelf, surface water pCO_2 significantly above atmospheric values have been reported (Anderson et al. 2009, 2011; Pipko et al. 2011; Semiletov et al. 2007; Semiletov 1999), with much higher values near the outflow of the Kolyma River (Semiletov 1999; Semiletov et al. 2007). The high pCO_2 values can be attributed primarily to the remineralization of organic matter introduced from the coastal erosion and Siberian Rivers (Anderson et al. 1990; Cauwet and Sidorov 1996; Semiletov 1999; Semiletov et al. 2007, 2011).

In the Beaufort Sea, seawater pCO_2 appears to be highly variable (~150–350 μatm) in the western parts of the shelf (Bates 2006; Murata and Takizawa 2003), with low values observed in areas with high proportions of freshwater and sea-ice melt (Bates 2006; Murata and Takizawa 2003; Cooper et al. 2005). In the eastern Beaufort Sea shelf, summertime surface seawater pCO_2 values were generally low close to equilibrium with the atmosphere (Fransson et al. 2009).

The central Canada Basin has not been extensively sampled for the marine carbon cycle due to heavy ice-coverage. The AOS expedition in 1994 found that surface waters under sea-ice had seawater pCO_2 values lower than the atmosphere (Jutterström and Anderson 2010). Several sections across the central basin studied later have shown similar results (Jutterström and Anderson 2010). In the early 2000s, very low seawater pCO_2 values were observed in the Canada Basin off the Chukchi Sea shelf (Bates 2006) and in the Makarov Basin next to the Canada Basin (Fransson et al. 2009). However, more recently, Yamamoto-Kawai et al. (2009a) showed that some surface areas of the Canada Basin had seawater pCO_2 values close to equilibrium with the atmosphere in areas with significant sea-ice loss (particularly during the unprecedented sea-ice loss in 2007). After the major summertime sea-ice retreat observed in 2007, Cai et al. (2010) showed that ice-free (mostly southern) surface areas of the Canada Basin had seawater pCO_2 close to equilibrium with the atmosphere, reflecting uptake of CO_2 from the atmosphere and warming during the exposure of surface waters, whereas areas with heavy ice-cover (mostly the northern and western part) had lower pCO_2 (Fig. 8.2). These new data contrast with low pCO_2 values of 260–300 μatm observed in the summer of 1999 (Murata and Takizawa 2003) and the very low pCO_2 (<260 μatm) from the summer of 1994 (Jutterström and Anderson 2010) (Fig. 8.2).

8.5.2 Air-Sea CO_2 Flux

Estimating air-sea CO_2 fluxes in various areas of the Arctic Ocean has been an important goal over the last two decades. Early work based on DIC mass balance suggested that the entire Arctic Ocean absorbed atmospheric CO_2 at a rate of 24 Tg C year^{-1} (Anderson et al. 1998). Based on the reports prior to 2005, Cai et al. (2006), in a global ocean margin synthesis, assigned an overall atmospheric CO_2 uptake in the entire Arctic Ocean of 41 Tg C year^{-1}. Bates et al. (2005) and Bates (2006) estimated annual net air-to-sea CO_2 fluxes of 38 and 66 Tg C year^{-1}, respectively, for the Chukchi Sea and the entire Arctic Ocean margins. We thus adopt a conservative CO_2 uptake flux of 35 ± 6 Tg C year^{-1} in the Chukchi Sea and 10 Tg C year^{-1} in the Canadian Archipelago for the budget analysis in Fig. 8.3. Bates and Mathis (2009) summarized studies conducted prior to the 2007 major sea-ice loss event and concluded that the Arctic Ocean surface waters have the ability to absorb relatively large amounts of carbon dioxide (with a range of reported CO_2 sink rates ranging from 66 to 175 Tg C year^{-1} or 5–14 % of the global ocean CO_2 uptake). However, Cai et al. (2010) suggested that the Arctic Ocean basin could be a smaller sink for atmospheric CO_2. It is probably reasonable to assume that the recent smaller estimate in air-to-sea CO_2 flux in the basin areas reflects both an improvement in the coverage of survey areas and a changing status of CO_2 sinks and sources in the Arctic (Bates et al. 2011) (Table 8.4).

The CO_2 fluxes summarized in Table 8.3 are largely based on the recent syntheses of Bates and Mathis (2009) and Jutterström and Anderson (2010), and a reduced

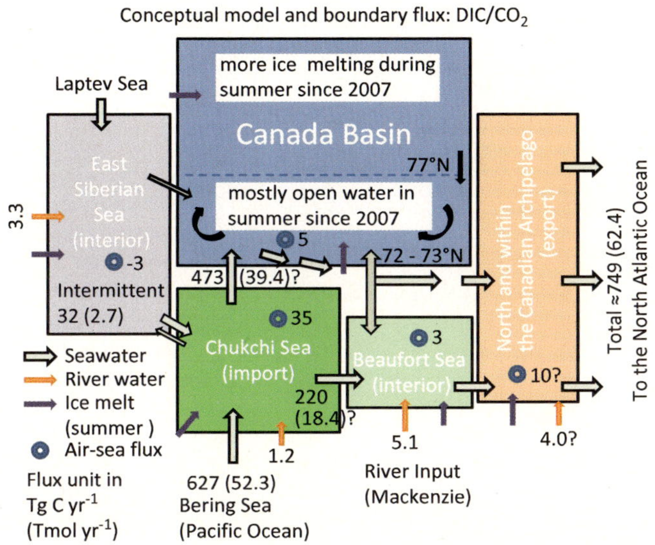

Fig. 8.3 Dissolved inorganic carbon (DIC) and air-sea CO_2 fluxes within the Pacific Arctic Region. The *larger hollow arrows* represent seawater DIC fluxes. The *orange color arrows* represent river water DIC fluxes. The *blue circles* indicate uptake of atmospheric CO_2 (values from Table 8.4 but in opposite sign as ocean gain is positive and loss is negative)

CO_2 flux estimate for the basin areas based on recent work of Cai et al. (2010). The accompanying chapter on C processes (Mathis et al. 2014, this volume) discusses the biogeochemical processes controlling the CO_2 fluxes and their climate sensitivity.

8.6 Impact of Seasonal Sea-Ice Cycle

The western Arctic Ocean margins experience extreme seasonal variations in insolation, sea-ice coverage, and freshwater inputs from Arctic rivers. Seasonal variations in sea-ice cover play an important role in shaping the water-masses and ecosystem. During the wintertime, sea-ice covers most of the shelf areas and the shelf water column is mostly well mixed. During the summer, a high Bering Strait inflow brings nutrient-rich Pacific Ocean water to the Chukchi Sea and beyond. Local sea-ice melt modifies this water to a relatively warm, fresher SML (Woodgate et al. 2005b).

In the southern Canada Basin, terrigenous freshwater often contributes as much as 10 % while sea-ice melt contributes about 5 % to the upper 50-m water column (Yamamoto-Kawai et al. 2009b). In recent summers with complete sea-ice melt at around 74°N, melt water contributes to freshwater inventory equally as the terrigenous input (Yamamoto-Kawai et al. 2009b). If we assume an average of

Table 8.4 Air-sea CO_2 exchange rates expressed in mmol m^{-2} day^{-1} and annual air-sea CO_2 exchange rates expressed in Tg C (or 10^{12} g C)

	Air-Sea flux (mmol m^{-2} day^{-1})	Annual flux (Tg C $year^{-1}$)	Reference
East Siberian Sea	−1 to −10.9	−1.2/−13	Semiletov et al. (2007)
	+0.3	+0.3	Nitishinsky et al. (2007)
	n/a	−5.9	Anderson et al. (1998)
Chukchi Sea	−12	−11	Murata and Takizawa (2003)
	−40 ± 22	−36 ± 6	Bates (2006)
	n/a	−46 ± 6	Bates (2006)
	n/a	−53 ± 14	Kaltin and Anderson (2005)
Beaufort Sea	n/a	−2.9	Anderson et al. (1998)
	−12	−2	Murata and Takizawa (2003)
Canadian Archipelago		−20	Bates and Mathis (2009) (by scaling to the Beaufort Sea)
Central basin	<−1 to −3	−6 to −19	Bates (2006)
		−5	Cai et al. (2010)
(PAR subtotal)		−50	This work (Fig. 8.3)
Arctic Ocean		−129 ± 65	Anderson et al. (1990)
		−70 ± 65	Anderson et al. (1994)
		−24 ± 17	Anderson et al. (1998)
		−41 ± 18	Anderson et al. (1998)
		−31	Kaltin and Anderson (2005)
		−66	Bates (2006)
		−66 to −199	Bates and Mathis (2009)

A negative air-sea CO_2 exchange rate indicates ocean uptake of CO_2 (i.e., CO_2 sink) (i.e., a loss term for the atmosphere). In Fig. 8.3, a negative value is converted to a positive one to show a gain for the ocean. Modified from (Bates and Mathis 2009)

6.0 ± 2 % (or 3 m) of melt water in the top 50-m water in half the area of the PAR (i.e., 3036×10^3 km^2, Table 8.1), then net freshwater flux to this area is 9.1×10^3 km^3 or 0.29 Sv during the sea-ice melt season. Moreover, based on a salinity-oxygen isotopic study, Yamamoto-Kawai et al. (2009b) concluded that the sea-ice melt component of surface water in the southern part of the Canada Basin has increased at a mean rate of 0.27 m/year in the top 50-m of the water column since 1987.

When sea ice is formed, crystals of water are produced and the salts are expelled, forming a concentrated brine solution. Theoretically, the cold brine solution leads to chemical oversaturation and precipitation of $CaCO_3$ (Richardson 1976). The precipitation will remove TA and DIC in a ratio of 2:1. Thus pCO_2 in the brine will increase. During recent fieldwork to the western Arctic Ocean, pCO_2 values observed in some ice melt ponds were up to 1,500 µatm (J. Mathis and N. Bates unpublished). The high pCO_2 in the brine can lead to a flux of CO_2 from the sea ice to the atmosphere or the water below, although the gas exchange velocity is likely very low and the inventory of the brine solution is small. With time the brine drains out of the sea ice to give a bulk sea-ice salinity of ~5 after one winter season to further decrease somewhat over coming seasons.

The chemical compositions of sea ice and brine have been studied in the field and laboratory. The composition was variable depending on the conditions that sea ice was exposed to in natural environments (Anderson and Jones 1985). High TA – DIC ratios have been shown in sea-ice melt waters collected north of Greenland (Rysgaard et al. 2007), presumably resulting from the dissolution of $CaCO_3$ crystals formed in the sea ice. It was further argued that, when the sea ice melts, the $CaCO_3$ dissolution would cause a pCO_2 undersaturation in the surface water and thus an uptake of CO_2 from the atmosphere (Rysgaard et al. 2007). The production of brine during sea ice formation and the transport of high density brine to deep depths, however, could also promote an uptake of CO_2 from the atmosphere as long as the surface water is undersaturated relative to the atmosphere (Anderson et al. 2004; Else et al. 2011).

Average values of sea-ice meltwater S, TA and DIC derived from the work of Rysgaard et al. (2007) are 5, 450 μmol/kg, and 400 μmol/kg respectively, which are equivalent to TA = 104.0 μmol/kg and DIC = 59.8 μmol/kg at 0 salinity (Cai et al. 2010). Taking the above sea-ice melt flux at face value, we derived a DIC flux at the level of 3.6 Tmol year^{-1} or 44 Tg C year^{-1} (with probably 33 % uncertainty). While this is not a trivial flux term, it is largely an internal or a seasonal source (during spring and summer) and sink (fall and winter). Therefore, for the annual net flux we do not count this part in the overall budget exercise given below. The current trend of increased summertime sea-ice melt will, however, require a reassessment of this issue.

8.7 Overall DIC Budget

DIC flux from the Pacific (627 TgC/year) and from the ESS (32 Tg C year^{-1}) to the western Arctic Ocean is estimated as 659 ± 21 Tg C year^{-1} based on the 2010 RUSALCA and the 2003–2004 SBI data (Table 8.2). We estimate DIC export flux to the Atlantic Ocean via the Canadian Archipelago by adopting a DIC value from an recent measurement that reported the average DIC value exported from the Archipelago into the Baffin Bay as 2,140 μmol kg^{-1} (Shadwick et al. 2011). We also assume a balance in the input and output of water, and thus, assign the total input water flux from the Pacific and the ESS to the export through the Archipelago (i.e., 0.90 Sv, see Fig. 8.1). Thus a total DIC export flux to the Atlantic Ocean is estimated as 749 Tg C year^{-1} with an unknown uncertainty probably larger than ±21 Tg C year^{-1}.

Most of the DIC flux into the region of the Chukchi and Beaufort Seas is from the Pacific Ocean whereas those from the ESS and from the atmosphere (38 Tg C year^{-1}) are also important. However, the total DIC fluxes from ocean currents and rivers and from atmospheric CO_2 uptake into the Chukchi and Beaufort Seas and the Canadian Archipelago (~704 Tg C year^{-1}) is slightly less than the export through the Archipelago (~749 Tg C year^{-1}). This difference is probably within the uncertainty of our estimates as we used the lower boundary of atmospheric CO_2 uptake rates and as DIC data from the exiting water (i.e., the Canadian Archipelago) are very limited. However, if we assume the values going into the budget analysis are accurate, then the total DIC input is less than the export by ~45 Tg C $year^{-1}$ or ~6 % of the total input or export. This result suggests that the PAR is slightly net heterotrophic, such that not only the OC generated in the highly productive Chukchi Sea is recycled within the system but also part of the OC from river and coastal erosion is respired there. Part of the OC production in the marginal seas must also be exported to the deep Arctic Ocean (Moran et al. 2005).

While high uncertainty is involved in deriving the above conclusion, it is consistent with our understanding of the physical and biogeochemical conditions in the Arctic Ocean. The Arctic Ocean is a land locked ocean with large inputs of organic and inorganic carbon from the rivers and coastal erosion. Newly produced marine OC is generally short-lived and recycled within the upper ocean, thus contributing only a small quantity to system-wide net production. Therefore, the conclusion that the western Arctic Ocean margin overall is net heterotrophic is consistent with our current understanding of global carbon cycle and biogeochemistry (Pipko et al. 2011; Cai 2011; Ducklow and McCallister 2004; Smith and Hollibaugh 1993). Admittedly, this conclusion is preliminary and requires both additional field observations and synthesis/modeling efforts. As warming progresses, we suggest that the Arctic Ocean likely will produce and thus export more DIC to the Atlantic Ocean. Whether this flux will turn the Arctic Ocean into a weaker CO_2 sink or even a CO_2 source to the atmosphere is currently unknown, and will depend on multiple factors controlling the rate of surface water pCO_2 increases versus that of the atmospheric CO_2 increase.

8.8 Summary

This chapter synthesizes inorganic and organic carbon fluxes across various boundaries in the Pacific Arctic Region (PAR). At ~50 Tg C $year^{-1}$, the PAR is a significant part of the global oceanic sink of atmospheric CO_2. Uptake of atmospheric CO_2, plus inputs from rivers and coastal erosion (including respired CO_2), enables the PAR to export more DIC to the Atlantic Ocean than it receives from the Pacific Ocean and other sources. Therefore, while there are large uncertainties in the flux estimates, with further warming, ice retreat, and transport of terrestrial OC (i.e., permafrost thawing and coastal erosion), the Arctic Ocean is likely to export more DIC to the Atlantic Ocean, contributing more DIC to the starting point of the global ocean deep water conveyor belt. It is unclear whether this trend will also lead to less CO_2 uptake from the atmosphere or even CO_2 release from the Arctic Ocean to the atmosphere in the future.

Acknowledgments We would like to thank Dr. Jackie M. Grebmeier and Dr. John A. Calder for their leadership in this effort. Review comments made by Lee Cooper and Arthur C-T. Chen greatly improved the quality of this chapter. Funding for synthesis activities was provided by the

Climate Program Office, R/CPO, of NOAA Arctic Research. W-JC acknowledges NOAA support through grant NA09OAR4310078 and NSF support through grants ARC-0909330 and PLR-1304337. NRB acknowledges NSF, NOAA and NASA support through NSF OPP-0732093, NSF OPP-1107457, NOAA NA080AR4310605, and NASA NNX10AG36G, respectively. DAH acknowledges NSF support through grants OPP-0822429 and OPP-0732082.

References

Aagaard K, Carmack EC (1989) The role of sea ice and other fresh water in the arctic circulation. J Geophys Res 94(C10):14485–14498. doi:10.1029/JC094iC10p14485

ACIA (2004) Impacts of a warming Arctic: Arctic climate impact assessment. Available at http://amap.no/acia/. Cambridge University Press, Cambridge

Anderson LG, Jones EP (1985) Measurements of total alkalinity, calcium, and sulfate in natural sea ice. J Geophys Res 90(C5):9194–9198. doi:10.1029/JC090iC05p09194

Anderson LG, Dyrssen D, Jones EP (1990) An assessment of the transport of atmospheric CO_2 into the Arctic Ocean. J Geophys Res 95(C2):1703–1711. doi:10.1029/JC095iC02p01703

Anderson LG, Olsson K, Skoog A (1994) Distribution of dissolved inorganic and organic carbon in the Eurasian basin of the Arctic Ocean. In: Johannessen OM, Muench RD, Overland JE (eds) The polar oceans and their role in shaping the global environment. American Geophysical Union, Washington, DC, pp 255–262

Anderson LG, Olsson K, Chierici M (1998) A carbon budget for the Arctic Ocean. Global Biogeochem Cycles 12:455–465

Anderson LG, Falck E, Jones EP, Jutterström S, Swift JH (2004) Enhanced uptake of atmospheric CO_2 during freezing of seawater: a field study in Storfjorden, Svalbard. J Geophys Res 109(C6):C06004. doi:10.1029/2003jc002120

Anderson LG, Jutterström S, Hjalmarsson S, Wåhlström I, Semiletov IP (2009) Out-gassing of CO_2 from Siberian Shelf seas by terrestrial organic matter decomposition. Geophys Res Lett 36(20):L20601. doi:10.1029/2009gl040046

Anderson LG, Tanhua T, Björk G, Hjalmarsson S, Jones EP, Jutterström S, Rudels B, Swift JH, Wåhlstöm I (2010) Arctic ocean shelf-basin interaction: an active continental shelf CO_2 pump and its impact on the degree of calcium carbonate solubility. Deep Sea Res I 57(7):869–879. doi:10.1016/j.dsr.2010.03.012

Anderson LG, Björk G, Jutterström S, Pipko I, Shakhova N, Semiletov I, Wåhlström I (2011) East Siberian Sea, an Arctic region of very high biogeochemical activity. Biogeosciences 8(6):1745–1754. doi:10.5194/bg-8-1745-2011

Andreev A, Chen C, Sereda N (2010) The distribution of the carbonate parameters in the waters of Anadyr Bay of the Bering Sea and in the western part of the Chukchi Sea. Oceanology 50(1):39–50. doi:10.1134/s0001437010010054

Arrigo KR, van Dijken GL, Pabi S (2008) Impact of a shrinking Arctic ice cover on marine primary production. Geophys Res Lett 35:L19603, doi:10.1029/2008GL035028

Bates NR (2006) Air-sea CO_2 fluxes and the continental shelf pump of carbon in the Chukchi Sea adjacent to the Arctic Ocean. J Geophys Res 111:C10013, 10010.10129/12005JC003083

Bates NR, Mathis JT (2009) The Arctic Ocean marine carbon cycle: evaluation of air-sea CO_2 exchanges, ocean acidification impacts and potential feedbacks. Biogeosciences 6:2433–2459

Bates NR, Best MHP, Hansell DA (2005) Spatio-temporal distribution of dissolved inorganic carbon and net community production in the Chukchi and Beaufort Seas. Deep Sea Res II 52(24–26):3303–3323. doi:10.1016/j.dsr2.2005.10.005

Bates NR, Cai W-J, Mathis JT (2011) The ocean carbon cycle in the western Arctic Ocean: distributions and air-sea fluxes of carbon dioxide. Oceanography 24(3):186–201. doi:10.5670/oceanog.2011.71

Borges AV, Dellile B, Frankignoulle M (2005) Budgeting sinks and sources of CO_2 in the coastal oceans: diversity of ecosystems counts. Geophys Res Lett 32:L14601, doi:10.1029/2005GL023053

Cai W-J (2011) Estuarine and coastal ocean carbon paradox: CO_2 sinks or sites of terrestrial carbon incineration? Annu Rev Mar Sci 3(1):123–145. doi:10.1146/annurev-marine-120709-142723

Cai W-J, Dai M, Wang Y (2006) Air-sea exchange of carbon dioxide in ocean margins: a province-based synthesis. Geophys Res Lett 33(12):L12603. doi:10.1029/2006gl026219

Cai W-J, Chen L, Chen B, Gao Z, Lee SH, Chen J, Pierrot D, Sullivan K, Wang Y, Hu X, Huang W-J, Zhang Y, Xu S, Murata A, Grebmeier JM, Jones EP, Zhang H (2010) Decrease in the CO_2 uptake capacity in an ice-free Arctic Ocean basin. Science 329(5991):556–559. doi:10.1126/science.1189338

Cauwet G, Sidorov I (1996) The biogeochemistry of Lena River: organic carbon and nutrients distribution. Mar Chem 53(3–4):211–227. doi:10.1016/0304-4203(95)00090-9

Chen C-TA, Borges AV (2009) Reconciling opposing views on carbon cycling in the coastal ocean: continental shelves as sinks and near-shore ecosystems as sources of atmospheric CO_2. Deep Sea Res II 56(8–10):578–590

Chen L, Gao Z (2007) Spatial variability in the partial pressures of CO_2 in the northern Bering and Chukchi seas. Deep Sea Res II 54:2619–2629

Chen CTA, Liu KK, Macdonald R (2003) Continental margin exchanges. In: Fasham MJR (ed) Ocean biogeochemistry: the role of the ocean carbon cycle in global change. Springer, New York, p 297

Codispoti LA, Richards FA (1963) Micronutrient distributions in the East Siberian and Laptev Seas during summer 1963. Arctic 21:67–83

Codispoti LA, Flagg C, Kelly V, Swift JH (2005) Hydrographic conditions during the 2002 SBI process experiments. Deep Sea Res II Top Stud Oceanogr 52(24–26):3199–3226

Codispoti LA, Flagg CN, Swift JH (2009) Hydrographic conditions during the 2004 SBI process experiments. Deep-Sea Res II 56:1144–1163

Cooper LW, Benner R, McClelland JW, Peterson BJ, Holmes RM, Raymond PA, Hansell DA, Grebmeier JM, Codispoti LA (2005) Linkages among runoff, dissolved organic carbon, and the stable oxygen isotope composition of seawater and other water mass indicators in the Arctic Ocean. J Geophys Res 110(G2):G02013. doi:10.1029/2005jg000031

Cooper LW, McClelland JW, Holmes RM, Raymond PA, Gibson JJ, Guay CK, Peterson BJ (2008) Flow-weighted values of runoff tracers ($\delta^{13}O$, DOC, Ba, alkalinity) from the six largest Arctic rivers. Geophys Res Lett 35:L18606. doi:10.1029/2008GL035007

Costard F, Gautier E, Brunstein D, Hammadi J, Fedorov A, Yang D, Dupeyrat L (2007) Impact of the global warming on the fluvial thermal erosion over the Lena River in Central Siberia. Geophys Res Lett 34(14):L14501. doi:10.1029/2007gl030212

Cota G, Pomeroy L, Harrison W, Jones E, Peters F, Sheldon WJ, Weingartner T (1996) Nutrients, primary production and microbial heterotrophy in the southeastern Chukchi Sea: arctic summer nutrient depletion and heterotrophy. Mar Ecol Prog Ser 135:247–258. doi:10.3354/meps135247

de Haas H, van Weering TCE, de Stigter H (2002) Organic carbon in shelf seas: sinks or sources, processes and products. Cont Shelf Res 22(5):691–717. doi:10.1016/s0278-4343(01)00093-0

Dou F, Yu X, Ping C-L, Michaelson G, Guo L, Jorgenson T (2010) Spatial variation of tundra soil organic carbon along the coastline of northern Alaska. Geoderma 154(3–4):328–335. doi:10.1016/j.geoderma.2009.10.020

Ducklow HW, McCallister SL (2004) The biogeochemistry of carbon dioxide in the coastal oceans. In: Robinson AR, Brink K (eds) The sea, vol 13. The global coastal ocean- multi-scale interdisciplinary processes. Harvard University Press, Boston, pp 269–315

Else BGT, Papakyriakou TN, Galley WMD RJ, Miller LA, Thomas H (2011) Wintertime CO_2 fluxes in an arctic polynya using eddy covariance: evidence for enhanced air-sea gas transfer during ice formation. J Geophys Res 116:C00G03. doi:10.1029/2010JC006760

Finlay J, Neff J, Zimov S, Davydova A, Davydov S (2006) Snowmelt dominance of dissolved organic carbon in high-latitude watersheds: implications for characterization and flux of river DOC. Geophys Res Lett 33(10):L10401. doi:10.1029/2006gl025754

Fransson A, Chierici M, Nojiri Y (2009) New insights into the spatial variability of the surface water carbon dioxide in varying sea ice conditions in the Arctic Ocean. Cont Shelf Res 29(10):1317–1328

Frey KE, Smith LC (2005) Amplified carbon release from vast West Siberian peatlands by 2100. Geophys Res Lett 32(9):L09401. doi:10.1029/2004gl022025

Goñi MA, Yunker MB, Macdonald RW, Eglinton TI (2005) The supply and preservation of ancient and modern components of organic carbon in the Canadian Beaufort Shelf of the Arctic Ocean. Mar Chem 93(1):53–73. doi:10.1016/j.marchem.2004.08.001

Guo L, Macdonald RW (2006) Source and transport of terrigenous organic matter in the upper Yukon River: evidence from isotope ($\delta^{13}C$, $\Delta^{14}C$, and $\delta^{15}N$) composition of dissolved, colloidal, and particulate phases. Glob Biogeochem Cycles 20:GB2011, doi:10.1029/2005GB002593

Guo L, Semiletov I, Gustafsson Ö, Ingri J, Andersson P, Dudarev O, White D (2004) Characterization of Siberian Arctic coastal sediments: implications for terrestrial organic carbon export. Global Biogeochem Cycles 18(1):GB1036. doi:10.1029/2003gb002087

Guo L, Ping C-L, Macdonald RW (2007) Mobilization pathways of organic carbon from permafrost to arctic rivers in a changing climate. Geophys Res Lett 34(13):L13603. doi:10.1029/2007gl030689

Guo L, Cai Y, Belzile C, Macdonald R (2012) Sources and export fluxes of inorganic and organic carbon and nutrient species from the seasonally ice-covered Yukon River. Biogeochemistry 107(1):187–206. doi:10.1007/s10533-010-9545-z

Gustafsson Ö, van Dongen BE, Vonk JE, Dudarev OV, Semiletov IP (2011) Widespread release of old carbon across the Siberian Arctic echoed by its large rivers. Biogeosci Discuss 8(1):1445–1461. doi:10.5194/bgd-8-1445-2011

Hansell DA, Kadko D, Bates NR (2004) Degradation of terrigenous dissolved organic carbon in the western Arctic Ocean. Science 304(5672):858–861. doi:10.1126/science.1096175

Hill V, Cota G (2005) Spatial patterns of primary production on the shelf, slope and basin of the Western Arctic in 2002. Deep Sea Res II 52(24–26):3344–3354. doi:10.1016/j.dsr2.2005.10.001

Holmes RM, McClelland JW, Peterson BJ, Shiklomanov IA, Shiklomanov AI, Zhulidov AV, Gordeev VV, Bobrovitskaya NN (2002) A circumpolar perspective on fluvial sediment flux to the Arctic Ocean. Global Biogeochem Cycles 16(4):1098. doi:10.1029/2001gb001849

Jorgenson MT, Brown J (2005) Classification of the Alaskan Beaufort Sea Coast and estimation of carbon and sediment inputs from coastal erosion. Geo-Mar Lett 25(2):69–80. doi:10.1007/s00367-004-0188-8

Jorgenson MT, Shur YL, Pullman ER (2006) Abrupt increase in permafrost degradation in Arctic Alaska. Geophys Res Lett 33(2):L02503. doi:10.1029/2005gl024960

Jutterström S, Anderson LG (2010) Uptake of CO_2 by the Arctic Ocean in a changing climate. Mar Chem 122(1–4):96–104. doi:10.1016/j.marchem.2010.07.002

Kaltin S, Anderson LG (2005) Uptake of Atmospheric carbon dioxide in Arctic shelf seas: evaluation of the relative importance of processes that influence pCO_2 in water transported over the Bering-Chukchi Sea shelf. Mar Chem 94:67–79

Lepore K, Moran SB, Grebmeier JM, Cooper LW, Lalande C, Maslowski W, Hill V, Bates NR, Hansell DA, Mathis JT, Kelly RP (2007) Seasonal and interannual changes in particulate organic carbon export and deposition in the Chukchi Sea. J Geophys Res 112(C10):C10024. doi:10.1029/2006jc003555

Letscher RT, Hansell DA, Kadko D (2011) Rapid removal of terrigenous dissolved organic carbon over the Eurasian shelves of the Arctic Ocean. Mar Chem 123(1–4):78–87. doi:10.1016/j.marchem.2010.10.002

Liu K-K, Atkinson L, Quiñones R, Talaue-McManus L (2010) Biogeochemistry of continental margins in a global context. In: Liu K-K, Atkinson L, Quiñones R, Talaue-McManus L (eds) Carbon and nutrient fluxes in continental margins: a global synthesis. Springer, Berlin

Lobbes JM, Fitznar HP, Kattner G (2000) Biogeochemical characteristics of dissolved and particulate organic matter in Russian rivers entering the Arctic Ocean. Geochim Cosmochim Acta 64(17):2973–2983. doi:10.1016/s0016-7037(00)00409-9

Macdonald RW, Anderson LG, Christensen JP, Miller LA, Semiletov IP, Stein R (2010) Polar margins: the Arctic 0. In: Liu K-K, Atkinson L, Quiñones R, Talaue-McManus L (eds) Carbon and nutrient fluxes in continental margins: a global synthesis. Springer, Berlin, pp 291–303

Mackenzie FT, Lerman A, Andersson AJ (2004) Past and present of sediment and carbon biogeochemical cycling models. Biogeosciences 1(1):11–32. doi:10.5194/bg-1-11-2004

Mathis JT, Pickart RS, Hansell DA, Kadko D, Bates NR (2007) Eddy transport of organic carbon and nutrients from the Chukchi Shelf: impact on the upper halocline of the western Arctic Ocean. J Geophys Res 112(C5):C05011. doi:10.1029/2006jc003899

Mathis JT, Grebmeier JM, Hansell DA, Hopcroft RR, Kirchman DL, Lee SH, Moran SB, Bates NR, VanLaningham S, Cross JN, Cai W-J (2014) Chapter 9: Carbon biogeochemistry of the western Arctic: primary production, carbon export and the controls on ocean acidification. In: Grebmeier JM, Maslowski W (eds) The Pacific Arctic region: ecosystem status and trends in a rapidly changing environment. Springer, Dordrecht, pp 223–268

McGuire AD, Anderson LG, Christensen TR, Dallimore S, Guo L, Hayes DJ, Heimann M, Lorenson TD, Macdonald RW, Roulet N (2009) Sensitivity of the carbon cycle in the Arctic to climate change. Ecol Monogr 79(4):523–555. doi:10.1890/08-2025.1

McLaughlin FA, Carmack EC, Macdonald RW, Melling H, Swift JH, Wheeler PA, Sherr BF, Sherr EB (2004) The joint roles of Pacific and Atlantic-origin waters in the Canada Basin, 1997–1998. Deep Sea Res I 51(1):107–128. doi:10.1016/j.dsr.2003.09.010

Michalak AM, Jackson RB, Marland G, Sabine CL (2011) A US carbon cycle science plan. Report of the carbon and climate working group for the U.S. Global Change Research Program. Washington, DC

Moran SB, Kelly RP, Hagstrom K, Smith JN, Grebmeier JM, Cooper LW, Cota GF, Walsh JJ, Bates NR, Hansell DA, Maslowski W, Nelson RP, Mulsow S (2005) Seasonal changes in POC export flux in the Chukchi Sea and implications for water column-benthic coupling in Arctic shelves. Deep Sea Res II 52(24–26):3427–3451. doi:10.1016/j.dsr2.2005.09.011

Murata A, Takizawa T (2003) Summertime CO_2 sinks in shelf and slope waters of the western Arctic Ocean. Cont Shelf Res 23:753–776

Nitishinsky M, Anderson LG, Hölemann JA (2007) Inorganic carbon and nutrient fluxes on the Arctic Shelf. Cont Shelf Res 27(10–11):1584–1599. doi:10.1016/j.csr.2007.01.019

Oechel WC, Vourlitis GL, Hastings SJ, Zulueta RC, Hinzman L, Kane D (2000) Acclimation of ecosystem CO_2 exchange in the Alaskan Arctic in response to decadal climate warming. Nature 406(6799):978–981, doi:http://www.nature.com/nature/journal/v406/n6799/suppinfo/406978a0_S1.html

Omar A, Johannessen T, Bellerby RGJ, Olsen A, Anderson LG, Kivin C (2005) Sea-ice and brine formation in Storfjorden: implications for the arctic wintertime air-sea CO_2 flux. American Geophysical Union, Washington, DC, ETATS-UNIS

Overland JE, Roach AT (1987) Northward flow in the Bering and Chukchi seas. J Geophys Res 92(C7):7097–7105. doi:10.1029/JC092iC07p07097

Perovich DK, Light B, Eicken H, Jones KF, Runciman K, Nghiem SV (2007) Increasing solar heating of the Arctic Ocean and adjacent seas, 1979–2005: attribution and role in the ice-albedo feedback. Geophys Res Lett 34(19):L19505. doi:10.1029/2007gl031480

Peterson BJ, Holmes RM, McClelland JW, Vörösmarty CJ, Lammers RB, Shiklomanov AI, Shiklomanov IA, Rahmstorf S (2002) Increasing river discharge to the Arctic Ocean. Science 298(5601):2171–2173. doi:10.1126/science.1077445

Pickart RS (2004) Shelfbreak circulation in the Alaskan Beaufort Sea: mean structure and variability. J Geophys Res 109(C4):C04024. doi:10.1029/2003jc001912

Pickart RS, Weingartner TJ, Pratt LJ, Zimmermann S, Torres DJ (2005) Flow of winter-transformed Pacific water into the Western Arctic. Deep Sea Res Part 52(24–26):3175–3198. doi:10.1016/j.dsr2.2005.10.009

Ping C-L, Michaelson GJ, Guo L, Jorgenson MT, Kanevskiy M, Shur Y, Dou F, Liang J (2011) Soil carbon and material fluxes across the eroding Alaska Beaufort Sea coastline. J Geophys Res 116(G2):G02004. doi:10.1029/2010jg001588

Pipko II, Semiletov IP, Tishchenko PY, Pugach SP, Christensen JP (2002) Carbonate chemistry dynamics in Bering Strait and the Chukchi Sea. Prog Oceanogr 55(1–2):77–94. doi:10.1016/s0079-6611(02)00071-x

Pipko II, Semiletov IP, Pugach SP, Wåhlström I, Anderson LG (2011) Interannual variability of air-sea CO_2 fluxes and carbon system in the East Siberian Sea. Biogeosciences 8(7):1987–2007. doi:10.5194/bg-8-1987-2011

Raymond PA, McClelland JW, Holmes RM, Zhulidov AV, Mull K, Peterson BJ, Striegl RG, Aiken GR, Gurtovaya TY (2007) Flux and age of dissolved organic carbon exported to the Arctic Ocean: a carbon isotopic study of the five largest arctic rivers. Global Biogeochem Cycles 21(4):GB4011. doi:10.1029/2007gb002934

Rember RD, Trefry JH (2004) Increased concentrations of dissolved trace metals and organic carbon during snowmelt in rivers of the Alaskan Arctic. Geochim Cosmochim Acta 68(3):477–489. doi:10.1016/s0016-7037(03)00458-7

Richardson C (1976) Phase relationships in sea ice as a function of temperature. J Glaciol 17:507–519

Roach AT, Aagaard K, Pease CH, Salo SA, Weingartner T, Pavlov V, Kulakov M (1995) Direct measurements of transport and water properties through the Bering Strait. J Geophys Res 100(C9):18443–18457. doi:10.1029/95jc01673

Rysgaard S, Glud RN, Sejr MK, Bendtsen J, Christensen PB (2007) Inorganic carbon transport during sea ice growth and decay: a carbon pump in polar seas. J Geophys Res 112:C03016, doi:10.1029/2006jc003572

Sarmiento JL, Wofsy S (1999) A U.S. carbon cycle science plan. Report of the carbon and climate working group for the U.S. Global Change Research Program. Washington, DC

Savelieva NI, Semiletov IP, Vasilevskaya LN, Pugach SP (2000) A climate shift in seasonal values of meteorological and hydrological parameters for Northeastern Asia. Prog Oceanogr 47(2–4):279–297. doi:10.1016/s0079-6611(00)00039-2

Semiletov IP (1999) Aquatic sources and sinks of CO_2 and CH_4 in the polar regions. J Atmos Sci 56(2):286–306. doi:10.1175/1520-0469(1999)056<0286:asasoc>2.0.co;2

Semiletov IP, Dudarev O, Luchin V, Charkin A, Shin K-H, Tanaka N (2005) The East Siberian Sea as a transition zone between Pacific-derived waters and Arctic shelf waters. Geophys Res Lett 32(10):L10614. doi:10.1029/2005gl022490

Semiletov IP, Pipko II, Repina I, Shakhova NE (2007) Carbonate chemistry dynamics and carbon dioxide fluxes across the atmosphere-ice-water interfaces in the Arctic Ocean: Pacific sector of the Arctic. J Mar Syst 66(1–4):204–226. doi:10.1016/j.jmarsys.2006.05.012

Semiletov IP, Pipko II, Shakhova NE, Dudarev OV, Pugach SP, Charkin AN, McRoy CP, Kosmach D, Gustafsson O (2011) On the biogeochemical signature of the Lena River from its headwaters to the Arctic Ocean. Biogeosci Discuss 8(2):2093–2143. doi:10.5194/bgd-8-2093-2011

Serreze M, Francis J (2006) The Arctic amplification debate. Clim Chang 76(3):241–264. doi:10.1007/s10584-005-9017-y

Shadwick EH, Thomas H, Gratton Y, Leong D, Moore SA, Papakyriakou T, Prowe AEF (2011) Export of Pacific carbon through the Arctic Archipelago to the North Atlantic. Cont Shelf Res 31(7–8):806–816. doi:10.1016/j.csr.2011.01.014

Smith SV, Hollibaugh JT (1993) Coastal metabolism and the oceanic organic-carbon balance. Rev Geophys 31(1):75–89

Spencer RGM, Aiken GR, Butler KD, Dornblaser MM, Striegl RG, Hernes PJ (2009) Utilizing chromophoric dissolved organic matter measurements to derive export and reactivity of dissolved organic carbon exported to the Arctic Ocean: a case study of the Yukon River, Alaska. Geophys Res Lett 36(6):L06401. doi:10.1029/2008gl036831

Springer AM, McRoy CP (1993) The paradox of pelagic food webs in the northern Bering Sea–III. Patterns of primary production. Cont Shelf Res 13(5–6):575–599. doi:10.1016/0278-4343(93)90095-f

Stein R, Macdonald RW (2004) The organic carbon cycle in the Arctic Ocean. Springer, New York

Striegl RG, Aiken GR, Dornblaser MM, Raymond PA, Wickland KP (2005) A decrease in discharge-normalized DOC export by the Yukon River during summer through autumn. Geophys Res Lett 32(21):L21413. doi:10.1029/2005gl024413

Striegl RG, Dornblaser MM, Aiken GR, Wickland KP, Raymond PA (2007) Carbon export and cycling by the Yukon, Tanana, and Porcupine rivers, Alaska, 2001–2005. Water Resour Res 43(2):W02411. doi:10.1029/2006wr005201

Takahashi T, Sutherland SC, Wanninkhof R, Sweeney C, Feely RA, Chipman DW, Hales B, Friederich G, Chavez F, Sabine C, Watson A, Bakker DCE, Schuster U, Metzl N, Yoshikawa-Inoue H, Ishii M, Midorikawa T, Nojiri Y, Körtzinger A, Steinhoff T, Hoppema M, Olafsson J, Arnarson TS, Tilbrook B, Johannessen T, Olsen A, Bellerby R, Wong CS, Delille B, Bates NR, de Baar HJW (2009) Climatological mean and decadal change in surface ocean pCO_2, and net sea-air CO_2 flux over the global oceans. Deep-Sea Res II 56(8–10):554–577. doi:10.1016/j.dsr2.2008.12.009

van Dongen BE, Semiletov I, Weijers JWH, Gustafsson Ö (2008) Contrasting lipid biomarker composition of terrestrial organic matter exported from across the Eurasian Arctic by the five great Russian Arctic rivers. Global Biogeochem Cycles 22(1):GB1011. doi:10.1029/2007gb002974

Vonk JE, Sánchez-García L, Semiletov I, Dudarev O, Eglinton T, Andersson A, Gustafsson Ö (2010) Molecular and radiocarbon constraints on sources and degradation of terrestrial organic carbon along the Kolyma paleoriver transect, East Siberian Sea. Biogeosciences 7(10):3153–3166. doi:10.5194/bg-7-3153-2010

Waelbroeck C, Monfray P, Oechel WC, Hastings S, Vourlitis G (1997) The impact of permafrost thawing on the carbon dynamics of tundra. Geophys Res Lett 24(3):229–232. doi:10.1029/97gl00071

Walsh JJ (1988) On the nature of continental shelves. Academic, San Diego

Walsh JJ, Dieterle DA, Muller-Karger FE, Aagaard K, Roach AT, Whitledge TE, Stockwell D (1997) CO_2 cycling in the coastal ocean. II. Seasonal organic loading of the Arctic Ocean from source waters in the Bering Sea. Cont Shelf Res 17(1):1–36. doi:10.1016/0278-4343(96)00021-0

Weingartner TJ, Cavalieri DJ, Aagaard K, Sasaki Y (1998) Circulation, dense water formation, and outflow on the northeast Chukchi shelf. J Geophys Res 103(C4):7647–7661. doi:10.1029/98jc00374

Wheeler PA, Gosselin M, Sherr E, Thibaultc D, Kirchman DL, Benner R, Whitledge TE (1996) Active cycling of organic carbon in the central Arctic Ocean. Nature 380(6576):697–699

Woodgate RA, Aagaard K, Weingartner TJ (2005a) Monthly temperature, salinity, and transport variability of the Bering Strait through flow. Geophys Res Lett 32(4):L04601. doi:10.1029/2004gl021880

Woodgate RA, Aagaard K, Weingartner TJ (2005b) A year in the physical oceanography of the Chukchi Sea: moored measurements from autumn 1990–1991. Deep Sea Res II 52:3116–3149

Woodgate RA, Weingartner T, Lindsay R (2010) The 2007 Bering Strait oceanic heat flux and anomalous Arctic sea-ice retreat. Geophys Res Lett 37(1):L01602. doi:10.1029/2009gl041621

Wu P, Wood R, Stott P (2005) Human influence on increasing Arctic river discharges. Geophys Res Lett 32(2):L02703. doi:10.1029/2004gl021570

Yamamoto-Kawai M, McLaughlin FA, Carmack EC, Nishino S, Shimada K (2009a) Aragonite undersaturation in the Arctic Ocean: effects of ocean acidification and sea ice melt. Science 326(5956):1098–1100. doi:10.1126/science.1174190

Yamamoto-Kawai M, McLaughlin FA, Carmack EC, Nishino S, Shimada K, Kurita N (2009b) Surface freshening of the Canada Basin, 2003–2007: river runoff versus sea ice meltwater. J Geophys Res 114:C00A05. doi:10.1029/2008JC005000

Chapter 9
Carbon Biogeochemistry of the Western Arctic: Primary Production, Carbon Export and the Controls on Ocean Acidification

Jeremy T. Mathis, Jacqueline M. Grebmeier, Dennis A. Hansell, Russell R. Hopcroft, David L. Kirchman, Sang H. Lee, S. Bradley Moran, Nicholas R. Bates, Sam VanLaningham, Jessica N. Cross, and Wei-Jun Cai

Abstract The Arctic Ocean is an important sink for atmospheric carbon dioxide (CO_2) with a recent estimate suggesting that the region accounts for as much as 15 % of the global uptake of CO_2. The western Arctic Ocean, in particular is a strong ocean sink for CO_2, especially in the Chukchi Sea during the open water

---

J.T. Mathis (✉)
Pacific Marine Environmental Laboratory, National Oceanic and Atmospheric Administration, 7600 Sand Point Way, Seattle, WA, USA

School of Fisheries and Ocean Sciences, University of Alaska Fairbanks, Fairbanks, AK, USA
e-mail: jeremy.mathis@noaa.gov

J.M. Grebmeier
Chesapeake Biological Laboratory, University of Maryland Center for Environmental Science, 146 Williams Street, Solomons, MD 20688, USA
e-mail: jgrebmei@umces.edu

D.A. Hansell
Rosenstiel School of Marine and Atmospheric Science, University of Miami, Miami, FL, USA

R.R. Hopcroft • S. VanLaningham • J.N. Cross
School of Fisheries and Ocean Sciences, University of Alaska Fairbanks, Fairbanks, AK, USA

D.L. Kirchman
School of Marine Science and Policy, University of Delaware, Lewes, DE, USA

S.H. Lee
Department of Oceanography, Pusan National University, Busan, South Korea

S.B. Moran
Graduate School of Oceanography, University of Rhode Island, Narragansett, RI, USA

N.R. Bates
Bermuda Institute of Ocean Sciences (BIOS), Ferry Reach, Bermuda

W.-J. Cai
School of Marine Science and Policy, The University of Delaware, Newark, DE, USA

J.M. Grebmeier and W. Maslowski (eds.), *The Pacific Arctic Region: Ecosystem Status and Trends in a Rapidly Changing Environment*, DOI 10.1007/978-94-017-8863-2_9,

season when rates of primary production can reach as high as 150 g C m^{-2}. The Arctic marine carbon cycle, the exchange of CO_2 between the ocean and atmosphere, and the fate of carbon fixed by marine phytoplankton appear particularly sensitive to environmental changes, including sea ice loss, warming temperatures, changes in the timing and location of primary production, changes in ocean circulation and freshwater inputs, and even the impacts of ocean acidification. In the near term, further sea ice loss and other environmental changes are expected to cause a limited net increase in primary production in Arctic surface waters. However, recent studies suggest that these enhanced rates of primary production could be short lived or not occur at all, as warming surface waters and increases in freshwater runoff and sea ice melt enhance stratification and limit mixing of nutrient-rich waters into the euphotic zone. Here, we provide a review of the current state of knowledge that exists about the rates of primary production in the western Arctic as well as the fate of organic carbon fixed by primary produces and role that these processes play in ocean acidification in the region.

Keywords Carbon cycle • Net community production • Grazing • Ocean acidification

9.1 Introduction

The Pacific sector of the western Arctic Ocean (Fig. 9.1) is now in a state of rapid transition, with potentially significant economic, social and environmental consequences. These changes are best exemplified by the marked warming and reduction in the thickness, extent, and seasonal persistence of ice cover witnessed in instrumental records over the last 30 years (e.g., Maslanik et al. 2007; Serreze et al. 2007, 2009). Sea-ice reduction is occurring more rapidly than predicted by global climate models (Stroeve et al. 2007; Polyak et al. 2010; Holland et al. 2010) and the observed changes in sea-ice cover are impacting the physical, chemical, and biological environments of established ecosystems as well as the terrestrial and atmospheric boundary conditions (e.g., via coastal erosion and ocean-atmosphere CO_2 flux). There is a growing consensus that the Arctic is passing 'tipping' points and exceeding ecological thresholds (Grebmeier et al. 2006a; Wassmann et al. 2008; Andersen et al. 2009; Sommerkorn and Hassol 2009). Therefore, it is now more critical than ever to synthesize existing and emerging datasets to provide context for the regional carbon system as the environment transitions to a new state.

The Arctic Ocean and adjacent polar continental shelf seas (Fig. 9.1) play an important role in the cycling of carbon, nutrients, and radiatively-important gases such as CO_2 (e.g., McGuire et al. 2010). Due to complex interactions and feedbacks (Fig. 9.2), this region is particularly sensitive to atmosphere-ocean-sea-ice physical forcing (Overland and Wang 2005; Wang et al. 2005), and ocean ecosystem changes such as the "greening" of the Arctic (Arrigo et al. 2008) associated with warming

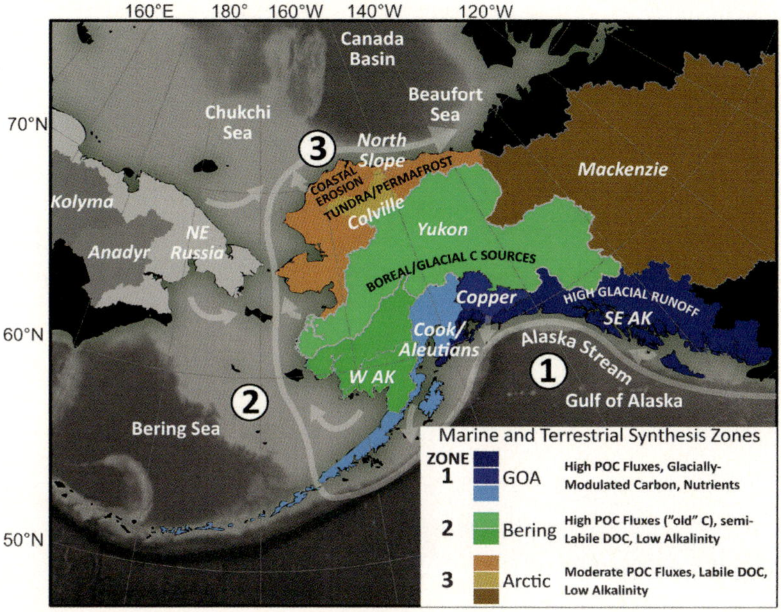

Fig. 9.1 Map of the Pacific Arctic Region (PAR) showing the major drainage basins and the associated biogeochemical properties that impact the carbon cycle. *1* Gulf of Alaska; *2* Bering Sea; and *3* Chukchi Sea. The *color bars* in the *inset* illustrate the various terrestrial drainage basins around North America and their predominant biogeochemical characteristics

temperatures (e.g., Serreze et al. 2009) and sea-ice loss (Moritz and Perovich 1996; Grebmeier and Whitledge 1996; Manabe and Stouffer 2000). For example, average temperatures in the Arctic have increased over the last century at nearly twice the global average, while sea ice extent has decreased on average by 2.7 % per decade (Cavalieri et al. 2003; Rothrock and Zhang 2005; Stroeve et al. 2005). However, since 2007, the pace of sea-ice decline has accelerated beyond model predictions (Stroeve et al. 2007; Holland et al. 2006; Winton 2006; Maslanik et al. 2007; Shimada et al. 2007) with summer sea-ice extent declining by more than 20–25 % (an additional loss of ~1.2 million km^2) coupled to a rapid loss of multi-year ice which has led to a reduction in overall ice volume. It is now predicted that the Arctic could be ice-free in the summertime within a few decades.

The importance of this system stems from the vast reservoirs of terrestrial and marine carbon in the Pacific-Arctic Region (PAR) (McGuire et al. 2006, 2009, 2010). Freshwater inputs (Peterson et al. 2002; Wu et al. 2005; Serreze and Francis 2006; Waddington and Roulet 1997), fluxes of dissolved organic carbon (DOC; Dittmar and Kattner 2003; Benner et al. 2004), particulate organic carbon (POC; Guo et al. 2004) and dissolved inorganic carbon (DIC) to the PAR shelves are disproportionately large (Opsahl et al. 1999) compared to other

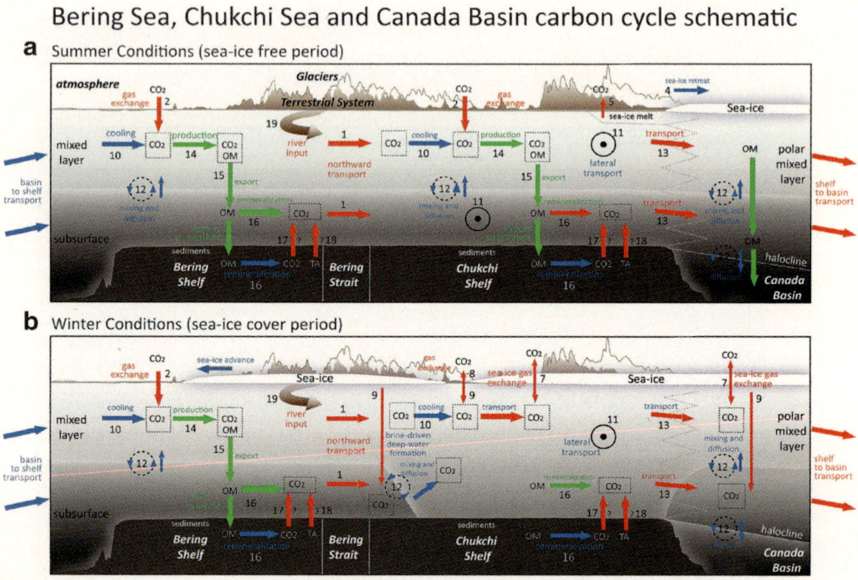

Fig. 9.2 Schematic of processes potentially influencing the inorganic carbon cycle and air-sea CO_2 gas exchange in the western Arctic Ocean from the Bering Sea shelf through the Bering Strait, across the Chukchi Sea and northwards into the Canada Basin (*left* to *right* on Figure) on "inflow" shelves of the Arctic (e.g., Barents and Chukchi seas). The *two panels* represent physical and biological processes likely operating during the summertime sea-ice free period (Panel **a**), and during the wintertime sea-ice covered period (Panel **b**). The processes are denoted by numeral with the caveat that the size of *arrow* does not necessarily reflect magnitude of flux, transport or transformation of CO_2. The processes include: *1* northward transport of DIC; *2* air-sea gas exchange; *3* warming; *4* exposure of surface water to the atmosphere due to sea-ice retreat and melting; *5* localized air-sea gas exchange from surface water highly influenced by sea-ice melt; *7* air-sea gas exchange through sea-ice; *8* winter air-sea gas exchange in leads and Polynyas; *9* inorganic carbon flux due to brine-rejection during deep-water formation in fall and winter. *10* cooling of surface waters during northward transport on Atlantic or Pacific Ocean waters into the Arctic Ocean; *11* between shelf transport of water and carbon; *12* redistribution of inorganic carbon between mixed layer and subsurface due to vertical diffusion and vertical entrainment/detrainment due to mixing; *13* shelf-basin exchanges of inorganic carbon (i.e., DIC) and organic carbon due to generalized circulation and eddy mediated transport; *14* net uptake of CO_2 due to phytoplankton photosynthesis or new production; *15* export flux of organic matter (OM) or export production; *16* remineralization of organic matter back to CO_2 either in subsurface waters or in sediments; *17* release of CO_2 from sediments; *18* release of alkalinity from sediments due to anaerobic processes in sediments, and; *19* river runoff input (Adapted from Bates and Mathis 2009)

basins (McGuire et al. 2009), and the responses of these carbon stocks and fluxes to projected climate change is highly uncertain under projected rapid changes in sea-ice cover and increased fluxes of glacial and fluvial source waters. Warming, sea-ice loss, "greening" of the Arctic and changes in physical circulation, stratification and freshwater inputs are important potential drivers on the carbon cycle but there are multiple synergistic processes at play (Bates and Mathis 2009; see Fig. 9.2).

The loss of sea-ice has the potential to increase the CO_2 sink in the region (e.g., Jutterström et al. 2010; Bates et al. 2010) to globally significant rates (~5–12 % of global; Bates and Mathis 2009). However, these shelves are highly variable in their CO_2 sink or source status. Adding complexity to such assessments is that the 2007 sea-ice loss event may have reset the Arctic's capacity to absorb CO_2 (Cai et al. 2010b) and that "greening" of the Arctic, and other physical changes, such as warming have opposite impacts on CO_2 sink or source potential of the Arctic Ocean (Bates and Mathis 2009). Furthermore, the uptake of anthropogenic CO_2, river/glacial inputs and sea-ice melt also appear to be amplifying the impact of ocean acidification on the saturation states of calcium carbonate ($CaCO_3$) minerals in the Pacific-Arctic region (e.g., Steinacher et al. 2009; Bates and Mathis 2009; Yamamoto-Kawai et al. 2009; Mathis et al. 2011a, b). It is thus critical to gain a better understanding of the interactions and feedbacks of the ocean carbon cycle to freshwater, sea-ice dynamics, climate and environmental change in this region. But significant gaps and uncertainties remain that limit our understanding of the PAR carbon sources and sinks and the linkages and feedbacks between carbon and climate (Anderson and Kaltin 2001).

The gateway fluxes of carbon as well as the atmosphere-terrestrial-ocean exchanges are discussed in the previous chapter. Here, the focus will be on the internal water column processes that impact the distribution and cycling of carbon in the PAR. The discussion starts with primary production and its contribution to the dissolved and particulate pools of carbon as well as the associated fluxes and fate of each. This is followed by a discussion of microbial remineralization of organic carbon in the water column and the role that sediments play in processing and storing carbon. Finally, we discuss the controls and potential impacts of ocean acidification in the PAR.

9.2 Primary Production

Over the past several decades, higher temperatures have decreased the extent of sea-ice cover as well as its thickness in the Arctic Ocean. Because of this, the overall amount of perennial sea ice, especially in the western Arctic Ocean, has been reduced (Perovich and Richter-Menge 2009). These recent changes in climate and ice conditions could change the patterns and the total amount of carbon production from phytoplankton (Fig. 9.2), with unknown consequences to higher trophic levels. However, there is some controversy as to whether changing climate conditions enhance or reduce the overall production in the Arctic Ocean. Since the responses of phytoplankton production to current environmental conditions vary by region, three main regions of the western Arctic Ocean—the northern Bering Sea, Chukchi Sea, and the deep Canada Basin are considered separately.

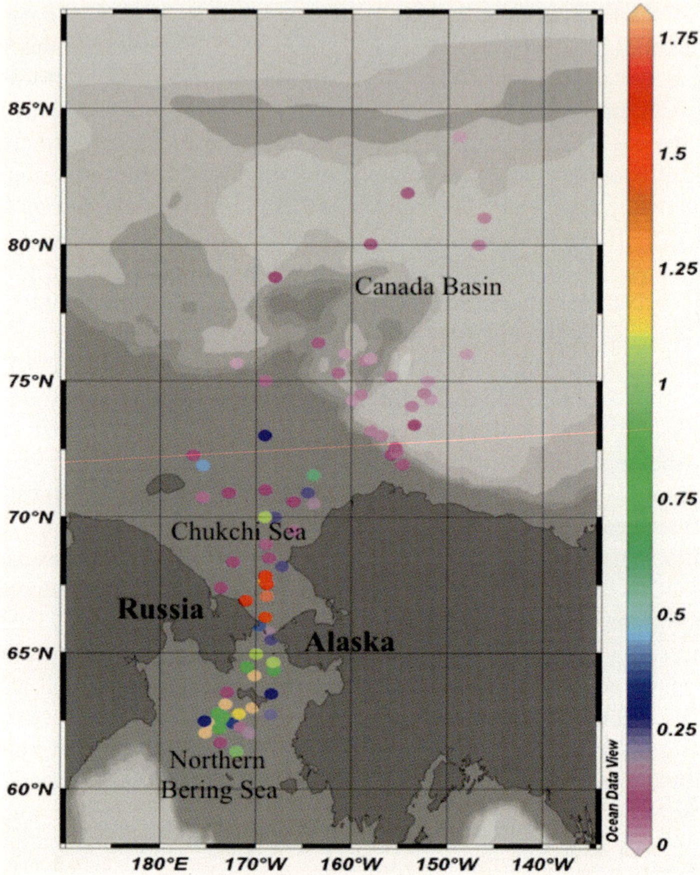

Fig. 9.3 Carbon uptake rates (g C m^{-2} day^{-1}) integrated to the 1 % light depth in the PAR (Data from Lee and Whitledge 2005; Lee et al. 2007, 2010, 2011)

9.2.1 Northern Bering Sea

The northern Bering Sea (Fig. 9.1) is a seasonally ice-covered shelf that is strongly influenced by the advection of cold, nutrient-rich Pacific water from the edge of the deep Bering Sea basin (Springer 1988; Grebmeier et al. 2006a). Recently measured inorganic carbon uptake rates of phytoplankton ranged from 0.1 to 3.9 g C m^{-2} day^{-1} (mean ± S.D. = 1.1 ± 1.2 g C m^{-2} day^{-1}) and from 0.2 to 2.0 g C m^{-2} (mean ± S.D. = 0.9 ± 0.6 g C m^{-2} day^{-1}) (south and north of St. Lawrence Island), respectively. These data were recorded in 2007 (Fig. 9.3), based on a 15-h photoperiod (Hansell and Goering 1990) and hourly carbon uptake rates. For comparative purposes although measurement methods are different among the studies, the annual primary production was estimated assuming a 120-day growing season in this region (Hansell et al. 1993; Sambrotto et al. 1984; Mathis et al. 2010). The estimated annual phytoplankton production was only 120 g C m^{-2} based on an average daily carbon production rate of about 1.0 g C m^{-2} day^{-1} during the whole cruise period from mid-May to June, 2007. These estimated annual rates are two to three times lower than those in previous studies in the northern Bering Sea done more than a decade ago (250–480 g C m^{-2}—see Hansell and Goering 1990; Springer and McRoy 1993; Springer et al. 1996). Recently, Grebmeier et al. (2006b) found that geographic displacement of marine mammal populations coincided with a reduction of benthic prey populations in the region from 1988 to 2004. They hypothesized that ecosystem change and declining productivity are reducing food supply to benthic prey, thus affecting apex predators (Grebmeier et al. 2006b). The recent lower total primary production and subsequent decline of benthic biomass might partly be due to a decrease in the phytoplankton biomass transported from lower latitudes, because the amount of phytoplankton production largely depends on the carbon biomass. Since northward flowing waters in the northern Bering Sea originate from the Pacific Ocean (Danielson et al. 2006), a decline of phytoplankton biomass transport as a seed population of phytoplankton from lower latitudes in the Bering Sea could cause a decrease in phytoplankton production in the higher latitudes of the Bering Sea.

9.2.2 *Chukchi Sea*

Northward flow of water through the narrow, shallow (<50 m) Bering Strait carries three water masses or branches, into the Chukchi Sea (Coachman et al. 1975): Anadyr Water (AW) in the west, Bering Shelf (BS) water in the central system and the Alaska Coastal Current (ACC) in the east. AW supplies the Chukchi continental shelf with high nutrients that promote abundant phytoplankton growth throughout the summer and transports oceanic zooplankton onto the northern shelf (Sambrotto et al. 1984; Springer et al. 1989; Springer and McRoy 1993). The location and direction of the three water masses moving through the Bering Strait have a strong influence on the physical conditions, nutrient concentrations, and phytoplankton communities observed in this important gateway to the Arctic Ocean (Springer and McRoy 1993; Lee et al. 2007). For example, Lee et al. (2007) found two distinctly different size compositions of phytoplankton communities in the different water masses in the Chukchi Sea. Large phytoplankton (>20 μm) contributed about 42 % of total phytoplankton biomass in the ACW, whereas they are dominant (about 94 %) in AW (Lee et al. 2007).

In previous studies, Sambrotto et al. (1984) estimated an annual total primary production of 324 g C m^{-2}, based on nitrate uptake and an *f*-ratio, while Hansell et al. (1993) calculated 576–720 g C m^{-2} for annual total production in the Anadyr flow north of the Bering Strait based on new production in the region. Based on ^{14}C uptake and chlorophyll-a concentrations in the region, Springer and McRoy (1993) estimated an annual production of 470 g C m^{-2}. These rates are unusually large considering the high latitude Arctic location (Sambrotto et al. 1984). However, Lee et al. (2007) recently found much lower rates in the region (Fig. 9.3), based on data from 2002 to 2004. The daily carbon uptake rates in these years ranged from 0.1 g C m^{-2} day^{-1} to 3.6 g C m^{-2} day^{-1} with a mean of 0.7 g C m^{-2} day^{-1} (S.D. = ±0.9 g C m^{-2} day^{-1}) (Lee et al. 2007). The estimates of production in the southern part of the Chukchi Sea during late July to August from previous studies were 1.7 g C m^{-2} day^{-1} with a range from 0.4 to 4.7 g C m^{-2} day^{-1} (Korsak 1992) and 1.6 g C m^{-2} day^{-1} with a range from 0.2 to 5.5 g C m^{-2} day^{-1} (Zeeman 1992). In the southern part of the Chukchi Sea, Lee et al. (2007) found 0.6 g C m^{-2} day^{-1} with a range from 0.1 to 1.5 g C m^{-2} day^{-1}, which is less than half of the values from the previous studies. The daily production in the southern Chukchi Sea measured by this study was somewhat lower than the rate (0.8 g C m^{-2} day^{-1}) on the northeastern shelf region reported by Hill et al. 2005. Although the production range (0.1–1.0 g C m^{-2} day^{-1}) from Hameedi (1978) is similar to the range of Lee et al. (2007), production was measured in the marginal ice zone of the northern Chukchi Sea during summer, rather than open water. The physical and chemical structures and thus production could be different in the water column of these two contrasting environments.

Based on a 100-day growing season, the recent annual total production of phytoplankton was between 10 and 150 g C m^{-2} with a mean of 73 g C m^{-2} for the southern Chukchi Sea, while the mean annual total production for the whole Chukchi Sea including the northwestern part was somewhat lower (55 g C m^{-2}) (Lee et al. 2007). In comparison, the estimated averages of annual total primary production rates for the whole Chukchi Sea are 148 and 170 g C m^{-2} from Zeeman (1992) and Korsak (1992), respectively. Although the average values of their production rates are comparable to the highest values in the study of Lee et al. (2007), the recent average annual production is 2 or 3 times lower than the previous estimates. The average production based on an interval of 120 days is estimated at 144 g C m^{-2} y^{-1} by Lee et al. (2007) which is still lower than that estimated by Hansell et al. (1993) (576–720 g C m^{-2} y^{-1}) or Sambrotto et al. (1984) (324 g C m^{-2} y^{-1}). The recent lower rates might be the result of seasonal, annual, and/or geographical variations in primary productivity in the Chukchi Sea. Those variations are well known in this area and are mostly attributed to different water masses and thus nutrient concentrations (Springer and McRoy 1993; Springer 1988; Hansell and Goering 1990). The lower production might be an indication of the recent decline in primary production of lower latitude regions in the Bering Sea. Gregg and Conkright (2002) found that chlorophyll-a concentrations decreased in the North Pacific from 1979–1986 to 1997–2000 period, suggesting that the decline in productivity may be a region-wide occurrence. Recently, small size phytoplankton (<2 μm) thrive, but larger cells decline in the Canada Basin because of freshened surface waters in recent years. In the Chukchi Sea, Woodgate et al. (2005) found that the ACC through the Bering Strait is recently fresher and warmer than previous years. In addition, Kaltin and Anderson (2005) suggested that much of the primary production previously reported in the Bering and Chukchi seas is a result of regeneration. Therefore, changes in species composition of phytoplankton might cause the recent lower primary production in the Chukchi Sea. More seasonal and annual measurements will be necessary to determine whether the recently detected decrease in phytoplankton productivity results from a common seasonal or intra-annual variation or an overall trend in the northern Bering Sea and the Chukchi Sea.

9.2.3 Deep Canada Basin

The Canada Basin (Fig. 9.1) represents one of the deepest parts of the Arctic Ocean and is covered with sea ice for most of the year. Pelagic primary production has been estimated rarely in this region (Cota et al. 1996; Gosselin et al. 1997; Chen et al. 2003; Lee and Whitledge 2005; Fig. 9.3–Lee et al. 2010). Opposite to the Bering and Chukchi Sea, recent rates of total primary production in the Canada Basin are higher than those previously measured (Apollonio 1959; English 1961; Pautzke 1979).

In general, chlorophyll-a at surface is very low (<0.2 mg Chl a m^{-3}), whereas the relatively higher concentrations (<1 mg Chl a m^{-3}) of the deep chlorophyll maximum layer is prominent between 40 and 60 m water depths depending on season (Lee and Whitledge 2005; Lee et al. 2010). The maximum photosynthetic rates were co-located with the chlorophyll maximum (Lee and Whitledge 2005). Total primary production rates in the open water ranged from 79 to 145 mg C m^{-2} day^{-1}, with a mean of 106 mg C m^{-2} from mid-August to early September 2002. These rates are much lower than those estimated in the eastern Canadian Arctic (Frobisher Bay and northern Baffin Bay), which were 227–450 mg C m^{-2} (Grainger 1975; Harrison et al. 1982), but comparable to those in the Canada Basin (Cota et al. 1996). The rates below snow covered sea ice are much lower than those in open waters due to low light intensity. Gosselin et al. (1997) found that the mean daily carbon uptake rate of phytoplankton was 35 mg C m^{-2} day^{-1} under a sea ice thickness of about 2.0-m in the Canada Basin from 26 July to 26 August 1994. The total primary production below sea ice of 2.3 m thickness ranged from 2.6 to 26.8 mg C m^{-2} day^{-1}, with a mean of 11.3 mg C m^{-2} in a study from 16 August to 5 September by Lee and Whitledge (2005). In comparison, the rates under thinner sea ice (1.5 m) ranged from 20.4 to 178.3 mg C m^{-2} day^{-1} in the Canada Basin from 27 June to 26 July 2005, with a mean of 59.5 mg C m^{-2} day^{-1} (Lee et al. 2010). This was significantly higher than found under thicker sea ice in 2002 (Lee and Whitledge 2005). Higher averaged particulate organic carbon and chlorophyll-a (POC/Chl-a) and particulate organic nitrogen (PON/Chl-a) of phytoplankton in 2005 compared to 2002 indicate that phytoplankton were less light limited in 2005 (Lee et al. 2010).

Since there are seasonal and interannual changes in the photosynthetic rates in the Arctic Ocean (English 1961; Pautzke 1979), annual primary production must be estimated with caution. However, for a comparison purpose, the annual total carbon production can be roughly estimated with the assumption of a 120-day growing season in the Arctic (Subba Rao and Platt 1984; Gosselin et al. 1997; Lee and Whitledge 2005). The annual total carbon production rate of phytoplankton under a mean sea-ice thickness of 1.5 m can range from 2.5 to 21.4 g C m^{-2}, with a mean of 7.1 g C m^{-2} in the Canada Basin in 2005 (Lee et al. 2010). This mean rate under sea ice is comparable with that in the open water column (8.9 g C m^{-2}) (Cota et al. 1996). Compared with a previous study, the under-ice productivity in 2005 is about one order of magnitude greater than that estimated by English (1961) (about 0.6 g C m^{-2}) under 2.5–3.0 m pack ice in the central Arctic Ocean several decades ago. If portions of ice algal production and the release of extracellular carbon are added to the total (Gosselin et al. 1997), the annual rates of total primary production under-ice in the Canada Basin will be higher than the estimation of Lee et al. (2010).

The decreasing sea-ice thickness in the Canada Basin (Rothrock et al. 2003; Nghiem et al. 2007; Perovich and Richter-Menge 2009) might be favorable for increased phytoplankton growth under the sea ice because light intensity governed by different sea-ice thicknesses is one of the most important factor affecting phytoplankton production in the Canada Basin (Lee and Whitledge 2005; Lee et al. 2010). However, phytoplankton adapted to low-light conditions under sea ice would show different effects if the sea ice cover melted away in the Arctic Ocean. In fact, nutrients were shown to be the main limiting factors for phytoplankton production rates at the surface in open waters, whereas light was the major limiting factor at the chlorophyll maximum layer just above the pycnocline in the Canada Basin (Lee and Whitledge 2005). Codispoti et al. (2009) also found the biological production rates to be low because of limited availability of nutrients in their SBI (Western Arctic Shelf-Basin Interaction) studies. More seasonal and annual data under a variety of environmental conditions in different regions should be obtained to improve the understanding of primary production processes in the Arctic and marine ecosystem responses to recent ongoing changes in sea-ice conditions in the Arctic Ocean.

The varying degree and spatial distribution of primary production in the Pacific-Arctic Region plays a strong role in determining the concentrations of organic matter and export of particles from the surface. Each of these processes and their impact on carbon distribution in the water column will be discussed independently.

9.3 DOC Production

An estimated 9 Pg of carbon reside in the Arctic Ocean as DOC (McGuire et al. 2009), or <2 % of the 662 Pg global DOC inventory (Hansell et al. 2009). An interesting feature in the system, however, is that DOC in the surface ocean has a very strong terrigenous signature (Wheeler et al. 1996; Opsahl et al. 1999), due to the Arctic Ocean being a relatively small basin receiving disproportionate fractions of global river runoff (~10 %) and terrigenous DOC (tDOC) export (~13 %) (Stein and Macdonald 2004). Ten to twenty percent of global vegetative carbon, and up to half the global inventory of upper-soil organic carbon, resides in the Arctic watersheds (Dixon et al. 1994; Zhulidov 1997; McGuire et al. 2009). Because of Arctic water entrainment into the deep North Atlantic, a potential (but unquantified) route exists for this terrigenous DOC to be transported to and sequestered in the ocean interior (Benner et al. 2005), unlike other regions of the global ocean. Hence, consideration of the terrigenous pool is included in most studies of Arctic DOC.

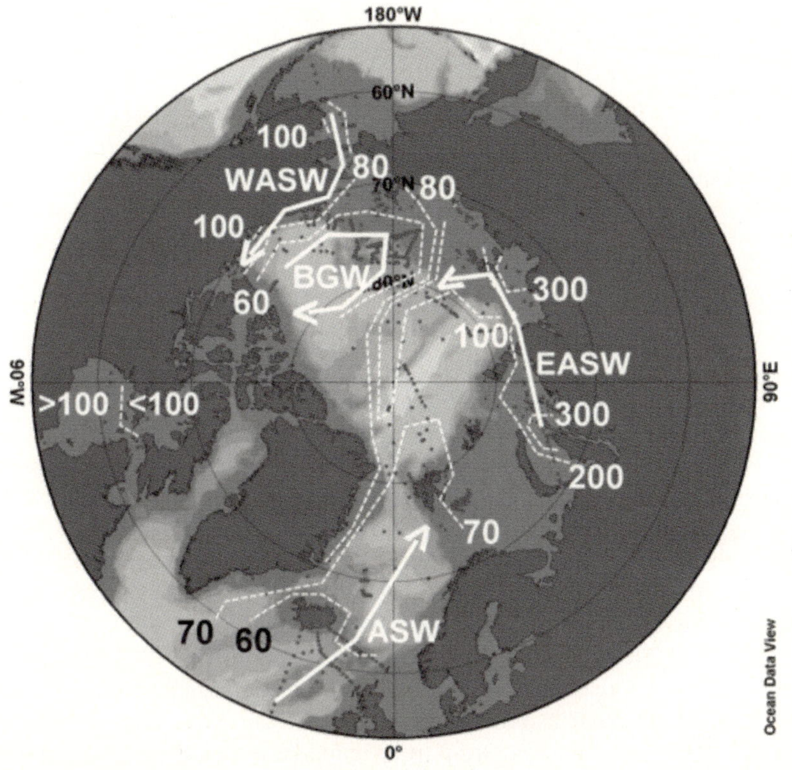

Fig. 9.4 Contoured distributions of DOC concentrations in the surface layer of the Arctic Ocean (measurement locations are indicated with *filled marks*). Also shown are water masses and circulations of particular relevance to DOC distributions, including Beaufort Gyre Water (BGW), Western Arctic Shelf Water (WASW), Eastern Arctic Shelf Water (EASW), and Atlantic Surface Water (ASW) (The data are collected from the personal data archives of the Arctic carbon community)

This section briefly considers controls on DOC concentrations in the western Arctic Ocean. Emphasis is on the allochthonous (terrigenous) inputs and marine consumption of tDOC. Excellent reviews of DOC in the Arctic carbon cycle, including fluvial and marine influences, can be found in Anderson (2002), Dittmar and Kattner (2003), and Amon (2004). Those works include consideration of organic composition, colored dissolved organic matter (CDOM), and the eastern Arctic, issues not addressed here.

9.3.1 Spatial Variability

The distribution of DOC in the surface layer of the Arctic Ocean is shown in Fig. 9.4. DOC concentrations are low in marine waters entering the Arctic: <60 μmol kg^{-1} in surface North Atlantic waters (Fig. 9.4) and similar values in the far North Pacific (Hansell et al. 2009). DOC is added to these marine waters as they invade the Arctic

Ocean shelves, both by autochthonous and allochthonous processes (Shin and Tanaka 2004; Amon 2004; Mathis et al. 2007). The DOC concentrations are highest near the outflows of the major rivers on the shelves of the Eastern Arctic, with concentrations >300 μmol kg^{-1} observed over relatively large areas, having been enriched with DOC of a terrigenous source (tDOC). Remnants of those tDOC-enriched waters leave the shelf to form the Transpolar Drift (Letscher et al. 2011), thus delivering elevated DOC waters to the central Arctic basins; note the elevated DOC (>100 um/kg) aligned with the dateline in Fig. 9.4. These high concentrations, and the responsible shelf-to-basin export of tDOC, were noted by Wheeler et al. (1996) and Bussmann and Kattner (2000). Similar enrichments occur on the western shelves, but to a lesser degree due to lower initial tDOC concentrations in the western rivers and lower flow (Cooper et al. 2008). Concentrations can be >100 μmol kg^{-1} in the interior of Norton Sound, reflecting drainage of the Yukon River. Similar concentrations exist near the mouth of the Mackenzie River. Note, though, that DOC concentrations are much more elevated further into the mouths of all rivers, relative to concentrations shown in Fig. 9.4 (e.g., DOC >300 μmol kg^{-1} near the mouth of the Mackenzie River; Guéguen et al. 2005). Based on these distributions, it appears that the tDOC surviving shelf circulation of 2–5 years in the eastern Arctic (Schlosser et al. 1994; Ekwurzel et al. 2001; Karcher and Oberhuber 2002) is exported to the deep Arctic basins, with eventual export (after another 2–5 years) (Letscher et al. 2011) via Fram Strait as the East Greenland Current (DOC > 80 μmol kg^{-1} along the east coast of Greenland). Western tDOC, in contrast, is largely transported along-shelf for eventual export through the Canadian Archipelago. Some of the western tDOC is mixed into the Beaufort Gyre for decadal circulation and mineralization (Hansell et al. 2004), resulting in concentrations of <60 μmol kg^{-1}, the lowest observed in the surface Arctic Ocean.

As can be inferred from the distributions in Fig. 9.4 and as suggested by Opsahl et al. (1999), the fraction of bulk DOC present as tDOC varies widely throughout the system. For example, at the shelf break in the Chukchi Sea during spring 2002 (prior to major sea ice melt), DOC concentrations (normalized to a salinity of 33, indicative of Pacific water) reached ~90 μmol kg^{-1} at a salinity of 29. If there was no tDOC present and if additional marine DOC input is small, the normalized DOC concentration would equal that in the marine fraction alone (the Pacific waters had ~65 μmol kg^{-1}). The difference in concentrations is the tDOC present; in this case, $90 - 65 = 25$ μmol kg^{-1}, or ~38 % of the total DOC was tDOC. In contrast, in higher salinity (essentially purely marine) surface waters, where measured DOC was ~65 μmol kg^{-1}, none of the DOC was terrigenous. All waters sampled in the Chukchi system fell within this range (0–38 %). Kattner et al. (1999) estimated that ~60 % of the DOC in the surface layer of the Laptev Sea is of terrigenous origin. As these systems are constantly mixing low salinity, high tDOC fluvial waters with high salinity, low tDOC marine waters, the contribution of tDOC will vary as a function of salinity and time elapsed since the tDOC entered the marine system.

9.3.2 The Use of DOC/Salinity Relationships

Because of this constant mixing of distinct waters, DOC-salinity relationships are commonly employed in studying DOC dynamics of the system. The apparent linearity in the correlation between the two variables led a number of studies to conclude that tDOC is highly conserved as it is mixed into the Arctic Ocean. tDOC does appear to be conserved upon initial mixing with marine waters, showing neither quick removal by biological mineralization nor by flocculation (Amon 2004; Alling et al. 2010). Once the material is delivered to the shelf system though, and enough time passes (a few years), then the non-conservative nature of tDOC becomes evident (Hansell et al. 2004; Cooper et al. 2005). Salinity-DOC regressions observed in the western Arctic are shown in Fig. 9.5. During late spring, prior to major sea ice melt, the negative correlation demonstrates a clear linearity (Fig. 9.5a). The marine end member is Pacific water of salinity near 33 and DOC ~65 μmol kg^{-1} (waters at higher salinity are subsurface Atlantic waters, which are not involved in the mixing discussed here), while the freshwater end member is river water (effective salinity of ~0). The western Arctic rivers, particularly the Mackenzie and the Yukon, drive the low salinity end of the regression (relevant river end member characterizations available in Cooper et al. 2008). Note that unlike coastal systems in the lower latitudes, where the low salinity/high DOC meteoric waters are present along the coast and marine waters are offshore, in the western Arctic the low salinity waters (in this regression) are in the Beaufort Gyre (over the deep Canada Basin) and the higher salinity waters are over the shelf. While the low salinity waters hold the most tDOC, they are also (in these data) the oldest waters (time since entering the shelf system).

Interpretation of the data in Fig. 9.5 is demonstrated in Fig. 9.6. The immediate (initial) mixing of river and marine end members results in a linear and negative correlation. The zero-salinity intercept indicates the tDOC concentrations in the source river. As tDOC is removed by biological and photo-oxidation, the relationship depicted by the dashed curve emerges. The entirety of the dashed curve, as an instantaneously complete mixing curve, can never be observed though because the river water, as it ages, is continuously mixed with marine water, thereby constantly diluting the riverine signal while surface currents carry it away. The remnant of the regression (i.e., the observable portion of the theoretical dashed line is labeled "observed" in Fig. 9.6a) is what we observe in nature (Fig. 9.5a). It is only over a short salinity range that the linear regression survives. Extrapolation of that short line back to the zero-salinity intercept (arrowed, dotted line in Fig. 9.6) estimates the tDOC (normalized to zero salinity) still present in the riverine fraction. The difference between the initial river value and the final observed value (determined by extrapolation) is the amount of tDOC that has been removed. Transient tracers of time, such as the $^{228}Ra/^{226}Ra$ ratio (Kadko and Muench 2005), are employed to determine decay rate constants for tDOC, ranging from 0.06 to 0.097 $year^{-1}$ in the western Arctic to 0.24 $year^{-1}$ in the eastern Arctic (Letscher et al. 2011).

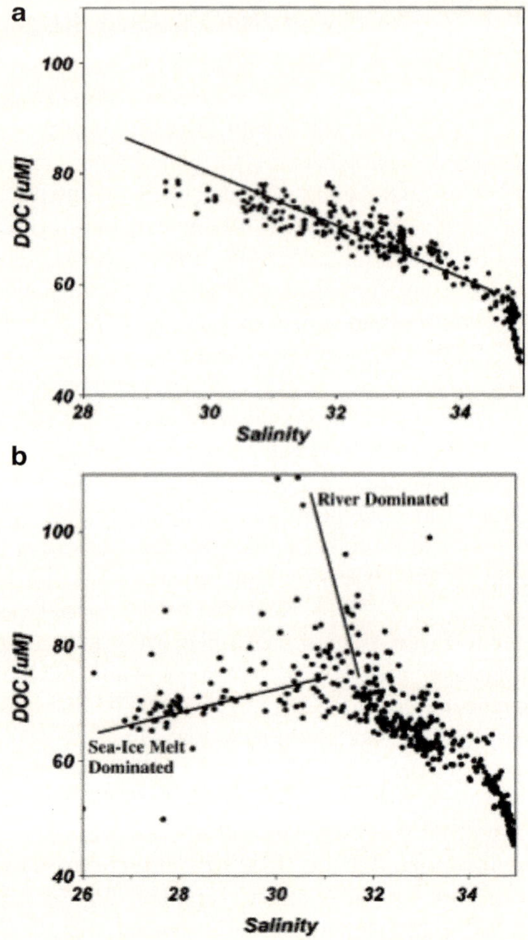

Fig. 9.5 Scatter plots of salinity and DOC (µmoles kg^{-1}) identifying mixing curves: (**a**) spring 2002; and (**b**) summer 2002. Shown with apparent *mixing lines*

The relationship shown in Fig. 9.5a exists during winter and spring, when the system is relatively simple. As summer develops, other processes can dominate the signal. High river runoff, high primary production, and strong sea ice melt make the system much more complex. Some of these impacts are shown in Fig. 9.5b, which is the DOC-salinity relationship in the same western Arctic system as in Fig. 9.5a, but during summer. Some parts of the regression in Fig. 9.5a are evident (such as at salinity >32), but lower salinity waters are highly impacted by the large input of river water associated with seasonal thawing of the watershed and by sea ice melt. The newly added river water is evident as elevated DOC (>100 µmol kg^{-1} in Fig. 9.5b), the regression of which against salinity would intercept zero-salinity at very high DOC concentrations. The melting of sea ice similarly lowers the salinity, but as its DOC concentrations are low (Mathis et al. 2005), the DOC concentrations are lowered, such that a regression against salinity has a very low zero-salinity intercept.

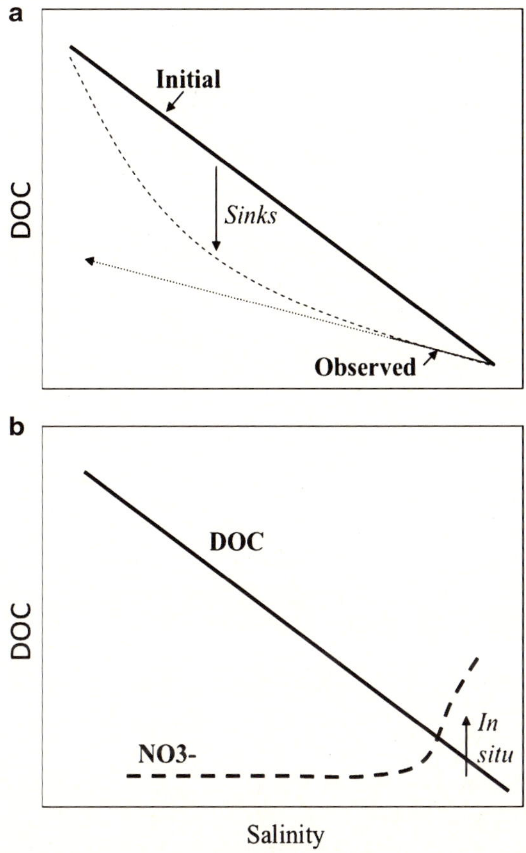

Fig. 9.6 Diagrammatic rendering of the relationship between western Arctic DOC concentrations and salinity in the surface layer as a function of (**a**) time since introduction of tDOC to the ocean system and (**b**) marine nutrient abundance. In (**a**) the initial relationship, when tDOC is first introduced to the ocean, is linear over the full salinity range. With time, tDOC is removed and the correlation (which is not actually observable in nature over the full salinity range due to rapid mixing) becomes curvilinear (*dashed line*). The observable portion of this curve (i.e., Fig. 9.5a), including the full marine end member and some minor dilution with freshwater, is labeled "observed". The zero intercept of the linear regression of the observed correlation (*dotted line*) provides an estimate for the tDOC concentration remaining in the freshwater fraction. In (**b**), the potential impact of marine nutrients on the "observed" correlation is demonstrated. Nutrients are enriched in the marine end of the salinity range (associated with Pacific water) and impoverished in the freshwater end. As such, net community production and net marine DOC production can only impact the DOC/salinity relationship in the high salinity water, increasing DOC at a given high salinity (*upward arrow*)

Evaluating regressions between DOC and salinity must be done with full understanding of the conflicting processes impacting the relationships.

The impacts described above focus on those occurring at the low salinity end of the regressions. Impacts from variations within the marine end member need to be considered as well. One important consideration is the impact on the regressions of the *in situ* production of marine DOC within marine waters. Net DOC production occurs at ~10 % of net community production (NCP) in the western Arctic (Mathis et al. 2007), consistent with findings elsewhere (Hansell and Carlson 1998). NCP, in turn, is dependent on the amount of new nutrients in a system. In the western Arctic, new nutrients are at very low concentrations in low salinity waters (such as in the surface layer of the Beaufort Gyre), so *in situ* production probably has the least impact there (where tDOC is still a large signal). The highest concentrations of new nutrients are in the Pacific water, so to the extent those nutrients are consumed, seasonal DOC accumulation will result. This situation is depicted in Fig. 9.6b, where the DOC/salinity relationship should be taken as that observed in the system (i.e., as in Fig. 9.5a, but not as in the full salinity scale, depicted in Fig. 9.6a). Nitrate is very low at the low salinity end, but elevated in the marine end. *In situ* production occurs during spring and summer when light is adequate, giving another reason for using the early spring data (Fig. 9.5a) rather than summer data (Fig. 9.5b). One further unknown is the extent to which the DOC produced *in situ* each summer resists degradation through the succeeding winter, such that there is a long-term accumulation of the material. If there is interannual accumulation, then the marine end of the regression (depicted in Fig. 9.6b) will be elevated relative to its true initial condition. This will reduce the slope of the regression, giving a lowered estimate for tDOC in the riverine fraction.

A final process that may impact the regressions is sea ice formation. Amon (2004) covered this topic well, so it will only be touched on here. As sea ice forms, high density brine is formed from the extruded salt. The brine is enriched with DOC as well, so the fate of the brine solution is important in considering the impact of this DOC on the system. If the brine sinks below the pycnocline, the process removes DOC from the surface layer. If it sinks only into the mixed layer, but not below or into the pycnocline, then the DOC is retained and the initial conditions are perhaps re-established with subsequent ice melt and vertical overturn of the surface layer. Both Schauer (1997) and Amon (2004) estimated that the salinity increase in brine produced in the eastern Arctic was inadequate for raising the density such that those waters break through the pycnocline. If brine impacted waters become decoupled from the region of ice melt, then a more unidirectional impact is expected. This issue needs to be further investigated.

9.3.3 Dynamical Characterization of tDOC—Inputs & Sinks

Large standing stocks of organic carbon in high-latitude soils and peatlands account for as much as 50 % of global soil carbon (Dixon et al. 1994; Zhulidov 1997). Total input of tDOC to the Arctic Ocean is 25–36 $\times 10^{12}$ gC/year, with the Mackenzie (1.4×10^{12} gC/year) and Yukon (1.7×10^{12} gC/year) rivers dominating input to the western Arctic (Raymond et al. 2007), together adding ~10 % of the total tDOC input to the Arctic Ocean. The tDOC yield of a watershed (g C m^{-2} $year^{-1}$) is dependent on the water yield (m^3 m^{-2} $year^{-1}$) (Raymond et al. 2007), such that an increase

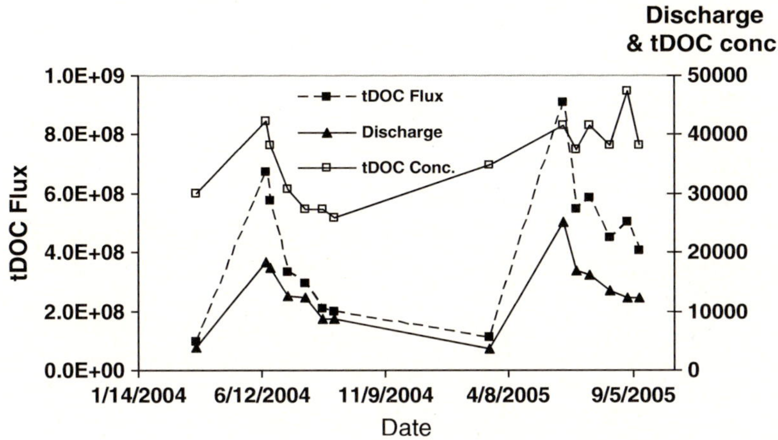

Fig. 9.7 Mackenzie River discharge ($m^3\ s^{-1}$), tDOC concentration ($\mu mol\ kg^{-1} \times 100$) and tDOC flux (mol C day^{-1}) during summer seasons of 2004 and 2005 (Data taken from Raymond et al. 2007)

in the hydrologic cycle over Arctic watersheds (Savelieva et al. 2000; Peterson et al. 2002) may well increase the tDOC export from those systems (Walvoord and Striegl 2007). Interannual variability in watershed runoff, then, will largely control variability in DOC export. The highest DOC concentrations in Arctic rivers commonly coincide with the spring freshet (Fig. 9.7), when highest discharge occurs (Finlay et al. 2006; Holmes et al. 2008; Spencer et al. 2008). Compositional changes occur in the DOC pool of high latitude rivers over the seasonal cycle (Neff et al. 2006; Guo et al. 2007; Striegl et al. 2007), reflecting the changing magnitude of sources.

The material that is exported is very young compared to the bulk marine DOC (Benner et al. 2004; Guo et al. 2007; Raymond et al. 2007). Approximately 50 % of tDOC exported during the arctic spring thaw is 1–5 years old, ~25 % is 6–10 years in age, and 15 % is 11–20 years old (Raymond et al. 2007). As lability (or reactivity) negatively correlates with age (Raymond and Bauer 2000), each DOC age cohort should exhibit a unique removal constant, with the youngest material being removed most rapidly upon export to the coastal system. That multiple fractions of varying lability comprise tDOC in rivers and coastal systems has been suggested (Hopkinson et al. 2002; Moran et al. 1999; Raymond and Bauer 2000). Holmes et al. (2008) found in Alaskan rivers that DOC exported during the spring freshet was readily removed, while DOC present during lower flow summer periods was more resistant. These distinct labilities suggest that tDOC removal in the ocean may best be described by a multi-compartment model (Letscher et al. 2011).

Lignins, a class of macromolecules indicating terrestrial plant residue, are a commonly employed tracer of tDOC in the Arctic Ocean. Opsahl et al. (1999) reported gymnosperm vegetation as its major source in the Arctic. Its concentration correlates well with riverine chromophoric dissolved organic matter (Spencer et al. 2009) and it has been located in deep North Atlantic waters (Benner et al. 2005), though at very low concentrations. The extent to which lignins conservatively trace tDOC once in the ocean is unknown. The sources of tDOC are more than gymnosperms, and the degree to which lignins represent each age class of tDOC has not been evaluated.

The marine removal of Arctic tDOC has only recently been considered (Hansell et al. 2004; Cooper et al. 2005; Alling et al. 2010; Letscher et al. 2011), but its reactivity makes the pool relevant to the ocean carbon cycle. Hansell et al. (2004) reported an approximate carbon mass balance in the removal of tDOC, such that its removal resulted in a like amount of inorganic carbon production in the system. Anderson et al. (2009) interpreted elevated pCO_2 over the Siberian shelf as due to the mineralization of terrigenous organic matter. The surface Arctic Ocean is presently a sink to atmospheric CO_2 (Bates et al. 2006), so an acceleration of tDOC export (and mineralization) in the future could work to partially neutralize that sink. Increased coastal erosion with reduced ice cover could result in accelerated release of particulate and dissolved organic matter presently locked in those soils (Semiletov et al. 2011), furthering this neutralization.

9.4 Export Flux of Particulate Organic Carbon

Wassmann et al. (2003) reported a comprehensive review of the vertical flux of POC in the Arctic Ocean. The major points from their work were that the vertical export of biogenic carbon is poorly resolved based on existing measurements in the Arctic Ocean, and that POC export is derived primarily from the adjacent margins that undergo seasonal decreases in sea-ice extent. They reported that the vertical export POC in the upper Arctic Ocean is highly variable, ranging from 1.5 to 14 g C m^{-2} $year^{-1}$ at ~200 m water depth. Vertical POC export is typically episodic, depending upon shelf production that surround the Arctic Ocean and in polynyas. Variations in sea-ice extent, light and stratification all impose constraints on the vertical export of biogenic matter, while remineralization of POC flux due to zooplankton grazing results in extensive attenuation of the total vertical POC export.

Wassmann et al. (2003) further reported that the deeper regions of the Arctic Ocean are characterized by a greater retention of carbon in the twilight zone, however marginal ice zones (MIZ) can also show this tendency. Over-wintering of large, long-lived zooplankton such as copepods and appendicularians appear to play an important role in vertical flux attenuation. In the decades to come, the dramatic northwards retreat of the MIZ due to global warming will result in a migration of the MIZ over a wider area throughout the year, represented by an extensive, stratified region that stretches from the shelves into the deep Arctic Ocean. As a consequence, primary production and vertical export of biogenic matter are hypothesized to increase will increase in the peripheral band of the contemporary permanent ice cover.

Since the review by Wassmann et al. (2003), publications appeared that reported results from major field studies of POC export in the upper Arctic Ocean, including the Shelf Basin Interaction II (SBI-II) study in the Chukchi Sea (Lalande et al. 2007a; Lepore et al. 2006; Moran et al. 2005), the CABANERA project in the Northern

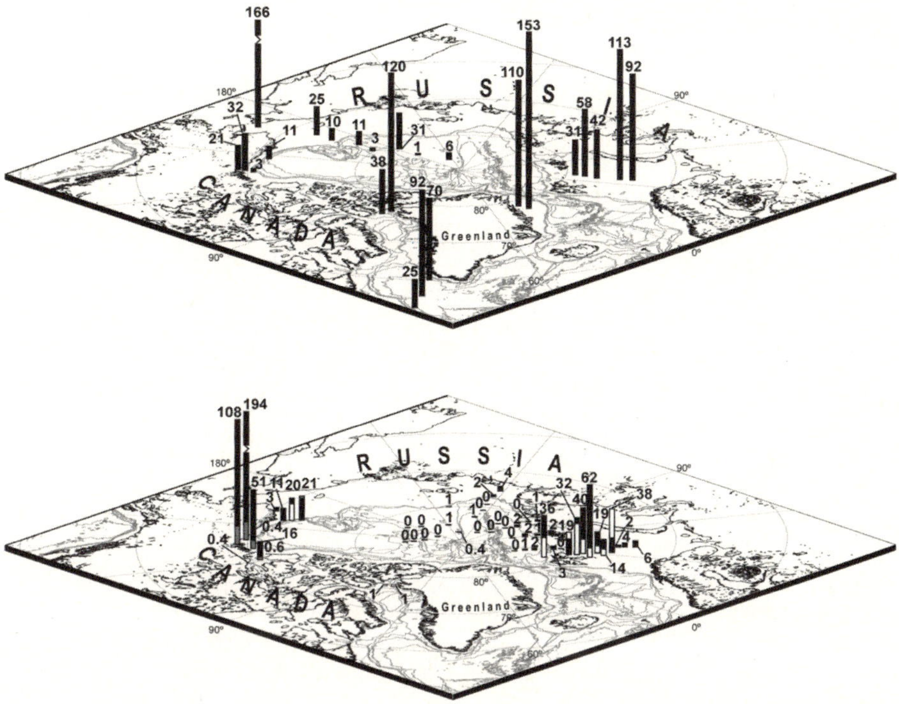

Fig. 9.8 Regional distribution of POC export fluxes (mmol C m^{-2} day^{-1}) in the Arctic Ocean and sub-polar Labrador Sea determined using sediment traps and $^{234}Th/^{238}U$ disequilibrium in the Chukchi Sea (Lalande et al. 2007b; Lepore et al. 2006; Moran et al. 2005), the northern Barents Sea (Lalande et al. 2008), the Mackenzie shelf (Amiel and Cochran 2008), the Laptev, northern Baffin Bay and Beaufort Seas (Lalande et al. 2009), and the eastern Arctic (Cai et al. 2010a). These results represent the major field programs of POC export flux in the upper Arctic Ocean since summarized by Wassmann et al. (2003). *Black bars* indicate POC fluxes calculated from $^{234}Th/^{238}U$ disequilibrium; *white bars* results from free-floating sediment traps; *grey bars* represent POC fluxes calculated for the marine component of the total POC flux in the Mackenzie shelf (Amiel and Cochran 2008). Seasonally averaged fluxes are plotted for the Chukchi Sea (Lalande et al. 2007a; Lepore et al. 2006; Moran et al. 2005), and regionally averaged POC fluxes are indicated for the Mackenzie shelf (Amiel and Cochran 2008)

Barents Sea (Lalande et al. 2008), the Canadian Arctic Shelf Exchange Study (CASES) over the Mackenzie Shelf (Amiel and Cochran 2008), the comparative study of POC export in the Laptev, northern Baffin Bay and Beaufort Seas (Lalande et al. 2009) and the recent eastern Arctic expedition ARK-XXII/2 (Cai et al. 2010b). These studies are summarized in the following sections and the results illustrated in Fig. 9.8. Note that in the following discussion, because of varying water depth and topography in the different regions, there may be either accumulation or transport of POC, and this may change through the annual cycle. For this reason, there is an inherent uncertainty in assessing the fate of exported POC in these budget computations.

9.4.1 Regional Case Studies

9.4.1.1 Chukchi Sea: The Shelf Basin Interaction Study (SBI-II)

As part of the 2002 Shelf-Basin Interaction (SBI) process study, measurements of the seasonal variations in the export flux of POC were reported for the upper waters of the Chukchi Sea (Moran et al. 2005). POC fluxes were quantified by determination of $^{234}Th/^{238}U$ disequilibrium and POC/^{234}Th ratios in large (>53 μm) aggregates collected using in-situ pumps. Samples were collected at 35 stations on two cruises, one in predominantly ice-coved conditions during the spring (May 6–June 15) and the other in predominantly open water during the summer (July 17–August 26). Enhanced particle export was observed in the shelf and slope waters, particularly within Barrow Canyon, and there was a marked increase in particle export at all stations during the summer (July–August) relative to the spring (May–June). ^{234}Th-derived POC fluxes exhibit significant seasonal and spatial variability, averaging 3 ± 5 mmol C m^{-2} day^{-1} (range = 0.03–22 C mmol m^{-2} day^{-1}) in the spring and increasing ~4-fold in the summer to an average value of 11 ± 9 mmol C m^{-2} day^{-1} (range = 1–39 mmol C m^{-2} day^{-1}). The fraction of primary production exported from the upper waters increases from ~15 % in the spring to ~32 % in the summer. By comparison, DOC accumulation associated with net community production represented ~6 % of primary production (~2 mmol C m^{-2} day^{-1}). The majority of shelf and slope stations indicate a close agreement between POC export and benthic C respiration in the spring, whereas there is an imbalance between POC export and benthic respiration in the summer. An implication of this study is that up to ~20 % of summer production (~6 ± 7 mmol C m^{-2} day^{-1}) may be seasonally exported off-shelf in this productive shelf/slope region of the Arctic Ocean.

In 2004, seasonal and interannual changes in POC export and deposition were estimated in the shelf-slope region of the Chukchi Sea using measurements of ^{234}Th-^{238}U disequilibria and the POC/^{234}Th ratio in large (>53-μm) particles (Lepore et al. 2006). These export fluxes were used in conjunction with rates of primary productivity and benthic carbon respiration to construct a POC budget for this shelf-slope region. Samples were collected along a series of shelf-basin transects in the spring (May–June) and summer (July–August) of 2004. These stations were previously occupied during the ice covered (spring) and open water (summer) seasons of 2002, allowing for an interannual comparison of export flux. In contrast to 2002, when open water POC fluxes were significantly higher than in the ice-covered period, POC export fluxes in 2004 were similar during the spring (average = 20 ± 25 mmol C m^{-2} day^{-1}) and summer (average = 20 ± 15 mmol C m^{-2} day^{-1}). The high POC fluxes measured during the spring are attributed to a plankton bloom, as evidenced by exceptionally high primary productivity (average = 124 ± 88 mmol C m^{-2} day^{-1}). The shelf-slope budget of particulate organic carbon indicates that 10–20 % of primary productivity was exported below 50 m but was not consumed during benthic carbon respiration or burial and oxidation in underlying sediments. Notwithstanding the regional and temporal patchiness inherent in these studies, a water column-sediment budget of ^{234}Th indicates that particulate material accumulates in shelf sediments on a seasonal basis.

As part of the SBI study, the ^{234}Th-derived POC export fluxes were compared to fluxes quantified using drifting sediment traps in the spring (May 15–June 23) and summer (July 17–August 26) (Lalande et al. 2007a, b). Measurements were obtained at five stations in the Chukchi Sea during the spring cruise and four stations during the summer cruise along Barrow Canyon (BC) and along a parallel shelf-to-basin transect from East Hanna Shoal (EHS) to the Canada Basin. ^{234}Th and POC fluxes obtained with *in situ* pumps and drifting sediments traps agreed to within a factor of 2 for 70 % of the measurements. POC fluxes were also measured with sediment traps at 50 m along BC were also similar in both seasons (31 ± 9 mmol C m^{-2} day^{-1} and 29 ± 14 mmol C m^{-2} day^{-1}, respectively), but were approximately twice as high as POC fluxes measured with *in situ* pumps. Sediment trap POC fluxes measured along the EHS transect also increased from spring to summer (3 ± 1 mmol C m^{-2} day^{-1} and 13 ± 6 mmol C m^{-2} day^{-1}, respectively), and these fluxes were similar to the POC fluxes obtained with *in situ* pumps. Discrepancies in POC export fluxes measured using *in situ* pumps and sediment traps may be reasonably explained by differences in the estimated POC/^{234}Th ratios that arise from differences between the techniques, such as time-scale of measurement and size and composition of the collected particles. Despite this variability, *in situ* pump and sediment trap-derived POC fluxes were only significantly different at a highly productive station in BC during the spring.

9.4.1.2 Mackenzie Shelf: Canadian Arctic Shelf Exchange Study (CASES)

Amiel and Cochran (2008) assessed both terrestrial and marine POC fluxes in the Mackenzie Shelf as part of the CASES study. Water column deficits of ^{234}Th relative to ^{238}U in the Mackenzie Shelf, Cape Bathurst Polynya, and Amundsen Gulf were used to estimate sinking fluxes of POC. The ^{234}Th fluxes were converted to marine and terrestrial POC fluxes using the POC/Th ratio of filterable particles >70 μm and δ^{13}C measurements to estimate the fraction of marine and terrestrial POC. In June/July 2004, the largest ^{234}Th deficits (0–100 m: 56–95 dpm m^{-2}) were observed in the Mackenzie outer shelf. Deficits in the upper 100 m ranged 3–59 dpm m^{-2} in the Cape Bathurst Polynya. The δ^{13}C values of POC in the >70 μm particles filtered using in situ pumps ranged from −25.1 ‰ to −28 ‰. A two-end member mixing model with marine POC = −21.4 ‰ and terrestrial POC = −28 ‰ indicates that terrestrial POC is most evident at the Mackenzie Shelf stations but is present throughout the region. The fraction of marine POC ranged from 0 % to 59 % in the area in June/July 2004, with highest values in the Cape Bathurst Polynya. Fluxes of marine POC in the polynya averaged ~5 mmol C m^{-2} day^{-1} at 50 m in June 2004 and increased to ~12 mmol C m^{-2} day^{-1} in July. Comparable fluxes were observed at 100 m in June but values decreased to ~6 mmol C m^{-2} day^{-1} at 100 m in July. These fluxes are greater than estimates of organic carbon remineralization and burial in sediments of the polynya (~3 mmol C m^{-2} day^{-1}). Amiel and Cochran (2008) suggested that POC may be exported out of the area, effectively remineralized by microbial activity in the twilight zone, or incorporated into biomass.

9.4.1.3 Laptev Sea, Northern Baffin Bay and the Beaufort Sea Shelves

Lalande et al. (2009) deployed moored sediment traps in 2005–2006 in the Beaufort Sea, Northern Baffin Bay, and the Laptev Sea to compare the annual variability in POC export flux and to evaluate the factors regulating the annual cycle of POC export over these continental shelves. POC fluxes estimated over the annual cycle at 200 m ranged from 0.4 to 1.3 mmol C m^{-2} day^{-1} with the highest export in Northern Baffin Bay and the lowest export over the shelf in the Beaufort Sea. Each annual cycle exhibited an increase in POC export a few weeks prior to, during, or immediately following sea-ice melt, though varying temporal patterns in export were observed over the remainder of the annual cycle. Enhanced primary production, discharge of the Lena River, and resuspension events contributed to periods of elevated POC export over the Laptev Sea slope. Greater POC fluxes in Northern Baffin Bay reflected periods of elevated primary production in the North Water polynya. In the Beaufort Sea, sediment resuspension contributed to most of the large export events. Lalande et al. (2009) suggested that the outer shelf of the Laptev Sea may sustain the largest increase in POC export in the coming years, due to the large reduction in ice cover and the possible increase in the Lena River discharge. Moreover, the large difference in forcing among the regions investigated reinforces the importance of monitoring POC fluxes in the different oceanographic regimes that characterize the Arctic shelves to assess the response of the Arctic Ocean carbon cycle to interannual variability and climate change.

9.4.1.4 Eastern and Central Arctic Ocean: Polarstern ARK-XXII/2 Expedition

There are still very few direct measurements of POC export under the permanently ice-covered central Arctic Ocean (Moran et al. 1997). During the *Polarstern* ARK-XXII/2 expedition to the eastern and central Arctic Ocean (28 July to 7 October in 2007), a high-resolution study of POC export was conducted using $^{234}Th/^{238}U$ disequilibrium (Cai et al. 2010a). Depth profiles of total ^{234}Th in the upper 100–200 m were collected at 36 stations in the eastern and central Arctic Ocean and the Barents, Kara, and Laptev seas. Samples were processed using a small volume MnO_2 co-precipitation method with addition of a yield tracer. The ^{234}Th deficit with respect to ^{238}U was found to be evident throughout the upper 100 m over the Arctic shelves. In comparison, the deficit was confined to the upper 25 m in the central Arctic Ocean. Below 25 m, secular equilibrium was approached between ^{234}Th and ^{238}U. Moreover, in association with the surface ^{234}Th deficit, the total chlorophyll concentration was generally found to be enhanced, indicating that *in-situ* production and export of biogenic particles are the main mechanism for upper ocean ^{234}Th removal in the central Arctic Ocean. ^{234}Th-derived POC fluxes were determined with a steady state model and pump-normalized POC/^{234}Th ratios on total suspended particles collected at 100 m. Results showed enhanced POC export over the Arctic shelves. On average, POC export fluxes over the various Arctic shelves were 3 ± 2 mmol m^{-2} day^{-1} (the Barents Sea), 0.5 ± 1 mmol m^{-2} day^{-1} (the Kara Sea) and 3 ± 2 mmol m^{-2} day^{-1} (the Laptev Sea), respectively. In comparison, the central Arctic Ocean was characterized by the lowest POC export flux ever reported, 0.2 ± 1 mmol m^{-2} day^{-1} (1SD, n=26). This value is very low compared to prior estimates, and is also about one order of magnitude lower than the POC export fluxes reported in other oligotrophic oceans. A ThE ratio (^{234}Th-derived POC export/primary production) of 2 % or 6 % in the central Arctic Ocean was estimated, depending on the definition of primary production. The low ThE ratio indicates that like other oligotrophic regimes, the central Arctic Ocean is characterized by low POC export relative to primary production, i.e., a tightly coupled food web. This recent study implies that the current role of the central Arctic Ocean in C sequestration is still very limited (Anderson et al. 2003).

9.4.2 Conclusions

Several field programs conducted in the Arctic Ocean over the past decade have reported new estimates of POC export fluxes based on the determination of $^{234}Th/^{238}U$ disequilibria in the upper waters of the permanently ice-covered interior basins, the seasonally ice-covered shelf and slope waters, and polynya regimes. These studies consistently indicate particle export from the upper few hundred meters of the water column occurring on a time-scale of days to months. $^{234}Th/^{238}U$ disequilibria observed in the interior basin and shelf-slope environments are attributed primarily to particle settling associated with variations in local primary and secondary productivity. In shelf regimes, resuspension of abiogenic bottom sediments also contributes to $^{234}Th/^{238}U$ disequilibria. As is the case with sediment traps, such processes may lead to biased estimates of the vertical POC export flux.

Taken as a whole, these recent Arctic studies indicate a marked regional variability in upper ocean POC export flux, ranging from 0.3 to 7 mmol C m^{-2} day^{-1} in the central Arctic Ocean to 0.7–45 mmol C m^{-2} day^{-1} in shelf and polynya regimes. A key issue for future studies is to determine the extent of seasonal variability in POC export, which is likely to be significant when considering the high nutrient concentrations and strong seasonal gradients in light and ice-cover that characterizes the Arctic Ocean. These observations provide the basis for what is likely to be a considerable increase in understanding the factors controlling the temporal and spatial variability in POC export flux and remineralization depths in the Arctic Ocean and adjacent seas.

9.5 Grazing

As in other oceans, both protists and metazoans are believed to be important grazers of both phytoplankton and heterotrophic microbes in the PAR domain. In general, our impressions of the relative importance of different taxonomic groups is based on extrapolation of their importance in similar ecosystems, modeling exercises, or better-known information such aspects of community composition and biomass. One of the difficulties in extrapolating results from other studies to the PAR is the mosaic of habits it encompasses (Hopcroft et al. 2008): the northern Bering Sea and much of the Chukchi operates as a cold extension of the Pacific Ocean fauna (Hopcroft et al. 2010), while other parts of the Chukchi and Beaufort seas have a fauna of more Arctic character (Darnis et al. 2008; Hopcroft et al. 2010), and the Canada Basin itself has a decidedly oceanic Arctic fauna (Ashjian et al. 2003; Lane et al. 2008; Kosobokova and Hopcroft 2010). Interwoven with these is a patchwork of polynyas with enhanced productivity and grazing rates (Deibel and Daly 2007).

The actual number of studies directly measuring grazing in the PAR domain is limited for both the protisan microzooplankton (Sherr et al. 1997, 2008) and metazoans (Deibel et al. 2005; Campbell et al. 2009), although several indirect measurements of community grazing (as vertical flux) exist (i.e., Forest et al. 2007; Seuthe ct al. 2007). Additional grazing work was been conducted at the southern PAR fringe during PROBES (i.e., Dagg et al. 1982), but this work—like that in BEST/BSIERP (E. Lessard, RG Campbell, unpublished)—has concentrated primarily on Pacific oceanic species.

Near the shelf break, of the Chukchi Sea, the dominant four copepod taxa (*Calanus glacialis, Calanus hyperboreus, Metridia longann and Pseudocalanus spp.*) remove ~13 % day^{-1} of the primary production in spring and 28 % day^{-1} in summer, with wide confidence intervals (Campbell et al. 2009). These copepods removed mostly phytoplankton during the spring bloom, but preyed preferentially on the microzooplankton community at all times (Campbell et al. 2009). The microzooplankton themselves were dominated by aloricate ciliates and heterotrophics dinoflagelllates, both with relatively low growth rates, that consumed on average 22 % of primary production, much less than typical for other oceanographic environments, including the Barents Sea. Although the bulk of the imbalance must be exported to the benthos, or advected further into the basins (Grebmeier et al. 2009), the grazing impact of larvaceans was not considered.

Larvaceans are important filter-feeders in the PAR communities (Lane et al. 2008), as well as elsewhere in the Chukchi Sea (Hopcroft et al. 2010), especially in the marginal ice zone of both the shelf and basins (R. Hopcroft, unpublished data). The high grazing rate of larvaceans allows them to consume as much as 20 % of the daily primary production in the St Lawrence Island polynya (Deibel et al. 2005) much greater than estimated for the copepods in this case as well as in other Arctic polynyas they considered (Deibel and Daly 2007). While food concentration does typically increase the rates of copepod reproduction (Plourde et al. 2005) and development (Ringuette et al. 2002) in Arctic habitats, larvacean growth rates are typically high compared to copepods (Hopcroft et al. 1998). This is true even for arctic species (Choe and Deibel 2009), proving advantageous in exploiting ephemeral habitats of high-food concentration.

The quantitative impact of other planktonic metazoans—grazers, scavengers, and predators—in the transfer and cycling of carbon in the PAR region is largely unknown. A significant amount of carbon must be lost as it is moved to higher planktonic trophic levels. Nonetheless, the cold temperatures characteristic of the Arctic result in relatively low rates of respiration (e.g., Ikeda et al. 2001) at all planktonic trophic levels, resulting in relatively high biomass given the relatively low annual rates of primary production. Although the period of high primary production may be limited, many of these animals accumulate relatively large stores of lipids during such times that carry them through the winter (often in diapause) and to fuel subsequent reproduction (e.g., Hirche and Kattner 1993). In contrast, the more predatory species may have relatively less seasonality in their prey supply and feeding strategy.

9.6 Benthic Carbon Cycling

Ice algae and open water production in the spring are key components of the seasonal phytoplankton bloom and the timing of the export of this material to the underlying benthos is critical to growth properties of the infaunal organisms. Zooplankton populations have a limited impact early in the spring due to slow growth rates at low seawater temperatures, thus low zooplankton populations allow most of the ice edge production to settle ungrazed to the sediment (Bluhm and Gradinger 2008). However, with continued sea ice retreat and associated seawater warming in the Arctic, zooplankton populations will likely grow fast enough to graze a large fraction of the new production (Coyle and Pinchuk 2002), even with the potential of increased open water production with reduced sea ice extent. Currently early season phytoplankton blooms are typical in the northern Bering Sea and into the Chukchi Sea. In the western regions under the influence of high nutrient Anadyr Water, zooplankton populations are only weakly coupled to water column production, resulting in low recycling in the water column and the fraction of primary production exported to the underlying sediments is high (Grebmeier and Barry 2007; Grebmeier 2012). Comparatively, the Alaska Coastal Water is a lower nutrient and water column production regime, thus most of the seasonal production and subsequent micro- and meso-zooplankton growth and microbial transformations utilize a large percentage of the water column organic carbon production, thus reducing the fraction of material exported to the sediments, thereby limiting benthic populations and benthic carbon cycling (Grebmeier 2012; Grebmeier et al. 2006a).

Spatial patterns of sediment community oxygen consumption (SCOC), sometime called sediment oxygen demand, identify regions of organic carbon deposition throughout the Pacific-influenced shelf region and can be used as a surrogate for variations of export production (Grebmeier 2012; Grebmeier and Barry 2007).

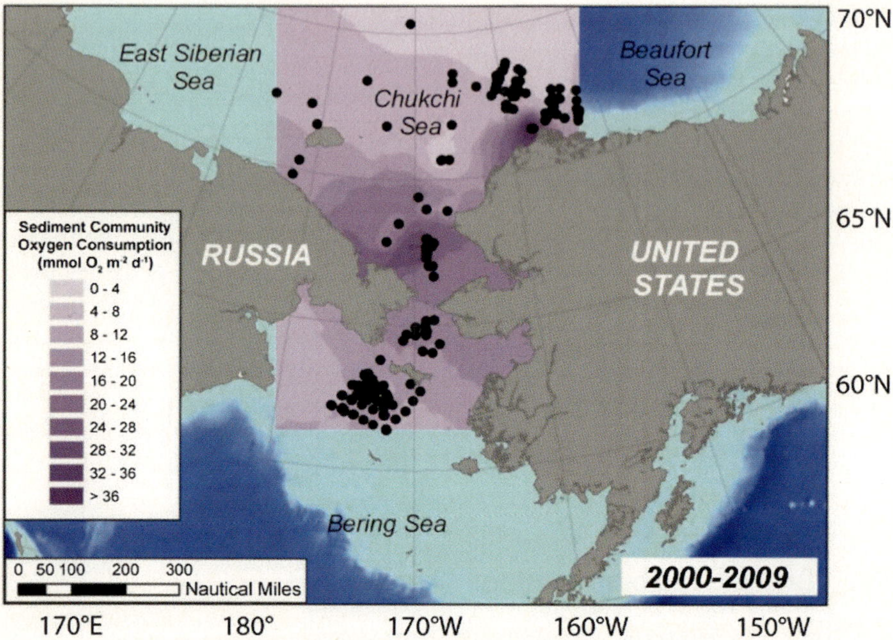

Fig. 9.9 Distribution of sediment community oxygen consumption (positive influx values) in the Bering and Chukchi seas from 2000 to 2009

We converted oxygen to carbon using a 1:1 stoichiometric relationship between oxygen and carbon utilization and a carbon respiration coefficient of 0.85 after Renaud et al. (2007b) for comparative purposes with carbon export values in the previous section of this chapter. Looking just at the last decade (2000 onward), we continue to observe specific "hot-spots" of high SCOC in regions southwest of St. Lawrence Island, in the Chirikov Basin north of St. Lawrence Island, in the southern Chukchi Sea and at the head of Barrow Canyon (Fig. 9.9) compared to previous decades (see Grebmeier 2012). SCOC ranges from ~0.3 to 40 mmol $O_2\,m^{-2}\,day^{-1}$ (84–335 mg C $m^{-2}\,day^{-1}$) in the northern Bering Sea just south and north of St. Lawrence Island to 20–50 mmol $O_2\,m^{-2}\,day^{-1}$ (167–418 mg C $m^{-2}\,day^{-1}$) in the southern Chukchi Sea and upper Barrow Canyon. SCOC declines to 1–10 mmol $O_2\,m^{-2}\,day^{-1}$ (8–84 mg C $m^{-2}\,day^{-1}$) on the outer Chukchi and western Beaufort shelves, with the lowest values from 0.1 to 5 mmol $O_2\,m^{-2}\,day^{-1}$ (<1–42 mg C $m^{-2}\,day^{-1}$) on the northern Chukchi and western Beaufort slope and Arctic basin regions (see Grebmeier et al. 2006a for SCOC methodology).

The lower SCOC rates nearshore in the northern Bering, Chukchi, East Siberian, and western Beaufort Seas, except Barrow Canyon, indicate reduced organic carbon export and deposition to the benthos (Grebmeier and Barry 2007; Grebmeier 2012). These values of SCOC converted to carbon values are comparable in range to export production determined from both the thorium method and sediment traps (Lalande et al. 2007b; Lepore et al. 2006, this chapter). For example, high SCOC in the southern Chukchi Sea and upper Barrow Canyon "hotspots" is consistent with elevated carbon export flux rates determined by the thorium method (Lepore et al. 2006), although lower than those measured using the floating sediment trap method in the summer period only (Lalande et al. 2007b). When SCOC is greater than the water column carbon export production, the import of organic carbon from upstream productive zones is suggested.

Past studies also indicate that sediment respiration in the productive area of the PAR is dominated by benthic macroinfauna (animals that live inside the sediment >1 mm in size) and that rates of microbial respiration are low by comparison (Grebmeier and McRoy 1989; Henriksen et al. 1993; Devol et al. 1997; Clough et al. 2005a, b; Grebmeier and Barry 2007). The dominance of macroinfauna in community respiration is assumed for the shelf and upper slope of the Pacific-influenced waters, although small meiofauna (>63 μm–0.5 mm) become more dominant in the deep slope and basin as macroinfauna diminish (Clough et al. 2005a, b; Pirtle-Levy 2006). Clough et al. (2005a, b) found that the role of macrofauna was more important than meiofauna to sediment respiration at depths shallower than 500 m in the Chukchi Sea during the summer. They also noted that benthic macroinfaunal biomass explained 75 % of the variability in benthic respiration rates, similar to the >50 % influence of benthic macroinfauna observed by Grebmeier and McRoy (1989) under the high carbon export areas of the northern Bering and Chukchi seas. Recent data also indicate that epifaunal animals (animals that live on the sediment surface) can increase overall benthic community oxygen consumption and thus their oxygen consumption should be considered when developing a total benthic carbon cycling budget (Ambrose et al. 2001; Piepenburg et al. 1995; Renaud et al. 2007a). Thus, whole core SCOC rates, although generally indicative of organic carbon supply to the benthos, are considered minimal values since they do not include either megafaunal animal impacts or anaerobic processes, both of which would increase total sediment oxygen consumption and overall carbon mineralization rates.

Similarly, macrofauna was found to be a dominant driver for higher sediment oxygen consumption in the Canadian Beaufort Sea during studies undertaken during the Canadian Arctic Shelf Exchange Study (CASES) program. Studies by Renaud et al. (2007a) found that 60 % of the oxygen demand during the productive spring period in 2005 was driven by macrofaunal activity, with only a minor portion through microbial and meiofaunal activities. Sediment-community carbon demand in fall 2003 and summer 2004 was 1.3 ± 0.3 to 8.0 ± 4.3 mmol C m^{-2} day^{-1}, with lowest levels measured in the Amundsen Gulf. Levels were generally higher on the shelf than at the shelf edge and highest at Cape Bathurst (Renaud et al. 2007a). This study estimated that the benthic carbon demand during the period of highest sediment oxygen demand was around 0.21 g C m^{-2} day^{-1}, about half the benthic organic carbon uptake of the "hotspot" areas in the southern Chukchi Sea and head of Barrow Canyon (Grebmeier et al. 1989; Moran et al. 2005). They also found epibenthos (primarily echinoderms) were important drivers of sediment respiration.

Benthic macroinfaunal biomass from 30 to 60 g C m^{-2} has been reported in the southern Chukchi Sea and upper Barrow Canyon productive regions (Grebmeier 2012; Grebmeier et al. 2006a). The implication of these high benthic biomass areas is that carbon storage in benthic infaunal biomass is likely an important component of the overall carbon budget in this region. Additionally, rates of high sediment carbon uptake of up to 30 mmol C m^{-2} day^{-1} at the head of Barrow Canyon indicate increased export to the sediments of organic carbon during the summer period (Lepore et al. 2006). This increase in carbon export in upper Barrow Canyon during the summer is likely due to both *in situ* phytoplankton production and deposition of particulate organic material from upstream production sites in the south, which would make Barrow Canyon an important region of carbon transport from the Chukchi Sea to the Arctic Basin. Notably, Renaud et al. (2007a, b) working in the Beaufort Sea found that not enough *in situ* particulate carbon was present in their study area to support the benthic carbon uptake rates, indicating an outside source of carbon was needed as was observed in the northern Chukchi Sea.

9.6.1 Sediment Nutrient Efflux

Spatial evaluation of ammonium and silicate efflux from 2000 onwards (encompassing the 2007–2009 International Polar Year) indicate a relationship between regions of high export production as indicated by SCOC (Fig. 9.9) and nutrient efflux (Figs. 9.10 and 9.11). Highest ammonium (Fig. 9.10) and silicate (Fig. 9.11) efflux occurred in the hotspot areas of carbon export (indicated by positive SCOC influx, Fig. 9.9) southwest of St. Lawrence Island (SLI), the north-central Chirikov Basin north of SLI, the southeast Chukchi Sea and upper Barrow Canyon, all known areas of high pelagic-benthic coupling and carbon export (Grebmeier 2012). The high ammonium effluxes (indicated by negative values of up to ~7 mmol NH_4 m^{-2} day^{-1}) occurred in regions of high carbon export as indicated by high SCOC (positive influx of up to ~47 mmol O_2 m^{-2} day^{-1}; Fig. 9.9) in similar locations and coincident with high populations of infaunal bivalves and amphipods. In particular, silicate efflux is greatest in the SE Chukchi Sea and upper Barrow Canyon benthic "hotspot" areas (Fig. 9.10) where SCOC is also the greatest, indicative of the high carbon supply to the sediments and benthic carbon recycling processes in this region.

SCOC, ammonium and silicate efflux are probably influenced by bioturbation processes within the upper sediments by infaunal bivalves, polychaetes and crustaceans that dominate in this region as indicated for regions in the southern Chukchi Sea. In addition, microbial recycling of organic carbon as well as nutrients excreted by infaunal organisms can be contributors to benthic carbon cycling in these areas. Bioturbation by infaunal organisms can influence the rates of oxygen uptake and nutrient regeneration, thus the location of "hotspots" of benthic carbon cycling are directly related to infaunal biomass and population structure. Since the region is characterized by high infaunal population levels, bioturbation by the dominant infauna as well as large benthic feeding marine mammals, such as walrus are likely critical for maintaining ecosystem structure and function on the productive northern Bering and Chukchi shelves. Any changes in physical forcing that influence

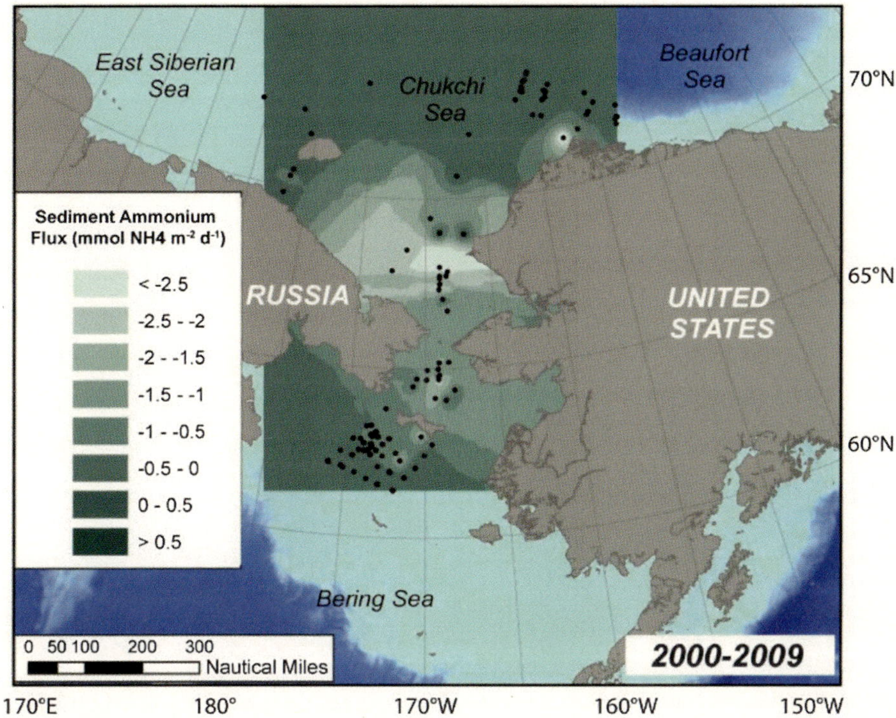

Fig. 9.10 Efflux of ammonium from sediments (negative values) in the northern Bering and Chukchi seas from 2000 to 2009

upper water column biological processes (nutrient availability, primary production, zooplankton production, export production) will directly influence sediment oxygen uptake, nutrient release and overall benthic carbon cycling that can feed back to both upper water column and higher trophic level productivity.

9.7 Contribution of Heterotrophic Bacteria to Carbon Cycling

Heterotrophic bacteria and the rest of the microbial loop usually process a large fraction of ^{14}C-primary production primary production and thus contribute substantially to the carbon cycle of the oceans. Bacterially-mediated carbon fluxes through dissolved organic material (DOM) account for half or more of primary production as measured by the ^{14}C method in most aquatic ecosystems (Ducklow 2000), based on data from biomass production estimates and size fractionation experiments that examine respiration or DOC consumption by the bacteria-size fraction. Since bacteria are the main users of DOC, data on bacterial activity can be used to

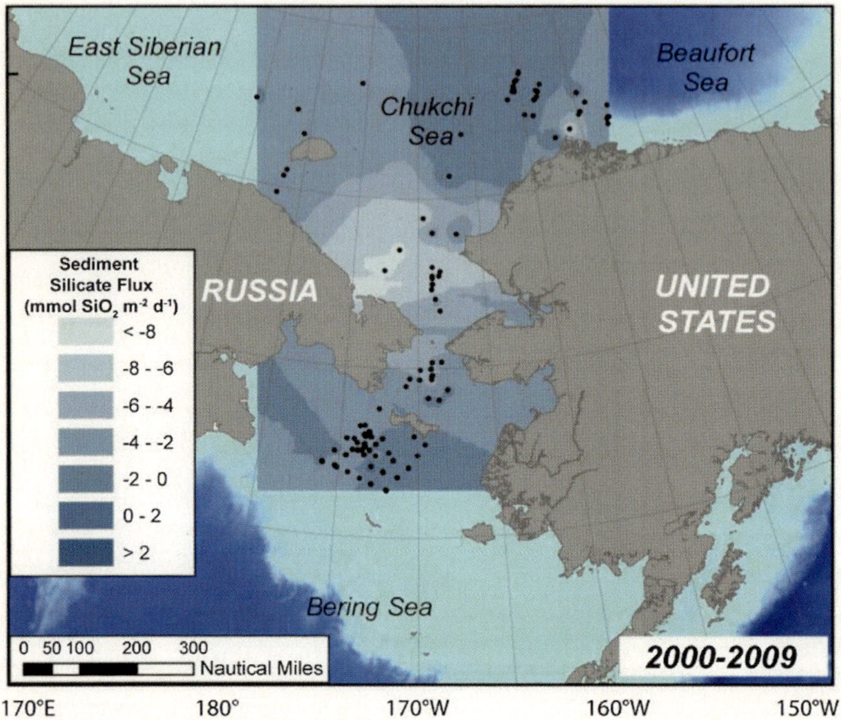

Fig. 9.11 Efflux of silicate from sediments (negative values) in the northern Bering and Chukchi seas from 2000 to 2009

estimate DOC fluxes, especially of the labile components of the DOC pool with turnover times on the scales of days to weeks. In some oceanic regions, these fluxes and respiration by bacteria and other heterotrophic organisms appear to exceed primary production rates over substantial time and space scales, raising questions about balancing carbon fluxes in these regions (Duarte and Regaudie-de-Gioux 2009; Robinson 2008).

Early studies suggested that bacterial activity would be low in perennially cold waters like the Arctic Ocean (Pomeroy and Deibel 1986). This low activity was thought to allow more carbon to be channeled to higher trophic levels and to be exported to benthic communities. Studies in the Ross Sea, Antarctica supported the cold temperature hypothesis and found that ratios of bacterial biomass production to ^{14}C-primary production were low (Ducklow et al. 2001; Ducklow 1999), suggesting low fluxes through the DOM pool. In the Arctic Ocean, however, one study found variable but often high ratios of bacterial to primary production (Rich et al. 1997).

The SBI program provided an opportunity to examine bacterial properties and DOC fluxes in the Chukchi Sea and nearby regions of the Canada Basin (Grebmeier and Harvey 2005). The review here will focus on two related questions: (i) what fraction of total respiration can be attributed to heterotrophic bacteria?; (ii) what fraction of primary production is processed by heterotrophic bacteria, as estimated from biomass production and growth efficiencies estimates? The term "heterotrophic bacteria" accurately describes the microbes involved in the processes discussed in this review because the abundances of autotrophic Archaea and cyanobacteria are low in these surface waters (Cottrell and Kirchman 2009; Garneau et al. 2009; Kirchman et al. 2007).

9.7.1 Respiration by Heterotrophic Bacteria

The standard size fractionation approach was used to examine respiration by the bacterial size fraction (<0.8 μm) in the summer of 2004 in the Chukchi Sea and Canada Basin (Kirchman et al. 2009a). The small size fraction usually includes a large part (>80 %) of total bacterial abundance, but excludes particle-bound microbes, which can be abundant and active in low salinity coastal waters impacted by river inputs (Garneau et al. 2009).

The size fractionation experiments indicated that heterotrophic bacteria accounted for a highly variable fraction of total respiration in the Chukchi Sea and Canada Basin (Kirchman et al. 2009a). The percentage was 40–60 % in shallow shelf waters of the Chukchi Sea and decreased to 3–26 % in slope waters. Respiration by the bacterial size fraction exceeded total respiration at two stations in the Canada Basin. The overall average of all experiments indicated that heterotrophic bacteria account for about 25 % of total respiration in these Arctic waters, which is smaller than typically seen in low-latitude oceans.

9.7.2 Biomass Production by Heterotrophic Bacteria and Phytoplankton

Estimates of biomass production by heterotrophic bacteria can be used to explore the transfer of carbon, other elements and energy from DOM, through heterotrophic bacteria and the rest of the microbial loop, and onto higher trophic levels. Since bacterial abundance and biomass is constant on short time scales, bacterial production is roughly equal to transfer of carbon to higher trophic levels after viral lysis is accounted for; the few available estimates indicate that lysis is low in the Arctic for unknown reasons (Steward et al. 1996). The production estimates also can be used to examine community growth rates and generation times, a fundamental property of biological populations. In the Arctic, growth rates range from 0.03 to 0.15 day^{-1}, equivalent to generation times of 5 days to over 3 weeks (Kirchman et al. 2009a). These growth parameters are slower and longer than those estimated in low latitude oceans (Kirchman et al. 2009b).

Biomass production rates also can be used to examine the fraction of primary production processed by heterotrophic bacteria. This approach has many well-known problems. Criticism has focused on bacterial production estimates, but ^{14}C-based estimates of primary production have also been questioned (Quay et al. 2010) and are difficult to interpret for the Arctic (see below). Still, it is technically much easier to measure biomass production than total respiration and respiration by the bacteria-size fraction, especially in cold waters where rates are low and oxygen concentrations are high. Consequently, the data set on production estimates is much larger than that on respiration. These estimates suggest high variability over time and space in relationships between bacterial biomass production and ^{14}C- primary production and by implication, the routing of carbon through the DOM pool from phytoplankton.

In general, bacterial production correlated with ^{14}C-primary production ($r=0.53$; $p<0.001$; $n=89$) for rates integrated over the euphotic zone in the SBI study area (Kirchman et al. 2009a), but the ratio of the two production estimates (BP:PP) varied substantially during the SBI expeditions. It was significantly higher in 2002 than in 2004 and higher in May–June than in July–August of both years. Both bacterial production and primary production tended to decrease in transects from shelf waters to the basin, but since primary production decreased more so, the BP:PP ratio tended to be highest in the basin. In waters such as on the shelf with high rates of primary production (>100 mgC m^{-2} day^{-1}), the BP:PP ratio was low (0.06 ± 0.01; $n=55$) whereas it was high (0.79 ± 0.21; $n=34$) in waters such as the basin with low rates of primary production (<100 mgC m^{-2} day^{-1}). The implication of these data for shelf-basin exchange is discussed below.

9.7.3 Growth Efficiency in the Arctic Ocean

The BP:PP ratios do not include respiration and thus cannot be used alone to explore questions about how much primary production is routed through DOC, bacteria and the microbial loop. The bacterial growth efficiency (BGE) is also needed. We can calculate bacterial respiration (BR) from bacterial production (BP) estimates using the equation:

$$BR = BP(1 - BGE) / BGE$$

The few estimates of BGE for Arctic waters vary greatly (see references in Kirchman et al. 2009a) with an overall average of 6.9 ± 9.0 % for the entire SBI data set. This average does not differ statistically from the oceanic average of 15 % (Del Giorgio and Cole 2000). There was no significant effect of temperature on BGE in experiments conducted during the SBI expeditions, consistent with the lack of a clear temperature effect on BGE in previous studies (see references in Kirchman et al. 2009a).

9.7.4 *Implications for Shelf-Basin Exchange*

The BGE values can be combined with the biomass production data to provide another estimate for how much of total respiration and of total primary production consumption can be attributed to heterotrophic bacteria. Assuming BGE to be 7 %, the production data suggest that about 80 % of ^{14}C-primary production in the western Arctic is consumed by heterotrophic bacteria in waters with high rates of primary production (Kirchman et al. 2009a). When ice effects on ^{14}C primary production are ignored, consumption by heterotrophic bacteria amounts to <50 % of ^{14}C primary production, significantly less than see in low-latitude oceans (Kirchman et al. 2009b). In contrast, bacterial respiration estimated by this approach exceeds 100 % in waters with low primary production rates. This does seem possible in some Arctic regions (e.g. ice-covered waters), but there are known problems with the data. In a discussion of possible errors, Kirchman et al. (2009a) concludes that the ^{14}C method probably underestimates true rates of both net and gross primary production. A recent study also found that ^{14}C-based estimates of primary production were much lower than estimates from other approaches (Quay et al. 2010).

The data still can be used to explore the possible export of organic carbon from productive shelf waters to less productive waters such as those in the basin. The high rates of bacterial production, bacterial respiration and total respiration in waters with low primary production can be sustained only by organic carbon exported from waters with high primary production. In addition to general shelf-basin exchange, organic carbon could be exchanged between ice-free and ice-covered shelf regions with high and low rates of primary production, respectively. These regions could have adequate nutrients but be light limited if covered by ice. The production data can be used to identify these regions.

The organic carbon most likely exported is that in the DOC pool. The amount is too large to be accounted for by particulate organic carbon, whereas it is small compared to the total integrated DOC pool; the daily rate given by Kirchman et al. (2009a) is <10 % of total DOC. DOC produced in productive waters an exported to less productive waters helps to explain the spatial and temporal variability in net and community production as measured by light–dark incubations in the Arctic (Cottrell et al. 2006).

9.8 Ocean Acidification

It has been widely shown that the uptake of anthropogenic CO_2 by the oceans (e.g., Sabine et al. 2004; Sabine and Feely 2007) has a significant effect on marine biogeochemistry by reducing seawater pH (Feely et al. 2009; Caldiera and Wickett 2003) and the saturation states (Ω) of important calcium carbonate ($CaCO_3$) minerals (Feely et al. 2004; Orr et al. 2005; Caldiera and Wickett 2005) through a process termed "ocean acidification". Seawater exhibiting undersaturated conditions (i.e., $\Omega < 1$) are potentially corrosive for biogenic $CaCO_3$ minerals such as aragonite, calcite and high-Mg calcite. The reduction of $CaCO_3$ mineral saturation states in the surface ocean and along continental margins could have potentially negative consequences for benthic and pelagic calcifying organisms, and entire marine ecosystems (Fabry et al. 2008). Of even greater concern is the rate at which ocean acidification and $CaCO_3$ mineral saturation state suppression are progressing, particularly in the high latitude PAR (Byrne et al. 2010; Fabry et al. 2009) where mixing processes and colder temperatures naturally precondition the water column to have lower pH and Ω values compared to more temperate ocean environments.

Recent observations in the sub-arctic North Pacific Ocean (Mathis et al. 2011b) have already revealed areas of seasonal $CaCO_3$ mineral Ω suppression. Aragonite undersaturation has been observed throughout the water column during certain times of the year, and models project widening areas of aragonite undersaturation for longer durations during the next several decades (Steinacher et al. 2009). Undersaturation has potentially negative consequences for the region because the expansive continental shelf in regions such as the eastern Bering Sea sustains a commercially valuable fishery (Cooley and Doney 2009; Cooley et al. 2009) that produces approximately 47 % of the US fish catch by weight.

As observed at several open-ocean time-series, the uptake of anthropogenic CO_2 has already decreased surface water pH by 0.1 units. IPCC scenarios, based on present-day CO_2 emissions, predict a further decrease in seawater pH by 0.3–0.5 units over the next century and beyond (Caldiera and Wickett 2003). Ocean acidification and decreased pH reduces the saturation states of calcium carbonate minerals such as aragonite and calcite, with many studies showing decreased $CaCO_3$ production by calcifying fauna (Buddemeier et al. 2004; Fabry et al. 2008) and increased $CaCO_3$ dissolution. The PAR is particularly vulnerable to ocean acidification due to relatively low pH as well as low temperature and salinity (implying low calcium ion concentration) of polar waters compared to other waters (Orr et al. 2005; Steinacher et al. 2009) together with the low buffer capacity of sea-ice melt water (Yamamoto-Kawai et al. 2009). In the high latitude PAR, the uncoupling of primary production and grazing leads to high export rates of organic matter to the bottom waters and the sediments. When this organic matter is remineralized back into CO_2, it naturally decreases pH and suppresses carbonate mineral saturation states. However, the presence of anthropogenic CO_2 in the water column has caused bottom waters over some parts of the PAR shelves to become undersaturated in carbonate minerals (mostly aragonite, but in some locations calcite undersaturations have been observed) (e.g., Mathis et al. 2011b; Anderson et al. 2011).

9.8.1 The Bering Sea

The eastern shelf of the Bering Sea (Fig. 9.1) is a highly dynamic area that is influenced by a number of terrestrial and marine processes (Fig. 9.12) that impact seawater carbonate chemistry with considerable spatial, seasonal and interannual variability

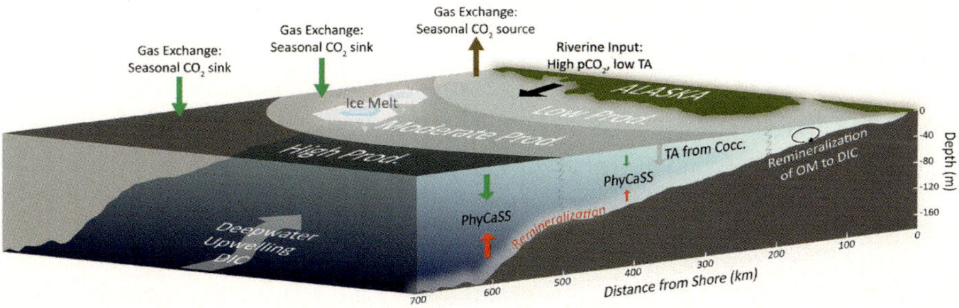

Fig. 9.12 Generalized description of the processes affecting the carbonate chemistry of the eastern Bering Sea shelf. The influx of runoff from the coast delivers water with high pCO_2, low TA, and moderate concentrations of dissolved organic matter (OM). The high pCO_2 of the water creates a seasonal source of CO_2 to the atmosphere while reducing carbonate mineral saturation states. Offshore, the *upper water column* is dominated by sea ice melt in late spring and summer that creates a highly satisfied surface layer where primary production is controlled by the confluence of coastal waters rich in micronutrients and basin water replete in macronutrients. Seasonally high rates of NCP lead to a rapid drawdown of CO_2 at the surface creating a strong seasonal sink for atmospheric CO_2. In 2009, coccolithophore (Cocc.) blooms were observed in the intermediate shelf waters and lowered TA concentrations at the surface. The varying degree of export production at the surface determined the amount of remineralization that occurred at depth which ultimately controlled saturation states. This PhyCaSS interaction can be observed to varying degrees across the shelf (Adapted from Mathis et al. 2011b)

in the saturation states of the two most biogenically important $CaCO_3$ minerals, aragonite ($\Omega_{aragonite}$) and calcite ($\Omega_{calcite}$) (Mathis et al. 2011a). The springtime retreat of sea ice, coupled with warming and seasonally high rates of freshwater discharge create distinctive horizontal and vertical zones over the shelf, each with their own unique characteristics (Stabeno et al. 1999). The onset of stratification in surface waters stimulates an intense period of phytoplankton primary production (PP), particularly over the middle region of the shelf where the confluence of macronutrient-rich Bering Sea water and coastal water replete in micronutrients is highest (Aguilar-Islas et al. 2007). In this region, historically referred to as the "*green belt*", rates of net community production (NCP) can exceed 480 mg C m^{-2} day^{-1} (see PP section in this chapter) while average rates across the shelf are ~330 mg C m^{-2} day^{-1}, making the eastern Bering Sea shelf one the most productive regions in the global ocean (Sambrotto et al. 2008; Mathis et al. 2010).

On the eastern Bering Sea shelf, a seasonal divergence in pH and Ω is observed between surface and subsurface waters, driven primarily by the biology of the system (Mathis et al. 2010). During the spring phytoplankton bloom, high rates of NCP effectively remove CO_2 from the surface waters creating a strong seasonal disequilibrium with the atmosphere (e.g. Bates et al. 2010), but also increasing pH and Ω values by ~0.1 and ~1 respectively (Mathis et al. 2011a). However, in subsurface waters the opposite is observed, with pH and Ω values decreasing significantly (~0.3 and ~0.2, respectively; Mathis et al. 2011a, b). Much of the organic matter that is produced during the spring phytoplankton bloom is exported vertically out of the mixed layer. By mid-summer, the water-column becomes highly stratified and bottom waters are effectively isolated from surface waters over much of the shelf. The vertical export of organic matter and its subsequent seasonal remineralization at depth induces a significant build-up of CO_2 in bottom waters (i.e., pCO_2 increases) and concurrent suppression of $CaCO_3$ mineral Ω values (Mathis et al. 2010, 2011b). The seasonal divergence of pH and Ω in surface and subsurface waters has been described in terms of a "Phytoplankton-Carbonate Saturation State" (PhyCaSS; Bates et al. 2009). In 2008, subsurface waters of the eastern Bering Sea shelf became undersaturated with respect to aragonite (but not calcite; Mathis et al. 2011b). It has also been shown that the addition of anthropogenic CO_2 to the ocean augments this natural seasonal interaction between ocean biology and seawater carbonate chemistry, tipping subsurface waters below the saturation state threshold ($\Omega_{aragonite} = 1$) for aragonite.

9.8.2 The Western Arctic Ocean

In the Arctic Ocean, potentially corrosive waters are found in the subsurface layer of the central basin (Jutterström and Anderson 2005; Yamamoto-Kawai et al. 2009; Chierici and Fransson 2009; Anderson et al. 2010), on the Chukchi/Beaufort Sea shelf (Bates and Mathis 2009; Mathis et al. 2012) and in outflow waters of the Arctic found on the Canadian Arctic Archipelago shelf (Azetsu-Scott et al. 2010). On the Chukchi Sea, shelf waters corrosive to $CaCO_3$ occur seasonally in the bottom waters with unknown impacts to benthic organisms. The seasonally high rates of summertime phytoplankton primary production in the Chukchi Sea drives a downward export of organic carbon, which is remineralized back to CO_2 which in turn increases seawater pCO_2 (and decreasing pH) of subsurface waters. Such a seasonal biological influence on the pH of subsurface waters amplifies existing impacts of ocean acidification induced by the uptake of anthropogenic CO_2 over the last century (Bates and Mathis 2009). Given the scenarios for pH changes in the Arctic, the Arctic Ocean and adjacent Arctic shelves, including the western Arctic, will be increasingly affected by ocean acidification, with potentially negative implications for shelled benthic organisms as well as those animals that rely on the shelf seafloor ecosystem.

9.9 Summary

The Arctic marine carbon cycle, including primary production, carbon transformations, and export production are sensitive to sea ice loss, warming temperatures, changes in the timing and location of primary production, freshwater inputs, and ocean acidification. A key issue for future studies is to determine the extent and variability in DOC and POC production, which can vary widely with seasonal gradients in light, ice-cover and riverine input to the Arctic Ocean and marginal seas. Further studies are needed to relate the biogeochemical impacts influencing the marine carbon cycle and remineralization, including organic carbon cycling in the water column, microbial and plankton processes, and in the underlying sediments to broader transitions in climate conditions. Recent studies indicate that ocean acidification and decreased pH are reducing the saturation states of calcium carbonate minerals and this process may have a detrimental impact on calcifying fauna. Further studies of the key processes within the marine carbon cycle will facilitate a better understanding of climate change impacts as the Arctic ecosystem undergoes a rapid transition to a potentially new state.

Acknowledgments We thank all of our colleagues who have conducted the work that we have summarized and explored here. This synthesis effort was supported by the National Science Foundation (PLR-1107997) through a grant to JTM.

References

Aguilar-Islas AM, Hurst MP, Buck KN, Sohst B, Smith GJ, Lohan MC, Bruland KW (2007) Micro- and macronutrients in the southeastern Bering Sea: insight into iron-replete and iron-depleted regimes. Prog Oceanogr 73:99–126

Alling V, Sanchez-Garcia L, Porcelli D, Pugach S, Vonk JE, van Dongen B, Mörth C-M, Anderson LG, Sokolov A, Andersson P, Humborg C, Semiletov I, Gustafsson O (2010) Nonconservative behavior of dissolved organic carbon across the Laptev and East Siberian Seas. Glob Biogeochem Cycles 24:GB4033. doi:10.1029/2010GB003834

Ambrose WG Jr, Tilney LM, Clough PR, Beer L (2001) Role of echinoderms in benthic remineralization in the Chukchi Sea. Mar Biol 139:937–949

Amiel D, Cochran JK (2008) Terrestrial and marine POC fluxes derived from 234Th distributions and δ13C measurements on the Mackenzie Shelf. J Geophys Res 113:C03S06

Amon RMW (2004) The role of dissolved organic matter for the organic carbon cycle in the Arctic Ocean. In: Stein RS, Macdonald RW (eds) The organic carbon cycle in the Arctic Ocean. Springer, New York, pp 83–99

Andersen T, Carstensen J, Hernandez-García E, Duarte CM (2009) Ecological thresholds and regime shifts: approaches to identification. Trends Ecol Evol 24(1):49–57

Anderson LG (2002) DOC in the Arctic Ocean. In: Hansell DA, Carlson CA (eds) Biogeochemistry of marine dissolved organic matter. Academic, San Diego, pp 665–683

Anderson LG, Kaltin S (2001) Carbon fluxes in the Arctic Ocean—potential impact by climate change. Polar Res 20(2):225–232

Anderson LG, Jones EP, Swift JH (2003) Export production in the central Arctic Ocean as evaluated from phosphate deficit. J Geophys Res 108(C6):3199. doi:10.1029/2001JC001057

Anderson LG, Jutterström S, Hjalmarsson S, Wåhlström I, Semiletov IP (2009) Out-gassing of CO_2 from Siberian Shelf seas by terrestrial organic matter decomposition. Geophys Res Lett 36:L20601. doi:10.1029/2009GL040046

Anderson LG, Tanhua T, Björk G, Hjalmarsson S, Jones EP, Jutterström S, Rudels B, Swift JH, Wåhlström I (2010) Arctic Ocean shelf—basin interaction, an active continental shelf CO_2 pump and its impact on the degree of calcium carbonate dissolution. Deep-Sea Res I 57:869–879. doi:10.1016/j.dsr.2010.03.012

Anderson LG, Björk G, Jutterström S, Pipko I, Shakhova N, Semiletov IP, Wåhlström I (2011) East Siberian Sea, an Arctic region of very high biogeochemical activity. Biogeosciences 8:1745–1754. doi:10.5194/bg-8-1745-2011

Apollonio S (1959) Hydrobiological measurements on IGY Drifting Station Bravo. EOS Trans AGU 40:316–319

Arrigo K, van Dijken G, Pabi S (2008) Impact of shrinking Arctic ice cover on marine primary production. Geophys Res Lett 35:L19603

Ashjian CJ, Campbell RG, Welch HE, Butler M, Keuren DV (2003) Annual cycle in abundance, distribution, and size in relation to hydrography of important copepod species in the western Arctic Ocean. Deep-Sea Res I 50:1235–1261

Azetsu-Scott K, Clarke A, Falkner K, Hamilton J, Jones EP, Lee C, Petrie B, Prinsenberg S, Starr M, Yeats P (2010) Calcium carbonate saturation states in the waters of the Canadian Arctic Archipelago and the Labrador Sea. J Geophys Res 115:C11021. doi:10.1029/2009JC005917

Bates NR, Mathis JT (2009) The Arctic Ocean marine carbon cycle: evaluation of air-sea CO_2 exchanges, ocean acidification impacts and potential feedbacks. Biogeosciences 6:2433–2459

Bates NR, Moran SB, Hansell DA, Mathis JT (2006) An increasing CO_2 sink in the Arctic Ocean due to sea-ice loss. Geophys Res Lett 33:C10013

Bates NR, Mathis JT, Cooper L (2009) The effect of ocean acidification on biologically induced seasonality of carbonate mineral saturation states in the Western Arctic Ocean. J Geophys Res Oceans 114:C11007

Bates NR, Mathis JT, Jefferies MA (2010) Air-sea CO_2 fluxes on the Bering Sea shelf. Biogeosci Discuss 7:1–44

Benner R, Benitez-Nelson B, Kaiser K, Amon RMW (2004) Export of young terrigenous dissolved organic carbon from rivers to the Arctic Ocean. Geophys Res Lett 31(5):L05305. doi:10.1029/2003GL019251

Benner R, Louchouarn P, Amon RMW (2005) Terrigenous dissolved organic matter in the Arctic Ocean and its transport to surface and deep waters of the North Atlantic. Glob Biogeochem Cycles 19(2):GB2025

Bluhm BA, Gradinger R (2008) Regional variability in food availability for Arctic marine mammals. Ecol Appl 18:S77–S96

Buddemeier RW, Keypas JA, Aronson RB (2004) Coral reefs and global climate change: potential contributions of climate change to stresses on coral reef ecosystems. Pew Center on Climate Change, Arlington, VA. Available at http://www.pewclimate.org/global-warming-in-depth/all_reports/coral_reefs/, 44 pp

Bussmann I, Kattner G (2000) Distribution of dissolved organic carbon in the central Arctic Ocean: the influence of physical and biological properties. J Mar Syst 27:209–219

Byrne RH, Mecking S, Feely RA, Liu Z (2010) Direct observations of basin-wide acidification of the North Pacific Ocean. Geophys Res Lett 37:L02601. doi:10.1029/2009GL040999

Cai P, Rutgers Van Der Loeff M, Stimac I, Nöthig E-M, Lepore K, Moran SB (2010a) Low export flux of particulate organic carbon in the central Arctic Ocean as revealed by $^{234}Th/^{238}U$ disequilibrium. J Geophys Res 115:C10037. doi:10.1029/2009JC005595

Cai W-J, Chen L, Chen B, Gao Z, Lee SH, Chen J, Pierrot D, Sullivan K, Wang Y, Hu X, Huang W-J, Zhang Y, Xu S, Murata A, Grebmeier JM, Jones EP, Zhang H (2010b) Decrease in the CO_2 uptake capacity in an ice-free Arctic Ocean basin. Science 329:556. doi:10.1126/science.1189338

Caldiera K, Wickett ME (2003) Anthropogenic carbon and ocean pH. Nature 425(6956):365

Caldiera K, Wickett ME (2005) Ocean model predictions of chemistry changes from carbon dioxide emissions to the atmosphere and ocean. J Geophys Res 110(C9):C09S04. doi:10.1029/2004JC002671

Campbell RG, Sherr EB, Ashjian CJ, Sherr BF, Hill V, Stockwell DA (2009) Mesozooplankton prey preference and grazing impact in the Western Arctic Ocean. Deep-Sea Res II 56:1274–1289

Cavalieri DJ, Parkinson CL, Vinnikov KY (2003) 30-year satellite record reveals contrasting Arctic and Antarctic decadal sea ice variability. Geophys Res Lett 30(18):1970

Chen M, Huang Y, Cai P, Guo L (2003) Particulate organic carbon export fluxes in the Canada Basin and Bering Sea as derived from $^{234}Th/^{238}U$ disequilibria. Arctic 56:32–44

Chierici M, Fransson A (2009) Calcium carbonate saturation in the surface water of the Arctic Ocean: undersaturation in freshwater influenced shelves. Biogeosciences 6:2421–2432. doi:10.5194/bg-6-2421-2009

Choe N, Deibel D (2009) Statolith diameter as an age indicator in the planktonic tunicate Oikopleura vanhoeffeni: variability in age-specific growth patterns in Conception Bay, Newfoundland. J Exp Mar Biol Ecol 375:89–98
Clough LM, Renaud PE, Ambrose WG Jr (2005a) Impacts of water depth, sediment pigment concentration, and benthic macrofaunal biomass on sediment oxygen demand in the western Arctic Ocean. Can J Fish Aquat Sci 62:1756–1765
Clough LM, Renaud PE, Ambrose WG Jr (2005b) Sediment oxygen demand, infaunal biomass, and sediment pigment concentration in the western Arctic Ocean. Can J Fish Aquat Sci 62:1756–1765
Coachman LK, Aagaard K, Tripp RB (1975) Bering Strait: the regional physical oceanography. University of Washington Press, Seattle, 172 pp
Codispoti LA, Flagg CN, Swift JH (2009) Hydrographic conditions during the 2004 SBI process experiments. Deep-Sea Res II 56:1144–1163
Cooley SR, Doney SC (2009) Anticipating ocean acidification's economic consequences for commercial fisheries. Environ Res Lett 4:024007. doi:10.1088/1748-9326/4/2/024007
Cooley SR, Kite-Powell HL, Doney SC (2009) Ocean acidification's potential to alter global marine ecosystem services. Oceanography 22(4):172–181
Cooper LW, Benner R, McClelland JW, Peterson BJ, Holmes RM, Raymond PA, Hansell DA, Grebmeier JM, Codispoti LA (2005) Linkages among runoff, dissolved organic carbon, and the stable oxygen isotope composition of seawater and other water mass indicators in the Arctic Ocean. J Geophys Res 110:G02013. doi:10.1029/2005JG000031
Cooper LW, McClelland JW, Holmes RM, Raymond PA, Gibson JJ, Guay CK, Peterson BJ (2008) Flow-weighted values of runoff tracers (δ18O, DOC, Ba, alkalinity) from the six largest Arctic rivers. Geophys Res Lett 35:L18606. doi:10.1029/2008GL035007
Cota GF, Pomeroy LR, Harrison WG, Jones EP, Peters F, Sheldon WM, Weingartner TR (1996) Nutrients, primary production and microbial heterotrophy in the southeastern Chukchi Sea: Arctic summer nutrient depletion and heterotrophy. Mar Ecol Prog Ser 135:247–258
Cottrell MT, Kirchman DL (2009) Photoheterotrophic microbes in the Arctic Ocean in summer and winter. Appl Environ Microbiol 75:4958–4966
Cottrell MT, Malmstrom RR, Hill V, Parker AE, Kirchman DL (2006) The metabolic balance between autotrophy and heterotrophy in the western Arctic Ocean. Deep-Sea Res 53:1831–1844
Coyle KO, Pinchuk AI (2002) Climate-related differences in zooplankton density and growth on the inner shelf of the southeastern Bering Sea. Prog Oceanogr 55:177–194
Dagg MJ, Vidal J, Whitledge TE, Iverson RL, Goering JJ (1982) The feeding, respiration, and excretion of zooplankton in the Bering Sea during a spring bloom. Deep-Sea Res 29:45–63
Danielson S, Aagaard K, Weingartner T, Martin S, Winsor P, Gawarkiewicz G, Quadfasel D (2006) The St. Lawrence polynya and the Bering shelf circulation: new observations and a model comparison. J Geophys Res 111:C09023
Darnis G, Barber DG, Fortier L (2008) Sea ice and the onshore-offshore gradient in pre-winter zooplankton assemblages in southeastern Beaufort Sea. J Mar Syst 74:994–1011
Deibel D, Daly KL (2007) Zooplankton processes in Arctic and Antarctic polynyas. In: Smith WO Jr, Barber DG (eds) Arctic and Antarctic polynyas. Elsevier, Amsterdam, pp 271–322
Deibel D, Saunders PA, Acuna JL, Bochdansky AB, Shiga N, Rivkin RB (2005) The role of appendicularian tunicates in the biogenic carbon cycle of three Arctic polynyas. In: Gorsky G, Youngbluth MJ, Deibel D (eds) Response of marine ecosystems to global change: ecological impact of appendicularians. Gordon and Breach, Paris, pp 327–356
Del Giorgio PA, Cole JJ (2000) Bacterial energetics and growth efficiency. In: Kirchman DL (ed) Microbial ecology of the oceans. Wiley, New York, pp 289–325
Devol AH, Codispoti LA, Christensen JP (1997) Summer and winter denitrification rates in western Arctic shelf sediments. Cont Shelf Res 17:1029–1050
Dittmar T, Kattner G (2003) The biogeochemistry of the river and shelf ecosystem of the Arctic Ocean: a review. Mar Chem 83:103–120
Dixon RK, Brown S, Houghton RA, Solomon AM, Trexler MC, Wisniewski J (1994) Carbon pools and flux of global forest ecosystems. Science 263:185–190

Duarte CM, Regaudie-de-Gioux A (2009) Thresholds of gross primary production for the metabolic balance of marine planktonic communities. Limnol Oceanogr 54:1015–1022

Ducklow HW (1999) The bacterial component of the oceanic euphotic zone. FEMS Microbiol Ecol 30:1–10

Ducklow H (2000) Bacterial production and biomass in the oceans. In: Kirchman DL (ed) Microbial ecology of the oceans. Wiley, New York

Ducklow H, Carlson C, Church M, Kirchman D, Smith D, Steward G (2001) The seasonal development of the bacterioplankton bloom in the Ross Sea, Antarctica 1994–1997. Deep-Sea Res II 48:4199–4221

Ekwurzel B, Schlosser P, Mortlock RA, Fairbanks RG, Swift JH (2001) River runoff, sea ice meltwater, and Pacific water distribution and mean residence times in the Arctic Ocean. J Geophys Res 106:9075–9092

English TS (1961) Some biological oceanographic observations in the central North Polar Sea Drift Station Alpha, 1957–1958. Research Paper 13. Arctic Institute of North America, 80 pp

Fabry VJ, Seibel BA, Feely RA, Orr JC (2008) Impacts of ocean acidification on marine fauna and ecosystem processes. ICES J Mar Sci 65:414–432

Fabry VJ, McClintock JB, Mathis JT, Grebmeier JM (2009) Ocean acidification at high latitudes: the bellwether. Oceanography 22(4):160–171

Feely RA, Sabine CL, Lee K, Berelson W, Keypas J, Fabry VJ, Millero FJ (2004) Impact of anthropogenic CO_2 on the $CaCO_3$ system in the oceans. Science 305(5682):362–366

Feely RA, Doney SC, Cooley SR (2009) Ocean acidification: present conditions and future changes in a high-CO_2 world. Oceanography 22(4):36–47

Finlay J, Neff J, Zimov S, Davydova A, Davydov S (2006) Snowmelt dominance of DOC in high-latitude watersheds: implications for characterization and flux of river DOC. Geophys Res Lett 33:L10401

Forest A, Sampei M, Hattori H, Makabe R, Sasaki H, Fukuchi M, Wassmann P, Fortier L (2007) Particulate organic carbon fluxes on the slope of the Mackenzie Shelf (Beaufort Sea): physical and biological forcing of shelf-basin exchanges. J Mar Syst 68:39–54

Garneau M-È, Vincent WF, Terrado R, Lovejoy C (2009) Importance of particle-associated bacterial heterotrophy in a coastal Arctic ecosystem. J Mar Syst 75:185–197

Gosselin M, Levasseur M, Wheeler PE, Horner RA, Booth BC (1997) New measurements of phytoplankton and ice algal production in the Arctic Ocean. Deep-Sea Res II 44:1623–1644

Grainger EH (1975) A marine ecology study in Frobisher Bay, Arctic Canada. In: Billingsly LW, Cameron TWM (eds) Energy flow—its biological dimensions. A summary of the IBP in Canada, 1964–1974. Canadian Committee for the IBP Royal Society of Canada, Ottawa, pp 261–266

Grebmeier GM (2012) Shifting patterns of life in the Pacific Arctic and sub-Arctic Seas. Annu Rev Mar Sci 4:16.1–16.16

Grebmeier JM, Barry JP (2007) Benthic processes in polynyas. In: Smith WO Jr, Barber DG (eds) Polynyas: windows to the world, Elsevier oceanography series 74. Elsevier, Amsterdam, pp 363–390

Grebmeier JM, Harvey HR (2005) The Western Arctic Shelf–Basin Interactions (SBI) project: an overview. Deep Sea Res II 52:3109–3115

Grebmeier JM, McRoy CP (1989) Pelagic-benthic coupling on the shelf of the northern Bering and Chukchi Seas. III. Benthic food supply and carbon cycling. Mar Ecol Prog Ser 53:79–91

Grebmeier JM, Whitledge TE (1996) Arctic system science ocean–atmosphere-ice interactions biological initiative in the Arctic: shelf-basin interactions workshop, ARCSS/OAII Report Number 4. University of Washington, Seattle, 39 pp

Grebmeier JM, Cooper LW, Feder HM, Sirenko BI (2006a) Ecosystem dynamics of the Pacific-influenced Northern Bering and Chukchi Seas in the Amerasian Arctic. Prog Oceanogr 71:331–361

Grebmeier JM, Overland JE, Moore SE, Farley EV, Carmack E, Cooper LW, Frey KE, Helle JH, McLaughlin F, McNutt SL (2006b) A major ecosystem shift in the Northern Bering Sea. Science 311:1461–1464

Grebmeier JM, Feder HM, McRoy CP (1989) Pelagic-benthic coupling on the shelf of the northern Bering and Chukchi Seas. Mar Ecol Prog Ser 51:253–268

Grebmeier JM, Harvey HR, Stockwell DA (2009) The Western Arctic Shelf–Basin Interactions (SBI) project, volume II: an overview. Deep-Sea Res II 56:1137–1143

Gregg WW, Conkright ME (2002) Decadal changes in global ocean chlorophyll. Geophys Res Lett 29:1730. doi:10.1029/2002GL014689

Guéguen C, Guo L, Tanaka N (2005) Distributions and characteristics of colored dissolved organic matter in the Western Arctic Ocean. Cont Shelf Res 25:1195–1207

Guo L, Smiletov I, Gustafsson O, Ingri J, Andersson P, Dudarev O, White D (2004) Characterization of Siberian Arctic estuarine sediments: implications for terrestrial organic carbon export. Glob Biogeochem Cycles 18(1):GB1036

Guo L, Ping C-L, Macdonald RW (2007) Mobilization pathways of organic carbon from permafrost to arctic rivers in a changing climate. Geophys Res Lett 34:L13603. doi:10.1029/2007GL030689

Hameedi MJ (1978) Aspects of water column primary productivity in the Chukchi Sea during summer. Mar Biol 45:37–46

Hansell DA, Carlson CA (1998) Net community production of dissolved organic carbon. Glob Biogeochem Cycles 12:443–453

Hansell DA, Goering JJ (1990) Pelagic nitrogen flux in the northern Bering Sea. Cont Shelf Res 10:501–519

Hansell DA, Whitledge TE, Goering JJ (1993) Patterns of nitrate utilization and new production over the Bering-Chukchi shelf. Cont Shelf Res 13:601–627

Hansell DA, Kadko D, Bates NR (2004) Degradation of terrigenous dissolved organic carbon in the western Arctic Ocean. Science 304(5672):858–861

Hansell DA, Carlson CA, Repeta DJ, Schlitzer R (2009) Dissolved organic matter in the ocean: new insights stimulated by a controversy. Oceanography 22:52–61

Harrison WG, Platt T, Irwin B (1982) Primary production and nutrient assimilation by natural phytoplankton populations of the eastern Canadian Arctic. Can J Fish Aquat Sci 39:335–345

Henriksen K, Blackburn TH, Lomstein BA, McRoy CP (1993) Rates of nitrification, distribution of nitrifying bacteria and inorganic N fluxes in northern Bering-Chukchi shelf sediments. Cont Shelf Res 13:629–651

Hill V, Cota GF, Stockwell D (2005) Spring and summer phytoplankton communities in the Chukchi and eastern Beaufort seas. Deep-Sea Res II 52:3369–3385

Hirche H-J, Kattner G (1993) Egg production and lipid content of Calanus glacialis in spring: indication of a food-dependent and food-independent reproductive mode. Mar Biol 104:615–622

Holland MM, Bitz CM, Tremblay B (2006) Future abrupt reductions in the summer Arctic sea ice. Geophys Res Lett 33(23):L23503

Holland MM, Serreze MC, Stroeve JC (2010) The sea ice mass budget of the Arctic and its future change as simulated by coupled climate models. Clim Dyn 34(2–3):185–200

Holmes RM, McClelland JW, Raymond PA, Frazer BB, Peterson BJ, Stieglitz M (2008) Lability of DOC transported by Alaskan rivers to the Arctic Ocean. Geophys Res Lett 35:L03402. doi: 10.1029/2007GL032837

Hopcroft RR, Roff JC, Bouman HA (1998) Zooplankton growth rates: the larvaceans Appendicularia, Fritillaria and Oikopleura in tropical waters. J Plankton Res 20:539–555

Hopcroft RR, Bluhm BA, Gradinger RR (2008) Arctic Ocean synthesis: analysis of climate change impacts in the Chukchi and Beaufort Seas with strategies for future research, 2nd edn. North Pacific Research Board, Anchorage

Hopcroft RR, Kosobokova KN, Pinchuk AI (2010) Zooplankton community patterns in the Chukchi Sea during summer 2004. Deep-Sea Res II 57:27–39

Hopkinson CS Jr, Vallino JJ, Nolin A (2002) Decomposition of dissolved organic matter from the continental margin. Deep-Sea Res II 49:4461–4478

Ikeda T, Kanno Y, Ozaki K, Shinada A (2001) Metabolic rates of epipelagic marine copepods as a function of body mass and temperature. Mar Biol 139:587–596

Jutterström S, Anderson LG (2005) The saturation of calcite and aragonite in the Arctic Ocean. Mar Chem 94:101–110

Jutterström S, Anderson LG, Bates NR, Bellerby R, Johannessen T, Jones EP, Key RM, Lin X, Olsen A, Omar AM (2010) Arctic Ocean data in CARINA. Earth Syst Sci Data 2:71–78

Kadko D, Muench R (2005) Evaluation of shelf-basin interaction in the western Arctic by use of short-lived radium isotopes: the importance of mesoscale processes. Deep-Sea Res II 52:3227–3244

Kaltin S, Anderson LG (2005) Uptake of atmospheric carbon dioxide in Arctic shelf seas: evaluation of the relative importance of processes that influence pCO2 in water transported over the Bering-Chukchi Sea shelf. Mar Chem 94(1–4):67–79

Karcher MJ, Oberhuber JM (2002) Pathways and modification of the upper and intermediate waters of the Arctic Ocean. J Geophys Res 107:2.1–2.13

Kattner G, Lobbes JM, Fitznar HP, Engbrodt R, Nöthig E-M, Lara RJ (1999) Tracing dissolved organic substances and nutrients from the Lena River through Laptev Sea (Arctic). Mar Chem 65:25–39

Kirchman DL, Elifantz H, Dittel A, Malmstrom RR, Cottrell MT (2007) Standing stocks and activity of archaea and bacteria in the western Arctic Ocean. Limnol Oceanogr 52:495–507

Kirchman DL, Hill V, Cottrell MT, Gradinger R, Malmstrom RR, Parker A (2009a) Standing stocks, production, and respiration of phytoplankton and heterotrophic bacteria in the western Arctic Ocean. Deep Sea Res II 56:1237–1248

Kirchman DL, Moran XAG, Ducklow H (2009b) Microbial growth in the polar oceans—role of temperature and potential impact of climate change. Nat Rev Microbiol 7:451–459

Korsak MN (1992) Primary production of organic matter. In: Nagel PA (ed) Results of the third joint US-USSR Bering and Chukchi Seas Expedition (BERPAC), summer 1988. U.S. Fish and Wildlife Service, Washington, DC, pp 215–218

Kosobokova KN, Hopcroft RR (2010) Diversity and vertical distribution of mesozooplankton in the Arctic's Canada Basin. Deep-Sea Res II 57:96–110

Lalande C, Grebmeier JM, Wassman P, Cooper LW, Flint MV, Sergeeva VM (2007a) Export fluxes of biogenic matter in the presence and absence of seasonal sea ice cover in the Chukchi Sea. Cont Shelf Res 27:2051–2065

Lalande C, Lepore K, Cooper LW, Grebmeier JM, Moran SB (2007b) Export fluxes of particulate organic carbon in the Chukchi Sea: a comparative study using $^{234}Th/^{238}U$ disequilibria and drifting sediment traps. Mar Chem 103:185–196

Lalande C, Moran SB, Wassmann P, Grebmeier JM, Cooper LW (2008) ^{234}Th-derived particulate organic carbon fluxes in the northern Barents Sea with comparison to drifting sediment trap fluxes. J Mar Syst 73:103–113

Lalande C, Forest A, Barber DG, Gratton Y, Fortier M (2009) Variability in the annual cycle of vertical particulate organic carbon export on Arctic shelves: contrasting the Laptev Sea, Northern Baffin Bay, and the Beaufort Sea. Cont Shelf Res 29:2157–2165

Lane PVZ, Llinás L, Smith SL, Pilz D (2008) Zooplankton distribution in the western Arctic during summer 2002: hydrographic habitats and implications for food chain dynamics. J Mar Res 70:97–133

Lee SH, Whitledge TE (2005) Primary production in the deep Canada Basin during summer 2002. Polar Biol 28:190–197

Lee SH, Whitledge TE, Kang SH (2007) Recent carbon and nitrogen uptake rates of phytoplankton in Bering Strait and the Chukchi Sea. Cont Shelf Res 27:2231–2249

Lee SH, Stockwell D, Whitledge TE (2010) Uptake rates of dissolved inorganic carbon and nitrogen by under-ice phytoplankton in the Canada Basin in summer 2005. Polar Biol 33:1027–1036

Lee SH, Joo Joo HM, Yun MS, Whitledge TE (2011) Recent phytoplankton productivity of the northern Bering Sea during early summer in 2007. Polar Biol 35:83–98. doi:10.1007/s00300-011-1035-9

Lepore K, Moran SB, Grebmeier JM, Cooper LW, Lalande C, Maslowski W, Hill V, Bates NR, Hansell DA, Mathis JT, Kelly RP (2006) Seasonal and interannual changes in POC export and deposition in the Chukchi Sea. J Geophys Res 112:C10024. doi:10.1029/2006JC003555

Letscher RT, Hansell DA, Kadko D (2011) Rapid removal of terrigenous dissolved organic carbon over the Eurasian shelves of the Arctic Ocean. Mar Chem 123:78–87

Manabe S, Stouffer RJ (2000) Study of abrupt climate change by a coupled ocean atmosphere model. Quat Sci Rev 19:285–299

Maslanik JA, Drobo S, Fowler C, Emery W, Barry R (2007) On the Arctic climate paradox and the continuing role of atmospheric circulation in affecting sea ice conditions. Geophys Res Lett 34:L03711

Mathis JT, Hansell DA, Bates NR (2005) Strong hydrographic controls on spatial and seasonal variability of dissolved organic carbon in the Chukchi Sea. Deep-Sea Res II 52:3245–3258

Mathis JT, Hansell DA, Kadko D, Bates NR, Cooper LW (2007) Determining net dissolved organic carbon production in the hydrographically complex western Arctic Ocean. Limnol Oceanogr 52:1789–1799

Mathis JT, Cross JN, Bates NR, Lomas ML, Moran SB, Mordy CW, Stabeno P (2010) Seasonal distribution of dissolved inorganic carbon and net community production on the Bering Sea Shelf. Biogeosciences 7:1769–1787. doi:10.5194/bg-7-1769-2010

Mathis JT, Cross JN, Bates NR (2011a) Coupling primary production and terrestrial runoff to ocean acidification and carbonate mineral suppression in the Eastern Bering Sea. J Geophys Res 116:C02030. doi:10.1029/2010JC006453

Mathis JT, Cross JN, Bates NR (2011b) The role of ocean acidification in systemic carbonate mineral suppression in the Bering Sea. Geophys Res Lett 38:L19602. doi:10.1029/2011GL048884

Mathis JT, Byrne RH, McNeil CL, Pickart RP, Juranek L, Liu S, Ma J, Easley RA, Elliot MW, Cross JN, Reisdorph SC, Morison J, Lichendorph T, Feely RA (2012) Storm-induced upwelling of high pCO_2 waters onto the continental shelf of the western Arctic Ocean and implications for carbonate mineral saturation states. Geophys Res Lett 39:L07606. doi:10.1029/2012GL051574

McGuire AD, Chapin FS, Walsh JE, Wirth C (2006) Integrated regional changes in arctic climate feedbacks: implications for the global climate system. Annu Rev Environ Resour 31:61–91

McGuire AD, Anderson LG, Christensen TR, Dallimore S, Guo L, Hayes DJ, Heimann M, Lorenson TD, Macdonald RW, Roulet N (2009) Sensitivity of the carbon cycle in the Arctic to climate change. Ecol Monogr 79(4):523–555

McGuire AD, Hayes DJ, Kicklighter DW, Manizza M, Zhuang Q, Chen M, Follows MJ, Gurney K, McClelland JW, Melillo JM, Peterson BJ, Prinn RG (2010) An analysis of the carbon balance of the Arctic basin from 1997 to 2006. Tellus 62(5):455–474

Moran SB, Ellis KM, Smith JN (1997) $^{234}Th/^{238}U$ disequilibria in the central Arctic Ocean: implications for particulate organic carbon export. Deep-Sea Res II 44:1593–1606

Moran MA, Sheldon WM Jr, Sheldon JE (1999) Biodegradation of riverine dissolved organic carbon in five estuaries of the southeastern United States. Estuaries 22(1):55–64

Moran SB, Kelly RP, Hagstrom K, Smith JN, Grebmeier JM, Cooper LW, Cota GF, Walsh JJ, Bates NR, Hansell DA, Maslowski RP (2005) Seasonal changes in POC export flux in the Chukchi Sea and implications for water column-benthic coupling in Arctic shelves. Deep-Sea Res II 52:3427–3451

Moritz RE, Perovich DK (1996) Arctic System Science: ocean–atmosphere-Ice Interactions. SHEBA. Surface Heat Budget of the Arctic Ocean Science Plan. ARCSS/OAII Report Number 5. University of Washington, Seattle, 60 pp

Neff JC, Finlay JC, Zimov SA, Davydov SP, Carrasco JJ, Schuur EAG, Davydova AI (2006) Seasonal changes in the age and structure of dissolved organic carbon in Siberian rivers and streams. Geophys Res Lett 33:L23401. doi:10.1029/2006GL028222

Nghiem SV, Rigor IG, Perovich DK, Clemente-Colón P, Weatherly JW, Neumann G (2007) Rapid reduction of Arctic perennial sea ice. Geophys Res Lett 34:L19504. doi:10.1029/2007GL031138

Opsahl S, Benner R, Amon RW (1999) Major flux of terrigenous dissolved organic matter through the Arctic Ocean. Limnol Oceanogr 44:2017–2023

Orr JC, Fabry VJ, Aumont O, Bopp L, Doney SC, Feely RA, Gnanadesikan A, Gruber N, Ishida A, Joos F, Key RM, Lindsay K, Maier-Reimer E, Matear R, Monfray P, Mouchet A, Najjar RG, Plattner GK, Rodgers KB, Sabine CL, Sarmiento JL, Schlitzer R, Slater RD, Totterdell IJ,

Weirig MF, Yamanaka Y, Yool A (2005) Anthropogenic ocean acidification over the twenty-first century and its impact on calcifying organisms. Nature 437(7059):681–686

Overland JE, Wang M (2005) The Arctic climate paradox: the recent decrease of the Arctic Oscillation. Geophys Res Lett 32(6):L06701

Pautzke CG (1979) Phytoplankton primary production below Arctic Ocean pack ice: an ecosystems analysis. Ph.D. thesis, University of Washington, Seattle, WA

Perovich DK, Richter-Menge JA (2009) Loss of sea ice in the Arctic. Annu Rev Mar Sci 1:417–441

Peterson BJ, Holmes RM, McClelland JW, Vorosmarty CJ, Shiklomanov IA, Shiklomanov AI, Lammers RB, Rahmsdorf S (2002) Increasing river discharge to the Arctic Ocean. Science 298:2171–2173

Piepenburg D, Blackburn TH, Dorrien CF, Gutt J, Hall POJ, Hulth S, Kendall MA, Opalinski KW, Rachor E, Schmid MK (1995) Partitioning of benthic community respiration in the Arctic (northwestern Barents Sea). Mar Ecol Prog Ser 118:199–213

Pirtle-Levy R (2006) A shelf-to-basin examination of food supply for Arctic benthic macrofauna and the potential biases of sampling methodology. Masters thesis, University of Tennessee, Knoxville, TN, USA. Available at/http://etd.utk.edu/2006/Pirtle-LevyRebecca.pdf

Plourde S, Campbell RG, Ashjian CJ, Stockwell DA (2005) Seasonal and regional patterns in egg production of Calanus glacialis/marshallae in the Chukchi and Beaufort Seas during spring and summer, 2002. Deep-Sea Res II 52:3411–3426

Polyak L, Alley RB, Andrews JT, Brigham-Grette J, Cronin TMN, Darby DA, Dyke AS, Fitzpatrick JJ, Funder S, Holland M, Jennings AE, Miller GH, O'Regan M, Savelle J, Serreze M, St John K, White JWC, Wolff E (2010) History of sea ice in the Arctic. Quat Sci Rev 29(15–16):1757–1778

Pomeroy LR, Deibel D (1986) Temperature regulation of bacterial activity during the spring bloom in Newfoundland coastal waters. Science 233:359–361

Quay PD, Peacock C, Björkman K, Karl DM (2010) Measuring primary production rates in the ocean: enigmatic results between incubation and non-incubation methods at Station ALOHA. Glob Biogeochem Cycles 24:GB3014

Raymond PA, Bauer JE (2000) Bacterial consumption of DOC during transport through a temperate estuary. Aquat Microb Ecol 22:1–12

Raymond PA, McClelland JW, Holmes RM, Zhulidov AV, Mull K, Peterson BJ, Striegl RG, Aiken GR, Gurtovaya TY (2007) Flux and age of dissolved organic carbon exported to the Arctic Ocean: a carbon isotopic study of the five largest arctic rivers. Glob Biogeochem Cycles 21:GB4011. doi:10.1029/2007GB002934

Renaud PE, Morata N, Ambrose WG, Bowie JJ, Chiuchiolo A (2007a) Carbon cycling by seafloor communities on the eastern Beaufort Sea shelf. J Exp Mar Biol Ecol 349(2):248–260

Renaud PE, Riedel A, Michel C, Morata N, Gosselin M, Juul-Pedersen T, Chiuchiolo A (2007b) Seasonal variation in benthic community oxygen demand: a response to an ice algal bloom in the Beaufort Sea, Canadian Arctic? J Mar Syst 67:1–12

Rich J, Gosselin M, Sherr E, Sherr B, Kirchman DL (1997) High bacterial production, uptake and concentrations of dissolved organic matter in the Central Arctic Ocean. Deep-Sea Res II 44:1645–1663

Ringuette M, Fortier L, Fortier M, Runge JA, Belancer S, Laroucher P, Weslawski J-M, Kwasniewski S (2002) Advanced recruitment and accelerated population development in Arctic calanoid copepods of the North Water. Deep-Sea Res II 49:5081–5100

Robinson C (2008) Heterotrophic bacterial respiration. In: Kirchman DL (ed) Microbial ecology of the Oceans. Wiley-Blackwell, Hoboken, pp 299–334

Rothrock DA, Zhang J (2005) Arctic Ocean sea ice volume: what explains its recent depletion? J Geophys Res 110(C1):C01002

Rothrock DA, Zhang J, Yu Y (2003) The arctic ice thickness anomaly of the 1990s: a consistent view from observations and models. J Geophys Res 108(C3):28-1–28-10

Sabine CL, Feely RA (2007) The oceanic sink for carbon dioxide. In: Reay D, Hewitt N, Grace J, Smith K (eds) Greenhouse gas sinks. CABI Publishing, Oxfordshire, pp 31–49

Sabine CL, Feely RA, Gruber N, Key RM, Lee K, Bullister JL, Wanninkhof R, Wong CS, Wallace DWR, Tilbrook B, Millero FJ, Peng T-H, Kozyr A, Ono T, Rios AF (2004) The oceanic sink for anthropogenic CO_2. Science 305(5682):367–371

Sambrotto RN, Goering JJ, McRoy CP (1984) Large yearly production of phytoplankton in the western Bering Strait. Science 225:1147–1150

Sambrotto RN, Mordy C, Zeeman SI, Stabeno PJ, Macklin SA (2008) Physical forcing and nutrient conditions associated with patterns of Chl a and phytoplankton productivity in the southeastern Bering Sea during summer. Deep Sea Res II 55:1745–1760

Savelieva NI, Semiletov IP, Vasilevskaya LN, Pugach SP (2000) A climate shift in seasonal values of meteorological and hydrological parameters for Northeastern Asia. Prog Oceanogr 47(2–4):279–297

Schauer U (1997) Impact of eastern Arctic shelf waters on the Nansen Basin intermediate layers. J Geophys Res 102:3371–3382

Schlosser P, Bauch D, Fairbanks R, Bönisch G (1994) Arctic river-runoff: mean residence time on the shelves and in the halocline. Deep-Sea Res I 41:1053–1068

Semiletov IP, Pipko II, Shakhova NE, Dudarev OV, Pugach SP, Charkin AN, McRoy CP, Kosmach D, Gustafsson Ö (2011) Carbon transport by the Lena River from its headwaters to the Arctic Ocean, with emphasis on fluvial input of terrestrial particulate organic carbon vs. carbon transport by coastal erosion. Biogeosciences 8:2407–2426

Serreze MC, Francis JA (2006) The Arctic amplification debate. Climate Change 76:241–264

Serreze MC, Holland MM, Stroeve J (2007) Perspectives on the Arctic's shrinking sea ice cover. Science 315:1533–1536

Serreze MC, Barrett AP, Stroeve JC, Kindig DN, Holland MM (2009) The emergence of surface-based Arctic amplification. Cryosphere 3(1):11–19

Seuthe L, Darnis G, Riser CW, Wassmann P, Fortier L (2007) Winter-spring feeding and metabolism of Arctic copepods: insights from faecal pellet production and respiration measurements in the southeastern Beaufort Sea. Polar Biol 30:427–436

Sherr EB, Sherr BF, Fessenden L (1997) Heterotrophic protists in the Central Arctic Ocean. Deep-Sea Res II 44:1665–1682

Sherr EB, Sherr BF, Hartz AJ (2008) Microzooplankton grazing impact in the Western Arctic Ocean. Deep-Sea Res II 56:1264–1273

Shimada K, Kamoshida T, Itoh M, Nishino S, Carmack E, McLaughlin F, Zimmermann S, Proshutinsky A (2007) Pacific Ocean inflow: influence on catastrophic reduction of sea ice cover in the Arctic Ocean. Geophys Res Lett 33(8):L08605

Shin K-H, Tanaka N (2004) Distribution of dissolved organic matter in the eastern Bering Sea, Chukchi Sea (Barrow Canyon) and Beaufort Sea. Geophys Res Lett 31:L24304. doi:10.1029/2004GL021039

Sommerkorn M, Hassol S (2009) Arctic climate feedbacks: global implications. World Wildlife Fund report. In: Sommerkorn M, Hassol S (eds) World Wildlife Fund report, Washington D.C., USA, 100 pp

Spencer RGM, Aiken GR, Wickland KP, Striegl RG, Hernes PJ (2008) Seasonal and spatial variability in dissolved organic matter quantity and composition from the Yukon River basin, Alaska. Glob Biogeochem Cycles 22:GB4002

Spencer RGM, Aiken GR, Butler KD, Dornblaser MM, Striegl RG, Hernes PJ (2009) Utilizing chromophoric dissolved organic matter measurements to derive export and reactivity of dissolved organic carbon exported to the Arctic Ocean: a case study of the Yukon River. Geophys Res Lett 36:L06401. doi:10.1029/2008GL036831

Springer AM (1988) The paradox of pelagic food webs on the Bering-Chukchi continental shelf. Ph.D. dissertation. University of Alaska Fairbanks, Fairbanks, 232 pp

Springer AM, McRoy CP (1993) The paradox of pelagic food webs in the northern Bering Sea-III. Patterns of primary production. Cont Shelf Res 13:575–599

Springer AM, McRoy CP, Turco KR (1989) The paradox of pelagic food webs in the northern Bering Sea: II. Zooplankton communities. Cont Shelf Res 9:359–386

Springer AM, McRoy CP, Flint MV (1996) The Bering Sea Green Belt: shelf-edge processes and ecosystem production. Fish Oceanogr 5:205–223

Stabeno PJ, Schumacher JD, Ohtani K (1999) The physical oceanography of the Bering Sea. In: Loughlin TR, Ohtani K (eds) Dynamics of the Bering Sea: a summary of physical, chemical, and biological characteristics, and a synopsis of research on the Bering Sea. AK-SG-99-03, 1–28. North Pacific Marine Science Organization (PICES), University of Alaska Sea Grant, Fairbanks

Stein R, Macdonald RW (2004) Arctic Ocean organic carbon accumulation and its global significance. In: Stein R, Macdonald RW (eds) The organic carbon cycle in the Arctic Ocean. Springer, Berlin, pp 315–322

Steinacher M, Joos F, Frolicher TL, Platter G-K, Doney SC (2009) Imminent ocean acidification of the Arctic projected with the NCAR global coupled carbon-cycle climate model. Biogeosciences 6:515–533

Steward GF, Smith DC, Azam F (1996) Abundance and production of bacteria and viruses in the Bering and Chukchi Seas. Mar Ecol Prog Ser 131:287–300

Striegl RG, Dornblaser MM, Aiken GR, Wickland KP, Raymond PA (2007) Carbon export and cycling by the Yukon, Tanana, and Porcupine rivers, Alaska, 2001–2005. Water Resour Res 43(2):W02411

Stroeve JC, Serreze MC, Fetterer F, Arbetter T, Meier W, Maslanik J, Knowles K (2005) Tracking the Arctic's shrinking ice cover: another extreme September minimum in 2004. Geophys Res Lett 32(4):L04501

Stroeve J, Holland MM, Meier W, Scambos T, Serreze M (2007) Arctic sea ice decline: faster than forecast. Geophys Res Lett 34(9):L09501

Subba Rao DV, Platt T (1984) Primary production of Arctic waters. Polar Biol 3:191–210

Waddington JM, Roulet NT (1997) Groundwater flow and dissolved carbon movement in a boreal peatland. J Hydrol 191:122–138

Walvoord MA, Striegl RG (2007) Increased groundwater to stream discharge from permafrost thawing in the Yukon River basin: potential impacts on lateral export of carbon and nitrogen. Geophys Res Lett 34(12):L12402

Wang J, Ikeda M, Zhang S, Gerdes R (2005) Linking the northern hemisphere sea-ice reduction trend and the quasi-decadal arctic sea-ice oscillation. Clim Dyn 24(2–3):115–130

Wassmann P, Bauerfeind E, Fortier M, Fukuchi M, Hargrave B, Moran B, Noji T, Nöthig E-M, Olli K, Peinert R, Sasaki H, Shevchenko VP (2003) Particulate organic carbon flux to the Arctic Ocean sea floor. In: Stein R, Macdonald RW (eds) The organic carbon cycle in the Arctic Ocean. Springer, Berlin, pp 101–138

Wassmann P, Carroll J, Bellerby RGJ (2008) Carbon flux and ecosystem feedback in the northern Barents Sea in an era of climate change: an introduction. Deep Sea Res II 55(20–21):2143–2153

Wheeler PA, Gosselin M, Sherr E, Thibault D, Kirchman DL, Benner R, Whitledge TE (1996) Active cycling of organic carbon in the central Arctic Ocean. Nature 380:697–699

Winton M (2006) Does the Arctic sea ice have a tipping point? Geophys Res Lett 33(23):L23504

Woodgate RA, Aagaard K, Weingartner TJ (2005) Monthly temperature, salinity and transport variability of the Bering Strait through flow. Geophys Res Lett 32(4):L04601

Wu PL, Wood R, Stott P (2005) Human influence on increasing arctic river discharges. Geophys Res Lett 32(2):L02703

Yamamoto-Kawai M, McLaughlin FA, Carmack EC, Nishino S, Shimada K (2009) Aragonite undersaturation in the Arctic Ocean: effects of ocean acidification and sea ice melt. Science 326(5956):1098–1100

Zeeman SI (1992) The importance of primary production and CO2. In: Nagel PA (ed) Results of the third joint US-USSR Bering and Chukchi seas expedition (BERPAC), summer 1988. U.S. Fish and Wildlife Service, Washington, DC, pp 39–49

Zhulidov AV (1997) Atlas of Russian wetlands: biogeography and metal concentrations. National Hydrology Research Institute, Environment Canada, Saskatoon, Saskatchewan, Canada, 312 pp

Chapter 10
Biodiversity and Biogeography of the Lower Trophic Taxa of the Pacific Arctic Region: Sensitivities to Climate Change

R. John Nelson, Carin J. Ashjian, Bodil A. Bluhm, Kathleen E. Conlan, Rolf R. Gradinger, Jacqueline M. Grebmeier, Victoria J. Hill, Russell R. Hopcroft, Brian P.V. Hunt, Hyoung M. Joo, David L. Kirchman, Ksenia N. Kosobokova, Sang H. Lee, William K.W. Li, Connie Lovejoy, Michel Poulin, Evelyn Sherr, and Kelly V. Young

Abstract The lower trophic level taxa underpin the marine ecosystems of the Pacific Arctic Region (PAR). Recent field observations indicate that range shifts, and changes in the relative abundance of particular taxa have occurred within the last decade. Here we provide a region wide survey of the diversity and distribution of viruses, bacteria, archaea, auto- and heterotrophic protists, as well as metazoan zooplankton and benthic organisms in the PAR. Our aim is to provide a foundation for the assessment of the changes within the lower trophic level taxa of the PAR and to document such change when possible. Sensitivities to the effects of climate change are also discussed. Our vision is to enable data-based predictions regarding ecological succession in the PAR under current climate scenarios, and to deepen our understanding regarding what the future holds for higher trophic level organisms and the carbon cycle.

Keywords Microbes • Phytoplankton • Zooplankton • Benthos • Biogeography • Biodiversity • Sea ice

---

R.J. Nelson (✉)
Department of Biology, Centre for Biomedical Research, University of Victoria, Victoria, BC, Canada
e-mail: jnelson@uvic.ca

C.J. Ashjian
Woods Hole Oceanographic Institution, Woods Hole, MA, USA

B.A. Bluhm • R.R. Gradinger • R.R. Hopcroft
School of Fisheries and Ocean Sciences, University of Alaska Fairbanks, Fairbanks, AK, USA

K.E. Conlan • M. Poulin
Canadian Museum of Nature, Ottawa, ON, Canada

J.M. Grebmeier
Chesapeake Biological Laboratory, University of Maryland Center for Environmental Science, 146 Williams Street, Solomons, MD 20688, USA
e-mail: jgrebmei@umces.edu

J.M. Grebmeier and W. Maslowski (eds.), *The Pacific Arctic Region: Ecosystem Status and Trends in a Rapidly Changing Environment*, DOI 10.1007/978-94-017-8863-2_10,

10.1 General Introduction

The way in which microbes, zooplankton, and benthic and sympagic (sea-ice associated) invertebrates of the Pacific Arctic Region (PAR) respond to environmental change has important implications for overall ecosystem function. Potential biotic responses to change include altered phenology, abundance, community composition, and species range shifts; some of these responses have been observed within the last decade (Grebmeier et al. 2006b; Li et al. 2009). Besides changes in arctic taxa, the PAR is also vulnerable to establishment of southern organisms from both the Atlantic and Pacific oceans.

The life span of the lower trophic-level organisms ranges from days (microbes) to decades (benthic infauna). The former will respond quickly to physical forcing and serve as sensitive indicators of ecosystem change although the detection of ecological trends by the study of these short-lived organisms may be confounded by their strong seasonal dynamics. Macrobenthic infauna are spatially constrained and have long life spans, and thus integrate local climate signals over longer periods of time. Study of the benthic infauna has therefore provided us with some of the strongest indication of ecosystem change in the PAR to date (Sect. 10.4.2).

The first objective of this work is to assess past and current change in the lower trophic level taxa in the PAR. The second objective is to provide context with

V.J. Hill
Department of Ocean, Earth and Atmospheric Sciences,
Old Dominion University, Norfolk, VA, USA

B.P.V. Hunt
Department of Earth, Ocean and Atmospheric Sciences,
University of British Columbia, Vancouver, BC, Canada

H.M. Joo
Korea Polar Research Institute, Incheon, South Korea

D.L. Kirchman
School of Marine Science and Policy, University of Delaware, Lewes, DE, USA

K.N. Kosobokova
Shirshov Institute of Oceanology, Russian Academy of Sciences,
Moscow, Russian Federation

S.H. Lee
Department of Oceanography, Pusan National University, Busan, South Korea

W.K.W. Li
Fisheries and Oceans Canada, Dartmouth, NS, Canada

C. Lovejoy
Department of Biology, Université Laval, Quebec, QC, Canada

E. Sherr
College of Earth, Ocean, and Atmospheric Sciences, Oregon State University,
Corvallis, OR, USA

K.V. Young
Fisheries and Oceans Canada, Sidney, BC, Canada

regard to understanding current sensitivities and vulnerabilities to climate forcing. Taken together these two objectives will enable data-based predictions regarding ecological succession in the PAR under current climate scenarios, and to deepen our understanding regarding what the future holds for higher trophic level organisms and the marine carbon cycle.

10.2 Phytoplankton in the PAR

10.2.1 Introduction

Populations of subarctic and arctic phytoplankton are controlled by the extreme seasonality of the polar environment. Primary production is dependent on two basic elements: light and nutrients. The annual cycle of phytoplankton growth starts with ice algae at the ice-water interface, stimulated by increasing irradiance as snow cover melts and day length increases (Gradinger 2009). As the sea ice thins and melt ponds expand, under ice light transmission increases to upwards of 50 % (Light et al. 2008) and phytoplankton blooms under the ice are observed (Arrigo et al. 2012; Gosselin et al. 1997). The season progresses with intense blooms along the retreating ice edge across both the shelf and basin regions. Diatoms, primarily *Thalassiosira* and *Chaeroceros* spp. (Aizawa et al. 2005; Suzuki et al. 2002), dominate the community composition at this time. Much of this carbon is sequestered to the benthos through a highly efficient biological pump (Campbell et al. 2009; Moran et al. 2005; Sherr et al. 2009; Wehrmann et al. 2011). In some areas, this strong benthic-pelagic coupling supports an extremely high benthic biomass that is utilized by higher trophic-level taxa, including seabirds, walruses and gray whales (e.g., Grebmeier et al. 2006a; see also Sect. 10.4). Strong stratification in the summer restricts the vertical replenishment of surface nutrients, promoting the growth of smaller celled flagellates which rely on recycled nutrients (Hill et al. 2005; Sukhanova et al. 1999).

Sea surface warming trends in the Arctic Ocean have been most pronounced in the Pacific region with recent summertime temperature anomalies of + 2.5 °C (Steele et al. 2008). Summertime sea ice coverage has decreased across the Arctic since 1979, with total extent declining by 2.8–7.2 % per decade per month (Serreze et al. 2007). An alteration of the food web through the introduction of species adapted to a more temperate climate, changes in the location and timing of phytoplankton blooms forced by variation in sea surface temperatures (SST), or alterations to ice dynamics, may shift ecological processes in the Bering and Chukchi seas. It is postulated that the timing of phytoplankton growth relative to water temperature plays a critical role in determining the pathways of reduced carbon biomass. Walsh and McRoy (1986) hypothesized that blooms occurring in warm water would be consumed by copepods whose foraging and growth increase in efficiency at higher water temperatures (Huntley and Lopez 1992). Blooms occurring earlier in cold water would not be effectively grazed by copepods, but fall to the bottom to support benthic organisms (Hunt et al. 2002). However, recent observations of warming and

Table 10.1 Total number of taxa and genera of phytoplankton and sea-ice unicellular eukaryotes recorded from Bering Strait and the Alaskan Beaufort Sea, including genus dominance with species number and occurrence (%) of dominant taxa (Poulin et al. 2011; M. Poulin, unpublished data)

Taxon	# taxa	# genera	Genus dominance	Dominant taxa
Diatoms	331	91		
Centrics	99	29	*Chaetoceros* (34)	*Attheya septentrionalis* (>50 %)
			Thalassiosira (21)	
Pennates	232	62	*Navicula* (52)	*Nitzschia frigida* (80 %)
			Nitzschia (36)	*Cylindrotheca closterium* (>65 %)
			Pinnularia (19)	*Thalassionema nitzschioides* (>60 %)
			Entomoneis (10)	*Fragilariopsis cylindrus* (>50 %)
				Fragilariopsis oceanica (>50 %)
				Navicula pelagica (>50 %)
Dinoflagellates	74	28	*Protoperidinium* (21)	
Chlorophytes	13	11		
Others	25	36		
Total	443	166		

cooling phases in the Bering Sea have indicated the need for revision to this hypothesis (Coyle et al. 2011). Although the details regarding the interplay between temperature and pelagic-benthic coupling remain to be worked out, what is clear is that modification of benthic communities caused by changes in the sinking of carbon from the euphotic zone will have profound impacts on the ability of the system to support mammals (e.g. polar bears, gray whales and walruses) that are a critical part of the cultural and social fabric of the Arctic. This section describes the biodiversity and biogeography of phytoplankton within the PAR, with the intent of understanding and predicting future changes.

10.2.2 Phytoplankton and Sea Ice Algae: An Overview

At the pan-arctic scale a total of 2,106 single-celled eukaryotic taxa, including 1,874 phytoplankton and 1,027 sea ice taxa have been reported (Poulin et al. 2011). The highest diversity of both phytoplankton and sea-ice eukaryotes was recorded in Canadian waters. In the Bering Strait and the Alaskan Beaufort Sea, pelagic diatoms accounted for 75 % of the total number of unicellular eukaryotes, whereas 17 % is contributed by sea ice taxa (Table 10.1). Landfast and pack ice in the Alaskan Beaufort Sea are predominantly colonized by pennate diatoms which accounted for 77 % of all unicellular eukaryotes recorded, with the micro-sized fraction (i.e. 20–200 μm) representing 96 %.

Marine phytoplankton in the Beaufort Sea and Canada Basin are predominantly large-celled diatoms and dinoflagellates which accounted for 58 % and 21 %, respectively, of all 555 microscopic forms recorded (Table 10.2). Arctic landfast and pack ice in this sector are mainly colonized by pennate diatoms which accounted

Table 10.2 Total number of taxa and genera of phytoplankton and sea-ice unicellular eukaryotes recorded from the Canadian Beaufort Sea and Canada Basin, including genus dominance with species number and occurrence (%) of dominant taxa (Poulin et al. 2011; M. Poulin, unpublished data)

Taxon	# taxa	# genera	Genus dominance	Dominant taxa
Diatoms	324	82		
Centrics	100	24	*Chaetoceros* (44)	*Attheya septentrionalis* (>80 %)
			Thalassiosira (17)	*Melosira arctica* (>75 %)
Pennates	224	58	*Navicula* (54)	*Cylindrotheca closterium* (95 %)
			Nitzschia (30)	*Fragilariopsis cylindrus* (>75 %)
			Pinnularia (16)	*Nizschia frigida* (>60 %)
			Entomoneis (10)	*Navicula directa* (>50 %)
Dinoflagellates	115	34	*Protoperidinium* (24)	
			Gyrodinium (18)	
			Gymnodinium (17)	
Chlorophytes	13	13		
Others	103	71		
Total	555	200		

for 69 % of all 257 unicellular eukaryotes recorded, with the micro-sized fraction (i.e. 20–200 μm) representing 90 % of all taxa observed. See also Sect. 10.5.2 for more detail on ice associated taxa.

Dominant arctic taxa are mostly associated with cold waters and the presence of available sea ice habitat. *Nitzschia frigida* Grunow can be considered a sentinel species for the occurrence of first-year sea ice, while *Fragilariopsis cylindrus* (Grunow ex Cleve) Frenguelli is often associated with first-year sea ice but is also known to occur in cold pelagic waters. *Melosira arctica* Dickie is associated with the under surface of the sea ice forming long threads consisting of huge number of colonies, and *Attheya septentrionalis* (Østrup) Crawford is correlated to cold pelagic waters. In addition, while diatoms dominate in regions with 50–90 % ice cover, flagellates are often more abundant than diatoms under thicker ice cover (Booth and Horner 1997; Gosselin et al. 1997). As water temperatures, and sea ice dynamics and characteristics change, we might expect alterations in the distribution and abundance of these arctic taxa which has unknown consequences for ecosystem function. In particular smaller taxa are predicted to thrive in a warmer more stratified Arctic Ocean (Li et al. 2009).

10.2.3 Latitudinal Variation of Phytoplankton Biodiversity and Community Composition in the Western Arctic Ocean

Temporal succession in phytoplankton taxa is observed from the dark, cold winter into the warm, stratified summer (Hill et al. 2005; Sukhanova et al. 1999, 2009). During winter, abundance in the water column is extremely low, and it is postulated that some arctic taxa overwinter in the sea ice, seeding the water column in the spring as the ice melts. The primary spring bloom is usually dominated by large, rapidly growing diatom species. Typically, the large taxa of the bloom are followed by smaller flagellate taxa, which rely on recycled nutrients, that increase in abundance at the nutricline (Hill et al. 2005; Sukhanova et al. 2009).

The sea ice melt stimulates phytoplankton growth, and occurs first in the Bering Sea, which is also the best studied region within the PAR (Shirshov 1982). Smaller phytoplankton (<10 μm) are generally characteristic of the Bering Sea, although larger diatoms (>10 μm) can contribute substantially to the biomass (Shiomoto 1999; Sukhanova et al. 1999). In the Bering Sea basin small size phytoplankton (<10 μm) dominate in both the summer and winter, with diatoms dominating the community structure (Shiomoto 1999). The assemblage is characterized by pennate diatoms such as *Fragilariopsis pseudonana* (Hasle) Hasle, *Neodenticula seminae* (Simonsen and Kanaya) Akiba and Yanagisawa, and *Pseudo-nitzschia* spp. Results from a long-term sediment trap study in the basin showed that *Neodenticula seminae* represented 25–90 % of the vertical diatom flux, with resting spores of *Chaetoceros* spp. becoming very abundant (Kurihara and Takahashi 2002).

Aizawa et al. (2005) reported that phytoplankton communities associated with the spring bloom on the Bering Shelf were dominated by centric diatoms belonging to the genera *Thalassiosira* Cleve and *Chaetoceros* Ehrenberg, namely *T. gravida* Cleve, *T. trifulta* Fryxell, *T. conferta* Hasle, *C. debilis* Cleve and *C. contortus* Schütt. Several pennate diatoms were also constituent members of such blooms, e.g., *Pseudo-nitzschia seriata* (Cleve) H. Peragallo, *Fragilariopsis cylindriformis* (Hasle) Hasle and *F. oceanica* (Cleve) Hasle (Broerse et al. 2003). In summer, *Chaetoceros* spp. and elongate centric diatoms belonging to *Leptocylindrus* Cleve, *Proboscia* Sundström, *Guinardia* H. Peragallo and *Rhizosolenia* Brightwell were also abundant (Sukhanova et al. 1999). In the surface waters overlying the deeper basins, dominant diatoms belong to both *Chaetoceros* subgenera *Phaeoceros* Gran and *Hyalochaete* Gran, and *Thalassiosira* in spring, whereas smaller diatoms *Neodenticula seminae* and *Fragilariopsis cylindriformis* are more abundant in summer (Sukhanova et al. 1999; Aizawa et al. 2005). Within the Bering Sea several frontal systems result in nutrient enrichment of the euphotic zone, and lead to higher phytoplankton biomass when compared to the rest of the middle shelf area. These fronts occur around the Pribilof Islands, resulting in summertime phytoplankton biomass and abundance values that equal those of the spring bloom over the middle shelf (Flint et al. 2002). The summer community here is dominated by larger sized phytoplankton (>10 μm) such as *Chaetoceros socialis* Lauder, *C. debilis*, *C. diadema* (Ehrenberg) Gran, *C. furcillatus* Bailey, *Thalassiosira hyalina* (Grunow) Gran, and *Fragilariopsis oceanica* in numbers double that of the adjacent middle shelf. Evidence exists for phytoplankton community structure shifts in the Bering Sea, with the recent appearance of coccolithophorid blooms and increased contributions of *Phaeocystis* taxa during the summer (Merico et al. 2003).

Three water masses with origins in the Pacific Ocean enter the Chukchi Sea though Bering Strait (Coachman and Aagaard 1988). These have been identified as (i) Anadyr Water (AW) in the west, (ii) Bering Shelf (BS) water in the center

(iii) Alaska Coastal Water (ACW) in the east (see Maslowski et al., this volume, for more detail). The location and direction of these flows dictate the physical conditions and nutrient concentrations on the Chukchi Shelf. The seasonal progression of the phytoplankton shows a similar pattern to the Bering Sea, with spring and summer assemblages driven by local patterns in warming, ice melt, stratification and nutrient supply (Sergeeva et al. 2010). However, the assemblages are found to differ from east to west across the shelf, with high chlorophyll *a* and larger cell sizes observed in the ACW (Robie et al. 1992; Lee et al. 2007). Throughout the area phytoplankton communities associated with the seasonal ice zone are characterized by a mixture of small cell-sized (<5 μm) taxa belonging to the prasinophytes and chrysophytes, and larger diatom and haptophyte cells (>5 μm) (Hill et al. 2005). Diatoms dominate communities with chlorophyll *a* concentrations characteristic of bloom conditions. Small sized phytoplankton cells, most likely prasinophytes, were found over the slope and basin areas where the ice cover was still continuous. Post-bloom conditions in the ACW and BS water masses are associated with low nutrient and biomass concentrations in the upper water column along with small celled prasinophytes, haptophytes and diatoms, while at the subsurface chlorophyll maximum diatoms were dominant. However, in the AW summertime assemblages are still dominated by large phytoplankton (>20 μm, Lee et al. 2007), with total chlorophyll *a* concentrations higher than those found in the BS or ACW (Robie et al. 1992).

Further north, on the shelf break of the Chukchi Sea (70 °N–72 °N), large centric diatoms (>10 μm) predominate the pelagic environment, whereas pennate diatoms were more abundant in the bottom ice microalgal communities (Booth and Horner 1997; Gosselin et al. 1997; von Quillfeldt et al. 2003). In the Beaufort Sea, microflagellates are the most abundant taxa in the winter when chlorophyll *a* is near the lower limit of detection (Horner and Schrader 1982). Ice algae in the spring are characterized by diatoms with *Nitzschia frigida* and *N. cylindrus* (Grunow ex Cleve) Hasle accounting for ~50 % of the population, small flagellates and cryptomonads making up the remainder (Horner and Schrader 1982). In spring the water column is dominated by pennate diatoms such as *Cylindrotheca* Rabenhorst, *Navicula* Bory, and *Nitzschia* Hassall.

A transect from the Bering Sea, through the Chukchi Sea to the Canada Basin in the summer of 2008 showed that large sized phytoplankton cells were present in the seasonal ice zone, whereas the most abundant microalgae belonged to the pico- and nano-sized fraction at the surface in open waters and also at the subsurface chlorophyll maximum (Joo et al. 2012). The second most dominant taxa were variable but most commonly belonged to *Thalassiosira*, *Chaetoceros* and unidentified pico- and nanophytoplankton such as *Dinobryon belgica* Meunier and *Cryptomonas* sp. (Fig. 10.1).

10.2.4 Synechococcus

One of the more intriguing distribution patterns for western arctic autotrophs is the occasional presence of the coccoid cyanobacterium *Synechococcus* Nägeli. This genus has been characterized as intolerant to cold water (Murphy and Haugen 1985),

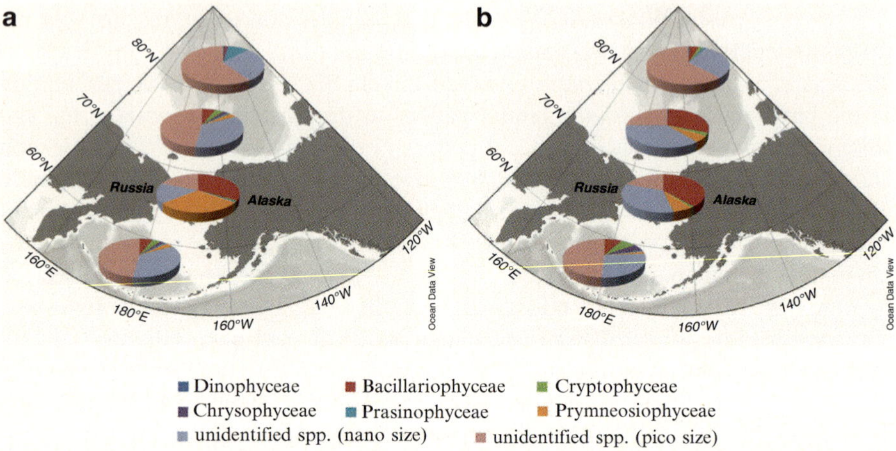

Fig. 10.1 Relative abundances of phytoplankton assemblages in the North Pacific Ocean and the western Arctic seas in summer 2008 ((**a**) surface, (**b**) depth of the subsurface chlorophyll maximum). The pie charts show relative abundance of the major taxa: Dinophyceae, Bacillariophyceae, Cryptophyceae, Chrysophyceae, Prasinophyceae, Prymnesiophyceae, unidentified nanophytoplankton, unidentified picophytoplankton (Joo et al. 2012)

occurring only at temperate and tropical latitudes and not growing at temperatures less than ~5 °C (Waterbury et al. 1986). In the western Arctic, *Synechococcus* has been observed at locations where Pacific waters are or have been present (Cottrell and Kirchman 2009; Ashjian et al. 2010) thus the distribution of *Synechococcus* in the PAR is driven by the relative quantity of Pacific water in the Chukchi and Beaufort seas, and is annually variable.

Virtually no *Synechococcus* were observed near Barrow in 2002, 2006 and 2008, when little Pacific water entered the region, in contrast to the situation in 2004, 2005, 2007 and 2009 when Pacific waters were a dominant feature of the hydrographic environment (Cottrell and Kirchman 2009; Tremblay et al. 2009; Ashjian et al. 2010). The highest cell abundances ($>10 \times 10^4$ cells/L) were observed in 2007, a year in which warmer Pacific water was present in the Chukchi Sea and near Point Barrow during the summer (Cottrell and Kirchman 2009; Woodgate et al. 2010). Surprisingly, *Synechococcus* was also abundant near Point Barrow in the winter of 2008, in similar numbers to those observed during summer (Cottrell and Kirchman 2009), which suggests that active growth may have persisted into the winter despite a water temperature near the freezing point (−1.8 °C). A similar situation exists in the eastern Arctic, where *Synechococcus* is reported from regions under the influence of warmer Atlantic waters such as the Greenland and Barents seas (Gradinger and Lenz 1989, 1995; Not et al. 2005).

The actual temperature at which *Synechococcus* was present in a number of these studies was lower than the 5 °C earlier thought required for growth (Waterbury et al. 1986), but much closer to the temperature of subpolar waters where *Synechococcus* can often be found (Li 2009). Perhaps *Synechococcus* can exist at lower temperatures but cannot survive for long without new inputs of Pacific or Atlantic waters (Gradinger and Lenz 1995). This explanation, however, is not consistent with the suggestion of Cottrell and Kirchman (2009), that *Synechococcus* are actively growing and utilizing dissolved organic matter in winter. An alternate explanation for their observation is that the winter of 2007–2008 may have been exceptional because of the high amounts of Pacific water that was brought into the Arctic during the anomalous summer of 2007. It is evident that the water temperature alone cannot be used as a metric to differentiate environments in which *Synechococcus* might be found.

10.2.5 Sensitivities to Habitat Changes

Few phytoplankton time series in the Arctic Ocean have the appropriate scope to address the issue of climate change, both because of insufficient temporal coverage to describe seasonality and because of inadequate spatial extent to describe trends (Ardyna et al. 2011). Therefore in order to assess climate related phytoplankton sensitivities to habitat changes over time, it is necessary to invoke a substitution of space for time, in which diverse habitats across space (arctic and non-arctic) are compared in order to gauge possible future changes in the Arctic. Projections of future states based on spatial macroecology can be successful if the mechanisms underlying the identified linkages between patterns are understood (Fisher et al. 2010).

The similarity of phytoplankton communities (diatoms, dinoflagellates, coccolithophorids) compared across geographic distances and environmental conditions of, for instance nutrient concentrations, suggests that species are broadly dispersed but local environmental selection controls community structure (Cermeno et al. 2010). More particularly in the Arctic, the similarity in picoeukaryote assemblages is strongly associated with water mass origin. That is, closely spaced assemblages are less similar if located in different water masses, but distantly spaced assemblages can be more similar if located in the same water mass (Hamilton et al. 2008).

At least four physical habitats which are likely to share a domain-wide response to climate variability have been identified in the Arctic (Carmack and Wassmann 2006): the seasonal ice zone, the riverine coastal domain, the Pacific-Arctic domain, and the pan-arctic shelf-break and margin domain. The contemporary distribution of *Synechococcus* across these contiguous domains (Fig. 10.2) is a striking demonstration of a strong environmental selection of marine microbes dispersed into these pan-arctic habitats from various source waters (see also Sect. 10.2.3). Extremely abundant populations of picocyanobacteria encounter a figurative wall of resistance as subarctic Pacific water moves through the Bering Sea into the Chukchi Sea, and also as subarctic Atlantic water moves around the Labrador Sea up to Davis Strait. Virtually no picocyanobacteria are found northward of the Chukchi Sea in the Beaufort Gyre, or northward of Davis Strait in Baffin Bay (also see Sect. 10.2.4). Patches of these cells, however, appear in Amundsen Gulf and Coronation Gulf near areas with terrigenous influence (Waleron et al. 2007; Tremblay et al. 2009). This pan-arctic perspective illustrates that any picocyanobacteria advected from subarctic waters and from freshwater, brackish and coastal environments probably manage to survive in open arctic

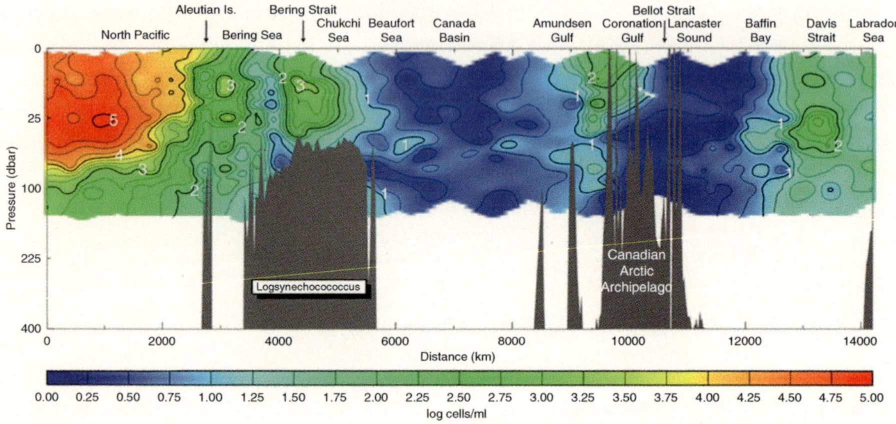

Fig. 10.2 Pan-arctic summer distribution of phycoerythrin-containing picocyanobacteria (W.K.W. Li unpublished data). The colors reflect the concentration of phycoerythrin-containing picocyanobacteria in log cell/ml as shown by horizontal scale according to depth (Y-axis) along a 14,000 km ocean transect from Victoria, British Columbia (km 0) to St. John's Newfoundland and Labrador (km 14,000)

waters but show little net growth (Gradinger and Lenz 1995; Waleron et al. 2007). It seems that in polar marine waters, picocyanobacteria have slow temperature-depressed and light-limited gross growth rates that are often exceeded by losses due to cell death, grazing, advection and mixing (Vincent 2000).

As climate changes and the upper Arctic Ocean receives more heat and fresh water, the habitats for phytoplankton become warmer and less saline, and consequently also more limited in nutrients because of greater stratification of the water column. These conditions favour picophytoplankton over nanophytoplankton and microplankton because small cells are more effective in acquiring nutrients and less susceptible to gravitational settling (Ardyna et al. 2011). In the future, if significant numbers of picocyanobacteria were to be found in the Arctic Ocean, we would have a strong basis to infer a shift in the habitat boundary for this group of phytoplankters. As yet, this does not seem to have happened. Instead, it is the indigenous picophytoplankton, namely the small eukaryotic algae such as the prasinophyte *Micromonas* Manton and Parke, that are increasing as the Arctic Ocean warms and freshens; concurrently, there has been a decrease in the abundance of nanophytoplankton (Li et al. 2009; Tremblay et al. 2009; Ardyna et al. 2011). If the present trend of increasing picophytoplankton to nanophytoplankton ratio continues, phytoplankton in the Arctic Ocean may change over time to appear more similar to that elsewhere in non-polar oceans. If so, the space-for-time substitution might be extended beyond size structure to other characteristics of the phytoplankton community such as phylogenetic composition or ecological functionality. This would allow us to infer broader aspects of ecological change in the Arctic from knowledge about the spatial macroecology of phytoplankton in the world oceans.

10.3 Heterotrophic Microbes in the PAR

10.3.1 Introduction

Heterotrophic microbes in the sea are encompassed by the prokaryotes, which includes two major domains of life, Bacteria and Archaea, and by single-celled eukaryotes, known as protists, which are in the third domain of life, Eukarya. Marine protists range from strictly photosynthetic algae, such as diatoms, to diverse mixotrophic phytoflagellates and dinoflagellates, ciliates and strict heterotrophs. Viruses, which infect both prokaryotic and eukaryotic microbes, are ubiquitous. All groups of microbes, plus viruses, are abundant and play vital roles in food webs in arctic and subarctic seas. In addition, there are communities of heterotrophic microbes in sea ice which are phylogenetically different from those found in the water column. While this much is certain, details of the diversity and distribution of heterotrophic microbes in western arctic and subarctic seas are sparse. The information summarized below is based on only relatively few studies of heterotrophic microbes and viruses, with limited geographic coverage.

10.3.2 Viruses

Extensive work in lower-latitude oceanic regions has demonstrated the importance of viruses in contributing to the top-down control of bacteria and phytoplankton (Suttle 2007; Breitbart et al. 2008; Tomaru et al. 2011). A more limited number of studies have focused on viruses in arctic and subarctic marine systems (Maranger et al. 1994; Steward et al. 1996, 2000; Yager et al. 2001; Middelboe et al. 2002; Hodges et al. 2005; Angly et al. 2006; Wells and Deming 2006; Wells et al. 2006; Payet and Suttle 2008). These studies suggest differences between arctic and lower-latitude oceanic regions. Bacterial and viral abundances are low in the Arctic Ocean (Payet and Suttle 2008), as are estimates of viral lysis based on the frequency of visibly infected cells (Steward et al. 1996). Low rates of viral lysis are consistent with fewer contacts between hosts and viruses due to low abundances and the higher viscosity of cold seawater, which limits the diffusivity of viruses and hosts by 60 % (Murray and Jackson 1992).

The diversity and types of viruses found in the Arctic may also differ. While viral communities in the Arctic Ocean include a majority of pan-ocean genotypes, arctic viruses, like arctic bacteria, appear to have somewhat lower diversity compared to viral communities in lower-latitude ocean regions (Angly et al. 2006). Temperate phage genes are relatively more abundant in arctic viruses than in lower-latitude viruses (Angly et al. 2006), and viral DNA is about threefold more abundant in bacterial metagenomes from the Arctic than in lower-latitude oceans (Cottrell and Kirchman 2012). Payet and Suttle (2008) detected two groups of viruses in the southern Beaufort Sea based on relative SYBR Green fluorescence, an index of amount of nucleic acid content, assessed by flow cytometry. A large population of viral particles with lower SYBR fluorescence was positively related to abundance of actively growing bacteria, while a smaller population of viral particles with higher SYBR fluorescence correlated with chlorophyll concentration. Payet and Suttle (2008) speculated that viruses with lower fluorescence represented smaller sized bacteriophages, while those with higher fluorescence represented phytoplankton-infecting viruses.

10.3.3 Bacterial Diversity

The diversity of bacteria in the oceans is becoming clearer as the sequencing of bacterial genes has become quicker, easier and more cost-effective. These sequence data are essential for exploring bacterial diversity in the oceans and other natural environments because bacteria and many other microbes lack distinguishing features that can be used for taxonomic studies. Also, most microbes cannot currently be isolated and grown in the laboratory where biochemical tests could be used to classify them. Instead, most studies of bacterial diversity focus on the 16S rRNA gene which is sequenced after PCR and cloning or by direct sequencing without cloning. A complementary approach is to enumerate bacteria in specific taxonomic groups microscopically by fluorescence in situ hybridization (FISH). Both approaches have been used to explore bacterial diversity in the Arctic Ocean.

Here we focus primarily on the euphotic zone of the Chukchi Sea, Beaufort Sea and the Canada Basin of the Arctic Ocean. Lovejoy et al. (2011) summarized results from 16S rRNA gene surveys carried out in other regions of the Arctic Ocean. The few studies of arctic benthic microbial communities (Niemann et al. 2006) and of the water column in subarctic regions (Gomez-Pereira et al. 2010) will not be considered here. Bacterial and archaeal communities of deep waters in the Canada Basin have been examined (Galand et al. 2010), but will not be discussed here in great depth.

The diversity and composition of arctic bacterial and archaeal communities are worth examining because community structure information may eventually help us understand the role of these microbes in carbon and nitrogen cycles, which in turn influence the structure of arctic food webs with effects up to higher trophic levels. Bacteria are also important for global sulfur cycling with important feedbacks on the arctic climate system (Motard-Côté 2010; Motard-Côté et al. 2011). Heterotrophic bacteria primarily recycle carbon; some data suggest that in the Arctic they process less organic carbon compared to lower-latitude oceans for reasons that are not entirely understood (Kirchman et al. 2009). The Arctic Ocean is also an important region for exploring general questions about microbial biogeography.

10.3.4 Bacterial and Archaeal Diversity Levels in the Arctic Ocean Versus Lower-Latitude Oceans

A basic component of diversity is richness, or the number of species making up the community of interest. Given the many problems in defining bacterial species, here we will use "phylotype" to refer to a bacterial taxon as having 16S rRNA genes that are ≥97 % similar. The species richness of large organisms tends to be lower in high-latitude ecosystems, and there is some evidence that numbers of bacterial taxa decrease with increasing latitude in the oceans (Pommier et al. 2007; Fuhrman et al. 2008) although not in soils (Fierer and Jackson 2006). The high Arctic Ocean was not explicitly considered in these previous studies. One study did find lower numbers of bacterial types in the Chukchi Sea than in Massachusetts coastal waters (Malmstrom et al. 2007). No comparative studies of archaeal diversity have been undertaken, although Archaea in general seem to be less diverse than Bacteria in most regions, including the Arctic Ocean (Galand et al. 2009a).

The recent application of high throughput next generation sequence technology to environmental genetics (Margulies et al. 2005) has yielded many more sequences than previously possible. Studies using this approach suggest that bacterial communities in coastal waters of the Arctic Ocean are less rich than lower-latitude marine waters, although the differences may not be great. In making this comparison, it is important to examine equal number of sequences for each data set. One study of samples from the Arctic Ocean found 400–600 phylotypes per 5,000 sequences (Kirchman et al. 2010) while 500–1,100 phylotypes per 5,000 sequences were found in samples from the English Channel (Gilbert et al. 2009). The estimates vary depending on location in the Arctic Ocean and on seasons in both the Arctic and English Channel, so it is difficult to make a direct comparison. The bacterial community in the Baltic Sea also may be more diverse than in the Arctic based on similar data (Andersson et al. 2009). All three studies were part of the International Census of Marine Microbes (ICoMM) program. Additional studies from this program and other high throughput sequencing studies will be published soon which will enable a more complete comparison of the Arctic with lower-latitude regions.

The types of microbes found in the Arctic Ocean can be explored at various taxonomic levels, ranging from domain down to phylotype or species. Here we will focus on higher taxonomic levels, as the forthcoming ICoMM reports and similar data are needed to examine taxonomic composition at the lower taxonomic levels.

Archaea may be more important in the Arctic Ocean than in other systems. There is some evidence of relatively more Archaea (as a percentage of total prokaryote abundance) in arctic surface waters than found in low-latitude oceans, as determined by FISH based studies (Kirchman et al. 2007). The relative importance of Archaea is correlated to particle abundance in arctic waters (Garneau et al. 2006), perhaps due to the influx of particle-rich river water (Wells et al. 2006). Changes in the community composition appear to following salinity gradients which are also

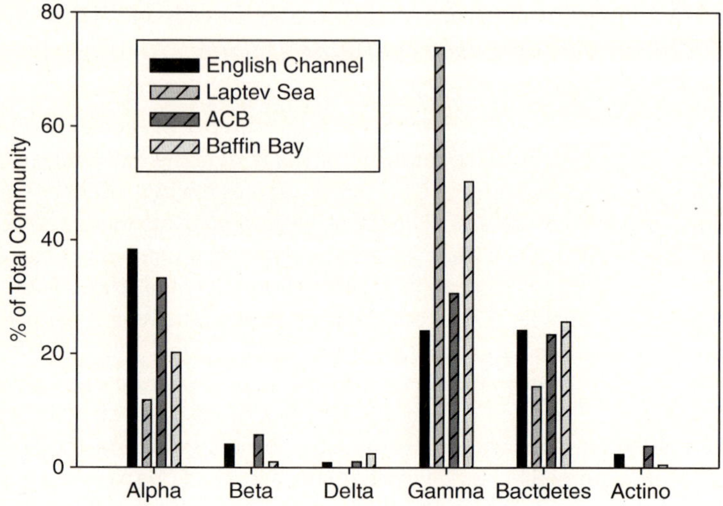

Fig. 10.3 Composition of bacterial communities in the Laptev Sea (Kellogg and Deming 2009) and Baffin Bay (Lovejoy et al. 2011), as determined by traditional clone library and Sanger sequencing. Data from the English Channel (2007 only) (Gilbert et al. 2009) and the ACB project (Kirchman et al. 2010) are from the pyrosequence method. The ACB data are mainly from the Chukchi and Beaufort seas. Alpha, Beta, Delta, and Gamma refer to subdivisions of Proteobacteria, while Bactdetes and Actino refer to Bacteroidetes and Actinobacteria. See Lovejoy et al. (2011) for a more complete list of studies examining bacterial community structure in the Arctic

influenced by river runoff (Galand et al. 2006). Archaea also appear to be more abundant in winter than in summer in the Arctic (Alonso-Sáez et al. 2008), which has also been noted in Antarctic waters (Church et al. 2003; Murray et al. 1998). The two most abundant archaeal groups in the ocean belong to the phyla Euryarchaeota and Marine Group I (MGI) Crenarchaeaota. The MGI has been recently elevated to phylum status, designated Thaumarcheaota (Brochier-Armanet et al. 2008; Spang et al. 2010). These MGI Archaea dominate the archaeal community over much of the Arctic (Galand et al. 2009b). The majority of Thaumarcheaota in the ocean appear to have the ammonia monooxygenase gene (*amoA*) involved in ammonium oxidation and nitrification (Kalanetra et al. 2009) and hence their abundance in the upper waters of the Arctic Ocean could be associated with high rates of nitrification. A recent study of Alaskan coastal waters found a correlation between *amoA* abundance and nitrification rates (Christman et al. 2011).

The major groups of bacteria found in low-latitude oceans are also seen in the Arctic Ocean, but there are some interesting differences even at the high taxonomic level (Fig. 10.3). Typically, Alphaproteobacteria, especially the SAR11 clade, dominate bacterial communities (Fuhrman et al. 2008), and these are common in the Arctic as well (Kirchman et al. 2010). However, the overall abundance of Alphaproteobacteria is greater in the English Channel than in the Arctic Ocean (Fig. 10.3). Instead, Gammaproteobacteria and Bacteroidetes are quite abundant in arctic waters, often accounting for >20 % of the total community, although there are exceptions (Lovejoy et al. 2011). These two groups are often <10 % of total bacterial diversity in open ocean waters such as the Sargasso Sea (Venter et al. 2004). There is much variability within the Arctic, however, some of which is due to sample size and PCR bias. Some of this variability is also due to location, depth and seasonality.

Communities in the oligotrophic arctic basins may be more similar to open ocean low-latitude communities than to bacterial communities in productive coastal arctic waters with higher proportions of Alphaproteobacteria (Lovejoy et al. 2011; Galand et al. 2010). Interestingly, ice-based bacterial communities are also relatively rich in Gammaproteobacteria and Bacteroidetes (Collins et al. 2010), and these groups may be important for the degradation of high molecular weight exopolymers associated with primary production (Elifantz et al. 2007; Kirchman 2002). If so, then the relative proportions of these major bacterial groups could be an indication of recent productivity and their assessment could be used to track ecological shifts, including those associated with climate change in the Arctic (Campeau et al. 2011).

10.3.5 Diversity and Distribution of Heterotrophic Protists

Heterotrophic protists include the phylogenetically diverse phototrophic pico- (0.2–2 μm) and nano- (2–20 μm) size flagellates, as well as the ciliates and heterotrophic dinoflagellates in the microzooplankton (20–200 μm). In open water columns, Radiolarians that include the Cantharis and Polycystinea are also present. Heterotrophic protists are a major component of all marine food webs, are the dominant eukaryotic microbes in deep water, and are a significant component of plankton biomass for most of the year in the upper mixed layer. In the Arctic, their trophic roles are similar to those in other oceanic systems. In the Bering Sea, nanoflagellates are presumably significant consumers of cyanobacteria, heterotrophic bacteria and small sized phytoplankton. In the western Arctic Ocean nanoflagellates are known to graze on the dominant prasinophyte *Micromonas* sp., as well as on bacteria (Sherr et al. 1997, 2003; Vaqué et al. 2008). In both the Bering Sea and western Arctic Ocean, protists in the microzooplankton, such as ciliates and heterotrophic dinoflagellates, are major grazers of diatoms and other phytoplankton (Olson and Strom 2002; Strom and Fredrickson 2008; Sherr et al. 2009). In turn these protists are food for mesozooplankton (Campbell et al. 2009). Little is known of the ecology of Cantharis and Polycystinea as they are infrequently recovered intact in bottle samples and are usually destroyed or ignored using standard net sampling techniques.

10.3.5.1 Diversity of Heterotrophic Protists Assessed by Microscopy

Most of the information on protist community composition in the Pacific arctic/subarctic region is from the Beaufort and Chukchi seas, and from the Bering Sea (Howell-Kubler et al. 1996; Sherr et al. 1997, 2003; Olson and Strom 2002; Strom and Fredrickson 2008). A diverse community of heterotrophic protists occurs in both the water column and sea ice in these regions. Ciliates and dinoflagellates in the plankton include morphospecies commonly found in other regions of the sea. Ice protists include ciliate morphotypes typical of benthic habitats, heterotrophic dinoflagellates, euglenids and amoebae, all of which are capable of consuming sea ice diatoms (see also Sects. 10.5 and 10.2.2, this volume).

Flagellate morphotypes. Heterotrophic flagellates occur in both water column and sea ice communities. These include colonial choanoflagellates which are sporadically abundant in the plankton in both the Chukchi and Bering seas (Sherr et al. 2009, unpublished data). The cercozoan *Cryothecomonas* Thomsen, Buck, Bolt and Garrison is also sporadically abundant in arctic waters and likely grazes on diatoms and other phytoplankton (Thaler and Lovejoy 2012). In the oligotrophic open Arctic Ocean and in the Beaufort Sea, *Dinobryon* spp., which are single-celled and colonial mixotrophic flagellates, occur sporadically but may sometimes be abundant (Sherr et al. 2013). Heterotrophic flagellates tentatively identified in the genus *Pirsonia* Schnepf, Drebes and Elbrächter have been observed parasitizing pelagic diatoms in the Bering Sea during summer diatom blooms (Sukhanova et al. 2006) these could also be uncultivated stramenopiles (Thaler and Lovejoy 2014). Putative flagellate parasites growing on *Fragilariopsis* spp. diatoms also occur during spring sea ice conditions in the Bering Sea (Sherr et al. 2013). Parasitic flagellates have previously been reported on diatoms in European waters (Tillmann et al. 1999).

Ciliate morphopecies Visually recognizable genera of ciliates in the Bering, Chukchi, and Beaufort seas include *Tintinnopsis* sp., *Ptychocyclis* sp. (e.g. Fig. 10.4b), *Codonellopsis* sp. and *Parafavella* sp. tintinnids, oligotrichous ciliates in the genera *Strombidium* Claparède and Lachmann and *Laboea* Lohmann … *Strobilidium* Schewiakoff, choreotrichous ciliates in the genera *Strobilidium*, *Leegaardiella* Lynn and Montagnes, and *Lohmanniella* Leegaard, and, less abundantly, haptorid species in the genera *Didinium* Poche and *Lacrymaria* Bory (Taniguchi 1984; Strom and Fredrickson 2008; Sherr et al. 2009, 2013). An unusual feature of ciliate communities in the Bering Sea is that some pelagic spirotrich ciliates are capable of ingesting large diatom cells and chains (Fig. 10.4a), this has not been observed in the Chukchi and Beaufort seas (Sherr et al. 2013). In sea ice communities ciliate species characteristic of the benthos, including scuticociliates and hypotrichs, are abundant (Sherr et al. 2009, 2013). In addition, ciliate endoparasites of krill have been reported in the Bering Sea (Capriulo et al. 1991).

Heterotrophic dinoflagellate morphospecies Large (50–150 μm) athecate dinoflagellates that resemble the heterotrophic species *Gyrodinium spirale* (Bergh) Kofoid and Swezy (e.g., Fig. 10.4c) and several *Gymnodinium* spp. have been noted since the studies of Bursa (1961, 1963) in the eastern Canadian Arctic and in coastal waters off Barrow, Alaska. In the Bering Sea such dinoflagellates have been observed ingesting large diatom cells and chains (Fig. 10.4d). Armored heterotrophic dinoflagellates in the genera *Protoperidinium* Bergh and *Dinophysis* Ehrenberg are common but are generally a minor component of heterotrophic dinoflagellate abundance and biomass. Smaller hetero- and mixotrophic dinoflagellates are ubiquitous but have not been taxonomically well characterized.

Fig. 10.4 Examples of heterotrophic protist morphotypes observed in western arctic and subarctic seas: (**a**) Pelagic strombidid ciliate which has ingested a long diatom chain, Bering Sea, (**b**) *Ptychocyclis* sp. tintinnid ciliate, Bering and Chukchi seas (**c**) *Gyrodinium* sp. athecate dinoflagellate, ubiquitous in arctic and subarctic regions, (**d**) *Gyrodinium* sp. similar to the cell pictured in (**c**), but which has ingested a long pennate diatom chain, Bering Sea. Scale bars = 50 μm. Micrographs of lugol-fixed samples collected during spring 2008–2010 Bering Ecosystem Study (BEST) cruises provided by E. and B. Sherr

10.3.5.2 Diversity of Heterotrophic Protists Assessed by Molecular Genetics

Assessment of the diversity and distribution of heterotrophic protists by way of environmental gene surveys began in 2002; data is limited, especially for the Western Pacific region, with no gene libraries published to date (Lovejoy et al. 2011). By comparison, data regarding community composition of heterotrophic protists by microscopic analysis covers a much broader range of arctic and subarctic systems, and includes data on cell abundances and biomass. However, taxonomic identification of protist species by microscopy is often less definitive as compared to identification by molecular genetics; frequently morphospecies can only be identified to genus or subgroup. Molecular techniques are more quantitative, and precise identification is possible when reference sequences exist. Even in the absence of reference sequences, environmental sequences can be compared to each other to provide reliable indication of the occurrence of the same species or ribotype over different regions and seasons (Lovejoy et al. 2007; Lovejoy and Potvin 2011). A current major challenge is to reconcile the morphological species information with environmental sequence information. Achieving this requires either cultivation of individual taxa or amplifying of the appropriate marker genes from single cells (yet to be done in the Arctic), both of which are time-consuming endeavors.

Another potential problem is that molecular diversity studies have mainly focused on the smallest plankton size fractions, which include pico- and nano-flagellate communities (Lovejoy et al. 2006; Terrado et al. 2009; Lovejoy and Potvin 2011), while microscopic analysis has primarily covered protists in the microzooplankton size class. Fortunately in this case, size fractionation methods are imperfect, so that larger cells frequently break during the prefiltration step and their genetic material is collected along with that of small cells. As a result, available arctic taxa genetic diversity data includes ribotype information on a broad range of protists, encompassing ciliates, heterotrophic dinoflagellates and Radiolaria, as well as small flagellates.

Among the small flagellates, genetic surveys have confirmed the widespread distribution of phylogenetically-diverse heterotrophic flagellates including choanoflagellates, Katblepharidia, *Telonemia,* cerozoans, diplomonads, and the diverse marine stramenopiles (MASTS). In contrast to the other major groups with at least some cultured representatives, MASTS are only known from their rRNA gene sequences and their ecology as heterotrophs has been deduced using FISH (Massana et al. 2004, 2006). These diverse heterotrophic flagellates range in size from $\leq$2 to 20 μm, and feed on both bacteria and other protists, including phytoplankton. Also among the uncultivated flagellates are the Picozoa (Not et al. 2007) that were originally thought to be photosynthetic; however recent work using single cell genomics has failed to identify genes required for photosynthesis (Yoon et al. 2011) and they are more likely heterotrophs. Many photosynthetic flagellates are also active grazers and at times are in the same functional category as the heterotrophic protists (Lovejoy et al. 2000; Zubkov and Terran 2008). These mixotrophs include members of the Chloroarchniophyceae, Haptophyceae, Cryptophyceae, Euglenophyceae and several major groups of stramenopiles (Pelagophyceae, Dictyochophyceae and Raphidophyceae), all of which have been observed in arctic seas (Lovejoy et al. 2006; Lui et al. 2009; Terrado et al. 2009; Lovejoy and Potvin 2011).

One surprising result of environmental 18S rRNA gene surveys has been the discovery of several major clades of uncultivated alveolates within, or related to, the parasitic Syndiniales. The Syndiniales fall into four or five major clades (Guillou et al. 2008). The most commonly recovered clade in the Arctic belongs to Syndiniales Group II, which contains the dinoflagellate parasitoid *Amoebophyra.* All known representatives of the Syndiniales are either parasitoids, parasitic or commensally dependent on a host and have complex life stages. The diversity of these protists suggests that they are fast evolving and many may be restricted to a single host (Guillou et al. 2008). Siano et al. (2011) reported that Syndiniales Group II protists were able to grow and infect dinoflagellates in both coastal and open water habitats in the Mediterranean Sea. In the Arctic these protists have been reported from all depths and seasons sampled to date when clone libraries were constructed using environmental DNA as a template (Terrado 2011).

Several ribotypes of dinoflagellates and ciliates have been recovered in almost all arctic 18S rRNA gene surveys. The majority of arctic dinoflagellate sequences fall into the poorly resolved Gymnodiniales, Prorocentrales, Peridiniales (GPP) group (Saldarriaga et al. 2004). A number of these sequences are found throughout the Arctic, notably two that cluster most closely to *Gyrodinium rubrum* and *G. helveticum*, which are both athecate heterotrophs. Several other sequences with closest matches to other environmental ribotypes falling within the GPP are found throughout the Arctic. Additional work using single cell amplification methods or fluorescence in situ hybridization (FISH) probing and microscopy is required to identify the organisms behind the ribotypes, although studies based on microscopy have found abundant gymnodinoid dinoflagellates in arctic and subarctic seas (Sherr and Sherr 2007; Sherr et al. 2009).

The most common ciliate sequences fall within the oligotrichs, with closest matches to other environmental sequences and a single sequence from a morphologically described species in the genus *Strombidium*. These fall into two clades, Strom A and Strom B (Lovejoy et al. 2006; Lovejoy and Potvin 2011). Other ciliate sequences have closest matches to tintinnids, although the morphological classification system used for Tintinnida and some naked oligotrichous ciliates does not follow the 18S rRNA gene phylogenies; this is either due to poor resolution using the 18S rRNA gene, or a misinterpretation of the phylogeny based on the presence or absence of lorica and the lorica form (Doherty et al. 2007). Ciliate sequences with closest matches to diverse predatory genera have been recovered from most samples in the North Pacific and Beaufort Sea regions where surveys have been carried out. Also, several Polysytina which are frequently encountered in the upper mixed layer, and one Acantharian which is always reported below 50 m, are known only from their ribotypes (Lovejoy et al. 2006; Terrado et al. 2009; Lovejoy and Potvin 2011).

10.3.5.3 Biogeographical and Depth Distribution of Heterotrophic Protists

Whether morphospecies identified by microscopic methods are identical across oceanic regions, or represent different strains or even species (Katz et al. 2005; Caron 2009; McManus and Katz 2009) has important implications for the 'everything is everywhere' debate. Recent work on marine ciliates has indicated that, while molecular genetic identification concurs with microscopic identification for some species, in other cases the gene sequences of similar ciliate morphospecies are not closely related (Katz et al. 2005; Doherty et al. 2007; McManus and Katz 2009). Some ciliate species are known to be confined to specific regions of the world ocean. For example, tintinnids in the genera *Ptychocyclis* (e.g., Fig. 10.3b) and *Parafavella* are restricted to boreal seas and in the genus *Tintinnopsis* to neritic marine systems (Pierce and Turner 1993). For other groups, such as the GPP dinoflagellates that are recovered from arctic waters and other seas, sequencing of the 18S rRNA gene and standard clustering techniques at 98 % similarity results in the grouping of very dissimilar morphospecies.

Such grouping implies that, for these dinoflagellates, there is insufficient information in the 18S rRNA gene to identify species unique or confined to arctic regions. In contrast, ribotypes belonging to a ciliate (Strom A), a Polysystinea and a chloroarachniophyte have only been reported in samples from the Arctic (Lovejoy and Potvin 2011). These ribotypes, along with the arctic *Micromonas* ecotype (Lovejoy et al. 2007), indicate that there may be a subset of protists restricted to the Arctic.

Bering Sea water enters the northwest Arctic Ocean with different characteristics in summer compared to winter. In addition, locally formed water masses belonging to the summer mixed layer and winter mixed layer lie atop the incoming Bering Sea water. Within these discrete water masses, the eukaryotic microbial communities are similar over the Canada Basin and Beaufort Sea (Lovejoy and Potvin 2011). However, the upper Bering Sea – Pacific photosynthetic community is very different from the communities in locally derived arctic water masses. Most notably, cyanobacteria are important primary producers and prey for heterotrophic flagellates in the Northwest Pacific. In contrast, cyanobacteria are rare or absent in the Canada Basin and Beaufort Sea (Li et al. 2009), although they have been reported in coastal waters of the North Slope of Alaska (Cottrell and Kirschman 2009; Ashjian et al. 2010; see also Sect. 10.2.4). It is unknown whether the Pacific-Bering heterotrophic community also changes on entering the Arctic, since there is a lack of molecular information on the heterotrophic protist community in incoming water. As summer ice retreats in the arctic basins, knowing how this Bering Sea community changes is critical to understanding how the Arctic Ocean food web would function in the future.

As compared to horizontal structure, there is even less information about how heterotrophic protists are distributed with depth. We know, based on a few depth profiles, that protist communities in arctic and subarctic seas may show variable abundance patterns over small spatial scales. For example, a profile of microzooplankton distribution with depth in the upper 20 m in Barrow Canyon in the summer of 2002 revealed distinct patterns of abundance of ciliates and heterotrophic dinoflagellates, with the greatest protist biomass in the upper 3 m rather than at the chlorophyll and diatom abundance maxima at 20 m (Table 10.3).

10.3.5.4 Heterotrophic Microbes: Future Research

Heterotrophic microbes are vitally important to biogeochemical cycles and ocean food webs in arctic and subarctic seas. Future research projects should examine the diversity and activity of Bacteria, Archaea, and heterotrophic protists. The ecology and microbial diversity of sea ice communities in particular is much in need of study (Comeau et al. 2013). Reconciling molecular and microscopic identification of eukaryotic microbes, and determining the extent of genetic variability within observed protist morphospecies, are important but challenging undertakings. In addition, several types of known or putative parasitic protists have been identified by molecular genetics or by microscopy, and this group is largely uncharacterized. Evaluating the distribution and abundance of protistan parasites and parasitoids in arctic marine systems is an important first step towards understanding the ecological role of this group. At present, data on the geographical and depth distribution of heterotrophic

Table 10.3 Depth profile of chlorophyll (Chl), diatom, ciliate and dinoflagellate abundance, and microzooplankton (MZP) biomass in the upper 20 m at a station in the Chukchi Sea off the northwestern tip of Alaska in Barrow Canyon (71.326 °N, 158.014 °W), bottom depth of 115 m, July 21 2002

Depth m	Chl *a* μg l^{-1}	Diatom cells ml^{-1}	Ciliate cells ml^{-1}	Heterotrophic dinoflagellate cells ml^{-1}	Total MZP biomass μgC l^{-1}
2.5	2.0	31	13	20	78
3.1	3.5	30	22	15	90
4.5	4.0	54	13	14	38
6.6	4.7	62	12	12	32
14.5	7.7	74	17	9	59
20	13.4	203	8	7	29

The chlorophyll and diatom cell abundance maxima at 20 m likely resulted from sinking of a spring diatom bloom. Sherr unpublished data from samples collected during a 2002 Shelf-Basins Interactions (SBI) project cruise

microbes is highly limited. Filling the knowledge gaps laid out above is critical to understanding how heterotrophic microbes in the Arctic will respond to sea ice retreat, increased freshwater and surface layer stratification, and warming arctic temperatures. How heterotrophic microbes respond to climate forcing has fundamental and far reaching implication for the future ecology of the Arctic.

10.4 Benthic Fauna of the PAR

10.4.1 Introduction

Benthic fauna are composed of both infauna, animals living within the bottom sediments, as well as epifauna, animals living on the surface of the seabed. Benthic animals are generally defined as micro-, meio-, and macrofauna, of size, <0.5 mm, 0.5–1 mm, and >1 mm, respectively. Also for the purposes of this chapter, epibenthic organisms are referred to as megafauna. In the shallow waters of the continental shelves of the PAR, benthic fauna have an important role as a foundational prey base for seabirds and marine mammals, and in carbon transformation processes. The high level of primary production in the PAR which is strongly influenced by Pacific water allows for substantial export of organic carbon to the underlying sea floor. This effect declines in intensity moving eastward towards the Beaufort Sea to Cape Bathurst, and offshore into the Canada Basin. In general, the importance of benthic fauna as prey items decreases with increased water depth while species diversity increases with depth until at least the shelf break. The number of identified species is 1,400–1,500 species on the productive Chukchi shelf and slope as well as in the geophysically more heterogeneous but less productive Beaufort Sea. However recent studies indicate all the species in the region have not been determined due to incomplete spatial sampling (Piepenburg et al. 2011). The important but poorly understood relationship between productivity and diversity has potential ramifications for the future success of higher trophic level taxa. Future changes in sea ice dynamics likely with a continued decrease in summertime areal coverage along with a general warming in the region will very likely change benthic ecosystem functioning.

10.4.2 Benthic Fauna of the Northern Bering, Chukchi, and Western Beaufort Seas

10.4.2.1 Environmental Setting

Primary production occurs in both the ice-covered and open waters of the Pacific-influenced northern Bering, Chukchi and western Beaufort seas. Annual primary production estimates (based on 120-day growth season) range from <50–800 g C m^{-2} d^{-1}; (Springer et al. 1996; Hill and Cota 2005). The highest values occur in the nutrient-rich Anadyr Water (AW) that transits northward into the Arctic on the western side of Bering Strait. Anadyr Water mixes with water from the northern Bering Sea Shelf to form Bering Sea Water (BSW) in the Chukchi Sea. By comparison, lower production occurs in the nutrient-poor Alaska Coastal Water (ACW) (Grebmeier et al. 2006a) which enters the Arctic through eastern Bering Strait. Tight pelagic-benthic coupling between seasonal water column production and short- and long-term benthic carbon transformation processes provide a "footprint" of persistent ecosystem events recorded in both the infaunal biomass (defined as >1 mm size fraction for all data presented in this section) and sediments (Dunton et al. 2005; Grebmeier et al. 2006a), as long lived benthic species can act as multi-year, long-term integrators of a variety of marine processes.

Most offshore sediments in the northern Bering, Chukchi and Beaufort seas are composed of soft substrates, such as mud or muddy sand with the exceptions of areas impacted by high currents, such as Bering Strait and Barrow Canyon where the substrate consists of mixed sand, mud and rocks of various size (Grebmeier et al. 2006a). The percentage of fine sediments (silt and clay) provides an indication of deposition zones as seen southwest of St. Lawrence Island in the northern Bering Sea, in the southeast Chukchi Sea as well as in Herald Valley and the slope and deep basin regions of the Canada Basin (Fig. 10.5). Studies indicate that organic carbon content is significantly correlated with silt and clay content in these sediments, although grain size is not the sole factor determining sediment organic carbon content (Grebmeier et al. 2006a).

10.4.2.2 General Biogeography

Macrobenthic infauna are normally stationary in marine sediments, thus their community patterns are directly impacted by export production from the overlying water column. Bivalves, amphipods, and polychaetes dominate the infaunal biomass south of St. Lawrence Island in the northern Bering Sea; amphipods and bivalves dominate in the central region from St. Lawrence Island to Bering Strait,

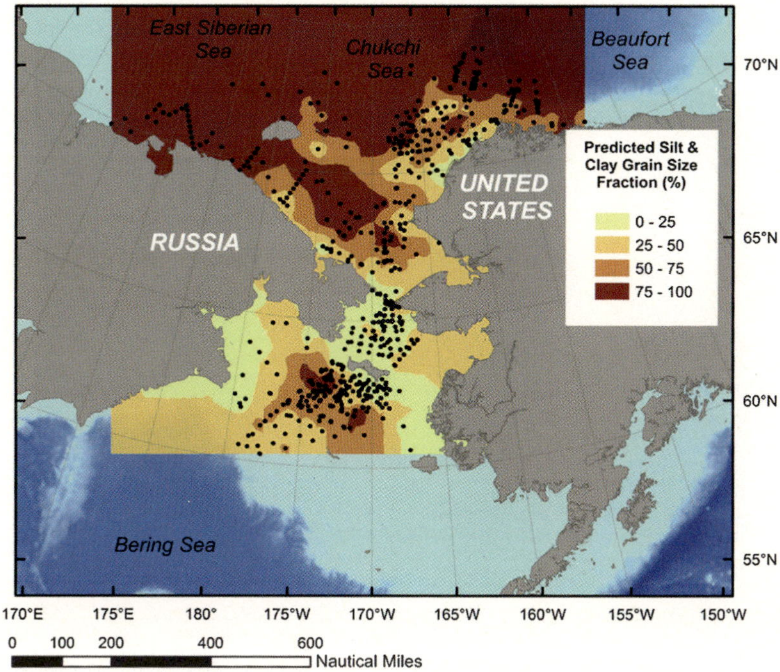

Fig. 10.5 Distribution of sediment surface silt and clay content in the northern Bering Sea to the Arctic Basin from 1973 to 2010 (provided by J. Grebmeier)

and bivalves and polychaetes dominate in the southern Chukchi Sea to the slope region of the Canada Basin and in the Alaskan Beaufort Sea (Fig. 10.6; Carey and Ruff 1977; Grebmeier et al. 2006a; Grebmeier 2012). Sediment grain-size composition is the major factor responsible for determining benthic community composition, including which benthic taxa are dominant, whereas organic carbon food supply is directly related to biomass itself (Grebmeier et al. 1988, 1989). Areas notable for high infaunal biomass are also sites of high sediment oxygen consumption and nutrient efflux, both of which are indicators of elevated organic carbon supply to the benthos (Fig. 10.6; also see Mathis et al. 2014, Figs. 9.9–9.11, this volume). Although infauna are food for many epifaunal species that are typically larger than their endobenthic prey and more motile, recent studies indicate that epibenthic faunal composition and biomass do not reflect the presence of a high biomass of infaunal prey, but instead are largely determined by sediment type and factors tied to latitude (Bluhm et al. 2009). Thus, while infauna respond to patterns of carbon export (Grebmeier 2012; Grebmeier et al. 2006a, b), the epibenthic fauna is not as closely coupled to carbon export (Bluhm et al. 2009) because the carbon export signal gets 'washed out' because of the mobility of some dominant taxa, such as crabs, that can undertake extensive migrations.

Infaunal macrobenthos High infaunal biomass is observed in persistent biomass "hot-spots" in the region ranging from southwest of St. Lawrence Island to Barrow

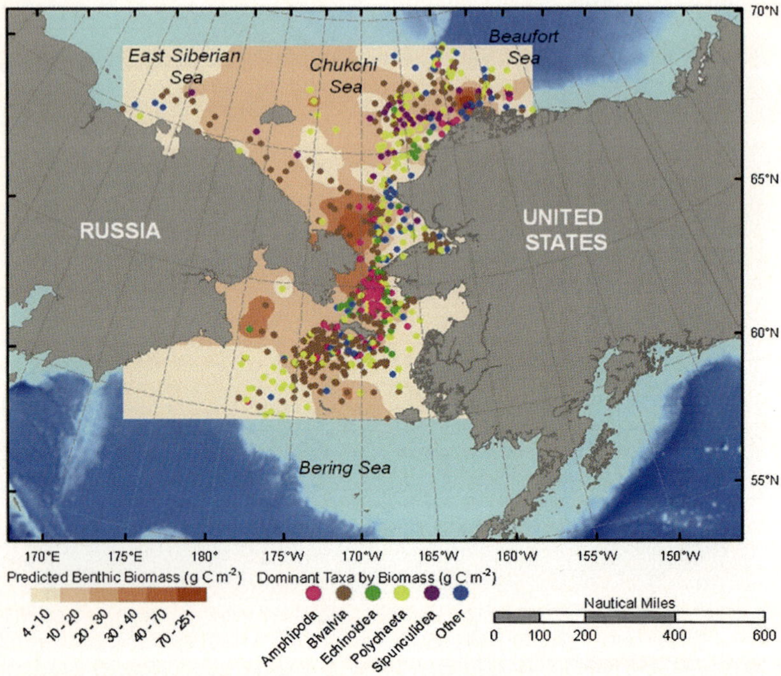

Fig. 10.6 Spatial distribution of infaunal benthic biomass and the most dominant taxa (based on infauna >1 mm size fraction) identified for each station in the northern Bering and Chukchi seas from 1973 to 2010 (With permission from Bluhm and Grebmeier 2011)

Canyon (Grebmeier et al. 2006a, 2012); such "hot spots" are indicative of high levels of organic carbon settling onto the sea floor (see also Mathis et al. 2014, this volume). Composition of the infaunal communities is not static. Currently the infauna southwest of St. Lawrence Island is dominated by the bivalves *Nuculana radiata* and *Ennucula tenuis* (Grebmeier 2012), while the bivalve *Macoma calcarea* was dominant before the 1980s (Lovvorn et al. 2003). This shift in bivalve prey species over the last decade has reduced the food quality for diving sea ducks of the region (Lovvorn et al. 2009).

Large quantities of organic carbon produced in the Anadyr Water passes northeastward through Anadyr Strait west of St. Lawrence Island into the Chirikov Basin, and northward through Bering Strait where it supports a rich benthic fauna on the western side of the system (Fig. 10.6). Strong bottom currents maintain fine sediments in suspension, resulting in coarse, sandy sediments in the Chirikov Basin (Fig. 10.5) that are inhabited by numerous interface-feeding ampeliscid amphipods (*Ampelisca macrocephala, Byblis* spp.) and bivalves (*Macoma calcarea*) which live in sandy muds (Fig. 10.6; Grebmeier et al. 1988, 2006a; Highsmith and Coyle 1992). These ampeliscid amphipods are a favored prey item of gray whales in that area. Recent amphipod declines in the northern Bering Sea, along with decreased sea ice extent, have resulted in a shift of gray whale feeding sites from the northern

Bering Sea to as far as north as Point Barrow (Moore et al. 2003, 2006, see Moore et al. 2014, this volume). These observations support the view that a large-scale ecosystem change is underway and underscore the importance of bottom-up control on higher trophic level taxa.

The nutrient- and phytoplankton-rich water that is transported northwestward through Bering Strait is a major driver of the high benthic faunal productivity of the south-central Chukchi Sea (Fig. 10.6). Macrobenthic infaunal biomass at sampling stations in the south-central Chukchi Sea ranges from 24 to 59 g C m^{-2} and exceeds 120 g C m^{-2} at the "hot-spot" just northwest of Bering Strait as observed by Sirenko and Gagaev (2007). This infaunal assemblage is dominated by tellinid and nuculid bivalves (*Macoma calcarea* and *Ennucula tenuis,* respectively) and ampeliscid and lysianassid amphipods (*Byblis* spp. and *Anonyx* sp., respectively). In contrast to the very productive western side of the system, benthic communities to the east, which are strongly influenced by Alaska Coastal Water, are more variable in composition and typically of very low biomass (<10 g C m^{-2}), though occasionally ranging up to higher values of 12–23 g C m^{-2} (also see Feder et al. 2007).

As it transits further north into the central Chukchi Sea, the Pacific water becomes progressively depleted of nutrients and phytoplankton. Perhaps not surprisingly then, infaunal biomass declines from the southern Chukchi Sea "hot-spot" to the central Chukchi Sea, with biomass dropping to <10 g C m^{-2} (Fig. 10.6, Sirenko and Gagaev 2007; Grebmeier et al. 2006a; Grebmeier 2012). Winter-transformed BSW flows through Herald Valley and Herald Trough (located in the western and central Chukchi Sea, respectively) and infaunal communities here are dominated by maldanid (*Maldane sarsi*), lumbrinerid (*Nicomache lumbricalis*) and nephtyid (*Nephtys ciliata*) polychaetes, brittle stars (*Ophiura sarsii*), and sipunculids (*Golfingia margaritacea*). It should also be noted that infaunal biomass increases to 10–20 g C m^{-2} in Herald Valley (Fig. 10.6), which indicates a current-driven focusing of high-quality organic material in this region.

In the US sector of the central Chukchi Sea, there is a hydrographic front off the Alaska coastline which extends to just north of Icy Cape (Feder et al. 1994a, b). Nutrient-poor ACW lies inshore of the front, while the relatively nutrient-rich BSW lies to the west and north of the front. Infaunal biomass values are higher north of the front (Sect. 10.4.2), which is probably the result of persistent deposition of organic carbon produced offshore and advected towards the Alaska coast (Feder et al. 1994a; Grebmeier et al. 2006a).

The distribution, abundance and biomass of infauna in the northern Chukchi Sea are related to water mass characteristics and current patterns (Feder et al. 1994a). Bivalves, polychaetes and sipunculids dominate the infaunal community of the northern Chukchi Sea, where average infaunal benthic biomass moderate at 5–15 g C m^{-2} (Fig. 10.6), with the highest values in the northeast sector and upper Barrow Canyon (20 and >100 g C m^{-2}, respectively). On the upper slope (200–1,000 m depth) and extending down into the Canada Basin, the benthic community becomes dominated by bivalves and polychaetes, with biomasses of <5 g C m^{-2} (Bluhm et al. 2005; Pirtle-Levy 2006; Grebmeier et al. 2006a). Foraminiferans >1 mm become dominant down slope into the basin (Grebmeier et al. 2006a). Virtually no information is available on meio- and microfauna for the northern Bering, Chukchi and western Beaufort seas; future studies are needed to address this information gap.

Epifaunal megabenthos Pacific species characterize the epibenthic fauna in the northern Bering Sea and southern part of the Chukchi Sea (Sirenko 2003). In the northeastern Bering and southeastern Chukchi seas echinoderms, mainly sea stars, tend to dominate the invertebrate epifaunal biomass (Feder et al. 2005). In the southeastern Chukchi Sea a number of predators, such as the sea star *Leptasterias polaris* that consumes both infauna and epifauna, and the large gastropods *Neptunea heros* and *Buccinum* spp. that mainly feed on infauna are found (Grebmeier et al. 2006a). In the coastal sandy sediments of the Gulf of Anadyr, the sand dollar *Echinarachnius parma* is abundant. In the central Gulf of Anadyr, which has finer sediments, the omnivorous brittle star *Ophiura sarsii,* as well as snow crabs (*Chionoecetes opilio)*, lyre crabs (*Hyas coarctatus*), hermit crabs (*Pagurus* spp.), gastropods (*Neptunea ventricosa* and *Margarites* sp.), and the basket star *Gorgonocephalus eucnemis* predominate (Sirenko and Koltun 1992). By contrast, the Anadyr Strait to the west of St. Lawrence Island, which has strong currents and mixed coarse pebbles and rock substrate, is dominated by suspension feeders, including sponges, barnacles and ascidians, as well as by sea urchins and brittle stars. North of St. Lawrence Island, the presence of abundant infauna prey (e.g., amphipods and bivalves) supports a substantial predatory epifauna assemblage, including hermit crabs, sea stars, crangonid shrimp, crabs (e.g. *Chionoecetes opilio*), and the large gastropod *Neptunea heros* (Feder and Jewett 1978; Grebmeier et al. 2006a). Closer to the coast sea anenomes, gastropods, bivalves, and sand dollars are common.

Further north into the Chukchi Sea, recent studies indicate that the epibenthic fauna is largely dominated by the brittle star *Ophiura sarsii* and the snow crab *Chionoecetes opilio,* while gastropods are the most species-rich group. High numbers of the holothurian *Myriotrochus rinkii* are also found (Bluhm et al. 2009). A rich assemblage of sedentary epifaunal organisms (sponges, soft corals and bryozoans) occur in the central Chukchi Sea in the shallow bank areas with coarse gravel sediments (Sirenko and Gagaev 2007). The inshore area in the central Chukchi Sea has a diverse mixture of mud, gravel and boulders (Fig. 10.5) and contains large numbers of epifaunal organisms, such as sponges, sea anemones, soft corals, barnacles, bryozoans, sea cucumbers and sea urchins (*Strongylocentrotus pallidus*) (Grebmeier et al. 2006a; Bluhm et al. 2009) Epifaunal amphipods are also numerous in both soft-bottom and gravel-boulder areas. Hermit crabs and gastropods (*Buccinum* spp.) are common and inhabit mud-gravel substrates. In addition, basket stars and sea stars occur inshore, while brittle stars and sea cucumbers are abundant in deeper water. Ascidians inhabit mud-gravel substrates off Barrow in extremely high densities. Echinoderms are numerous in the northeastern Chukchi Sea as well (Feder et al. 1994a; Ambrose et al. 2001); these include the sand dollar *Echinarachnius parma* nearshore and dense assemblages of brittle stars, that are dominated by the omnivore *O. sarsii* and the suspension-feeder *Ophiopholis aculeata.* Bluhm et al. (2009) found crustaceans to be the second most abundant and biomass-rich group in the Chukchi Sea, particularly at eastern stations under ACW, which contrasts with the trend suggested by Feder et al. (2005) of epifaunal crustacean dominance in the south and echinoderm dominance in the north.

The upper Barrow Canyon is a "hot spot" region in the northeastern Chukchi Sea, where the presence of a rich epifauna of suspension feeders (e.g., sponges, bivalves, barnacles, basket stars and tunicates) attached to rocks and cobble suggests the prevalence of strong currents (Ambrose et al. 2001; Grebmeier et al. 2006a). The suspension-feeding mussel *Musculus discors* is very abundant (with dry weight biomasses of up to 100 g C m^{-2}) in areas with interspersed silt, clay and gravel. This enrichment in benthic biomass at the head of Barrow Canyon (Fig. 10.6) coincides with extremely high sediment community oxygen consumption (Mathis et al. 2014, Fig. 9.9, this volume). The mud at the bottom of the canyon is covered with tubes of the deposit-feeding polychaetes, and a variety of other infauna and epifauna (Fig. 10.6).

Epifaunal shelf communities in the western Beaufort Sea are dominated by locally very abundant ophiuroid beds (Carey 1977; Frost and Lowry 1983; Ravelo et al. 2014) and more recently by snow crab (*Chionoecetes opilio*) that were found at some western >200 m shelf break locations (Rand and Logerwell 2011). Areas shallower than ~20 m are affected by ice scouring and river run-off which reduces epifaunal densities and diversity to a few amphipod and isopod taxa (Ravelo et al. 2014). Amidst these soft-bottom communities, the "Boulder Patch" in Stefansson Sound adjacent to the oil production areas of Prudhoe Bay, provides substrate for sponges, encrusting bryozoans, soft corals, tube worms, and other fauna (Dunton et al. 1982). Similarly, glacial erratics scattered across the eastern Alaskan Beaufort Sea provide small islands of hard substrate for sessile primarily filter-feeding taxa (Bluhm et al. pers. obs.).

The benthic fauna of the western Beaufort Sea shelf, which is hydrographically downstream of the northward flowing Pacific water, exploits both advected and locally produced organic carbon (Cooper et al. 2005; Grebmeier et al. 2006a). By comparison, the more riverine-influenced central and eastern Beaufort Sea has lower integrated water column and sediment chlorophyll inventories, and the sediment-surface organic carbon is mainly of terrestrial origin, as indicated by relatively depleted $\delta^{13}C$ values (Dunton et al. 2005, 2012; Grebmeier et al. 2006a). As a result infaunal benthic biomass is higher on the western Beaufort Sea shelf (10–20 g C m^{-2}) than in the organic carbon-limited systems to the east in the nearshore Beaufort Sea (<10 g C m^{-2}) (Fig. 10.6).

10.4.3 Benthic Invertebrate Patterns in the Canadian Beaufort Sea Shelf

10.4.3.1 Environmental Setting

The Canadian Beaufort Shelf (CBS) is the 65×10^3 km^2 central shelf of the Beaufort Sea, bounded by the Alaskan Beaufort Sea shelf (44×10^3 km^2) to the west and the Banks Island shelf (23×10^3 km^2) to the east. The Mackenzie Trough and Amundsen Gulf split this shelf system. The CBS is also referred to as the Mackenzie Shelf (Carmack and Macdonald 2002; Macdonald et al. 2004; Williams and Carmack 2008).

The Mackenzie River dominates the CBS in terms of the input of both freshwater (~330 km^3 $year^{-1}$ ± 25 %) and sediments (127 × 10^6 ± 6 × 10^6 Mt $year^{-1}$) (Macdonald et al. 1998; Walker et al. 2008). Relative to other rivers emptying into the Arctic Ocean, the Mackenzie ranks fourth for freshwater discharge and first for sediment import. Wind and wave effects modify the location of the Mackenzie plume, decrease stratification, disperse suspended sediment, modify light penetration and transport nutrients to the benthos (Melling 1993; Carmack and Macdonald 2002). Inshore, the benthos is strongly influenced by this estuarine system while further offshore, at water depths of 40–280 m, the benthos is bathed by low-salinity, nutrient-rich Pacific water and below this warmer, saltier Atlantic water (Carmack and Wassmann 2006).

Along the outer shelf and shelf-break, the benthos benefits from the eastwardly flowing Beaufort Undercurrent which provides a nutrient source for phytoplankton (Carmack and Macdonald 2002). This current and associated smaller-scale eddies also provide transport for pelagic stages of resident benthic fauna as well as for predators and new colonists (Carmack and Macdonald 2002). Upwelling of nutrient-rich water at the shelf edge especially around Cape Bathurst to the east (Williams and Carmack 2008), along the Mackenzie Trough (Williams et al. 2006) and in the smaller Kugmallit Valley (Williams et al. 2008) boosts production and thereby modifies the benthic communities locally (Conlan et al. 2008; Darnis et al. 2012). Mixing of diverse water types also enhances coastal productivity (Carmack and Macdonald 2008; Juul-Pedersen et al. 2008). Phytoplankton production is seasonally variable both in terms of cell size and taxonomic composition. In spring prior to ice break-up, sympagic algae provide some organic carbon to the benthos but this is a relatively small contribution (Macdonald et al. 1998; Renaud et al. 2007b; Morata et al. 2008; 2011; Riedel et al. 2008). Sea ice provides an upside-down substrate for some benthic meiofauna, larval stages of macrofauna and motile benthic macrofauna, which graze on sympagic microbes and primary producers in large numbers (up to 482,000 ind m^{-2} in spring) (Carey 1992).

Primary production peaks at 200 mg C m^{-2} d^{-1} in summer open water and declines to <10 mg C m^{-2} d^{-1} under ice cover (Carmack et al. 2004). Export of organic carbon from the water column to the benthos is variable, depending on stratification, currents, and grazing by meso- and microzooplankton. Particulate and dissolved organic matter generated by erosion and introduced by the Mackenzie River also provides seasonally variable organic carbon to the benthos of the CBS (Macdonald et al. 1998, 2004; O'Brien et al. 2006; Terrado et al. 2008; Walker et al. 2008). Surface-sediment organic carbon grades from terrestrial to oceanic with distance from shore (Macdonald et al. 2004) and eastward from the Mackenzie River (Morata et al. 2008). Deep-buried organic carbon in the sediment may be a long-term benthic food resource (Pirtle-Levy et al. 2009). Methane that is released from the seabed at numerous locations (Collett and Dallimore 1999) can be a nutrient source to benthic microbes but may also be a habitat-disturbing agent. However, the pingo-like features produced at some methane venting sites provide elevated habitats for the benthos, resulting in elevated species numbers (Jerosch et al. 2010).

During the winter, benthic organisms on the CBS live under a ceiling of landfast sea ice up to the 20 m isobath and under pack ice in deeper waters. A build-up of pack ice jammed against the fast ice (the stamukhi zone) dams the Mackenzie River outflow, resulting in a $\sim 12 \times 10^3$ km^2 "lake" of freshwater that forms a seasonal ceiling inshore (Macdonald et al. 1995; Carmack and Macdonald 2002). This influences the seasonal geochemistry of the shelf (Emmerton et al. 2008). Recurrent flaw leads around the 20–30 m isobath (Melling 1993) provide variable open water and thinner ice cover to 70 km offshore of the fast ice in late winter (Williams et al. 2008). Ice formation in the flaw leads causes release of dense brine which sinks down and mixes the seawater to 40–50 m depth (Carmack and Macdonald 2002). Where flushing is low, brine pooling on the seafloor may impact the benthos negatively (Kvitek et al. 1998). Ice scours the CBS to as deep as ~70 m at a rate of up to 8.2 events km^{-1} $year^{-1}$, reworking >75 % of the seabed at 22–25 m depth in the stamukhi zone (Lewis and Blasco 1990; Héquette et al. 1995). Such effects on the benthic fauna may be similar to that reported in the eastern Arctic (Conlan and Kvitek 2005) but the environmental regime and benthic community structure differ in the CBS. Ice scour also increases coastal erosion (Héquette and Barnes 1990), which modifies turbidity, sediment transport and carbon supply to the benthos. Shelf and slope slumping of deposited sediment and associated turbidity and deposition can locally disturb the benthos (Macdonald et al. 2004).

10.4.3.2 General Biogeography and Biodiversity

During the last Pleistocene glacial maximum, dating to 27 Ka, the global sea level was about 140 m lower than today, and large portions of the CBS (i.e., all areas at contemporary depths <140 m) had fallen dry (Macdonald et al. 2004). Still, the CBS sediment ~20 cm below the seafloor surface is permafrost (Allen et al. 1988). The CBS is currently (~1,850-present) less productive than it was between ~1,450 and 1,850 (Richerol et al. 2008). The benthos of the CBS is a mix of Atlantic- and Pacific-origin species with a small proportion of endemics (Dunton 1992) but its composition is in flux, a product of historic conditions and modern controls.

Up to 919 macro- and megafaunal species have been recorded on the CBS (Chapman and Kostylev 2005) with a total of 992 infaunal taxa enumerated for the entire Canadian Arctic (Archambault et al. 2010). Arthropods dominate at 37 % of total species, followed by annelids (26 %), molluscs (19.5 %) and echinoderms (4 %); these proportions are similar to the overall composition of the Arctic benthic fauna (Piepenburg et al. 2011). Simulated extrapolations to a theoretically infinite number of samples yield an expected number of 745 ± 47 molluscs+arthropods+echinoderm species (as compared to 455 observed to date) and 349 ± 17 annelids (as compared to 305 observed to date) (Table 10.4) on the Beaufort shelf (Canadian+Alaskan). This is an estimated 31–43 % of all arctic molluscs+arthropods+echinoderms and 37–49 % of all arctic annelids. Despite being less than half the size of the shelves of the adjacent Chukchi Sea and western Canadian Archipelago, the expected species richness of the Beaufort Shelf is 1.2–1.8 times

Table 10.4 Comparison of observed (S_{obs}), average taxonomic distinctness (Δ^+) and expected species richness (Chao2) of the major macro- and megafaunal phyla on the Beaufort Shelf compared to the shelves of the Chukchi Sea and the western Canadian Archipelago (Modified from Piepenburg et al. 2011)

Location	Size ($\times 10^3$ km^2)	No. stations	S_{obs} (MAE)[a]	S_{obs} (A)[b]	Δ^+ (MAE)	Δ^+ (A)	Chao2 ± SD (MAE)	Chao2 ± (A)
Chukchi[a]	464	131	401	188	74.4	55.1	443 ± 13	243 ± 21
Beaufort Shelf[b]	145	518	455	305	73.4	54.7	745 ± 47	349 ± 17
Western Canadian Archipelago[c]	605	337	364	185	74.0	54.1	482 ± 22	250 ± 25
Total Arctic[d]	9,787	4,452	1,467	668			2,040 ± 190[e]	816 ± 71[e]

Molluscs + arthropods + echinoderms (MAE) are treated as a single group and separately from annelids (A) due to different levels of taxonomic effort on these groups

[a]Area 13: Bering Strait E to Barrow and W to Proliv Pevek, 30 m to shelf edge

[b]Area 12: Barrow to Cape Bathurst, 30 m to shelf edge

[c]Area 11: Beaufort shelf E of Cape Bathurst, Amundsen Gulf, M'Clure Strait, Viscount-Melville Sound, M'Clintock Channel, 30 m to shelf edge

[d]Includes 14 of 19 Arctic shelf marine ecoregions; insufficient data for five ecoregions

[e]Based on data interpretation rather than a strict Chao2 calculation due to lack of sufficient data for some Arctic shelves

greater. This higher expected diversity may be partly due to the calculations being based on a higher number of samples on the Beaufort Shelf than on the other two shelves. However, this can also be attributed to high habitat variability on the Beaufort Shelf due to the Mackenzie River, shelf-edge influence, submarine glacial beaches and riverbeds, canyons and valleys, pingo-like features, methane vents, submerged drilling islands, ice-scour disturbance, and post-glacial evolutionary history (Cusson et al. 2007; Piepenburg et al. 2011; Conlan unpublished data).

Infaunal macrobenthos Macrofauna account for an average 25–69 % of the benthic carbon demand on the CBS (Renaud et al. 2007a). They are quick to respond to food pulses (Renaud et al. 2007b). Community composition on the CBS broadly grades ocean-ward (Conlan et al. 2008, unpublished data; Wacasey 1975). Inshore, excessive suspended material from the Mackenzie River plume selects for deposit feeders over suspension feeders, and the episodically high freshwater load selects for species that are physiologically flexible. Offshore, community structure and production are influenced by shelf topography which controls the degree of coupling with the pelagic environment.

Figure 10.7 illustrates the typical community composition of the dominant macrofauna (defined in this section as body size >0.5 mm) in various regions of the CBS as described in Conlan et al. (2008). Inshore, the community is dominated by large numbers of the small predatory polychaete *Micronephthys minuta* and lesser numbers of the deposit-feeding bivalve *Portlandia arctica*. Biomass close to the Mackenzie River plume is <1 g m^{-2} dry wt (Wacasey 1975) (equivalent to ~0.4 g C m^{-2} using conversions in Rowe 1983 and Ricciardi and Bourget 1999). On the outer shelf, the deep-burrowing bamboo worm *Maldane sarsi* dominates in abundance and biomass along with a variety of other polychaetes and crustaceans. This species and

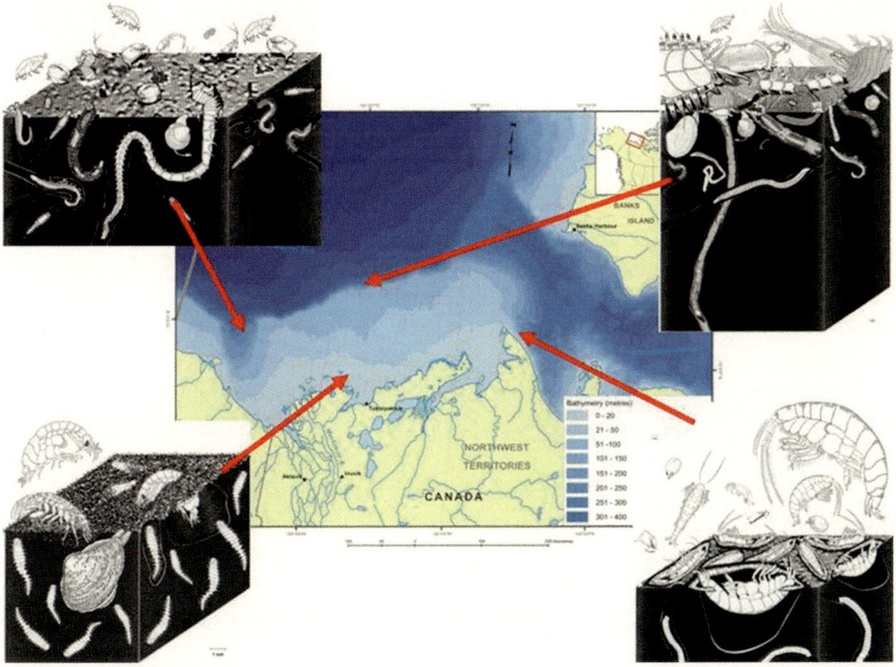

Fig. 10.7 Dominant macrofauna on the Canadian Beaufort Shelf. *Bottom left*: inshore near the Mackenzie delta (20 m) (the polychaete *Micronephthys minuta*, the bivalve *Portlandia arctica* and the amphipods *Anonyx* sp. and *Protomedeia* sp.). *Top left*: Mackenzie Trough (240 m) (the polychaetes *Micronephythys minuta*, *Prionospio* spp., *Terebellides stroemi* and *Pholoe minuta* and the crustaceans *Brachydiastylis nimia* (cumacean), myodocopid ostracods and phoxocephalid amphipods). *Top right*: continental slope at Kugmallit Valley (214 m) (the polychaetes *Maldane sarsi*, *Spiophanes* sp., *Tharyx* sp. and the large *Eunoe* sp. and *Pherusa* sp. (shown at the surface), the crustaceans *Brachydiastylis nimia,* myodocopid ostracods and phoxocephalid amphipods, the bivalves *Thyasira* sp. and *Yoldiella* sp. and the brittle star *Ophiacantha bidentata*). *Bottom right*: Cape Bathurst (38 m) (the amphipod *Ampelisca macrocephala* and the polychaete *Barantolla americana*). Community illustrations by Susan Laurie-Bourque (With permission from Conlan et al. 2008, modified)

other maldanid polychaetes provide an important benthic ecosystem service by regenerating nutrients at the surface, oxygenating deeper layers and mixing the sediment (Holte 1998; Levin et al. 1997; Dufour et al. 2007). Unlike the case for zooplankton (Darnis et al. 2008), the flaw lead and polynya system on the shelf does not leave a footprint on the benthos. Biomass of the macrofauna in the Mackenzie Trough is greater than on the surrounding shelf, but the community here is not distinct. Mean abundance across the shelf ranges from 828.1 ± 212.1 ind. m^{-2} on the slope to 3636.2 ± 1355.1 ind. m^{-2} in the flaw lead. Abundances under the fast ice, on the outer shelf and in Mackenzie Trough are intermediate.

Wind-driven upwelling at Cape Bathurst has a strong imprint on the underlying benthos. Macrofaunal abundance and biomass here is greater than elsewhere on the CBS shelf by 4–18-fold on average, to up to 18.0×10^3 ind m^{-2} (Conlan et al. 2008) and 71.4 g m^{-2} dry wt (Wacasey 1975) (equivalent to ~27.5 g C m^{-2} using conversions in Rowe 1983 and Ricciardi and Bourget 1999), levels which are comparable to those seen for the rich benthic community of the southeast Chukchi Sea (Grebmeier et al. 2006a; Feder et al. 2007). Species composition is also highly modified, being dominated by the suspension/surface deposit feeding amphipod *Ampelisca macrocephala* and the deposit feeding capitellid polychaete *Barantolla americana*. Although present elsewhere on the shelf, neither species achieves such high densities as at Cape Bathurst. *Ampelisca macrocephala* is a key component of the gray whale (*Eschrichtius robustus*) feeding areas of the Bering and Chukchi seas (Oliver et al. 1983; Highsmith et al. 2007) where pelagic-benthic coupling is also high (Highsmith and Coyle 1990; Grebmeier and Barry 1991; Feder et al. 1994b, 2007). The Eastern North Pacific population of gray whales feeds as far northwest as Cape Bathurst in the summer (Rugh and Fraker 1981) although most individuals forage in the Bering and Chukchi seas (Highsmith et al. 2007). However, the Bering Sea feeding area is declining in biomass and production and gray whales are foraging further north, on the Alaskan Beaufort Shelf (Grebmeier et al. 2006b, 2012; see also Moore et al. 2014, this volume). Possibly, therefore, more gray whales may reach Cape Bathurst in the future. Bowhead whales (*Balaena mysticetus*), which frequent Cape Bathurst in larger numbers than gray whales, can also feed benthically (Hazard and Lowry 1984; Würsig et al 1989). King and common eiders (*Somateria spectabilis* and *Somateria mollissima*) and long-tailed ducks (*Clangula hyemalis*) feed in large numbers on the benthos at Cape Bathurst (Dickson and Gilchrist 2002) as well.

Meio- and microfauna Meio- and microfauna account for 31–75 % of benthic carbon demand on the CBS (Renaud et al. 2007a). Bessière et al. (2007) found highest abundance and biomass on the shelf east of Cape Bathurst ($5.1 \pm 1.1 \times 10^6$ ind m^{-2}, 680.6 ± 262.7 mg C m^{-2} respectively). Much lower meiofaunal abundance and biomass occurred in the dense *Ampelisca* beds ($0.8 \pm 0.1 \times 10^6$ ind. m^{-2} and 63.0 ± 19.4 mg C m^{-2} respectively). High abundance and biomass values of meiofauna were also found in Franklin Bay. Abundance and biomass of meiofauna on the rest of the shelf and in Amundsen Gulf were 0.2 ± 0.04–$3.9 \pm 0.9 \times 10^6$ ind. m^{-2} and 24.9–240.6 ± 50.7 mg C m^{-2} respectively, the larger values being on the shelf edge and in Mackenzie Trough. Meiofaunal abundance correlated with the quantity of phytodetritus at the seafloor. In all cases, nematodes contributed 50–95 % of the abundance, while small polychaetes and copepods, being larger, contributed more biomass when abundant. In general, meiofaunal densities were within the range found on other arctic shelves.

Epibenthic megafauna Megafauna are most abundant on the CBS at 60–90 m depth and account for up to 41 % of the community carbon demand (Renaud et al 2007a). Although an inventory of the fauna exists (included in Piepenburg et al. 2011), regional trends in abundance, biomass and epifaunal composition are unpublished as yet; ongoing work will fill this gap over the next few years. The large scavenging isopods *Saduria entomon*, *S. sabini* and *S. sibirica* are a common feature of the CBS benthos (Kostylev and Chapman 2004). Other large epifauna are the soft coral *Gersemia rubiformis*, barnacle *Balanus crenatus*, brittle stars *Ophiocten sericeum* and *Ophiura robusta*, basket star *Gorgonocephalus arcticus*, sea stars (*Urasterias lincki*, *Leptasterias* sp. *Ctenodiscus* sp.), crinoid *Heliometra glacialis*, sea urchin *Strongylocentrotus droebachiensis* and sea cucumbers *Myriotrochus rinkii* and *Thyonidium* sp., and various anemones (Wacasey et al. 1977; Atkinson and Wacasey 1989; Kostylev and Chapman 2004). Epibenthic distribution is controlled by sediment type, food resources and freshwater influence. Benthos-feeding large predators such as beluga and bowhead whales and bearded, ringed and spotted seals exploit this resource (Dehn et al. 2007; Bluhm and Gradinger 2008).

10.4.4 Deep-Sea Benthos

The benthic fauna of the deep sea (Canada Basin) in the PAR, as within all arctic deep-sea basins, is shaped by the geological history and semi-isolation of the area. The Canada Basin, now filled with Atlantic-origin bottom waters (Rudels et al. 1994), was influenced by Pacific fauna until ~80 Ma when the deep-water connection to the Pacific Ocean closed (Marincovich et al. 1990). Currently the shallow Bering Strait presents a barrier for the northward transfer of Pacific deep-sea species, while the mid-arctic oceanic ridges do not prevent Atlantic fauna from dispersing even into the Canada Basin (Vinogradova 1997; Bluhm et al. 2011).

International Polar Year and Census of Marine Life efforts increased the previous inventory of eukaryotic fauna in the deep Arctic Ocean (Sirenko 2001) by ~300 to over 1,100 taxa ranging from the PAR to Fram Strait. The current faunal inventory is dominated by soft-bottom arthropods (>350 taxa), foraminiferans (~200), annelids (~200), and nematodes (>140) (Bluhm et al. 2011). Isolated hard-substrate habitats such as drop stones form patches of enhanced diversity (MacDonald et al. 2010). Typical for the deep sea, half of all taxa recorded so far were found at only one or two locations, including numerous recently described species and range extensions (e.g. Gagaev 2008; Sharma and Bluhm 2010). Considerable overlap in species composition with arctic shelf communities (Piepenburg et al. 2011) indicates that part of the deep-sea fauna originates from shelf fauna that survived Pleistocene glaciations by submergence into the deep, which was curiously confined to the Atlantic sector (Nesis 1984). In contrast to other deep-sea regions, no mid-depth (1,000–3,000 m) peak in species richness was observed; most studies showed instead a decrease in diversity and an increase in evenness with water depth. Distinct benthic communities are recognized within the slope, abyssal and plateau PAR regions; however, on a pan-arctic scale, regional differences in inventories are moderate (Bluhm et al. 2011).

Primary production over deep water in the PAR is low compared to the adjacent shelf due to the short photo-period, limited nutrients and multi-year ice cover (Lee and Whitledge 2005). Although local production is complemented by organic carbon advected from the highly productive Chukchi shelf (Moran et al. 2005; Grebmeier et al. 2006a), overall the arctic deep sea receives relatively low input of marine-derived organic matter. As a result, gradients in faunal abundance and biomass from the shelf to the abyss are steep and values typically decrease with water

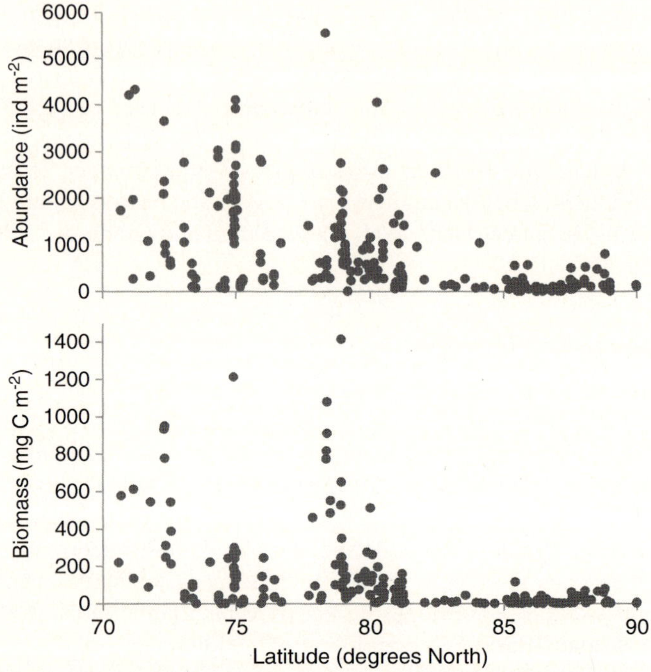

Fig. 10.8 Macrobenthic abundance (*upper plot*) and biomass (*lower plot*) as a function of latitude in the Arctic Ocean at depths >500 m (data from Bluhm et al. 2011)

depth. At depths >500 m, meiofaunal abundance ranges from <100 to >4,500 ind 10 cm^{-2} (mean around 1,200 ind 10 cm^{-2}), macrofaunal abundance ranges from <1 to almost 10,000 ind m^{-2} (mean around 930 ind m^{-2}), (Fig. 10.7, summarized in Bluhm et al. (2011)), and megafaunal abundance ranges from <1 to >200 ind m^{-2} (mean ~3 ind m^{-2}; MacDonald et al. 2010). In terms of both abundance and biomass, foraminiferans and nematodes generally dominate the meiofauna (Vanreusel et al. 2000), annelids, crustaceans and bivalves dominate the macrofauna (Clough et al. 1997; Bluhm et al. 2005; MacDonald et al. 2010), and echinoderms dominate the megafauna (MacDonald et al. 2010), much like in the adjacent shelf soft-sediment habitats (Conlan et al. 2008; Grebmeier 2012) (Fig. 10.8).

10.4.5 Effect of Climate Change on Benthic Fauna of the PAR

The PAR is undergoing significant change in climate, water-mass properties as well as sea-ice dynamics (Comiso 2003; Barber et al. 2008; Grebmeier et al. 2012). The environmental changes are very likely impacting the benthic fauna but due to limited accessibility to many regions of the PAR and lack of long-term monitoring programs, the actual documented shifts within benthic communities that are attributable to climate change are minimal, with the exceptions from the northern Bering Sea (Grebmeier et al. 2006b).

Based on observed physical changes, however, some projections can be made. Longer summertime open-water periods will likely increase light availability, wind mixing, upwelling at the shelf-break, and storm erosion and resuspension (Carmack and McLaughlin 2001; Carmack and Chapman 2003; Walker et al. 2008). Highly productive regions, such as the southern and northeast Chukchi Sea (Grebmeier 2012) and Cape Bathurst (see Sect. 10.4.3) may therefore expand in duration. In addition, the productivity in canyon regions and shelf edges, such as Barrow Canyon in the Chukchi Sea, and Mackenzie Trough and Kugmallit Valley shelf-edge areas in the Canadian Beaufort, which act as conduits for upwelling, may increase (Carmack and Wassmann 2006). Repeated upwelling events combined with ice ablation indeed increased the biological production including that of the benthos on the Beaufort Sea shelf in recent years (Tremblay et al. 2011). Such changes would favor large benthic predators such as gray whales, bearded seals, benthic fish and diving seabirds, common megafauna in the Chukchi Sea and parts of the Beaufort Sea (Bluhm and Gradinger 2008; Grebmeier et al. 2010).

Increases and changes in timing of flows from the Mackenzie River and other rivers in the PAR will affect input of terrigenous nutrients, sediments, and contaminants, as well as increase turbidity, alter local ice dynamics and the influence of freshwater on water-column salinity and stratification (Carmack and Wassmann 2006; Barber et al. 2008). This in turn will affect, and already has affected, primary production, species composition, grazer and microbial diversion and the extent to which organic carbon is available to the benthos (Li et al. 2009; Agustí et al. 2010). The possibility of greater diversion of organic carbon to pelagic consumers will negatively impact benthic fauna and their predators (Grebmeier et al. 2006b). Change in snowfall over sea ice will affect transmission of photosynthetically active radiation to primary producers (Mundy et al. 2007) and hence further reduce carbon available to the benthos. Increased storminess will alter bottom disturbance, upwelling, sediment transport, circulation patterns and bottom transport (Barber et al. 2008). In the marginal ice zone there can be increased productivity. As the marginal ice zone changes in space and time the benthos will be correspondingly impacted (Carmack and Macdonald 2002; Sejr et al. 2007).

Shifting water mass fronts will affect benthic dispersal, as well as the immigration and movements of benthic predators (Carmack and McLaughlin 2001). Altered benthic community composition, interactions, metabolic rates and life histories will change carbon cycling dynamics with attendant alteration of nutrient regeneration and CO_2 sequestering (Renaud et al. 2008; Grebmeier 2012; Mathis et al. 2014, this volume) and cascading effects on the pelagic ecosystem and large predators (e.g. Coyle et al. 2007). Such benthic changes may similarly modify contaminant pathways through the benthos to higher trophic levels (Macdonald et al. 2005). Ocean warming and greater human access to open water will favor immigration and introduction of non-indigenous species, some of which may become resident and modify benthic composition and processes (Cheung et al. 2009).

Out-competed shelf species may not be able to retreat to the slope and deep sea (Renaud et al. 2008). Commercial harvesting of mammals and fish may intensify, affecting the benthos through trawling disturbance and by-catch (Renaud et al. 2008).

Less sea ice, particularly deep-keeled multi-year ice will result in a contraction of the ice scour zone. Moderate ice scour can have a positive effect on benthic diversity (Conlan and Kvitek 2005). Thinning and reduction of the extent of landfast ice will have multiple effects on coastal erosion, the amount of ice scour, the extent of the flaw-lead system and stamukhi zone, the timing of ice break-up, catchment of freshwater, amount of primary production, use by apex predators and use for transportation and industrial activity, all of which are relevant to the benthos (Dumas et al. 2005).

Changes in seawater chemistry associated with uptake of anthropogenic CO_2 will affect the depth- and latitudinal range of such calcifying benthic fauna as foraminiferans, sponges, bryozoans, molluscs, spirorbids, barnacles, crabs, sea urchins, brittle stars, crinoids and brachiopods (Orr et al. 2005). Magnesium-calcite producers will respond first (Andersson et al. 2008). Synergistic effects of other climate-induced changes with acidification can accelerate and magnify effects on the benthos (Wood et al. 2011; Mathis et al. 2014, this volume).

In summary, the rapid sea ice and environmental changes being observed in the PAR, coincident with increased human activities, are likely to modify benthic faunal communities and associated carbon cycling, with both short- and long-term impacts to the ecosystem. There are still substantial knowledge gaps regarding the benthic fauna of the PAR, in particular at depths >3,000 m and in the epibenthic megafauna. Since a recent pilot study of the bathymetrically complex Chukchi Borderlands region including a pock mark, discovered a range of different faunal communities, locally high megafaunal abundance and species richness but a lack of seep fauna (MacDonald et al. 2010), this area also warrants further study. Finally, a deep-sea observing system, such as the HAUSGARTEN located in the eastern Fram Strait (Soltwedel et al. 2005), would be able to capture changes associated with the dramatic retreat of the perennial sea ice over the Chukchi Sea slope and adjacent Canada Basin.

10.5 Sea Ice Associated Diversity and Production in the PAR

10.5.1 Introduction

Sea ice occurs in the entire PAR with different regional characteristics. Coastal fast ice forms a relatively narrow band (a few km compared to hundreds of km in the Laptev Sea) along the entire Alaskan coastline from the Bering Sea to the Chukchi and Beaufort seas (Mahoney et al. 2007; Gradinger et al. 2009). Pack ice covers most of the PAR in winter, with first-year ice dominating the Bering Sea and increasing contributions of multi-year sea ice (MYI) in the Chukchi and Beaufort seas with latitude (Grebmeier et al. 2012). The relative contribution of MYI has decreased over the past decades (Maslanik et al. 2007).

The Bering Sea experiences the largest seasonal sea ice advance and retreat of any arctic and subarctic region, averaging about 1,700 km (Walsh and Johnson

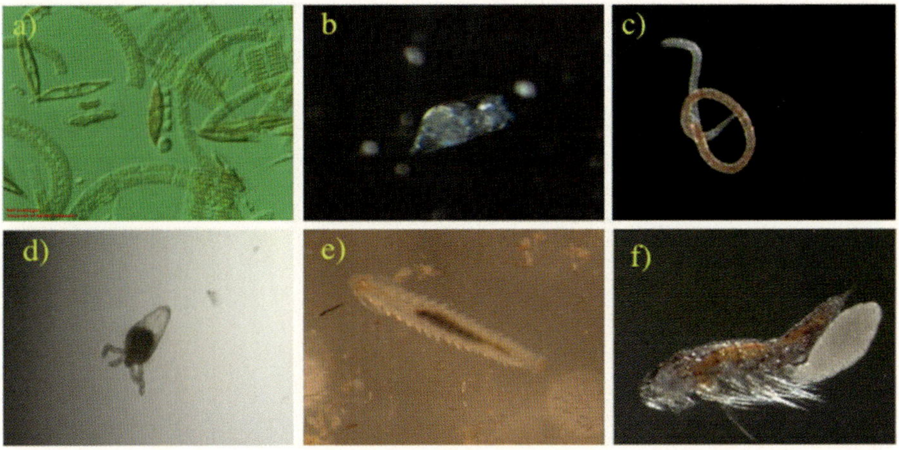

Fig. 10.9 Examples of sea ice biota: (**a**) Diatoms, (**b**) Rotifera, (**c**) Nematoda, (**d**) Hydroid, *Sympagohydra tuuli*, (**e**) Polychaeta, (**f**) Copepoda. Images from (**a**) Bering Sea pack ice, (**b–f**) Alaskan fast ice

1979) with large inter-annual variability (Niebauer 1983; Minobe 2002; Drinkwater et al. 2009). On average about 37 % of the Bering Sea is seasonally ice covered, mostly in the eastern shelf regions (Niebauer et al. 1999). In the Chukchi and Beaufort seas, ice coverage is 100 % for most of the winter months, except for some coastal flaw leads. Record low ice concentrations in the last decade during the September ice minimum have left the Beaufort and Chukchi seas ice-free well north of 75 °N (e.g. Stroeve et al. 2007).

Sea ice biota (Fig. 10.9) is present in any ice-covered ocean, not only in the Arctic and Antarctic, but also for example in the Baltic Sea and the Sea of Okhotsk. In any of these areas, strong ties exist between physical and chemical processes and sea ice characteristics and biological responses. The yearly cycle of physical processes, especially of light level and water mass stratification, are strongly related to seasonal sea ice extent (e.g. Sakshaug 2004). On a local scale ice structural features including snow depth, ice thickness and sediment load within the ice explain much of the tremendous small-scale patchiness of biological properties within sea ice (Gradinger et al. 2009). The information available for the following summary of ice-associated biota for the PAR is most limited for the Bering Sea.

10.5.2 Primary Producers: Diversity, Abundance and Activity

In the recent first pan-arctic assessment of ice algal diversity, species richness was determined to be lowest in the Bering Sea compared to the Chukchi and Beaufort seas as well as the Russian, Canadian and Scandinavian arctic regions (Poulin et al. 2011; Fig. 10.10). This finding, however, should not be interpreted as a true biological pattern but rather as a reflection of the poor observational record in the Bering Sea.

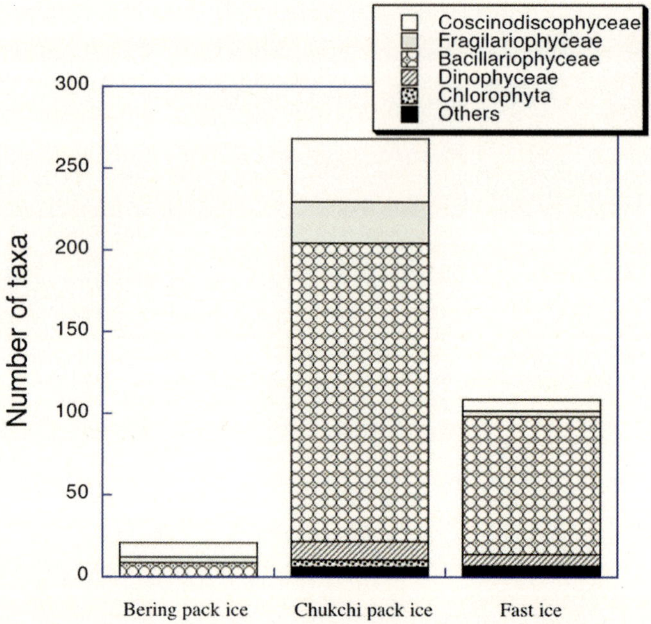

Fig. 10.10 Number of sea ice algal taxa documented from pack ice of the Bering and Chukchi/Beaufort seas and from Alaskan fast ice (Data from Poulin et al. 2011)

Only 20 ice algal species have been recorded from there so far, compared to 268 from the pack ice and 110 from the fast ice areas of the combined Chukchi and Beaufort seas. Based on Poulin et al.'s data (2011) pennate diatoms (Bacillariophycae) dominate in the entire PAR, followed by Coscinodiscophyceae and Fragilariophyceae. The total number of other taxa (including, e.g., Dinophyceae, Chlorophyta, and Cryptophyceae) varies between 0 (Bering Sea) and 20 (Chukchi/Beaufort seas). Many of the taxa reported in previous studies were not identified to species level and information on smaller taxa that are not readily identified by light microscopy (e.g. smaller flagellates in the pico- and nano-size classes) is completely absent from the PAR. See also Sect. 10.2.2 for more information regarding sea ice taxa.

Only a few studies have assessed the seasonality and magnitude of ice algal biomass and production in the PAR. On the PAR shelves with their regionally high phytoplankton productivity, ice algal production is only a very minor relative contribution to the total annual formation of new organic matter with absolute values typically less than 10 g C m^{-2} $year^{-1}$ for Bering, Chukchi and Beaufort sea sea ice (Gradinger 2009, Gradinger unpublished data) while annual phytoplankton productivity can exceed 400 g C m^{-2} $year^{-1}$ (Hill and Cota 2005). Although low in absolute values, ice algae can contribute as much as 50 % of the total primary production in the central Arctic because of low phytoplankton productivity (Gosselin et al. 1997). Regionally, maximum algal biomass accumulation in the sea ice is highly variable due to gradients in surface-water nutrient concentrations (Gradinger 2009).

In any region, the highest ice algal biomass is always found at the bottom of the first-year sea ice in a thin layer of only a few centimeters in vertical extent. In the PAR, these maximum values decrease primarily with latitude as a function of nutrient availability. The highest values exceed 600 μg Chl *a* L^{-1} in the Bering Sea (Gradinger et al. unpublished data); maximum values are substantially lower (less than 10 μg Chl *a* L^{1}) in the oligotrophic Canada Basin (Gradinger et al. 2005, 2010). Within coastal fast ice, Chl *a* concentrations can be similar to the highest pack-ice values given above (Gradinger et al. 2009).

Work done in fast ice provides the best available seasonal time series (e.g. Horner and Schrader 1982; Gradinger et al. 2009). Ice algal growth starts as soon as light intensity is sufficient to sustain algal growth, typically at photosynthetically active radiation levels below 1 μE m^{-2} s^{-1} (Mock and Gradinger 1999). Thick snow cover and high sediment loads both reduce light intensities and can delay or completely prevent the development of ice algal layers at the bottom of the fast ice (Gradinger et al. 2009). For the Bering Sea, the ice algal growth period was estimated at about 3 months (March to May) based on samples collected northeast of St. Lawrence Island (Alexander and Chapman 1981). At stations located further south between St. Lawrence Island and the Aleutians (McRoy and Goering 1974), ice algal pigment concentration (2.97 mg Chl *a* m^{-2}) was similar to the concomitant phytoplankton biomass integrated over a 30 m thick water column. In the Chukchi and Beaufort seas as well as other arctic sites, maximum biomass and productivity values of ice algae may exceed integrated values for the phytoplankton (Tremblay et al. 2006; Gradinger 2009).

Regarding the fate of ice algal production, direct ingestion of ice algae by either under-ice amphipods (Carey 1985; Werner and Gradinger 2002) or migrating zooplankters (Runge et al. 1991) is well documented in the arctic seas outside the PAR. Grazing by meiofauna inside the sea ice brine channel system does not appear to reduce the biomass by more than a few percent (Gradinger 1999; Nozais et al. 2001; Michel et al. 2006). Most of the ice algal biomass is released as one intense pulse when the ice melts, and can either seed the water column or sink to the sea floor (e.g., Goering and Iverson 1981; Gogorev and Okolodkov 1996; Michel et al. 2006). The importance of ice algal production as an early food pulse to benthic systems is widely acknowledged for subarctic and arctic seas including the PAR (Carey 1987; Michel et al. 2006). In addition, the high concentration of polyunsaturated fatty acids in ice algal material gives it high nutritional value (e.g. Sun et al. 2009).

10.5.3 Sea Ice Meiofauna Abundance and Diversity

Similar to ice algae, sea ice meiofauna (those animals living within the ice) occur in highest concentration within the bottom few centimeters of the sea ice (Bluhm et al. 2010). Their overall abundance patterns follow the above described nutrient – influenced ice algal biomass gradients; faunal abundance is highest on the nutrient-rich shelves and lowest in the oligotrophic gyre regions (Gradinger et al. 2010; Fig. 10.11). The highest abundance of sea ice meiofauna has been described within the arctic fast ice. All major taxa typically observed in arctic sea ice such as

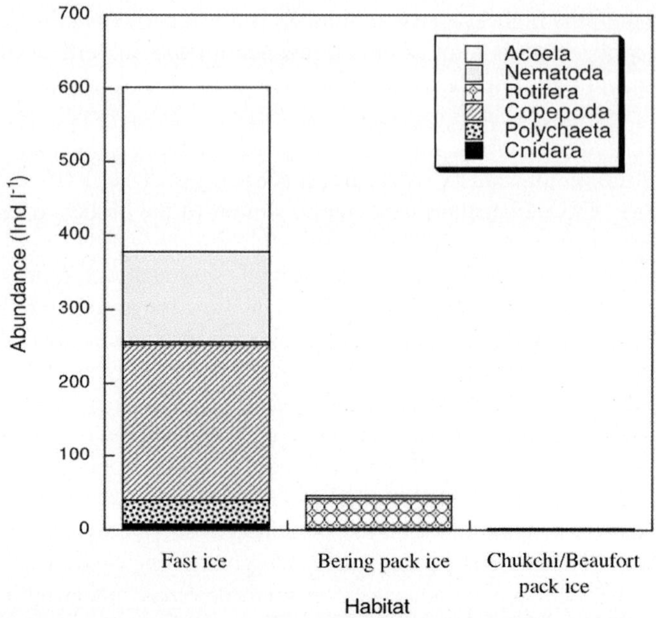

Fig. 10.11 Abundance (individuals L^{-1}) of sea ice meiofauna in the bottom 2–5 cm of sea ice from Alaskan coastal fast ice near Barrow (March 2002, data from Gradinger et al. 2009, mean of 4 replicates from 1 site), from pack ice in the Bering Sea (mean across 15 stations in spring 2008, Gradinger et al. unpublished data), and from offshore pack ice in the Chukchi/Beaufort Sea (mean across 13 number of stations in summer 2005, data from Gradinger et al. 2010)

nematodes, harpacticoid copepods, rotifers, and acoel flatworms (Bluhm et al. 2010) have also been observed in the sea ice of the PAR region. However, some ice-endemic meiofauna taxa (e.g. Nematoda) linked to the multi-year sea ice cover were absent from the Bering Sea (Gradinger et al. unpublished data). The number of known arctic meiofauna taxa is estimated at 30–50 (Bluhm et al. 2010) but on a species level, their pan-arctic distribution or that in the PAR area is not well known. Many of the ice associated taxa are herbivores, however carnivory was observed in the hydroid *Sympagohydra tuuli* (Sievert et al. 2009).

A unique fauna lives at the ice – water interface. North of Bering Strait, this under-ice fauna consists mainly of gammarid amphipods. Ice amphipod abundance on the Chukchi and Beaufort shelves remains unknown, but abundances over the slope and within Canada Basin are typically less than 10 individuals m^{-2} of sea ice (Gradinger et al. 2010). The majority of these amphipod species are endemic to arctic sea ice (Carey 1985) and include the gammarids *Apherusa glacialis*, *Onisimus glacialis, O. nanseni* and *Gammarus wilkitzkii*. *Apherusa glacialis* is the most herbivorous of those four species, while *G. wilkitzkii* feeds nearly entirely on ice-associated zooplankton or other amphipods (e.g. Poltermann 2001). In nearshore areas, the otherwise benthic gammarid *O. litoralis* exploits ice algal production in the spring period (Gradinger and Bluhm 2010). Although non-quantitative, baited-traps deployed under nearshore fast ice suggest this species is very abundant. Ice amphipods were, however, absent during 2008–2010 expeditions to the Bering Sea south of St. Lawrence Island (Gradinger et al. unpublished data). Instead, we documented patchy occurrence of euphausiids (*Thysanoessa raschii*) under the sea ice. We assume that those krill feed on the rich ice algal biomass similar to the behavior observed for Antarctic krill, *Euphausia superba* (Nicol 2006). The absence of under-ice amphipods in the Bering Sea might be attributed to the complete ice melt each year, and consequently the lack of a suitable summer substrate. The recently increased summer ice melt in the Chukchi and Beaufort seas could lead to a similar loss of ice-endemic biota in these areas. During a recent summer expedition to the Canada Basin, we observed higher abundances of under-ice amphipods and certain sea ice meiofauna taxa at pressure ridges than at the surrounding level ice. We suggested that amphipods might use those ridges as summer habitats to avoid the low-salinity surface layers under level pack ice (Gradinger et al. 2010). Pressure ridges might maintain seed populations of ice-endemic fauna during periods of enhanced summer melt; this observation raises questions with regard to the ecological implications of the observed decline in the percentage of multi-year ice in arctic waters.

10.5.4 Effects of Climate Change

Ongoing and future climate-related changes in environmental settings will alter not only biodiversity, but also functional relationships. The decreased proportion of MYI over the last decade (Maslanik et al. 2007) as well as recent decreases in total summer sea ice areal extent (Stroeve et al. 2007) puts ice-endemic species at the risk of extinction. Future climate-change scenarios predicting increased water temperatures and earlier ice melt suggest that primary productivity of phytoplankton might increase (Bluhm and Gradinger 2008; Popova et al. 2012) while production in sea ice will decrease, at least in some areas of the Arctic. Also, increased carbon cycling within the water column is anticipated with attendant reduced export of organic carbon to the sea floor (Carroll and Carroll 2003; Bluhm and Gradinger 2008). This could change growth and reproductive success of benthic biota (also see Sect. 10.4 this volume), particularly because ice algae are rich in polyunsaturated fatty acids that are of high food value for benthic organisms (e.g. Sun et al. 2009).

10.6 Biodiversity and Biogeography of Metazoan Zooplankton of the PAR

10.6.1 Introduction

Metazoan zooplankton are the main link between primary producers and higher trophic level taxa such as fishes, seabirds, and marine mammals and play a fundamental role in nutrient cycling (Buitenhuis et al. 2006) and carbon export to the benthos. The ecology of the seasonally and perennially ice covered seas of the PAR is less well understood than that of more temperate latitudes. This is due to the general remoteness of the region and the difficulties associated with access through ice-covered waters. The first records of zooplankton in the region date back to the late 1800s (e.g. Murdoch 1885). More comprehensive studies began to emerge in the 1930s (e.g., Johnson 1934; Stepanova 1937a, b; Bogorov 1939; Jaschnov 1940), followed intermittently by other studies in the region (e.g. Grainger 1965; Springer et al. 1989; Thibault et al. 1999; Ashjian et al. 2003; Hopky et al. 1994a; and refer to Table 1 for a comprehensive summary).

In the last decade, and particularly in 2007–2008 during the International Polar Year, consistent and comprehensive studies to understand the composition, abundance and distribution of the metazoan zooplankton fauna of the PAR have been carried out, some of which are still ongoing. Such recent efforts include: the Shelf Basin Interaction program on the Beaufort and Chukchi shelves (e.g. Plourde et al. 2005; Lane et al. 2008), the Hidden Ocean Expeditions (e.g. Hopcroft et al. 2005; Raskoff et al. 2005, 2010; Kosobokova and Hopcroft 2010), the Canadian Arctic Shelf Exchange Study (e.g. Darnis et al. 2008), Canada's Three Oceans program (Carmack et al. 2010), the Circumpolar Flaw Lead system study (Barber et al. 2010), the Beaufort Gyre Observing System/Joint Ocean Ice Study (Proshutinsky et al. 2009). Additional sampling activities have been conducted by Hokkaido University (2007–2008) in the Chukchi Sea, the Japan Agency for Marine-Earth Science Technology in the Chukchi and Bering seas, and from the Chinese Polar Institute's Icebreaker *Xue Long* (2003, 2008) in the Chukchi Sea and Canada Basin. Most recently, the Distributed Biological Observatory (Grebmeier et al. 2010, 2012) and the Circumpolar Biodiversity Monitoring Program (Gill et al. 2011) initiatives hold promise to promote international cooperative research with consistent sampling methodology, wide geographical coverage, and greater temporal sampling, all of which will accelerate the understanding of zooplankton biogeography and biodiversity in the PAR.

The deepest water in the PAR is in the Canada Basin (depth of 3,600–3,800 m) which is surrounded by the shallow shelf waters (<100 m) of the Chukchi Sea to the west and the Beaufort Sea to the south and east. The southernmost waters of the PAR, in the northern Bering Sea, are less that 100 m in depth as well. It is in these shallow shelf waters where the majority of the primary production occurs, and the majority of zooplankton biomass is thought to be found (Springer et al. 1989). Localized production enhancement can occur, for example in the Cape Bathurst Polynya of Amundsen Gulf (Arrigo and van Dijken 2004; Darnis et al. 2008), and there is high primary production associated with upwelling at the Mackenzie and Kugmallit canyons (Carmack et al. 2004; Sampei et al. 2011). Enhanced polynya and shelf production is associated with elevated zooplankton biomass, for example, Darnis et al. (2008) recorded biomass levels twice as high in the Cape Bathurst polynya than on the shelf slope. Other processes may also underlie local zooplankton accumulation. For example, transport and upwelling are both thought to be responsible for the accumulation of zooplankton near Barrow Canyon (Berline et al. 2008; Ashjian et al. 2010).

Table 10.5 Zooplankton species diversity in the PAR

Phylum	Group	# species
Annelida	Polychaete	5
Arthropoda	Amphipod	18
	Copepod	117
	Decapod	2
	Euphausiid	6
	Isopod	2
	Mysid	4
	Ostracod	7
Chaetognatha	Chaetognath	5
Cnidaria	Hydromedusa	33
	Siphonophore	6
	Scyphozoan	5
Ctenophora	Ctenophore	9
Mollusca	Cephalopod	2
	Pteropod	2
Chordata	Larvacean	5

Data from the following data sets were compiled to determine the numbers of species of all of the groups of zooplankton found in the PAR. Johnson 1934, 1956, 1963; Brodsky and Nikitin 1955; Virketis 1957, 1959; Pautzke 1979; McConnell 1977; Turco 1992a, b, Hopky et al. (1994a, b), Pinchuk 1993, 2006; Ashjian et al. 2003; Hopcroft et al. 2005; Darnis et al. 2008; Kosobokova and Hopcroft 2010; Walkusz et al. 2008, 2010; Kosobokova et al. 2011, Nelson (unpublished), Hopcroft (unpublished)

The Lomonosov Ridge with a sill depth of approximately 1,500 m, delimits the Eurasian and Amerasian basins of the Arctic Ocean. Other notable major geophysical features with potential to influence zooplankton biodiversity and biogeography are the Chukchi Plateau (depth 600–1,000 m) which is a northward extension of the Chukchi Shelf and the Northwind Ridge which rises up to approximately 500 m below the surface and runs roughly north-south along the western margin of the Canada Basin.

10.6.2 Species Diversity

The metazoan zooplankton of the PAR is represented by 230 known species from 7 phyla (Table 10.5). Copepods, of which 117 species have been identified in the PAR, are dominant throughout the region both in terms of biomass and abundance. In terms of taxonomic diversity no other group seriously rivals the copepods with the next most speciose orders being the hydrozoans with 32 species and the amphipods with 18 species. The biomass of gelatinous zooplankters, such as the larvacean *Oikopleura vanhoeffeni* have been observed at times to approach that of the copepods (Springer et al. 1989; Gorsky et al. 2005; Hopcroft et al. 2010) and there is general consensus that gelatinous zooplankters are under sampled and their ecological role is largely underappreciated (see Purcell et al. 2010; Raskoff et al. 2010). The biomass of fast swimming taxa, such as the amphipod *Themisto libellula,* may also be underestimated due to net avoidance (Dalpadado et al. 2001). Among the copepods the large-bodied calanoids, *Calanus glacialis*, *C. hyperboreus* and *Metridia longa*, are widespread and typically dominate zooplankton biomass in deeper waters. While copepod biomass is dominated by large bodied *Calanus* species, numerical dominance belongs to the small copepods such as *Microcalanus pygmaeus, Oithona similis, Triconia borealis*, and – on the shelves – *Pseudocalanus* spp. The large-bodied zooplankters, particularly the lipid-rich *Calanus* species, are important as a prey item for fishes, sea birds and baleen whales (Bradstreet et al. 1986; Lowry 1993).

10.6.3 Zooplankton Advection: Expatriate Analysis

The pattern and intensity of advection of Pacific and Atlantic zooplankters into the PAR must be considered when attempting to understand the ecology of this region of the Arctic. In addition, climate-forced change raises the distinct possibility of range extensions of southern zooplankters into the PAR with attendant ecological effects (see Occhipinti-Ambrogi 2007). To date there has been no documented contemporary reproductive establishment of an Atlantic or Pacific expatriate in PAR waters. The likelihood of a successful invasion of the PAR by a southern expatriate depends on both physical and biological factors any or all of which currently act to limit range extensions. The establishment of an Atlantic expatriate in the PAR may be less likely than a Pacific expatriate given the large expanse of arctic habitat that must be traversed prior to arriving in the PAR from the North Atlantic (see Hirche and Kosobokova 2007).

In the mid-1970s it was suggested that the Canada Basin hosts a unique deep-water zooplanktonic fauna of arctic endemics (Brodsky and Pavshtiks 1976). However, recent research using data collected throughout the entire water column with identical techniques and consistent taxonomic analysis, found no evidence of a zoogeographical barrier between the zooplankton fauna of the Eurasian and Canadian basins (Kosobokova et al. 2011). All deep-water and arctic endemic species are found on both sides of the Lomonosov Ridge suggesting effective exchange across the ridge. Notable differences in the species composition in the upper and midwater layers are related to the occurrence of Atlantic and Pacific expatriates advected into the Eurasian or Amerasian Arctic. Atlantic expatriates such as *Calanus finmarchicus*, are not observed far into the waters of the PAR (Kosobokova and Hopcroft 2010; Kosobokova et al. 2011), nor are Pacific expatriates observed on the Eurasian side of the Lomonosov Ridge (Kosobokova et al. 2011).

In contrast to the lack of Atlantic expatriates entering the PAR, huge numbers of Pacific zooplankters are advected directly into the PAR every year (Springer et al. 1989). Pacific zooplankters are commonly observed in the Chukchi Sea (see Hopcroft et al. 2010) and occasionally in the Canada Basin and Beaufort Sea (Grainger 1965; Hopcroft et al. 2005; Llinas et al. 2009); prominent Pacific expatriates include the copepods: *Eucalanus bungii, Neocalanus cristatus, Metridia pacifica, Pseudocalanus newmani, Neocalanus plumchrus, Neocalanus flemingeri, and Pseudocalanus mimus.* Because the bulk of the Pacific zooplankton currently does not make it out of the Chukchi Sea, their direct ecological impact is limited mainly to this region. However recent work suggests that euphausiids taken by baleen whales off Point Barrow may have been advected to the area from the Bering Sea (Berline et al. 2008; Ashjian et al. 2010). Any of the factors largely limiting Pacific zooplankton to the Chukchi Sea could change in the near future due to climate change; given the huge numbers of Pacific zooplankters advected annually into arctic waters, there is great potential for rapid and significant northward range expansions with associated potential for ecological transformation.

For many decades zooplankton sampling in the PAR has been sporadic and non-standardized, thus determination of trends in *abundance* of Pacific expatriate zooplankton in the PAR is problematic. It is possible however, to summarize historical records of presence and absence to get an idea of the areas of penetration for particular species and identify areas which may be future "hot spots" subject to invasion and establishment of Pacific expatriate zooplankters, with the caveat that sampling effort varies greatly across this domain. The presence of the Pacific expatriate species *Metridia pacifica, Neocalanus cristatus, Neocalanus plumchrus/flemingeri, Eucalanus bungii* in arctic waters is plotted in Fig. 10.12. All of these species have been regularly observed in the Chukchi Sea, and less frequently in the Beaufort Sea. *Neocalanus* species appear to be the least successful at penetrating into the basins and are restricted to shelf waters. In contrast, both *M. pacifica* and *E. bungii* have been reported far into the basin. Based on current speed in the Chukchi Sea (see Woodgate et al. 2005) and Canada Basin (see Timmermans et al. 2008) Pacific zooplankters found far into the Beaufort Sea or Canada Basin must have survived at least one arctic winter. Analysis of the population genetics of *Calanus glacialis* from the North Bering Sea with the 16S ribosomal RNA gene has been used to trace zooplankton advection from the Pacific Ocean (Nelson et al. 2009). Recent analysis of the genetics shows three fronts of penetration that roughly correspond to the three main water flow pathways through the Chukchi Sea (R.J. Nelson unpublished data). Future analysis of both the taxonomic composition and genetics of zooplankton collected in arctic waters will facilitate monitoring of the penetration and establishment of Pacific zooplankters in arctic waters.

10.6.4 Horizontal Zooplankton Community Structure

One of the earliest descriptions of the zooplankton biogeography of arctic regions of the PAR is that of Grainger (1965) who described three groups in Canadian waters of the PAR. Grainger's (1965) groups (adjusted to currently accepted nomenclature) were defined as (1): Widespread arctic – species found in both deep and

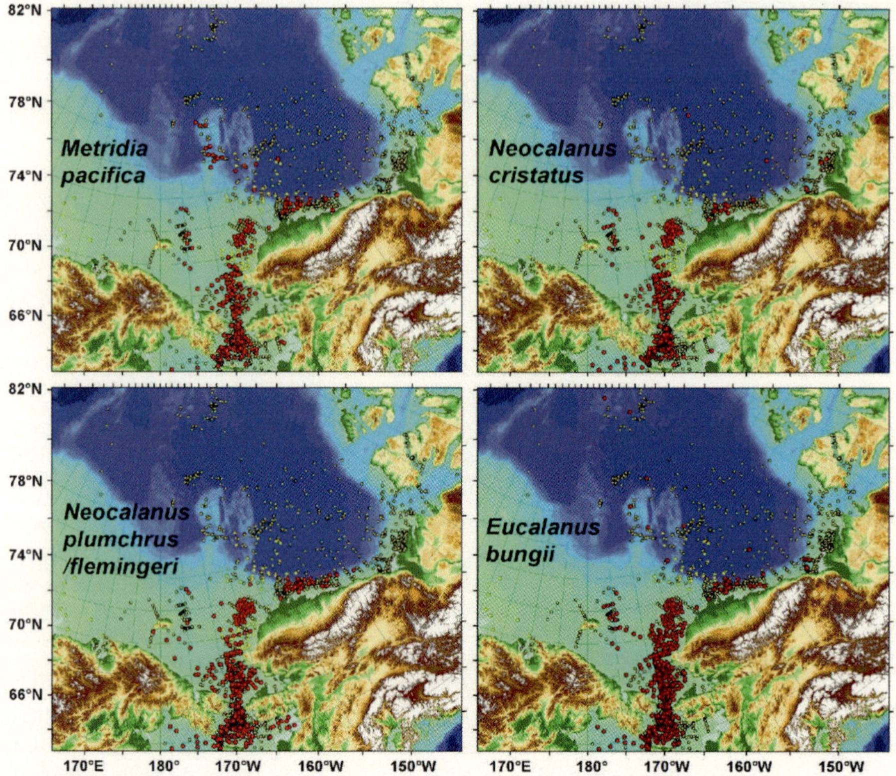

Fig. 10.12 Historical patterns of advection of Pacific zooplankters into the Arctic Ocean. The presence of *Metridia pacifica, Neocalanus cristatus*, *Neocalanus plumchrus/flemingeri* (studies carried out prior to 1989 did not consistently distinguish between these two congeners), and *Eucalanus bungii* in the Arctic Ocean is indicated by the *red symbols* placed at the location where they were detected (Note some locations are represented by only a single individual). *Yellow symbols* represent sites where zooplankton community composition data was collected (Data from the same sources as Table 10.5)

shelf waters. This group may be considered the archetypical PAR zooplankton assemblage and includes: the cnidarians *Aglantha digitale*, and *Aeginopsis laurentii*, the copepods *Calanus hyperboreus*, *C. glacialis*, *Pseudocalanus* spp., *Microcalanus pygmaeus, Paraeuchaeta glacialis, Metridia longa, Oithona similis*, and *Triconia* (= *Oncaea*) *borealis* and the larvaceans *Oikopleura vanhoeffeni* and *Fritillaria borealis*. (2): Offshore – which includes the ostracod *Boroecia maxima*, the copepods *Spinocalanus antarcticus*, *Gaetanus brevispinus*, *G. tenuispinus*, and *Heterorhabdus norvegicus*, and the chaetognath *Eukrohnia hamata*. (3): Coastal – including the cnidarians *Sarsia princeps, Euphysa flammea, Halitholus cirratus, Eumedusa birulai, Ptychogena lactea*, and *Obelia* spp., the copepods *Eurytemora herdmani* and *Limnocalanus macrurus, Acartia clausi and A. longiremis*. These groupings are still generally relevant with regard to present day zooplankton distributions in areas of the PAR not strongly influenced by the presence of Pacific expatriates.

More powerful detailed multivariate analyses are emerging with associated refinement to the general biogeographical distribution put forth by Grainger (1965). Darnis et al. (2008) found evidence for three pre-winter assemblages (shelf, polynya and slope) in the southeastern Beaufort Sea. Similarly Lane et al. (2008) discerned four zooplankton communities ranging from the inner shelf of the Chukchi Sea into deep water offshore. Working in the southern and western Chukchi Sea, Hopcroft et al. (2010) indentified six major assemblages of zooplankton, and each was closely tied to physical properties of the water from which they were collected.

Consistent with the result of Darnis et al. (2008) and Lane et al. (2008), more recent surveys provide further support for a consistent separation between the epipelagic zooplankton communities of the Canada Basin and shelf of the Beaufort Sea. This separation is largely driven by differences in the abundance of species that occur in both regions. In the basin, the cyclopoid copepod *Oithona similis* dominates, and the calanoid copepod *Microcalanus pygmaeus* and pteropod *Limacina helicina* are also abundant. Abundance of the copepod genus *Pseudocalanus* is low in the basin. In shelf waters, however, *Pseudocalanus* spp. makes up a much larger proportion of zooplankton abundance while the proportional contributions of *O. similis*, *M. pygmaeus* and *L. helicina* are lower than in the Basin. The large calanoid copepods *Calanus glacialis* and *C. hyperboreus* make a relatively small contribution to total abundance but taken together tend to dominate zooplankton biomass, contributing up to 70 % to total biomass (Hopcroft et al. 2005; Darnis et al. 2008; Kosobokova and Hopcroft 2010) both on the shelf and in the basin. Occasionally shelf community zooplankters are found in the basin due to offshore transport in eddies that originate along the shelf break (Carmack and Macdonald 2002; Spall et al. 2008; Llinas et al. 2009). Recently, Kosobokova et al. (2011) showed the generality of such neritic and oceanic species domains throughout the Arctic. Besides such large scale community structuring, zooplankton communities also have the potential to vary across much smaller spatial scales. For example in the vicinity of the Mackenzie River plume three distinct zooplankton communities can be detected in response to decreasing riverine influence (Walkusz et al. 2010).

10.6.5 Vertical Distribution of Zooplankton in the Deep Waters of the PAR

Studies of the vertical distribution of zooplankton in the deep waters of the PAR have been carried out from drifting ice platforms (Johnson 1963; Grainger 1965; Harding 1966; Scott 1968; Kosobokova 1980, 1982; Geynrikh et al. 1983) and ice breakers (Thibault et al. 1999; Kosobokova and Hirche 2000; Kosobokova and Hopcroft 2010; Kosobokova et al. 2011). Of the approximately 230 species of metazoan zooplankton inhabiting the PAR (see Table 10.5) more than half of them are known to occur in the deep basin (Kosobokova et al. 2011). This includes representatives of seven metazoan taxa: Cnidaria, Ctenophora, Mollusca, Annelida, Crustacea, Chaetognatha, and Larvacea, all predominantly oceanic species. Crustaceans strongly dominate in terms of the species number (70 %), and copepods are the most diverse group amongst them. The composition of epi- and mesopelagic zooplankton is relatively well characterized as compared to bathypelagic species. Of the 111 species identified in the Canada Basin by Kosobokova and Hopcroft (2010), 25 % were either observed for the first time or previously undescribed; the majority of these novel observations were made from samples collected from below 500 m.

As noted previously, the zooplankton community composition of the arctic shelves is structured by cross-shelf gradients in physical conditions and can be complex (Lane et al. 2008; Darnis et al. 2008; Hopcroft et al. 2010; Walkusz et al. 2010). There is a distinct difference between communities found in the shelf water and in the deep basin (see Sect. 10.6.3 this volume). Across the deep water however, horizontal structure is low and communities are primarily structured by depth (Kosobokova and Hopcroft 2010; Raskoff et al. 2010). Zooplankton abundance and biomass rapidly drops from maximal values in the epipelagic layer then declines slowly in the meso- and bathypelagic layers (e.g. Kosobokova and Hirche 2000; Ashjian et al. 2003; Kosobokova and Hopcroft 2010). Although the zooplankton is concentrated in the epipelagic layer more than half of the total metazoan zooplankton biomass is found in deeper water (Kosobokova and Hopcroft 2010). Species diversity increases from the surface to a maximum occurring within the Atlantic water layer at 500–1,000 m, followed by a decrease with further depth (Kosobokova et al. 2011).

Distinct zooplankton assemblages are observed with water depth. One of the first descriptions of this structure is that of Harding (1966). In the summer of 2005 Kosobokova and Hopcroft (2010) observed that the 0–50 m stratum was characterized by a high proportion of copepods and chaetognaths; with increasing depth, down to 1,000 m, the copepods decline to be replaced by a mix of cnidarians, ctenophores, ostracods, amphipods and decapods; below 1,000 m the proportion of copepods increases as do jellies, while chaetognaths decrease; in the arctic bottom water (2,000 m and deeper) copepods again decrease to be replaced by cnidarians, ctenophores, ostracods and amphipods. Within the entire water column copepods represent ~75 % of the total biomass, chaetognaths represent 13 % on average (ca. 50 % of non-copepod biomass), followed by cnidarians and ctenophores (4.6 %), ostracods (3.6 %), and other groups – notably pteropods, amphipods and larvaceans – contributed ~2 % or less (Kosobokova and Hopcroft 2010). These general patterns are largely conserved between basins (Kosobokova et al. 2011).

10.6.6 Long-Term Change

The way in which metazoan zooplankton respond to climate-forced change has the potential to effect a range of ecosystem processes. For example climate-induced reduction of ice cover duration on arctic shelves has been hypothesized to favor the population growth of the predominant large calanoid copepods (Ringuette et al. 2002) and *Pseudocalanus* (Darnis et al. 2008). This may improve the early survival and population levels of larval Arctic Cod, (*Boreogadus saida*) which prey primarily on the naupliar stages of *Pseudocalanus* (Michaud et al. 1996).

There is limited indication as yet on how gelatinous zooplankton may respond to changes in sea ice cover, although larvaceans will likely increase in importance (Gorsky et al. 2005; Deibel and Daly 2007). Other predatory groups may also increase (Mills 1995, 2001). Increased sea ice melt and run off has reduced Ω aragonite saturation in Canada Basin surface waters (Yamamoto-Kawai et al. 2009), and in some areas of the Beaufort Sea (Chierici and Fransson 2009), and Chukchi Sea shelves (Mathis et al. 2005, 2009) to levels potentially deleterious to aragonite-shelled zooplankton fauna such as *Limacina helicina*. In the Canada Basin, freshwater accumulation is resulting in a deepening of the nutricline, and consequently, the depth of the chlorophyll maximum (McLaughlin and Carmack 2010), favoring growth of pico- over nanophytoplankton (Li et al. 2009). Picoplankton are less efficiently grazed by most mesozooplankton species, such as the large calanoid copepods *Calanus hyperboreus,* and *C. glacialis* which are of key ecological importance (Bradstreet et al. 1986; Fortier et al. 1994). A shift in phytoplankton size structure may therefore result in decreased abundance of *Calanus*, and a shift towards small particle grazers.

Planktonic communities of the Chukchi and Beaufort seas may undergo climate related changes through increased transport and/or survival rates of Pacific species. It is generally believed that current arctic temperatures are too low for the long-term survival of Pacific zooplankton species, and that once transported into the Arctic Ocean, they are unable to establish viable populations (Smith and Schnack-Schiel 1990). Subsequently they simply become sources of carbon for the resident communities. While this generalization may apply to the large-bodied Pacific species, highly adapted to the seasonal cycles of the Pacific, the smaller-bodied forms (many of which are common to both the Bering and Chukchi seas) might remain viable in arctic waters because they respond more directly to current environmental conditions (e.g. Hopcroft and Kosobokova 2010). Additionally, as smaller species typically grow faster and have shorter generation times than larger species (Hirst and Bunker 2003), we might expect them to continually seed new production to the region.

Although we have some insight into what the future holds for zooplankton of the PAR there is much to be learned. Increased research activity in the PAR, is accelerating our understanding of zooplankton community ecology and biogeography. We are now on the cusp of having sufficient data to closely examine trends in abundance and distribution. Our hope is that this knowledge, coupled with insight into the linkages between ecology, physics and geochemistry, will allow for prediction of the future state of arctic marine ecosystems during this period of rapid climate change.

10.7 Summary

The environment of the PAR is changing; these changes impact the productivity and abundance of the lower trophic taxa and increase the likelihood of northward range extensions. This chapter documents the distribution and abundance of PAR microbes, zooplankton, and benthic organisms. When possible the effects of environmental changes such as increasing temperature, freshening of the water column and changes in ice dynamics are assessed.

Microbes are the foundation of food webs in the PAR and they appear to be effected by the environmental changes. Freshening of the water column increases stratification thus reduces surface nutrient availability. Such increased stratification seems to favor the smallest types of phytoplankton. This sort of shift may presage things to occur more widely in the Arctic Ocean with attendant effects on the higher trophic level taxa. However, it is difficult to extrapolate such relationships across the entire heterogeneous PAR. There is still much to be learned about the roles of even the most prominent taxa in biogeochemical cycles and food webs; the use of molecular identification of microbes is filling some of these knowledge gaps.

Recent studies indicate that environmental changes in the PAR are impacting benthic fauna. Notable among potential benthic impact drivers is the change in seawater chemistry associated with increased uptake of CO_2 which has potential to affect the productivity and distribution of calcifying organism including many of the benthos. Changes in the species composition and productivity of benthic organisms in the PAR will affect the roles played by this diverse fauna in carbon cycling and as a prey items. Due to a lack of long-term monitoring programs, documentation of shifts in benthic communities is minimal.

Besides the direct effects of environmental change, the arctic planktonic communities of the PAR, (especially those close to the Bering Strait gateway), will likely be increasingly impacted by the transport and activity of Pacific taxa. Current arctic seawater temperatures are considered too low for the long-term survival of Pacific zooplankton. However, smaller-bodied Pacific taxa with shorter generation times and facultative life histories, may prove adaptable and persist for longer and longer in arctic waters potentially changing food web dynamics.

Recent increased research activity in the PAR is improving our understanding of lower trophic community ecology and biogeography. Notable among this research is the Distributed Biological Observatory initiative which promotes and coordinates multidisciplinary studies in the PAR to build a change detection array of environmental and biological measurements. Multidisciplinary studies of the linkages between lower tropic species ecology, physics and geochemistry are enhancing our understanding and capacity to predict the future status of PAR marine ecosystems during this period of rapid change.

Acknowledgements Production of this chapter as well as new data acquisition and analysis reported here was supported by a variety of funding sources including: Fisheries and Oceans Canada, the US National Science Foundation (incl. grants OPP 0806295 to D. Kirchman, OPP

073201 to B. and E. Sherr, OPP 0732767 to R Gradinger, B. Bluhm and Katrin Iken, and OPP-0802290 to J. Grebmeier and L. Cooper), the National Oceanographic and Atmospheric Administration Office of Ocean Exploration and Research and Arctic Research Program, the Bureau of Ocean Energy Management, the Natural Sciences and Engineering Research Council of Canada (partially as a contribution to the Canadian Healthy Ocean Network and ArcticNet), the Alfred P. Sloan Foundation through the Census of Marine Life, the Canadian Museum of Nature, Natural Resources Canada, and Russian Foundation for Basic Research under Grant 13-04-00551. We thank Curt Suttle for providing recent papers and reviews of viruses in arctic and subarctic seas.

References

Agustí S, Sejr MK, Duarte CM (2010) Impacts of climate warming on polar marine and freshwater ecosystems. Polar Biol 33:1595–1598

Aizawa C, Tanimoto M, Jordan RW (2005) Living diatom assemblages from North Pacific and Bering Sea surface waters during summer 1999. Deep-Sea Res II 52:2186–2205

Alexander V, Chapman T (1981) The role of epontic algal communities in Bering Sea ice. In: Hood DW, Calder JA (eds) The Bering Sea shelf: oceanography and resources. U.S. Department of Commerce, pp 757–761

Allen DM, Michel FA, Judge AS (1988) The permafrost regime in the Mackenzie Delta, Beaufort Sea region, N.W.T. and its significance to the reconstruction of the palaeoclimatic history. J Quat Sci 3:3–13

Alonso-Sáez L, Sánchez O, Gasol JM, Balagué V, Pedrós-Alio C (2008) Winter-to summer changes in the composition and single-cell activity of near-surface Arctic prokaryotes. Environ Microbiol 10:2444–2454

Ambrose WG Jr, Clough LM, Tilney PR, Beer L (2001) Role of echinoderms in benthic remineralization in the Chukchi Sea. Mar Biol 139:937–949

Andersson AJ, Mackenzie FT, Bates NR (2008) Life on the margin: implications of ocean acidification on Mg-calcite, high latitude and cold-water marine calcifiers. Mar Ecol Prog Ser 373:265–273

Andersson AF, Riemann L, Bertilsson S (2009) Pyrosequencing reveals contrasting seasonal dynamics of taxa within Baltic Sea bacterioplankton communities. ISME J 4:171–181

Angly FE, Felts B, Breitbart M, Salamon P, Edwards RA, Carlson C, Chan A, Haynes M, Kelley S, Liu H, Mahaffy J, Mueller J, Nulton J, Olson R, Parsons R, Rayhawk S, Suttle C, Rohwer F (2006) The marine viromes of four oceanic regions. PLoS Biol 4(11):e368. doi:10.1371/journal.pbio.0040368

Archambault P, Snelgrove PVR, Fisher JAD, Gagnon J-M, Garbary DJ, Harvey M, Kenchington EL, Lesage V, Levesque M, Lovejoy C, Mackas DL, McKindsey CW, Nelson RJ, Pepin P, Piche L, Poulin M (2010) From sea to sea: Canada's three oceans of biodiversity. PLoS One 5(8):e12182

Ardyna M, Gosselin M, Michel C, Poulin M, Tremblay J-E (2011) Environmental forcing of phytoplankton community structure and function in the Canadian High Arctic: contrasting oligotrophic and eutrophic regions. Mar Ecol Prog Ser 442:37–57

Arrigo KR, van Dijken GL (2004) Annual cycles of sea ice and phytoplankton in Cape Bathurst polynya, southeastern Beaufort Sea, Canadian Arctic. Geophys Res Lett 31:L08304. doi:10.1029/2003GL018978

Arrigo KR, Perovich DK, Pickart RS, Brown ZW, van Dijken GL, Lowry KE, Mills MM, Palmer MA, Balch WM, Bahr F, Bates NR, Benitez-Nelson C, Bowler B, Brownlee E, Ehn JK, Frey KE, Garley R, Laney SR, Lubelczyk L, Mathis J, Matsuoka A, Mitchell BG, Moore GWK, Ortega-Retuerta E, Pal S, Polashenski CM, Reynolds RA, Schieber B, Sosik HM, Stephens M, Swift JH (2012) Massive phytoplankton blooms under Arctic sea ice. Science 336:1408

Ashjian CJ, Campbell RG, Welch HE, Butler M, Van Keuren D (2003) Annual cycle in abundance, distribution, and size in relation to hydrography of important copepod species in the western Arctic Ocean. Deep-Sea Res I 50:1235–1261

Ashjian CJ, Braund SR, Campbell RG, George JC, Kruse J, Maslowski W, Moore SE, Nicolson CR, Okkonen SR, Sherr BF, Sherr EB, Spitz YH (2010) Climate variability, oceanography, bowhead whale distribution, and Iñupiat subsistence whaling near Barrow, Alaska. Arctic 63:179–194

Atkinson EG, Wacasey JW (1989) Benthic invertebrates collected from the western Canadian Arctic, 1951 to 1985. Can Data Rep Fish Aquat Sci 745:1–132

Barber DG, Lukovich JV, Keogak J, Baryluk S, Fortier L, Henry GHR (2008) The changing climate of the Arctic. Arctic 61(Suppl 1):7–26

Barber DG, Asplin MG, Gratton Y, Lukovich J, Galley RJ, Raddatz RL, Leitch D (2010) The International Polar Year (IPY) Circumpolar Flaw Lead (CFL) system study: overview and the physical system. Atmos Ocean 48:225–243

Berline L, Spitz YH, Ashjian CJ, Campbell RG, Maslowski W, Moore SE (2008) Euphausiid transport in the Western Arctic Ocean. Mar Ecol Prog Ser 360:163–178

Bessière A, Nozais C, Brugel S, Demers S, Desrosiers G (2007) Metazoan meiofauna dynamics and pelagic-benthic coupling in the Southeastern Beaufort Sea, Arctic Ocean. Polar Biol 30:1123–1135

Bluhm BA, Gradinger R (2008) Regional variability in food availability for Arctic marine mammals. Ecol Appl 18(suppl):77–96

Bluhm BA, Grebmeier JM (2011) Biodiversity: status and trends of benthic organisms. In: Arctic report card: update for 2011. http://www.arctic.noaa.gov/report11/biodiv_benthic_organisms.html

Bluhm BA, MacDonald IR, Debenham C, Iken K (2005) Macro- and megabenthic communities in the high Arctic Canada Basin: initial findings. Polar Biol 28:218–231

Bluhm BA, Iken K, Mincks-Hardy S, Sirenko BI, Holladay BA (2009) Community structure of epibenthic megafauna in the Chukchi Sea. Aquat Biol 7:269–293

Bluhm BA, Gradinger RR, Schnack-Schiel SB (2010) Sea ice meio-and macrofauna. In: Thomas D, Dieckmann G (eds) Sea ice. Wiley-Blackwell, Oxford, UK, pp 357–394

Bluhm BA, Ambrose WG Jr, Bergmann M, Clough LM, Gebruk AV, Hasemann C, Iken K, Klages M, Macdonald IR, Renaud PE, Schewe I, Soltwedel T, Wlodarska-Kowalczuk M (2011) Diversity of the Arctic deep-sea benthos. Mar Biodiv 41:87–107

Bogorov VG (1939) The characteristics of seasonal phenomena in the plankton of the Arctic seas and their significance for ice forecastings. Zool J 18(5) (in Russian)

Booth B, Horner R (1997) Microalgae on the Arctic Ocean section, 1994: species abundance and biomass. Deep-Sea Res II 44:1607–1622

Bradstreet MSW, Finley KJ, Sekerak AD, Griffiths WB, Evans CR, Fabijian MF, Stallard HE (1986) Aspects of the biology of arctic cod (*Boreogadus saida*) and its importance in arctic marine food chains. Can Tech Rep Fish Aquat Sci 1491:1–193

Breitbart M, Middelboe M, Rohwer F (2008) Marine viruses: community dynamics, diversity and impact on microbial processes. In: Kirchman DL (ed) Microbial ecology of the oceans. Wiley, Hoboken, pp 443–479

Brochier-Armanet C, Bousseau B, Gribaldo S, Forterre P (2008) Mesophilic crenarchaeota: proposal for a third archaeal phylum the Taumarchaeota. Nat Rev Micobiol 6:245–252

Brodsky KA, Nikitin MN (1955) Observational data of the scientific research drifting station of 1950–1951. Hydrobiological work. Morskoy Transp 1:404–410 (in Russian)

Brodsky KA, Pavshtiks EA (1976) Plankton of the central part of the Arctic Basin. Pol Geogr 1:143–161 (trans: from Voprosy Geographii, 101:148–157)

Broerse ATC, Tyrrell T, Young R, Poulton AJ, Merico A, Balch W, Miller PI (2003) The cause of bright waters in the Bering Sea in winter. Cont Shelf Res 23:1579–1596

Buitenhuis E, Le Quere C, Aumont O, Beaugrand G, Bunker A, Hirst A, Ikeda T, O'Brien T, Piontovki S, Straile D (2006) Biogeochemical fluxes through mesozooplankton. Global Biogeochem Cycles 20:GB2003

Bursa AS (1961) The annual oceanographic cycle at Igloolik in the Canadian Arctic. II. The phytoplankton. J Fish Res Board Can 18:563–615

Bursa AS (1963) Phytoplankton in coastal waters of the Arctic Ocean at Point Barrow, Alaska. Arctic 16:239–262

Campbell RG, Sherr EB, Ashjian CJ, Plourde S, Sherr BF, Hill VJ, Stockwell D (2009) Mesozooplankton prey preference and grazing impact in the western Arctic Ocean. Deep-Sea Res II 56:954–963

Campeau AM, Li WKW, Carmack EC, Tremblay J-E, Carmack EC, Lovejoy C (2011) Arctic Ocean microbial community structure before and after the 2007 record sea ice minimum. PLoS One 6(11):e27492

Capriulo GM, Sherr EB, Sherr BF (1991) In Trophic behaviour and related community feeding activities of heterotrophic marine protists. In: Reid PC, Turley CM, Burkill PH (eds) Protozoa and their role in marine processes. Springer, Berlin, pp 219–265

Carey AG Jr (1977) Summarization of existing literature and unpublished data on the distribution, abundance, and life histories of benthic organisms (Beaufort Sea). Outer continental shelf energy program final report, Contract No 03-5-022-68, Task order 4. Volumes I–IV

Carey AG Jr (1985) Marine ice fauna: Arctic. In: Horner RA (ed) Sea ice biota. CRC Press, Boca Raton, pp 173–190

Carey AG Jr (1987) Particle flux beneath fast ice in the shallow southwestern Beaufort Sea, Arctic Ocean. Mar Ecol Prog Ser 40:247–257

Carey AG Jr (1992) The ice fauna in the shallow southwestern Beaufort Sea, Arctic Ocean. J Mar Syst 3:225–236

Carey AG Jr, Ruff RE (1977) Ecological studies of the benthos in the Western Beaufort Sea with special reference to bivalve molluscs. In: Dunbar MJ (ed) Polar oceans. Arctic Institute of North America, Calgary, pp 505–530

Carmack E, Chapman DC (2003) Wind-driven shelf/basin exchange on an Arctic shelf: The joint roles of ice cover extent and shelf-break bathymetry. Geophys Res Lett 30:1778

Carmack EC, Macdonald RW (2002) Oceanography of the Canadian Shelf of the Beaufort Sea: a setting for marine life. Arctic 55:29–45

Carmack EC, Macdonald R (2008) Water and ice-related phenomena in the coastal region of the Beaufort Sea: some parallels between native experience and western science. Arctic 61:265–280

Carmack EC, McLaughlin F (2001) Arctic Ocean change and consequences to biodiversity: a perspective on linkage and scale. Mem Natl Inst Polar Res Spec Issue 54:365–375

Carmack EC, Wassmann P (2006) Food webs and physical–biological coupling on pan-Arctic shelves: unifying concepts and comprehensive perspectives. Prog Oceanogr 71:446–477

Carmack EC, Macdonald RW, Jasper S (2004) Phytoplankton productivity on the Canadian Shelf of the Beaufort Sea. Mar Ecol Prog Ser 277:37–50

Carmack EC, McLaughlin FA, Vagle S, Melling H, Williams WJ (2010) Structures and property distributions in the three oceans surrounding Canada in 2007: a basis for a long-term ocean climate monitoring strategy. Atmos Ocean 48:211–224

Caron DA (2009) Past president's address: protistan biogeography: why all the fuss? J Eukaryot Microbiol 56:105–112

Carroll ML, Carroll J (2003) The Arctic Seas. In: Black KD, Shimmield GB (eds) Biogeochemistry of marine systems. CRC Press, Boca Raton, pp 127–156

Cermeno P, de Vargas C, Abrantes F, Falkowski PG (2010) Phytoplankton biogeography and community stability in the ocean. PLoS One 5:e10037. doi:10.1371/journal.pone.0010037

Chapman AS, Kostylev VE (2005) Distribution, abundance and diversity of benthic species from the Beaufort Sea and western Amundsen Gulf – a summary of data collected between 1951 and 2000. Geol Surv Can Open File 5498

Cheung WL, Lam VWY, Sarmiento JL, Kearney K, Watson R, Pauly D (2009) Projecting global marine biodiversity impacts under climate change scenarios. Fish Fish. doi:10.1111/j.14672979200800315x

Chierici M, Fransson A (2009) Calcium carbonate saturation in the surface water of the Arctic Ocean: undersaturation in freshwater influenced shelves. Biogeosciences 6:2421–2432

Christman GD, Cottrell MT, Popp BN, Gier E, Kirchman DL (2011) Abundance, diversity, and activity of ammonia-oxidizing prokaryotes in the coastal Arctic Ocean in summer and winter. Appl Environ Microbiol 77:2026–2034

Church MJ, Delong EF, Ducklow HW, Karner MB, Preston CM, Karl DM (2003) Abundance and distribution of planktonic archaea and bacteria in the waters west of the Antarctic Peninsula. Limnol Oceanogr 48:1893–1902

Clough LM, Ambrose WG Jr, Cochran JK, Barnes C, Renaud PE, Aller RC (1997) Infaunal density, biomass and bioturbation in the sediments of the Arctic Ocean. Deep-Sea Res II 44:1683–1704

Coachman LK, Aagaard K (1988) Transports through the Bering Strait: annual and interannual variability. J Geophys Res 93:15535–15539

Collett TS, Dallimore SR (1999) Hydrocarbon gases associated with permafrost in the Mackenzie Delta, Northwest Territories, Canada. Appl Geochem 14:607–620

Collins RE, Rocap G, Deming JW (2010) Persistence of bacterial and archaeal communities in sea ice through an Arctic winter. Environ Microbiol 12:1828–1841

Comeau AM, Philippe B, Thaler M, Gosselin M, Poulin M, Lovejoy C (2013) Protists in Arctic drift and land-fast sea ice. J Phycol 49:229–240. doi:10.1111/jpy.12026

Comiso JC (2003) Warming trends in the Arctic from clear-sky satellite observations. J Climate 16:3498–3510

Conlan KE, Kvitek RG (2005) Recolonization of ice scours on an exposed Arctic coast. Mar Ecol Prog Ser 286:21–42

Conlan KE, Aitken A, Hendrycks E, Melling H, McClelland C (2008) Distribution patterns of Canadian Beaufort Shelf macrobenthos. J Mar Syst 74:864–886

Cooper LW, Larsen IL, Grebmeier JM, Moran SB (2005) Rapid deposition of sea ice-rafted material to the Arctic Ocean benthos demonstrated using the cosmogenic tracer 7Be. Deep-Sea Res II 52:3452–3461

Cottrell MT, Kirchman DL (2009) Photoheterotrophic microbes in the Arctic Ocean in summer and winter. Appl Environ Microbiol 75:4958–4966. doi:10.1128/AEM.00117-09

Cottrell MT, Kirchman DL (2012) Virus genes in Arctic marine bacteria identified by metagenomic analysis. Aquat Microb Ecol 66:107–116

Coyle KO, Konar B, Blanchard A, Highsmith RC, Carroll J, Carroll M, Denisenko S, Sirenko BI (2007) Potential effects of temperature on the benthic macrofaunal community of the southeastern Bering Sea shelf: possible impacts of climate change. Deep-Sea Res II 54:2885–2905

Coyle KO, Eisner LB, Mueter FJ, Pinchuk AI, Janout MA, Cieciel KD, Farley EV, Andrews AG (2011) Climate change in the southeastern Bering Sea: impacts on pollock stocks and implications for the oscillating control hypothesis. Fish Oceanogr 20:139–156

Cusson M, Archambault P, Aitken A (2007) Biodiversity of benthic assemblages on the Arctic continental shelf: historical data from Canada. Mar Ecol Prog Ser 331:291–304

Dalpadado P, Borkner N, Bogstad B, Mehl S (2001) Distribution of *Themisto* (Amphipoda) spp. in the Barents Sea and predator-prey interactions. ICES J Mar Sci 58:876–895

Darnis G, Barber DG, Fortier L (2008) Sea ice and the onshore-offshore gradient in pre-winter zooplankton assemblages in southeastern Beaufort Sea. J Mar Syst 74:994–1011

Darnis G, Robert D, Pomerleau C, Link H, Archambault P, Nelson RJ, Geoffroy M, Tremblay J-E, Lovejoy C, Ferguson SH, Hunt BPV, Fortier L (2012) Current state and trends in Canadian Arctic marine ecosystems: II. Heterotrophic food web, pelagic-benthic coupling, and biodiversity. Clim Change 115:179–205

Dehn L-A, Sheffield GG, Follmann EH, Duffy LK, Thomas DL, O'Hara TM (2007) Feeding ecology of phocid seals and some walrus in the Alaskan and Canadian Arctic as determined by stomach contents and stable isotope analysis. Polar Biol 30:167–181

Dickson DL, Gilchrist HG (2002) Status of marine birds of the Southeastern Beaufort Sea. Arctic 55(Suppl 1):46–58

Deibel D, Daly KL (2007) Zooplankton processes in Arctic and Antarctic polynyas. In: Smith WO Jr, Barber D (eds) Polynyas: windows into polar oceans, vol 74, Elsevier oceanography series. Elsevier, Amsterdam, pp 271–322

Doherty M, Costas BA, McManus GB, Katz LA (2007) Culture-independent assessment of planktonic ciliate diversity in coastal Northwest Atlantic waters. Aquat Microb Ecol 48:141–154

Drinkwater KF, Mueter F, Friedland KD, Taylor M, Hunt GL, Hare J, Melle W (2009) Recent climate forcing and physical oceanographic changes in Northern Hemisphere regions: a review and comparison of four marine ecosystems. Prog Oceanogr 81:10–28

Dufour SC, White C, Desrosiers G, Juniper SK (2007) Structure and composition of the consolidated mud tube of *Maldane sarsi* (Polychaeta: Maldanidae). Estuar Coast Shelf Sci 78:360–368

Dumas J, Carmack E, Melling H (2005) Climate change impacts on the Beaufort shelf landfast ice. Cold Reg Sci Technol 42:41–51

Dunton K (1992) Arctic biogeography: the paradox of the marine benthic fauna and flora. Trends Ecol Evol 7:183–189

Dunton KH, Reimnitz E, Schonberg SV (1982) An Arctic kelp community in the Alaskan Beaufort Sea. Arctic 35:354–484

Dunton KH, Goodall JL, Schonberg SV, Grebmeier JM, Maidment DR (2005) Multi-decadal synthesis of benthic-pelagic coupling in the western arctic: role of cross-shelf advective processes. Deep-Sea Res II 52:3462–3477

Dunton KH, Schonberg SV, Cooper LW (2012) Food-web structure of the Alaskan nearshore shelf and estuarine lagoons of the Beaufort Sea. Estuar Coasts 35:416–435

Elifantz H, Dittel AI, Cottrell MT, Kirchman DL (2007) Dissolved organic matter assimilation by heterotrophic bacterial groups in the western Arctic Ocean. Aquat Microb Ecol 50:39–49

Emmerton CA, Lesack LFW, Vincent WF (2008) Nutrient and organic matter patterns across the Mackenzie River, estuary and shelf during the seasonal recession of sea-ice. J Mar Syst 74:741–755

Feder HM, Jewett SC (1978) Survey of the epifaunal invertebrates of southeastern Chukchi Sea and Kotzebue Sound. IMS report R78–1, Institute of Marine science, University of Alaska, Fairbanks, 124 pp

Feder HM, Foster NR, Jewett SC, Weingartner TJ, Baxter R (1994a) Mollusks in the northeastern Chukchi Sea. Arctic 47:145–163

Feder HM, Naidu AS, Jewett SC, Hameedi JM, Johnson WR, Whitledge TE (1994b) The northeastern Chukchi Sea: benthos-environmental interactions. Mar Ecol Prog Ser 111:171–190

Feder HM, Jewett SC, Blanchard A (2005) Southeastern Chukchi Sea (Alaska) epibenthos. Polar Biol 28:402–421

Feder HM, Jewett SC, Blanchard AL (2007) Southeastern Chukchi Sea (Alaska) macrobenthos. Polar Biol 30:261–275

Fierer N, Jackson RB (2006) The diversity and biogeography of soil bacterial communities. Proc Natl Acad Sci USA 103:626–631

Fisher JAD, Frank FT, Leggett WC (2010) Dynamic macroecology on ecological time scales. Global Ecol Biogeogr 19:1–15

Flint M, Sukhanova IN, Kopylov AI, Poyarkov SG, Whitledge T (2002) Plankton distribution associated with frontal zones in the vicinity of the Pribilof Islands. Deep-Sea Res II 49:6069–6093

Fortier L, Le Fèvre J, Legendre L (1994) Export of biogenic carbon to fish and to the deep ocean: the role of large planktonic microphages. J Plankton Res 16:809–839

Frost KJ, Lowry LF (1983) Demersal fishes and invertebrates trawled in the northeastern Chuckhi and western Beaufort Seas, 1976–77. NOAA technical report NMFS SSRF-764

Fuhrman JA, Steele JA, Hewson I, Schwalbach MS, Brown MV, Green JL, Brown JH (2008) A latitudinal diversity gradient in planktonic marine bacteria. Proc Natl Acad Sci USA 105:7774–7778

Gagaev SY (2008) *Sigambra healyae* sp. n., a new species of polychaete (Polychaeta: Pilargidae) from the Canada Basin of the Arctic Ocean. Russ J Mar Biol 34:73–75

Galand PE, Lovejoy C, Vincent WF (2006) Contrasting and diverse Archaea communities in the Mackenzie River and coastal Arctic Ocean. Aquat Microb Ecol 44:115–126

Galand PE, Casamayor EO, Kirchman DL, Lovejoy C (2009a) Unique archaeal assemblages in the Arctic Ocean unveiled by massively parallel tag sequencing. ISME J 3:860–869

Galand PE, Lovejoy C, Hamilton AK, Ingram RG, Pedneault E, Carmack EC (2009b) Archaeal diversity and a gene for ammonia oxidation are coupled to oceanic circulation. Environ Microbiol 11:971–980

Galand PE, Potvin M, Casamayor EO, Lovejoy C (2010) Hydrography shapes bacterial biogeography of the deep Arctic Ocean. ISME J 4:564–576

Garneau M-E, Vincent WF, Alonso-Sáez L, Gratton Y, Lovejoy C (2006) Prokaryotic community structure and heterotrophic production in a river-influenced coastal arctic ecosystem. Aquat Microb Ecol 42:27–40

Geynrikh AK, Kosobokova KN, Rudyakov YA (1983) Seasonal variations in the vertical distribution of some prolific copepods of the Arctic basin. Can Trans Fish Aquat Sci 4925:1–22

Gilbert JA, Field D, Swift P, Newbold L, Oliver A, Smyth T, Somerfield PJ, Huse S, Joint I (2009) The seasonal structure of microbial communities in the Western English Channel. Environ Microbiol 11:3132–3139

Gill M, 19 co-authors (2011) Arctic marine biodiversity monitoring plan, CAFF monitoring series report No. 3, April 2011, CAFF International Secretariat, Akureyri. ISBN 978-9979-9778-7-2

Goering JJ, Iverson RL (1981) Phytoplankton distribution in the southeastern Bering Sea shelf. In: Hood DW, Calder JA (eds) The eastern Bering Sea shelf: oceanography and resources, vol 2. OMPA/NOAA, University of Washington Press, Seattle, pp 933–946

Gogorev RM, Okolodkov YB (1996) Species composition of the planktonic and sea-ice algae in the Chukchi Sea and Lavrentiya Bay (Bering Sea, August 1991). Bot J 81:35–41

Gomez-Pereira PR, Fuchs BM, Alonso C, Oliver MJ, Van Beusekom JEE, Amann R (2010) Distinct flavobacterial communities in contrasting water masses of the North Atlantic Ocean. ISME J 4:472–487

Gorsky G, Youngbluth MJ, Deibel D (2005) Response of marine ecosystems to global change: ecological impact of appendicularians. Éditions Scientifiques, Paris, 435 pp. ISBN 2-8470-302-9-8

Gosselin M, Levasseur M, Wheeler PA, Horner RA, Booth BC (1997) New measurements of phytoplankton and ice algal production in the Arctic Ocean. Deep-Sea Res II 44:1623–1644

Gradinger R (1999) Integrated abundances and biomass of sympagic meiofauna from Arctic and Antarctic pack ice. Polar Biol 22:169–177

Gradinger R (2009) Sea-ice algae: major contributors to primary production and algal biomass in the Chukchi and Beaufort Seas during May/June 2002. Deep Sea Res II 56:1201–1212

Gradinger R, Bluhm BA (2010) Timing of ice algal grazing by the Arctic nearshore benhic amphipod *Onisimus litoralis*. Arctic 63:355–358

Gradinger R, Lenz J (1989) Picocyanobacteria in the high Arctic. Mar Ecol Prog Ser 52:99–101

Gradinger R, Lenz J (1995) Seasonal occurrence of picocyanobacteria in the Greenland Sea and central Arctic Ocean. Polar Biol 15:447–452

Gradinger RR, Meiners K, Plumley G, Zhang Q, Bluhm BA (2005) Abundance and composition of the sea-ice meiofauna in off-shore pack ice of the Beaufort Gyre in summer 2002 and 2003. Polar Biol 28:171–181

Gradinger RR, Kaufman MR, Bluhm BA (2009) Pivotal role of sea ice sediments in the seasonal development of near-shore Arctic fast ice biota. Mar Ecol Prog Ser 394:49–63

Gradinger R, Bluhm B, Iken K (2010) Arctic sea-ice ridges-safe havens for sea-ice fauna during periods of extreme ice melt? Deep Sea Res II 57:86–95

Grainger EH (1965) Zooplankton from the Arctic and adjacent Canadian waters. J Fish Res Board Can 22:543–564

Grebmeier JM (2012) Shifting patterns of life in the Pacific Arctic and Subarctic seas. Ann Rev Mar Sci 4:63–78

Grebmeier JM, Barry JP (1991) The influence of oceanographic processes on pelagic-benthic coupling in polar regions: a benthic perspective. J Mar Syst 2:495–518

Grebmeier JM, McRoy CP, Feder HM (1988) Pelagic-benthic coupling on the shelf of the northern Bering and Chukchi Seas. I. Food supply source and benthic biomass. Mar Ecol Prog Ser 48:57–67

Grebmeier JM, Feder HM, McRoy CP (1989) Pelagic-benthic coupling on the shelf of the northern Bering and Chukchi Seas. II. Benthic community structure. Mar Ecol Prog Ser 51:253–268

Grebmeier JM, Cooper LW, Feder HM, Sirenko BI (2006a) Ecosystem dynamics of the Pacific-influenced Northern Bering and Chukchi Seas in the Amerasian Arctic. Prog Oceanogr 71:331–361

Grebmeier JM, Overland JE, Moore SE, Farley EV, Carmack EC, Cooper LW, Frey KE, Helle JH, McLaughlin FA, McNutt SL (2006b) A major ecosystem shift in the northern Bering Sea. Science 311:1461–1464

Grebmeier JM, Moore SE, Overland JE, Frey KE, Gradinger RR (2010) Biological response to recent Pacific Arctic Sea ice retreats. Eos T Am Geophys Un 91:161–168

Grebmeier J, Pickart R, Ashjian C, Cooper L, Frey K, He J, Itoh M, Kedra M, Kikuchi T, Moore S, Nelson J, Vagle S (2012) Ecosystem observations in Barrow Canyon: a focus for the international distributed biological observatory (DBO), pp 68–72. [In Arctic Report Card 2012]. http://www.arctic.noaa.gov/reportcard/barrow_canyon.html

Guillou L, Viprey M, Chambouvet A, Welsh RM, Kirkham AR, Massana R, Scanlan DJ, Worden AZ (2008) Widespread occurrence and genetic diversity of marine parasitoids belonging to Syndiniales (Alveolata). Environ Microbiol 10:3349–3365

Hamilton AK, Lovejoy C, Galand PE, Ingram RG (2008) Water mass and biogeography of picoeukaryote assemblages in a cold hydrographically complex system. Limnol Oceanogr 53:922–935

Harding GCH (1966) Zooplankton distribution in the Arctic Ocean with notes on life cycles. M.Sc. thesis, McGill University

Hazard KW, Lowry LF (1984) Benthic prey in a bowhead whale from the northern Bering Sea. Arctic 37:166–168

Héquette A, Barnes PW (1990) Coastal retreat and shoreface profile variations in the Canadian Beaufort Sea. Mar Geol 91:113–132

Héquette A, Desrosiers M, Barnes PW (1995) Sea ice scouring on the inner shelf of the southeastern Canadian Beaufort Sea. Mar Geol 128:201–219

Highsmith RC, Coyle KO (1990) High productivity of northern Bering Sea benthic amphipods. Nature 344:862–864

Highsmith RC, Coyle KO (1992) Productivity of Arctic amphipods relative to gray whale energy requirements. Mar Ecol Prog Ser 83:141–150

Highsmith RC, Coyle KO, Bluhm BA, Konar B (2007) Gray whales in the Bering and Chukchi Seas. In: Estes JA, DeMaster DP, Doak DF, Williams TM, Brownell RL Jr (eds) Whales, whaling, and ocean ecosystems. University of California Press, Berkeley, Oakland

Hill VJ, Cota GF (2005) Spatial patterns of primary production in the Chukchi Sea in the spring and summer of 2002. Deep-Sea Res II 52:3344–3354

Hill VJ, Cota GF, Stockwell D (2005) Spring and summer phytoplankton communities in the Chukchi and Eastern Beaufort Sea. Deep Sea Res II 52:3369–3385

Hirche H-J, Kosobokova KN (2007) Distribution of *Calanus finmarchicus* in the northern North Atlantic and Arctic Ocean-Expatriation and potential colonization. Deep-Sea Res II 54:2729–2747

Hirst AG, Bunker AJ (2003) Growth of marine planktonic copepods: Global rates and patterns in relation to chlorophyll *a*, temperature, and body weight. Limnol Oceanogr 48:1988–2010

Hodges LR, Bano N, Hollibaugh JT, Yager PL (2005) Illustrating the importance of particulate organic matter to pelagic microbial abundance and community structure – an Arctic case study. Aquat Microb Ecol 40:217–227

Holte B (1998) The macrofauna and main functional interactions in the sill basin sediments of the pristine Holandsfjord, North Norway, with autecological reviews for some key-species. Sarsia 83:55–68

Hopcroft RR, Clarke C, Nelson RJ, Raskoff KA (2005) Zooplankton communities of the Arctic's Canada Basin: the contribution by smaller taxa. Polar Biol 28:197–206

Hopcroft RR, Kosobokova KN (2010) Distribution and egg production of *Pseudocalanus* species in the Chukchi Sea. Deep-Sea Res II 57:49–56

Hopcroft RR, Kosobokova KN, Pinchuk AI (2010) Zooplankton community patterns in the Chukchi Sea during summer 2004. Deep-Sea Res II 57:27–39

Hopky GE, Lawrence MJ, Chiperzak DB (1994a) NOGAP B2; Zooplankton data from the Canadian Beaufort Sea shelf, 1984 and 1985. Canadian Data Report of Fisheries and Aquatic Sciences, 922

Hopky GE, Lawrence MJ, Chiperzak DB (1994b) NOGAP B2; Zooplankton data from the Canadian Beaufort Sea shelf, 1986. Canadian Data Report of Fisheries and Aquatic Sciences, 912

Horner R, Schrader GC (1982) Relative contributions of ice algae, phytoplankton, and benthic microalgae to primary production in nearshore regions of the Beaufort Sea. Arctic 35:485–503

Howell-Kubler AN, Lessard EJ, Napp JM (1996) Springtime microprotozoan abundance and biomass in the southeastern Bering Sea and Shelikof Strait, Alaska. J Plankton Res 18:731–745

Hunt GL, Stabeno PJ, Walters G, Sinclair E, Brodeur RD, Napp JM, Bond NA (2002) Climate change and control of the southeastern Bering Sea pelagic ecosystem. Deep-Sea Res II 49:5821–5853

Huntley ME, Lopez MG (1992) Temperature dependant production of marine copepods: a global analysis. Am Nat 140:201–242

Jaschnov V (1940) Plankton productivity of the northern seas of the USSR. Moscovskoe Obshestvo Ispytatelei Prirody Press, Moscow, 85 p. (in Russian)

Jerosch KJ, Kostylev VE, Blasco SM (2010) Analysis of epi-benthos distribution and their possible association to Gary Knoll's pingo-like-features on the Canadian Beaufort Shelf: a small-scale case study. In: Calado H, Gil A (eds) Geographic technologies applied to marine spatial planning and integrated coastal zone management. pp 78–85. E-Book ISBN 978-972-8612-64-1

Johnson MW (1934) The production and distribution of zooplankton in the surface waters of the Bering Sea and Bering Strait, Part II Report of the oceanographic cruise U.S. Coast Guard Cutter Chelan – 1934, pp 45–82

Johnson MW (1956) The plankton of the Beaufort and Chukchi Sea areas of the Arctic and its relation to hydrography. Arctic Institute of North America, Montreal

Johnson MW (1963) Zooplankton collections from the high polar Basin with special reference to the Copepoda. Limnol Oceanogr 8:89–102

Joo HM, Lee SH, Jung SW, Dahms H-U, Lee JH (2012) Latitudinal variation of phytoplankton communities in the western Arctic Ocean. Deep-Sea Res II 81–84:3–17

Juul-Pedersen T, Michel C, Gosselin M (2008) Influence of the Mackenzie River plume on the sinking export of particulate material on the shelf. J Mar Syst 74:810–824

Kalanetra KM, Bano N, Hollibaugh JT (2009) Ammonia-oxidizing Archaea in the Arctic Ocean and Antarctic coastal waters. Environ Microbiol 11:2434–2445

Katz LA, McManus GB, Snoeyenbos-West OLO, Griffin A, Pirog K, Costas B, Foissner W (2005) Reframing the everything is everywhere debate: evidence for high gene flow and diversity in ciliate morphospecies. Aquat Microb Ecol 41:55–65

Kellogg CTE, Deming JW (2009) Comparison of free-living, suspended particle, and aggregate-associated bacterial and archaeal communities in the Laptev Sea. Aquat Microb Ecol 57:1–18

Kirchman DL (2002) The ecology of Cytophaga-Flavobacteria in aquatic environments. FEMS Microb Ecol 39:91–100

Kirchman DL, Eliphantz H, Dittel AI, Malmstrom RR, Cottrell MT (2007) Standing stocks and activity of archaea and bacteria in the western Arctic Ocean. Limnol Oceanogr 52:495–507

Kirchman DL, Moran XAG, Ducklow H (2009) Microbial growth in the polar oceans- role of temperature and potential impact of climate change. Nat Rev Micro 7:451–459

Kirchman DL, Cottrell MT, Lovejoy C (2010) The structure of bacterial communities in the western Arctic Ocean as revealed by pyrosequencing of 16S rRNA genes. Environ Microbiol 12:1132–1143

Kosobokova KN (1980) Seasonal changes of the vertical distribution and age composition of *Microcalanus pygmaeus, Oithona similis, Oncaea borealis* and *O. notopus* populations in the central Arctic basin. In: Vinogradov ME (ed) Biology of the central Arctic Basin. Nauka, Moscow, pp 167–182 (in Russian)

Kosobokova KN (1982) Composition and distribution of the biomass of zooplankton in the central Arctic Basin. Oceanology 22:744–750

Kosobokova K, Hirche H-J (2000) Zooplankton distribution across the Lomonosov Ridge, Arctic Ocean: species inventory, biomass and vertical structure. Deep-sea Res I 47:2029–2060. doi:10.1016/S0967-0637(00)00015-7

Kosobokova KN, Hopcroft RR (2010) Diversity and vertical distribution of mesozooplankton in the Arctic's Canada Basin. Deep-Sea Res II 57:96–110

Kosobokova KN, Hopcroft RR, Hirche H-J (2011) Patterns of zooplankton diversity through the depths of the Arctic's central basins. Mar Biodiv 41:29–50

Kostylev VE, Chapman AS (2004) Report on the Beaufort Sea benthic epifauna identified from seabed video collected on board CCGS Nahidik, 4–17 August 2004: Geological Survey of Canada, Open file report 4938, 53 p

Kurihara M, Takahashi K (2002) Long-term size variation and life cycle patterns of a predominant diatom *Neodenticula seminae* in the subarctic Pacific and the Bering Sea. B Plank Soc Japan 49:77–87

Kvitek RG, Conlan KE, Iampietro P (1998) Black pools of death: anoxic, brine-filled ice gouge depressions become lethal traps for benthic organisms in an Arctic embayment. Mar Ecol Prog Ser 162:1–10

Lane PVZ, Llinás L, Smith SL, Pilz D (2008) Zooplankton distribution in the western Arctic during summer 2002: hydrographic habitats and implications for food chain dynamics. J Mar Syst 70:97–133

Lee SH, Whitledge TE (2005) Primary and new production in the deep Canada Basin during summer 2002. Polar Biol 28:190–197

Lee SH, Whitledge TE, Kang S-H (2007) Recent carbon and nitrogen uptake rates of phytoplankton in Bering Strait and the Chukchi Sea. Cont Shelf Res 27:2231–2249. doi:10.1016/j.csr.2007.05.009

Levin L, Blair N, DeMaster D, Plaia G, Fornes W, Martin C, Thomas C (1997) Rapid subduction of organic matter by maldanaid polychaetes on the North Carolina slope. J Mar Res 55:595–611

Lewis CFM, Blasco SM (1990) Character and distribution of sea-ice and iceberg scours. In: Clark JI (ed) Workshop on ice scouring and design of offshore pipelines, Calgary, Alberta, April 18–19, 1990. Canada oil and gas lands administration and centre for cold ocean resources engineering

Li WKW (2009) Plankton populations and communities. In: Witman JD, Roy K (eds) Marine macroecology. University of Chicago Press, Chicago, pp 29–64

Li WKW, McLaughlin FA, Lovejoy C, Carmack EC (2009) Smallest algae thrive as the Arctic Ocean freshens. Science 326:539

Light BT, Grenfell TC, Perovich DK (2008) Transmission and absorption of solar radiation by Arctic sea ice during the melt season. J Geophys Res 113. doi:10.1029/2006JC003977

Llinas L, Pickart R, Mathis J, Smith S (2009) Zooplankton inside an Arctic Ocean cold core eddy; probable origin and fate. Deep-Sea Res 56:1290–1304

Lovejoy C, Potvin M (2011) Microbial eukaryotic distribution in a dynamic Beaufort Sea and the Arctic Ocean. J Plankton Res 33:431–444

Lovejoy C, Legendre L, Therriault J-C, Tremblay J-E, Klein B, Ingram RG (2000) Growth and distribution of marine bacteria in relation to nanoplankton community structure. Deep-Sea Res II 47:461–487

Lovejoy C, Massana R, Pedros-Alio C (2006) Diversity and distribution of marine microbial eukaryotes in the Arctic Ocean and adjacent seas. Appl Environ Microbiol 72:3085–3095

Lovejoy C, Vincent WF, Bonilla S, Roy S, Martineau M-J, Terrado R, Potvin M, Massana R, Pedros-Alio C (2007) Distribution, phylogeny and growth of cold-adapted picoprasinophytes in Arctic Seas. J Phycol 43:78–89

Lovejoy C, Galand PE, Kirchman DL (2011) Picoplankton diversity in the Arctic Ocean and surrounding seas. Mar Biodiv 41:5–12

Lovvorn JR, Richman SE, Grebmeier JM, Cooper LW (2003) Diet and body condition of spectacled eiders wintering in pack ice of the Bering Sea. Polar Biol 26:259–267

Lovvorn JR, Grebmeier JM, Cooper LW, Bump JK, Richman SE (2009) Modeling marine protected areas for threatened eiders in a climatically changing Bering Sea. Ecol Appl 19:1596–1613

Lowry LF (1993) Foods and feeding ecology. In: Burns JJ, Montague JJ, Cowles CJ (eds) The bowhead whale. Society for marine mammalogy, Special publication No. 2. Allen Press, Lawrence, pp 201–238

Lui H, Probert I, Uitz J, Claustre H, Aris-Brosou S, Frada M, Not F, de Vargas C (2009) Extreme diversity in noncalcifying haptophytes explains a major pigment paradox in open oceans. Proc Natl Acad Sci USA 106:12803–12808

Macdonald RW, Paton DW, Carmack EC, Omstedt A (1995) The freshwater budget and under-ice spreading of Mackenzie River water in the Canadian Beaufort Sea based on salinity and $^{18}O/^{16}O$ measurements in water and ice. J Geophys Res 100:895–919

Macdonald RW, Soloman SM, Cranston RE, Welch HE, Yunker MB, Gobeil C (1998) A sediment and organic carbon budget for the Canadian Beaufort Shelf. Mar Geol 144:255–273

Macdonald RW, Naidu AS, Yunker MB, Gobeil C (2004) The Beaufort Sea: distribution, sources, fluxes and burial of organic carbon. In: Stein R, Macdonald RW (eds) The organic carbon cycle in the Arctic Ocean. Springer, Berlin, pp 177–193

Macdonald RW, Harner T, Fyfe J (2005) Recent climate change in the Arctic and its impact on contaminant pathways and interpretation of temporal trend data. Sci Total Environ 342:5–86

MacDonald IR, Bluhm BA, Iken K, Gagaev S, Strong A (2010) Benthic macrofaunal and megafaunal assemblages in the Arctic deep-sea Canada Basin. Deep-Sea Res II 57:136–152

Mahoney A, Eicken H, Gaylord AG, Shapiro L (2007) Alaska landfast sea ice: links with bathymetry and atmospheric circulation. J Geophys Res 112. doi:10.1029/2006JC003559

Malmstrom RR, Cottrell MT, Kirchman DL (2007) Diversity and abundance of closely-related bacteria in the western Arctic Ocean. Aquat Microb Ecol 47:45–55

Maranger RJ, Bird DF, Juniper K (1994) Viral and bacterial dynamics in Arctic sea ice during the spring algal bloom near Resolute NWT, Canada. Mar Ecol Prog Ser 111:121–127

Margulies M, Egholm M, Altman WE, Attiya S, Bader JS, Bemben LA, Berka J, Braverman MS, Chen YJ et al (2005) Genome sequencing in microfabricated high-density picolitre reactors. Nature 437:376–380

Marincovich L Jr, Brouwers EM, Hopkins DM, McKenna MC (1990) Late Mesozoic and Cenozoic paleogeographic and paleoclimatic history of the Arctic Ocean Basin, based upon shallow-water marine faunas and terrestrial vertebrates. In: Gantz A, Johnson L, Sweeny JF (eds) The Arctic Ocean region. The geology of North America, vol. L, Geological Society of America, Boulder, pp 403–426

Maslanik JA, Fowler C, Stroeve J, Drobot S, Zwally J, Yi D, Emery W (2007) A younger, thinner Arctic ice cover: increased potential for rapid, extensive sea-ice loss. Geophys Res Lett 34. doi:10.1029/2007GL032043

Massana R, Castresana J, Balgue V, Guilou L, Romari K, Groisillier A, Valentin K, Pedros-Alio C (2004) Phylogenetic and ecological analysis of novel marine stramenopiles. Appl Environ Microbiol 70:3428–3535

Massana R, Guillou L, Terrado R, Forn I, Pedros-Alio C (2006) Growth of uncultured heterotrophic flagellates in unamended seawater incubations. Aquat Microb Ecol 45:171–180

Mathis JT, Hansell DA, Bates NR (2005) Strong hydrographic controls on spatial and seasonal variability of dissolved organic carbon in the Chukchi Sea. Deep-Sea Res II 52:3245–3258

Mathis JT, Hansell DA, Bates NR (2009) Inter annual variability of dissolved inorganic carbon distribution and net community production during the Western Arctic Shelf-Basin Interactions Project. Deep-Sea Res II 56:1213–1222. doi:10.1016/j.dsr2.2008.10.017

Mathis JT, Grebmeier JG, Hansell DA, Hopcroft RR, Kirchman DL, Lee SH, Moran SB, Bates NR, VanLaningham S, Cross JN, Cai WJ (2014) Chapter 9: Carbon biogeochemistry of the western Arctic: primary production, carbon export and the controls on ocean acidification. In: Grebmeier JM, Maslowski W (eds) The Pacific Arctic region: ecosystem status and trends in a rapidly changing environment. Springer, Dordrecht, pp 223–268

McConnell M (1977) An analysis of the zooplankton community structure of the western Beaufort Sea (Websec 1971). M.Sc. thesis, University of Rhode Island, 218 p

McLaughlin FA, Carmack EC (2010) Deepening of the nutricline and chlorophyll maximum in the Canada Basin interior, 2003–2009. Geophys Res Lett 37. doi:10.1029/2010GL045459

McManus GB, Katz LA (2009) Molecular and morphological methods for identifying plankton: what makes a successful marriage? J Plankton Res 31:1119–1129. doi:10.1093/plankt/fbp061

McRoy CP, Goering JJ (1974) The influence of ice on the primary productivity of the Bering Sea. In: Hood DW, Kelley EJ (eds) Oceanography of the Bering Sea. Institute of Marine Science, University of Alaska, Fairbanks, pp 403–421

Melling H (1993) The formation of a haline shelf front in wintertime in an ice-covered arctic sea. Cont Shelf Res 13:1123–1147

Merico A, Tyrrell T, Brown CW, Groom SB, Miller PI (2003) Analysis of satellite imagery for *Emiliania huxleyi* blooms in the Bering Sea before 1997. Geophys Res Lett 30:1337. doi:10.1029/2002GL016648

Michaud J, Fortier L, Rowe P, Ramseier R (1996) Feeding success and survivorship of Arctic cod larvae, *Boreogadus saida*, in the Northeast Water polynya (Greenland Sea). Fish Oceanogr 5:120–135

Michel C, Ingram RG, Harris LR (2006) Variability in oceanographic and ecological processes in the Canadian Arctic Archipelago. Prog Oceanogr 71:379–401

Middelboe M, Nielsen TG, Bjørnsen PK (2002) Viral and bacterial production in the North Water: in situ measurements, batch-culture experiments and characterization and distribution of a virus-host system. Deep-Sea Res II 49:5063–5079

Mills CE (1995) Medusae, siphonophores, and ctenophores as planktivorous predators in changing global ecosystems. Ices J Mar Sci 52:575–581

Mills CE (2001) Jellyfish blooms: are populations increasing globally in response to changing ocean conditions? Hydrobiologia 451:55–68

Minobe S (2002) Interannual to interdecadal changes in the Bering Sea and concurrent 1998/99 changes over the North Pacific. Prog Oceanogr 55:45–64

Mock T, Gradinger R (1999) Determination of Arctic ice algal production with a new and easy in situ incubation technique. Mar Ecol Prog Ser 177:15–26

Moore SE, Grebmeier JM, Davis JR (2003) Gray whale distribution relative to forage habitat in the northern Bering Sea: current conditions and retrospective summary. Can J Zool 81:734–742

Moore SE, Stafford KM, Mellinger DK, Hildebrand JA (2006) Listening for large whales in the offshore waters of Alaska. Bioscience 56:49–55

Moran SB, Kelly RP, Hagstrom K, Smith JN, Grebmeier JM, Cooper LW, Cota GF, Walsh JJ, Bates NR, Hansell DA, Maslowski W (2005) Seasonal changes in POC export flux in the Chukchi Sea and implications for water column-benthic coupling in Arctic shelves. Deep-Sea Res II 52:3427–3451

Morata N, Renaud PE, Brugel S, Hobson KA, Johnson BJ (2008) Spatial and seasonal variations in the pelagic-benthic coupling of the southeastern Beaufort Sea revealed by sedimentary biomarkers. Mar Ecol Prog Ser 371:47–63

Morata N, Poulin M, Renaud PE (2011) A multiple biomarker approach to tracking the fate of an ice algal bloom to the sea floor. Polar Biol 34:101–112

Motard-Côté J (2010) Dynamique bactérienne du diméthylsulfoniopropionate (DMSP) et du diméthylsulfure (DMS) en function des mases d'eau de surface dans l'est de l'Arctique canadien. M.Sc. thesis, Université Laval, 41 pp

Motard-Côté J, Scarratt M, Michaud S, Gratton Y, Rivkin R, Keats K, Gosselin M, Tremblay J-E, Kiene R, Lovejoy C (2011) Distribution and phylogenetic affiliation of dimethylsulfoniopropionate (DMSP)-degrading bacteria in northern Baffin Bay/Lancaster Sound. J Geophys Res Oceans 117:C00G11. doi:10.1029/2011JCX

Mundy CJ, Ehn J, Barber DG, Michel C (2007) Influence of snow cover and algae on the spectral dependence of transmitted irradiance through Arctic landfast first-year sea ice. J Geophys Res 112:C03007. doi:10.1029/2006jc003683

Murdoch J (1885) Marine invertebrates (exclusive of mollusks), pp 136–176. In: Report of the international polar expedition to Point Barrow, Alaska. US Government Printing Office, Washington, DC

Murphy LS, Haugen EM (1985) The distribution and abundance of phototrophic ultraplankton in the North Atlantic. Limnol Oceanogr 30:47–58

Murray AG, Jackson GA (1992) Viral dynamics – a model of the effects of size, shape, motion and abundance of single-celled planktonic organisms and other particles. Mar Ecol Prog Ser 89:103–116

Murray AE, Preston CM, Massana R, Taylor LT, Blakis A, Wu K, Delong EF (1998) Seasonal and spatial variability of bacterial and archaeal assemblages in the coastal waters near Anvers Island, Antarctica. Appl Environ Microbiol 64:2585–2595

Nelson RJ, Carmack EC, McLaughlin FA, Cooper GA (2009) Penetration of Pacific zooplankton into the western Arctic Ocean tracked with molecular population genetics. Mar Ecol Prog Ser 381:129–138

Nesis KN (1984) A hypothesis on the origin of western and eastern Arctic distribution of areas of marine bottom animals. Sov J Mar Biol 9:235–243

Nicol S (2006) Krill, currents, and sea ice: *Euphausia superba* and its changing environment. BioScience 56:111–120

Niebauer HJ (1983) Multiyear sea ice variability in the eastern Bering Sea: an update. J Geophys Res 88:2733–2742

Niebauer HJ, Bond NA, Yakunin LP, Plotnikov VV (1999) An update on the climatology and sea ice of the Bering Sea. In: Loughlin TR, Ohtani K (eds) The Bering Sea: a summary of physical, chemical and biological characteristics and a synopsis of research. North Pacific Marine Science Organization, PICES, Alaska Sea Grant, pp 22–83

Niemann H, Losekann T, De Beer D, Elvert M, Nadalig T, Knittel K, Amann R, Sauter EJ, Schluter MM, Klages M, Foucher JP, Boetius A (2006) Novel microbial communities of the Haakon Mosby mud volcano and their role as a methane sink. Nature 443:854–858

Not F, Massana R, Latasa M, Marie D, Colson C, Eikrem W, Pedros-Alio C, Vaulot D, Simon N (2005) Late summer community composition and abundance of photosynthetic picoeukaryotes in Norwegian and Barents Seas. Limnol Oceanogr 50:1677–1686

Not F, Valentin K, Romari K, Lovejoy C, Massana R, Töbe K, Vaulot D, Medlin LK (2007) Picobiliphytes: a marine picoplanktonic algal group with unknown affinities to other eukaryotes. Science 315:253–255

Nozais C, Gosselin M, Michel C, Tita G (2001) Abundance, biomass, composition and grazing impact of the sea-ice meiofauna in the North Water, northern Baffin Bay. Mar Ecol Prog Ser 217:235–250

O'Brien MC, Macdonald RW, Melling H, Iseki K (2006) Particle fluxes and geochemistry on the Canadian Beaufort Shelf: implications for sediment transport and deposition. Cont Shelf Res 26:41–81

Occhipinti-Ambrogi A (2007) Global change and marine communities: alien species and climate change. Marine Pollut Bull 55:342–352

Oliver JS, Slattery PN, Silberstein MA, O'Connor EF (1983) A comparison of gray whale, *Eschrichtius robustus*, feeding in the Bering Sea and Baja California. Fish Bull Calif 81:513–522

Olson MB, Strom SL (2002) Phytoplankton growth, microzooplankton herbivory and community structure in the southeast Bering Sea: insight into the formation and temporal persistence of an *Emiliania huxleyi* bloom. Deep-Sea Res II 49:5969–5990

Orr JC, Fabry VJ, Aumont O, Bopp L, Doney SC, Feely RA, Gnanadesikan A, Gruber N, Ishida A, Joos F, Key RM, Lindsay K, Maier-Reimer E, Matear R, Monfray P, Mouchet A, Najjar RG, Plattner G-K, Rodgers KB, Sabine CL, Sarmiento JL, Schlitzer R, Slater RD, Totterdell IJ, Weirig M-F, Yamanaka Y, Yool A (2005) Anthropogenic ocean acidification over the twenty-first century and its impact on calcifying organisms. Nature 437:681–686

Pautzke CG (1979) Copepoda collected from the Canada Basin Arctic Ocean; Fletcher's Ice Island (T-3) 1970–1972 and AIDJEX, 1975. In: Phytoplankton primary production below Arctic Ocean pack ice: an ecosystems analysis. Ph.D. thesis, University of Washington

Payet JP, Suttle CA (2008) Physical and biological correlates of virus dynamics in the southern Beaufort Sea and Amundsen Gulf. J Mar Syst 74:933–945

Piepenburg D, Archambault P, Ambrose WG Jr, Blanchard AL, Bluhm B, Carroll ML, Conlan KE, Cusson M, Feder HM, Grebmeier JM, Jewett SC, Lévesque M, Petryashev VV, Sejr MK, Sirenko BI, Wlodarska-Kowalczuk M (2011) Towards a pan-Arctic inventory of the species diversity of the macro- and megabenthic fauna of the Arctic shelf seas. Mar Biodiv 41:51–70

Pierce RW, Turner JT (1993) Global biogeography of marine tintinnids. Mar Ecol Prog Ser 94:11–26

Pinchuk A (1993) The distribution of zooplankton in the Bering Strait, July 1991. Exploration of the Fauna of the Seas (in Russian) 45(53):138–154

Pinchuk A (2006) Zooplankton in vertical stratified collections on board of the Russian R/V Lomonosov, program ANIIA-65. Chukchi Sea, August 1953–1954. http://dw.sfos.uaf.edu/rest/metadata/ArcOD/2006P1

Pirtle-Levy RS (2006) A shelf-to-basin examination of food supply for Arctic benthic macrofauna and the potential biases of sampling methodology. M.Sc. thesis, University of Tennessee, Knoxville. http://etd.utk.edu/2006/Pirtle-LevyRebecca.pdf

Pirtle-Levy R, Grebmeier JM, Cooper LW, Larsen IL (2009) Chlorophyll *a* in Arctic sediments implies long persistence of algal pigments. Deep-Sea Res II 56:1326–1338

Plourde S, Campbell RG, Ashjian CJ, Stockwell D (2005) Seasonal and regional patterns in egg production of *Calanus glacialis/marshallae* in the Chukchi and Beaufort Seas during Spring and Summer, 2002. Deep-Sea Res II 52:3411–3426

Poltermann M (2001) Arctic sea ice as feeding ground for amphipods – food sources and strategies. Polar Biol 24:89–96

Pommier T, Canbäck B, Riemann L, Boström KH, Simu K, Lundberg P, Tunlid A, Hagström Å (2007) Global patterns of diversity and community structure in marine bacterioplankton. Mol Ecol 16:867–880

Popova EE, Yool A, Coward AC, Dupont F, Deal C, Elliott S, Hunke E, Jin M, Steele M, Zhang J (2012) What controls primary production in the Arctic Ocean? Results from an intercomparison of five general circulation models with biogeochemistry. J Geophys Res 117:C00D12

Poulin M, Daugbjerg N, Gradinger R, Ilyash L, Ratkova T, von Quillfeldt C (2011) The pan-Arctic biodiversity of marine pelagic and sea-ice unicellular eukaryotes: a first-attempt assessment. Mar Biodiv 41:13–28

Proshutinsky A, Krishfield R, Barber D (2009) Preface to special section on Beaufort Gyre Climate System Exploration Studies: documenting key parameters to understand environmental variability. J Geophys Res 114:C00A08. doi:10.1029/2008JC005162

Purcell JE, Hopcroft RR, Kosobokova KN, Whitledge TE (2010) Distribution, abundance and predation effects of epipelagic ctenophores and jellyfish in the western Arctic Ocean. Deep-Sea Res II 57:127–135

Rand KM, Logerwell EA (2011) The first demersal trawl survey of benthic fish and invertebrates in the Beaufort Sea since the late 1970s. Polar Biol 34:475–488

Raskoff KA, Purcell JE, Hopcroft RR (2005) Gelatinous zooplankton of the Arctic Ocean: in situ observations under the ice. Pol Biol 28:207–217

Raskoff KA, Hopcroft RR, Kosobokova KN, Purcell JE, Youngbluth M (2010) Jellies under ice: ROV observations from the Arctic 2005 hidden ocean expedition. Deep Sea Res II 57:111–126

Ravelo AM, Konar B, Trefry JH, Grebmeier JM (2014) Epibenthic community variability in the northeastern Chukchi Sea. Deep-Sea Res II 102:119–131. doi:10.1016/j.dsr2.2013.07.017

Renaud PE, Morata N, Ambrose WG Jr, Bowie JJ, Chiuchiolo A (2007a) Carbon cycling by seafloor communities on the eastern Beaufort Sea shelf. J Exp Mar Biol Ecol 349:248–260

Renaud PE, Riedel A, Michel C, Morata N, Gosselin M, Juul-Pedersen T, Chiuchiolo A (2007b) Seasonal variation in benthic community oxygen demand: a response to an ice algal bloom in the Beaufort Sea, Canadian Arctic? J Mar Syst 67:1–12

Renaud PE, Carroll ML, Ambrose WG Jr (2008) Effects of global warming on Arctic sea-floor communities and its consequences for higher trophic levels. In: Duarte CM (ed) Impacts of global warming on polar ecosystems. Fundación BBVA, Bilbao

Ricciardi A, Bourget E (1999) Global patterns of macroinvertebrate biomass in marine intertidal communities. Mar Ecol Prog Ser 185:21–35

Richerol T, Rochon A, Blasco S, Scott DB, Schell TM, Bennett RJ (2008) Evolution of paleo sea-surface conditions over the last 600 years in the Mackenzie Trough, Beaufort Sea (Canada). Mar Micropaleontol 68:6–20

Riedel A, Michel C, Gosselin M, LeBlanc B (2008) Winter-spring dynamics in sea-ice carbon cycling in the coastal Arctic Ocean. J Mar Syst 74:918–932

Ringuette M, Fortier L, Fortier M, Runge JA, Bélanger S, Larouche P, Weslawski JM, Kwasniewski S (2002) Advanced recruitment and accelerated population development in Arctic calanoid copepods of the North Water. Deep-Sea Res 49:5081–5099

Robie WS, McRoy CP, Springer AM (1992) Phytoplankton biomass distribution in the Northern Bering Sea and Southern Chukchi Sea. In: Nagel PA (ed) Results of the third joint US–USSR Bering and Chukchi Seas expedition (BERPAC), Summer 1988, pp 123–127. US Fish and Wildlife Service, Washington, DC

Rowe GT (1983) Biomass and production of the deep-sea benthos, pp 97–121. In: Rowe GT (ed) The sea, vol 8. Wiley Interscience, New York

Rudels B, Jones EP, Anderson LG, Kattner G (1994) On the intermediate depth waters of the Arctic Ocean. Geophys Monogr 85:33–46

Rugh DJ, Fraker MA (1981) Gray whale (*Eschrichtius robustus*) sightings in Eastern Beaufort Sea. Arctic 34:186–187

Runge JA, Therriault J-C, Legendre L, Ingram RG, Demers S (1991) Coupling between ice micro-algal productivity and the pelagic, metazoan food web in southeastern Hudson Bay: a synthesis of results. Polar Rec 10:325–338

Sakshaug E (2004) Primary and secondary production in the Arctic seas, pp 57–81. In: Stein R, Macdonald RW (eds) The organic carbon cycle in the Arctic Ocean. Springer, New York

Saldarriaga JF, Taylor FJ, Cavalier-Smith T, Menden-Deuer S, Keeling PJ (2004) Molecular data and the evolutionary history of dinoflagellates. Eur J Protistol 40:85–111

Sampei M, Sasaki H, Makabe R, Forest A, Hattori H, Tremblay J-É, Gratton Y, Fukuchi M, Fortier L (2011) Production and retention of biogenic matter in the southeast Beaufort Sea during 2003–2004: insights from annual vertical particle fluxes of organic carbon and biogenic silica. Polar Biol 34:501–511

Scott DA (1968) Seasonal and vertical distributions of zooplankton in the Canadian Basin of the Arctic Ocean. M.Sc. thesis, University of Washington

Sejr MK, Nielsen TG, Rysgaard S, Risgaard-Petersen N, Sturluson M, Blicher ME (2007) Fate of pelagic organic carbon and importance of pelagic-benthic coupling in a shallow cove in Disko Bay, West Greenland. Mar Ecol Prog Ser 341:75–88

Sergeeva VM, Sukhanova IN, Flint M, Pautova LA, Grebmeier JM, Cooper LW (2010) Phytoplankton community in the Western Arctic in July–August 2003. Oceanology 50:184–197. doi:10.1134/S0001437010020049

Serreze MCH, Holland MM, Stroeve J (2007) Perspectives on the Arctic's shrinking sea-ice cover. Science 315:1533–1536

Sharma J, Bluhm BA (2010) Diversity of larger free-living nematodes from macrobenthos (>250 m) in the Arctic deep-sea Canada Basin. Mar Biodiv 41:455–465

Sherr EB, Sherr BF (2007) Heterotrophic dinoflagellates: a significant component of microzooplankton biomass and major grazers of diatoms in the sea. Mar Ecol Prog Ser 352:187–197

Sherr EB, Sherr BF, Fessenden L (1997) Heterotrophic protists in the central Arctic Ocean. Deep-Sea Res II 44:1665–1682

Sherr EB, Sherr BF, Wheeler PA, Thompson K (2003) Temporal and spatial variation in stocks of autotrophic and heterotrophic microbes in the upper water column of the central Arctic Ocean. Deep-Sea Res I 50:557–571

Sherr EB, Sherr BF, Hartz AJ (2009) Microzooplankton grazing impact in the Western Arctic Ocean. Deep-Sea Res II 56:1264–1273

Sherr EB, Sherr BF, Ross C (2013) Microzooplankton grazing impact in the Bering Sea during spring sea ice conditions. Deep Sea Res II 94:57–67

Shiomoto A (1999) Effect of nutrient on phytoplankton size in the Bering Sea basin. In: Loughlin TR, Ohtani K (eds) Dynamics of the Bering Sea. University of Alaska Sea Grant, Fairbanks, pp 323–340

Shirshov PP (1982) Plankton of the Arctic waters. Nauka Publishing, Moscow (in Russian)

Siano R, Alves-de-Souza C, Foulon E, El Bendif M, Simon N, Guillou L, Not F (2011) Distribution and host diversity of Amoebophryidae parasites across oligotrophic waters of the Mediterranean Sea. Biogeosciences 8:67–278

Sievert S, Anton-Erxleben F, Kiko R, Kramer M (2009) *Sympagohydra tuuli* (Cnidaria, Hydrozoa): first report from sea ice of the central Arctic Ocean and insights into histology, reproduction and locomotion. Mar Biol 156:541–554

Sirenko BI (2001) List of species of free-living invertebrates of Eurasian Arctic seas and adjacent deep waters. Explorations of the Fauna of the Seas 51:1–129

Sirenko BI (2003) First Pacific species of genus *Choneplax* (Mollusca, Polyplacophora). Ruthenica 13:33–39

Sirenko BI, Gagaev SY (2007) Unusual abundance of macrobenthos and biological invasions in the Chukchi Sea. Russ J Mar Biol 33:355–364

Sirenko BI, Koltun VM (1992) Characteristics of benthic biocenosis of the Chukchi and Bering seas. In: Results of the third joint US-USSR Bering and Chukchi seas expedition (BERPAC), Summer 1988, pp 251–261

Smith SL, Schnack-Schiel SB (1990) Polar zooplankton, pp 527–598. In: Smith WO Jr (ed) Polar oceanography, Part B, Chemistry, biology, and geology. Academic, New York

Soltwedel T, Bauerfeind E, Bergmann M, Budaeva N, Hoste E, Jaeckisch N, Juterzenka K, Matthießen J, Mokievsky V, Nöthig E, Quéric N, Sablotny B, Sauter E, Schewe I, Urban-Malinga B, Wegner J, Wlodarska-Kowalczuk M, Klages M (2005) HAUSGARTEN: multidisciplinary investigations at a deep-sea, long-term observatory in the Arctic Ocean. Oceanography 18:46–61

Spall MA, Pickart RS, Fratantoni PS, Plueddemann AJ (2008) Western Arctic shelfbreak eddies: formation and transport. J Phys Oceanogr 38:1644–1668

Spang A, Hatzenpichier R, Brochier-Armanet C, Rattel T, Tischler P, Spieck E, Streit W, Stahl DA, Wagner M, Schleper C (2010) Distinct gene set in two different lieages of ammonia-oxidizing archaea supports the phylum Thaumarchaeota. Trends Microbiol 18:331–340

Springer AM, McRoy CP, Turco KR (1989) The paradox of pelagic food webs in the northern Bering Sea – II. Zooplankton communities. Cont Shelf Res 9:359–386

Springer AM, McRoy CP, Flint MV (1996) The Bering Sea Green Belt: shelf edge processes and ecosystem production. Fish Oceanogr 5:205–223

Steele M, Ermold W, Zhang J (2008) Arctic Ocean surface warming trends over the past 100 years. Geophys Res Letts 35:L02614. doi:10.1029/2007GL031651

Stepanova VS (1937a) Biological indicators of currents in the northern Bering and southern Chukchi Seas. Issledovanija Morei SSSR 25:175–216 (in Russian)

Stepanova VS (1937b) Distribution of Bering Sea water in the Chukchi Sea (from data of the analysis of zooplankton from a cruise on the ice breaker Krasinin 1935). Trudy Arkticheskogo Nauchno-Issledovatel 0skogo Institute 82:113–143 (in Russian)

Steward GF, Smith D, Azam F (1996) Abundance and production of bacteria and viruses in the Bering and Chukchi Seas. Mar Ecol Prog Ser 131:287–300

Steward GF, Montiel JL, Azam F (2000) Genome size distributions indicate variability and similarities among marine viral assemblages from diverse environments. Limnol Oceanogr 45:1697–1706

Stroeve J, Holland MM, Meier W, Scambos T, Serreze M (2007) Arctic sea ice decline: faster than forecast. Geophys Res Lett 34. doi:10.1029/2007GL029703

Strom SL, Fredrickson KA (2008) Intense stratification leads to phytoplankton nutrient limitation and reduced microzooplankton grazing in the southeastern Bering Sea. Deep-Sea Res 55:1761–1774

Sukhanova IN, Semina HJ, Venttsel MV, Loughlin TR, Ohtani K (1999) Spatial distribution and temporal variability of phytoplankton in the Bering Sea. Dynamics of the Bering Sea. University of Alaska Sea Grant, Fairbanks, pp 453–484

Sukhanova IN, Flint MV, Whitledge TE, Stockwell DA, Rho TK (2006) Mass development of the planktonic diatom *Proboscia alata* over the Bering Sea shelf in the summer season. Oceanology 46:200–216

Sukhanova IN, Flint MV, Pautova LA, Stockwell D, Grebmeier JM, Sergeeva VM (2009) Phytoplankton of the western Arctic in the spring and summer of 2002: structure and seasonal changes. Deep-Sea Res II 56:1223–1236. doi:10.1016/j.dsr2.2008.12.030

Sun M-Y, Clough LM, Carroll ML, Dai J, Ambrose WG Jr, Lopez GR (2009) Different responses of two common Arctic macrobenthic species (*Macoma balthica* and *Monoporeia affinis*) to phytoplankton and ice algae: will climate change impacts be species specific? J Exp Mar Biol Ecol 376:110–121

Suttle CA (2007) Marine viruses – major players in the global ecosystem. Nat Rev Microbiol 5:801–812

Suzuki K, Minami C, Liu H, Saino T (2002) Temporal and spatial patterns of chemotaxonomic algal pigments in the subarctic Pacific and the Bering Sea during the early summer of 1999. Deep-Sea Res II 49:5685–5704

Taniguchi A (1984) Microzooplankton biomass in the arctic and subarctic Pacific Ocean in summer. Mem Nat Inst Polar Res Spec Iss 32:63–76

Terrado R (2011) Diversité et succession des protistes dans l'océan Arctique. Ph.D. thesis, Faculté des sciences et de génie, Doctorat interuniversitaire en océanographie Université Laval, Numéro unique: 27849

Terrado R, Lovejoy C, Massana R, Vincent WF (2008) Microbial food web responses to light and nutrients beneath the coastal Arctic Ocean sea ice during the winter-spring transition. J Mar Syst 74:964–977

Terrado R, Vincent WF, Lovejoy C (2009) Mesopelagic protists: diversity and succession in a coastal Arctic ecosystem. Aquat Microb Ecol 56:25–39

Thaler M, Lovejoy C (2012) Distribution and diversity of a protist predator *Cryothecomonas* (Cercozoa) in arctic marine waters. J Eukaryot Microbiol 59:291–299

Thaler M, Lovejoy C (2014) Environmental selection of marine stramenopile clades in the Arctic Ocean and coastal waters. Polar Biol 37:347–357

Thibault D, Head EJH, Wheeler PA (1999) Mesozooplankton in the Arctic Ocean in summer. Deep-Sea Res I 46:1391–1415

Tillmann U, Hesse K-J, Tillmann A (1999) Large-scale parasitic infection of diatoms in the Northfrisian Wadden Sea. J Sea Res 42:255–261

Timmermans M-L, Toole J, Proshutinsky A, Krishfield R, Plueddmann A (2008) Eddies in the Canada Basin, Arctic Ocean, observed from ice-tethered profilers. J Phys Oceanogr 38:133–145

Tomaru Y, Fujii N, Oda S, Toyoda K, Nagasaki K (2011) Dynamics of diatom viruses on the western coast of Japan. Aquat Microb Ecol 63:223–230

Tremblay J-É, Michel C, Hobson KA, Gosselin M, Price NM (2006) Bloom dynamics in early opening waters of the Arctic Ocean. Limnol Oceanogr 51:900–912

Tremblay G, Belzile C, Gosselin M, Poulin M, Roy S, Tremblay J-É (2009) Late summer phytoplankton distribution along a 3500 km transect in Canadian Arctic waters: strong numerical dominance of picoeukaryotes. Aquat Microb Ecol 54:55–70

Tremblay J-E, Bélanger S, Barter DG, Asplin M, Martin J, Darnis G, Fortier L, Gratton Y, Link H, Archambault P, Sallon A, Michel C, Williams WJ, Philippe B, Gosselin M (2011) Climate forcing multiplies biological productivity in the coastal Arctic Ocean. Geophys Res Lett 38:L18604

Turco K (1992a) Zooplankton taxa, abundance and biomass data. ISHTAR Data Report No. 6, Part 1 (1985–1987). Institute of Marine Science, University of Alaska Fairbanks, Fairbanks

Turco K (1992b) Zooplankton taxa, abundance and biomass data. ISHTAR Data Report No. 6, Part 2 (1988–1989). Institute of Marine Science, University of Alaska Fairbanks, Fairbanks

Vanreusel A, Clough L, Jacobsen K, Ambrose W, Jivaluk J, Ryheul V, Herman R, Vincx M (2000) Meiobenthos of the central Arctic Ocean with special emphasis on the nematode community structure. Deep-Sea Res I 47:1855–1879

Vaqué D, Guadayol Ò, Peters F, Felipe J, Angel-Ripoll L, Terrado R, Lovejoy C, Pedrós-Alió CS (2008) Seasonal changes in planktonic bacterivory rates under the ice-covered coastal Arctic Ocean. Limnol Oceanogr 53:2427–2438

Venter JC, Remington K, Heidelberg JF, Halpern AL, Rusch D, Eisen JA, Wu D, Paulsen I, Nelson KE, Nelson W, Fouts DE, Levy S, Knap AH, Lomas MW, Nealson K, White O, Peterson J, Hoffman J, Parsons R, Baden-Tillson H, Pfannkoch C, Rogers YH, Smith HO (2004) Environmental genome shotgun sequencing of the Sargasso Sea. Science 304:66–74

Vincent WF (2000) Cyanobacterial dominance in the polar regions. In: Whitton BA, Potts M (eds) The ecology of cyanobacteria. Kluwer, Netherlands, pp 321–340

Vinogradova NG (1997) Zoogeography of the abyssal and hadal zones. Adv Mar Biol 32:326–387

Virketis MA (1957) Some data on the zooplankton from the central part of the Arctic Basin (in Russian). Materials of scientific observations of the drift stations North Pole 3 and North Pole 4 1954/55 1, Moscow, pp 238–311 (in Russian)

Virketis MA (1959) Materials on the zooplankton of the central part of the Arctic Ocean (in Russian). Results of scientific observations on the drift stations North Pole 4 and North Pole 5, 1955–1956, Moscow, pp 132–206 (in Russian)

von Quillfeldt CH, Ambrose WG Jr, Clough LM (2003) High number of diatom species in first-year ice from the Chukchi Sea. Polar Biol 26:806–818

Wacasey JW (1975) Biological productivity of the southern Beaufort Sea: zoobenthic studies. Beaufort Sea Project Technical Report 12b. Department Environment, Victoria BC

Wacasey JW, Atkinson EG, Derick L, Weinstein A (1977) Zoobenthos data from the southern Beaufort Sea, 1971–1975. Fish Mar Serv Res Devel Tech Rep 41

Waleron M, Waleron K, Vincent WF, Wilmotte A (2007) Allochthonous inputs of riverine picocyanobacteria to coastal waters in the Arctic Ocean. FEMS Microbiol Ecol 5:356–365

Walker TR, Grant J, Cranford P, Lintern DG, Hill P, Jarvis P, Barrell J, Nozais C (2008) Suspended sediment and erosion dynamics in Kugmallit Bay and Beaufort Sea during ice-free conditions. J Mar Syst 74:794–809

Walkusz W, Paulic JE, Papst MH, Kwasniewski S, Chiba S, Crawford RE (2008) Zooplankton and ichthyoplankton data collected from the Chukchi and Beaufort Seas during the R/V Mirai cruise, September 2002. Can Data Rep Fish Aquat Sci 1211:1–34

Walkusz W, Paulic JE, Kwaniewski S, Williams W, Wong S, Papst M (2010) Distribution, diversity and biomass of summer zooplankton from the coastal Canadian Beaufort Sea. Polar Biol 33:321–355

Walsh JE, Johnson CM (1979) Analysis of Arctic sea ice fluctuations 1953–77. J Phys Oceanogr 9:580–591

Walsh JJ, McRoy CP (1986) Ecosystem analysis in the southeastern Bering Sea. Cont Shelf Res 5:259–288

Waterbury JB, Watson SW, Valois FW, Franks DG (1986) Biological and ecological characterization of the marine unicellular cyanobacterium synechococcus. In: Platt T, Li W (eds) Photosynthetic picoplankton. Can J Fish Aquat Sci Bull 214:71–120

Wehrmann LM, Risgaard-Petersen N, Schrum HN, Walsh E, Huh Y, Ikehara M, Peierre C, D'Hondt S, Ferdelman TG, Ravelo A, Takahashi K, Zarikian C (2011) Coupled organic and inorganic cycling in the deep subseafloor sediment of the northeastern Bering Sea Slope 9IODP EXP. 323. Chem Geol 284:251–261

Wells LE, Deming JW (2006) Modelled and measured dynamics of viruses in Arctic winter sea-ice brines. Environ Microbiol 8:1115–1121

Wells LE, Cordray M, Bowerman S, Miller LA, Vincent WF, Deming JW (2006) Archaea in particle-rich waters of the Beaufort Shelf and Franklin Bay, Canadian Arctic: clues to an allochthonous origin? Limnol Oceanogr 51:47–59

Werner I, Gradinger R (2002) Under-ice amphipods in the Greenland Sea and Fram Strait (Arctic): environmental controls and seasonal patterns below the pack ice. Mar Biol 140:317–326

Williams WJ, Carmack EC (2008) Combined effect of wind-forcing and isobath divergence on upwelling at Cape Bathurst, Beaufort Sea. J Mar Res 66:645–663

Williams WJ, Carmack EC, Shimada K, Melling H, Aagaard K, Macdonald RW, Ingram RG (2006) Joint effects of wind and ice motion in forcing upwelling in Mackenzie Trough, Beaufort Sea. Cont Shelf Res 26:2352–2366

Williams WJ, Melling H, Carmack EC, Ingram RG (2008) Kugmallit Valley as a conduit for cross-shelf exchange on the Mackenzie Shelf in the Beaufort Sea. J Geophys Res 113:C02007. doi:10.1029/2006JC003591

Wood HL, Spicer JI, Kendall MA, Lowe DM, Widdicombe S (2011) Ocean warming and acidification; implications for the Arctic brittlestar *Ophiocten sericeum*. Polar Biol 34:1033–1044

Woodgate RA, Aagaard K, Weingartner TJ (2005) Monthly temperature, salinity, and transport variability of the Bering Strait through flow. Geophys Res Lett 32. doi:10.1029/2004GL021880

Woodgate RA, Weingartner TR, Lindsay R (2010) The 2007 Bering Strait oceanic heat flux and anomalous Arctic sea-ice retreat. Geophys Res Lett 37. doi:10.1029/2009GL041621

Würsig B, Dorsey EM, Richardson WJ, Wells RS (1989) Feeding, aerial and play behaviour of the bowhead whale, *Balaena mysticetus*, summering in the Beaufort Sea. Aquat Mamm 15:27–37

Yager PL, Connelly TL, Mortazavi B, Wommack KE, Bano N, Bauer JE, Opsahl S, Hollibaugh JT (2001) Dynamic bacterial and viral response to an algal bloom at subzero temperatures. Limnol Oceanogr 46:790–801

Yamamoto-Kawai M, McLaughlin FA, Carmack EC, Nishino S, Shimada K (2009) Aragonite undersaturation in the Arctic Ocean: effects of ocean acidification and sea ice melt. Science 326:1098–1100

Yoon HS, Price DC, Stepanauskas R, Rajah VD, Sieracki ME, Wilson WH, Yang EC, Duffy S, Bhattacharya D (2011) Single-cell genomics reveals organismal interactions in uncultivated marine protists. Science 332:714–717

Zubkov MV, Terran GA (2008) High bacterivory by the smallest phytoplankton in the North Atlantic Ocean. Nature 455:224–226

Chapter 11
Marine Fishes, Birds and Mammals as Sentinels of Ecosystem Variability and Reorganization in the Pacific Arctic Region

Sue E. Moore, Elizabeth Logerwell, Lisa Eisner, Edward V. Farley Jr., Lois A. Harwood, Kathy Kuletz, James Lovvorn, James R. Murphy, and Lori T. Quakenbush

Abstract Extreme reductions in sea ice extent and thickness in the Pacific Arctic Region (PAR) have become a hallmark of climate change over the past decade, but their impact on the marine ecosystem is poorly understood. As top predators, marine fishes, birds and mammals (collectively, upper trophic level species, or UTL) must adapt via biological responses to physical forcing and thereby become sentinels to ecosystem variability and reorganization. Although there have been no coordinated long-term studies of UTL species in the PAR, we provide a compilation of information for each taxa as an ecological foundation from which future investigations can benefit. Subsequently, we suggest a novel UTL-focused research framework focused on measurable responses of UTL species to environmental variability as one way to

S.E. Moore (✉)
Fisheries Office of Science and Technology, National Oceanic and Atmospheric Administration, Seattle, WA, USA
e-mail: sue.moore@noaa.gov

E. Logerwell • L. Eisner • E.V. Farley Jr. • J.R. Murphy
Alaska Fisheries Science Center, National Oceanic and Atmospheric Administration, Seattle, WA, USA

L.A. Harwood
Canada Department of Fisheries and Oceans, Yellowknife, NT, Canada

K. Kuletz
USF&WS, Alaska Region, Anchorage, AK, USA

J. Lovvorn
Southern Illinois University, Carbondale, IL, USA

L.T. Quakenbush
Alaska Department of Fish and Game, Fairbanks, AK, USA

J.M. Grebmeier and W. Maslowski (eds.), *The Pacific Arctic Region: Ecosystem Status and Trends in a Rapidly Changing Environment*, DOI 10.1007/978-94-017-8863-2_11,

ascertain shifts in the PAR marine ecosystem. In the PAR, indigenous people rely on UTL species for subsistence and cultural foundation. As such, marine fishes, birds and mammals represent a fundamental link to local communities while simultaneously providing a nexus for science, policy, education and outreach for people living within and outside the PAR.

Keywords Arctic marine ecosystems • Marine fish • Marine mammals • Seabirds • Sea ducks

11.1 Introduction

The marine ecosystem of the Pacific Arctic Region (PAR: northern Bering, Chukchi and Beaufort seas) has undergone dramatic physical changes in the past decade. The iconic 'signal' of this change is the extreme sea ice minima observed each September since 2002, with dramatic seasonal retreats since September 2007 (Perovich 2011; Stroeve et al. 2011; Frey et al. 2014, this issue). This signal has been accompanied by a profound reduction in sea ice thickness (Maslanik et al. 2011), a shift in regional climatology marked by increased summer air temperatures that extend through autumn (Overland 2009; Overland et al. 2014, this issue), increases in both fresh water content (Proshutinsky et al. 2009) and ocean temperature (Shimada et al. 2006; Woodgate et al. 2012), and an overall ~20 % increase in primary production as compiled from satellite observations for the 1998–2009 period (Arrigo and van Dijken 2011). The recent discoveries of massive (>100 km) under-ice phytoplankton blooms on the continental shelf (Arrigo et al. 2012) and the deposition of significant ice algal biomass (average = 9 g C/m^2) in the central Arctic basins (Boetius et al. 2013) suggest that past estimates of net primary production in arctic waters is at least tenfold too low. Together these observations indicate that the PAR has experienced a rapid and directional change, with the potential to shift the marine ecosystem to a new state. Although biological responses to these physical changes are complex and therefore difficult to elucidate (Grebmeier 2012; Wassmann 2011), upper trophic level (UTL) species, including marine fishes, birds and mammals can act as sentinels to ecosystem alterations to which they must adapt to survive.

Predictions of the effects of physical forcing on the population dynamics of UTL species should be based on empirical data, ideally multi-decadal time series of both physical (e.g. sea ice variability) and biological factors (e.g., body condition, productivity, recruitment, survival, seasonal distribution and abundance). Unfortunately, such data rarely exist. Even when physical and biological data are available, unequivocally determining that physical factors are influencing the biological ones remains difficult (Krebs and Berteaux 2006). Fortunately, comparing seasonal patterns of distribution, abundance, diet and body condition across UTL species can provide clear evidence of ecosystem variability and reorganization. A striking example of this was empirical evidence for regime shifts in North Pacific and Bering Sea ecosystems in 1977 and 1989, based in part on examination of 69 biological time series (Hare and Mantua 2000). A notable feature in this analysis was the "relative clarity" of ecosystem reorganization found in the biological records compared to a "relative lack of clear changes expressed by indices of Pacific climate", causing the authors to conclude that monitoring ecosystems "may allow for an earlier identification of regime shifts than is possible from monitoring climate data alone". Indeed, seminal papers describing basin-wide shifts in zooplankton, shrimp, fish and seabird communities following the 1977 regime shift (e.g., Francis et al. 1998; Anderson and Piatt 1999) underscored the usefulness of multi-species approaches to identify ecosystem reorganization.

11.1.1 Ecological Scale

Ecological scale, described as the interface between population biology and ecosystem science (Levin 1992; Norris 2003), is fundamental to the use of UTL species as sentinels of ecosystem variability (e.g., Moore 2005; Moore and Huntington 2008). Simply stated, it is essential to consider the spatial and temporal scales at which each species 'samples' the environment to gain information on ecosystem state. The temporal ecological scale for fishes, marine birds and marine mammals extends from days to decades (to centuries for bowhead whales), while their spatial ecological scale extends from meters to thousands of kilometers, as defined by each species natural history (Fig. 11.1a). Upper trophic level species respond to ecosystem variability via non-linear complex trophic pathways reflective of their position along the ecological scale ladder, which in turn can be related to oceanographic processes at comparable spatial and temporal domains (Fig. 11.1b). Therefore, consideration of UTL species as ecosystem sentinels must begin with an overview of their ecology in the PAR; that is, what is known about the natural history of marine fishes, birds and mammals in relation to sea ice and other oceanographic features such as bathymetry, temperature, salinity, fronts and eddies. From that foundation, case studies are provided to show how reorganization in the PAR ecosystem might be identified from extrinsic and intrinsic responses of UTL species. There follows an evaluation of UTL species as sentinels of ecosystem variability and reorganization, including how aggregate indicators may strengthen the role of UTL as sentinels and suggestions for a thresholds-based research framework. The chapter concludes by emphasizing the need to integrate sampling of UTL species with oceanographic and community-based research and suggests that these species provide a fundamental link to subsistence-based human communities and often act as 'ambassadors' for public outreach and education with regard to climate change in the Arctic.

To our knowledge, this is the first attempt to consider how marine fishes, birds and mammals might collectively serve as indicators of ecosystem variability and

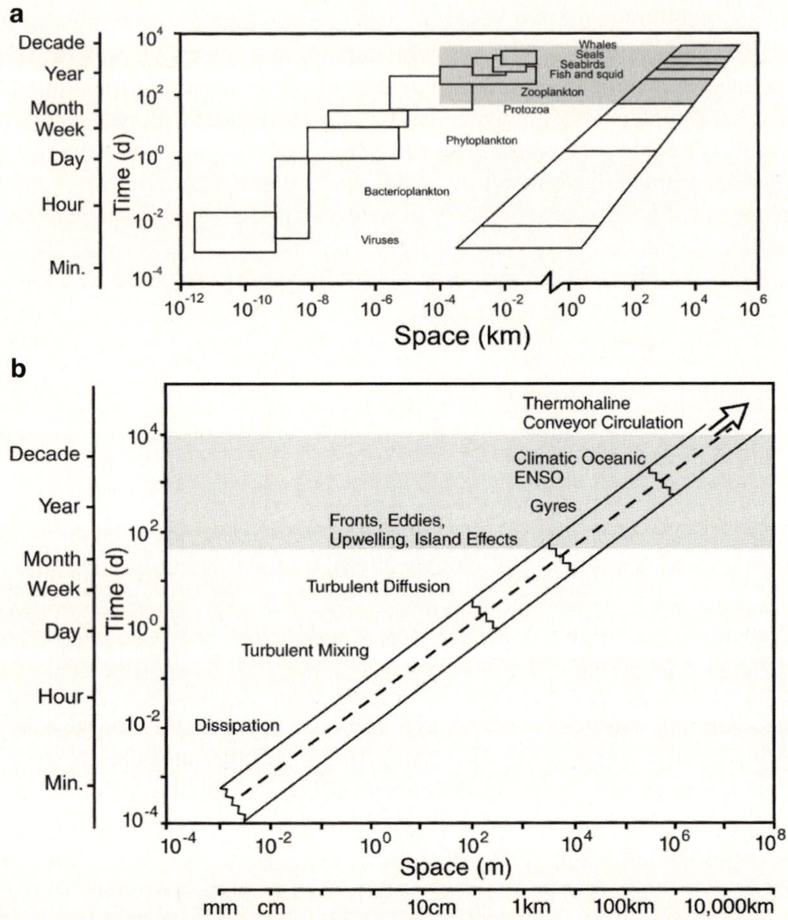

Fig. 11.1 The ecological scale for UTL species (*shaded area*) extends from months to decades and from 10 to 1,000s km, as defined by species' natural history (**a**) oceanographic features such as fronts, eddies, gyres, ENSO events and climatic features such as sea ice (**b**) affect availability and concentration of UTL prey, as well as the distribution and movements of the animals themselves (Modified from Moore 2005)

reorganization in the PAR. This work builds upon earlier contributions by many researchers, especially Frost and Lowry (e.g., 1984, 1981 and references therein), Springer et al. (e.g., 1984, 1986 and references therein) Piatt and Springer (2003) and Piatt et al. (1991, 2007 and references therein), who provided pioneering papers on trophic relationships among marine fishes, birds and mammals in the Alaskan Arctic. A comprehensive review of the literature or data available for all UTL species in the PAR is beyond the scope of this chapter. Rather, here we provide a framework and champion the case for inclusion of UTL species in multidisciplinary research of the PAR, during a period of rapid transition in the marine ecosystem.

11.2 Overview: Ecology of Upper Trophic Level (UTL) Species

The oceanography of the PAR is comparatively well described, with reviews of bathymetry, seasonal patterns of sea ice extent, water masses and current flow relative to the ecology of seabirds and whales provided in Piatt and Springer (2003), Moore et al. (2000) and Niebauer and Schell (1993). In brief, the PAR consists of a broad, shallow (40–70 m deep) continental shelf extending north from St. Matthew Island in the Bering Sea to the narrow shelf, slope and deep basin of the Beaufort Sea. Sea ice covers the region from October through June, although the ice free period now often extends into November (Perovich 2011; Frey et al. 2014, this issue). During the ice-covered period, an extensive and predictable system of ice-flaw zones and polynyas develop near islands and along the coast, providing open-water refugia for species that over-winter in the PAR (Niebauer and Schell 1993). During the ice-free period, three distinct water masses are recognized – the cold, saline and nutrient rich Anadyr Water (AW) to the west, the seasonally warm, low salinity Alaskan Coastal Water (ACW) that hugs the Alaska coast to the east and the central Bering Shelf Water (BSW) which often mixes with AW and ACW when passing through the shallow and turbulent Bering Strait (Weingartner et al. 2005). In-flow of these three water masses through Bering Strait defines the character of the Chukchi Sea and strongly influences the hydrography of the Beaufort Sea. There is a marked wind-driven seasonal cycle, with maximum and minimum northward flow in July and February, respectively (Aagaard et al. 1985; Woodgate et al. 2005). North of Bering Strait, currents are bathymetrically channeled, with AW crossing the Chukchi shelf and (largely) flowing into the Beaufort basin via Herald Canyon, BSW flowing north between Herald and Hanna shoals and ACW flowing along the Alaskan coast and into the Beaufort sea via Barrow Canyon (Pickart et al. 2010). Often there is little lateral mixing among these water masses such that fronts defined by marked differences in sea surface temperature (SST) can be identified on satellite images (e.g., Piatt and Springer 2003), providing a snapshot of the heterogeneity of the physical system.

With regard to prey for UTL species, much of the primary production in the PAR is coupled to the seafloor on the vast continental shelves, resulting in localized regions of very high benthic community biomass (e.g., Grebmeier 2012; Nelson et al. 2014, this volume). Conversely, much of the zooplankton biomass, including large copepods (*Calanus* and *Neocalanus* spp.) and euphausiids (mainly *Thysanoessa* spp.) are advected into the PAR from the Bering Sea; up to 1.8 M metrictons are carried into the Chukchi Sea annually by one estimate (Springer et al. 1989; Nelson et al. 2014, this volume). Like the water masses that transport them, patterns of zooplankton abundance and biomass are heterogeneous in the PAR and subject to change with the transport rate and penetration of Pacific water (Hopcroft et al. 2010; Eisner et al. 2013). It is important to recognize that, for UTL species, much of their benthic prey is locally derived, while pelagic prey is often transported into the system and so subject to biophysical processes in the North Pacific and southern Bering Sea.

Table 11.1 Mean catch per unit effort (CPUE in kg/ha) of common marine fish species found in the Bering, Chukchi and Beaufort Seas

		Bering Sea 2008[a]	Chukchi Sea 1990[b]	Beaufort Sea 2008[c]
Common name	Scientific name	CPUE (kg/ha)	CPUE (kg/ha)	CPUE (kg/ha)
Arctic cod	*Boreogadus saida*	1.04	3.02	6.12
Arrowtooth/Kamchatka flounder	*Atheresthes stomias/A. evermanni*	11.87	*	*
Bering flounder	*Hippoglossoides robustus*	0.45	0.18	0.11
Sculpins	Cottidae	4.22	0.76	0.03
Flathead sole	*Hippoglossoides elassodon*	10.81	*	*
Greenland turbot	*Reinhardtius hippoglossoides*	0.27	<0.01	*
Northern rock sole	*Lepidopsetta polyxystra*	41	*	*
Pacific cod	*Gadus macrocephalus*	8.65	0.12	*
Saffron cod	*Eleginus gracilis*	<0.01	0.39	*
Walleye pollock	*Theragra chalcogramma*	61.2	0.02	0.13
Yellowfin sole	*Limanda aspera*	42.4	*	*
Eelpouts	Zoarcidae	0.5	0.21	0.03

The CPUE reported for the Beaufort Sea is from unlined bottom trawl nets only. * = no catch of that species
[a]J. Conner, AFSC, personal communication
[b]Barber et al. (1997)
[c]Rand and Logerwell (2010)

11.2.1 Fishes and Crabs

There are more marine fish species in the boreal Pacific (south of approx. latitude 60 ° N) than in the Arctic (Allen and Smith 1988). Indeed, the Bering Sea supports one of the most productive commercial fisheries in the world (Table 11.1; Ianelli et al. 2012). The best-studied region is the southern Bering Sea, with comparatively few fishery-independent surveys for marine fishes in the northern Bering Sea (hereafter, NBS). Overall, available data show a marked decline in species richness and abundance along the latitudinal gradient extending north from the eastern Bering Sea shelf to the Chukchi Sea (Stevenson and Lauth 2012).

11.2.1.1 Northern Bering and Chukchi Seas

The Bering Sea-Aleutian Salmon International Survey (BASIS) was developed to investigate the biological response of salmon during a period of climate change (Helle 2009). The survey typically extends from the Alaska Peninsula to just north of Saint Lawrence Island (SLI), but in 2007 included the Bering Strait and north-eastern Chukchi Sea. Of the Pacific salmon, chum (*Oncorhynchusketa*) and pink (*O. gorbuscha*) have the broadest distribution and occur, as spawning populations, in the Arctic (Stephenson 2006; Irvine et al. 2009). A review of all known captures

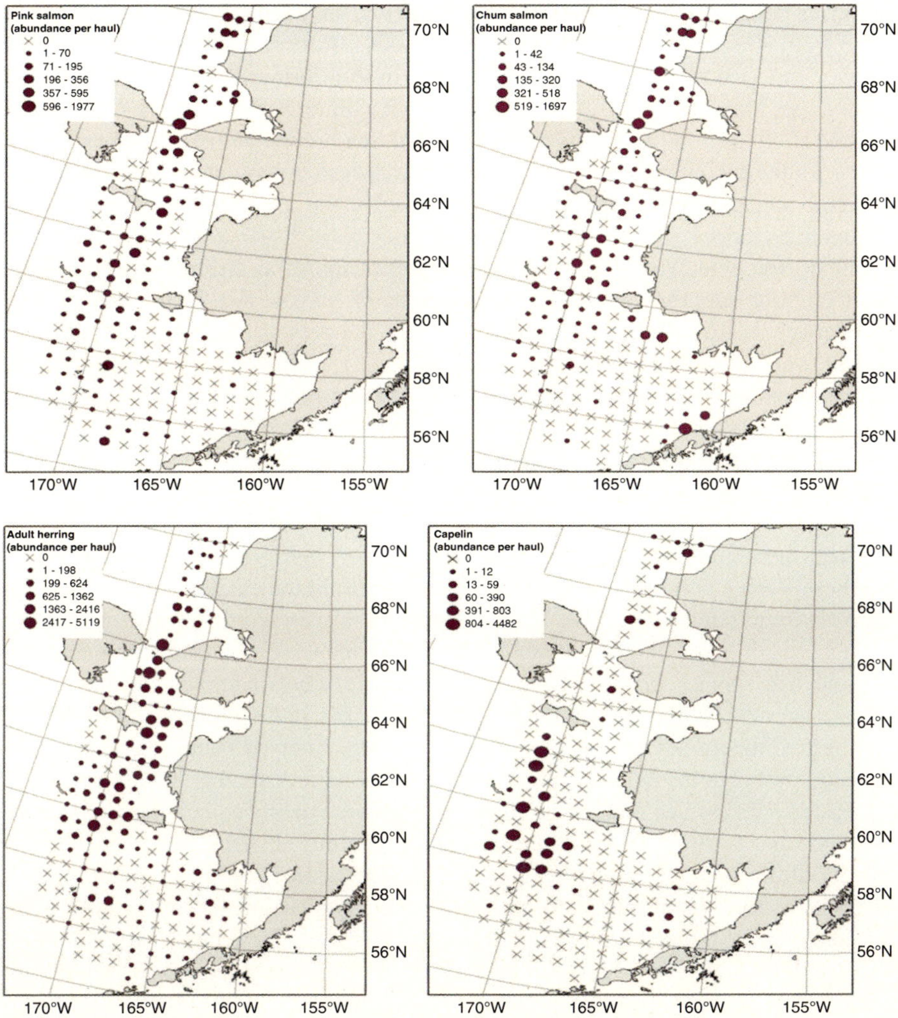

Fig. 11.2 Distribution of juvenile pink and chum salmon, Pacific herring and capelin from the 2007 BASIS survey

in the Canadian western Arctic (through 2003) documented the occurrence of other Pacific salmon species, including: chinook (*O. tshawytscha*), sockeye (*O. nerka*) and coho (*O. kisutch*). It has been suggested that salmon populations that occupy the Bering Sea are expanding their range into the Arctic, coincident with ocean warming (Moss et al. 2009), but more evidence is needed to confirm this hypothesis (Stephenson 2006).

During the 2007 BASIS survey, Moss et al. (2009) documented the distribution, diet and growth of juvenile pink and chum salmon from the northern Bering north into the Chukchi Sea. They found highest densities of salmon in the vicinity of Bering Strait (Fig. 11.2). Genetic stock analysis showed that juvenile chum salmon in Bering Strait were predominantly from N.E. Russia, whereas chum in the Chukchi were from Alaskan stocks (Kondzela et al. 2009). Information on diet and growth showed that juvenile pink and chum salmon in the Bering Strait and Chukchi Sea were larger than in the coastal Bering Sea, with salmon in the northern Chukchi the largest of all. Salmon in the northern Chukchi Sea preyed on fish, whereas salmon in the southern Chukchi Sea and Bering Strait preyed on zooplankton (euphausiids and crab megalope). Murphy et al. (2009) describe the stock-structured distribution of juvenile chinook salmon and showed that they were distributed as far north as the Chukchi Sea, with Yukon River fish found as far north as Bering Strait. Of note, large chinook salmon were caught by residents of Barrow, Alaska in summer 2010 (C. George, personal communication). Notably, the presence of maturing salmon in Arctic waters north of known salmon producing river drainages likely reflects straying and not colonization (Stephenson 2006). Unless or until new sustainable populations are established in Arctic river drainages, continued warming of the Arctic may simply result in increased straying rates in salmon.

Capelin (*Mallotus villosus*), Pacific herring (*Clupea pallasii*) and rainbow smelt (*Osmerus mordax*) are the dominant pelagic fish biomass in the northern Bering and Chukchi seas, along with juvenile walleye pollock (*Theragra chalcogramma*) in the northern Bering and Arctic cod (*Boreogadus saida*) in the Chukchi. These fish provide an important forage base for a wide range of piscivorous predators such as salmon, seabirds and marine mammals. Pacific herring were among the most abundant pelagic fish species in the Bering Sea in surface trawls sampled during the fall BASIS surveys (Fig. 11.2), and were found primarily in the colder saltier NBS and southern Chukchi seas (Eisner et al. 2013). Catch rates of Pacific herring during BASIS surveys were much higher north of 60 °N latitude than at lower latitudes of the eastern Bering Sea shelf (Fig. 11.3). Capelin were nearly equally abundant in the northern and southern Bering Sea and were generally distributed offshore within the middle domain, between 50 and 100 m bottom depth. Conversely, rainbow smelt are a nearshore anadromous species and were abundant in the NBS, but largely absent from the Chukchi Sea. Capelin and herring diets varied with latitude during the BASIS surveys (Fig. 11.4). In the southern Bering Sea (south of 60 °N), herring consumed primarily walleye pollock and euphausiids. In the NBS the relative abundance of pollock and euphausiids in their diet declined with amphipods and pelagic tunicates (*Oikopleura* spp.) gaining in importance. Capelin diets also changed dramatically south and north at 60 °N. To the south, capelin consumed primarily small copepods and euphausiids. To the north capelin consumed nearly exclusively large copepods. Finally, saffron and Arctic cod were absent south of SLI, but relatively abundant in the northern Chukchi Sea during the 2007 BASIS survey (Eisner et al. 2013). This contrasts with findings of Cui et al. (2009) who reported that Arctic cod, Bering flounder and snailfish (Liparidae) as the dominant species south of SLI, while Arctic staghorn sculpin (*Gymnocanthus tricuspis*), shorthorn sculpin (*Myoxocephalus scorpius*) and Arctic alligatorfish (*Ulcina olrikii*) were the dominant species north of the island from two bottom trawl surveys in 2006 and 2007.

Arctic cod are one of the most abundant fish in the Chukchi (and Beaufort) Sea (Frost and Lowry 1983; Barber et al. 1997; Norcross et al. 2010; Thedinga et al. 2013).

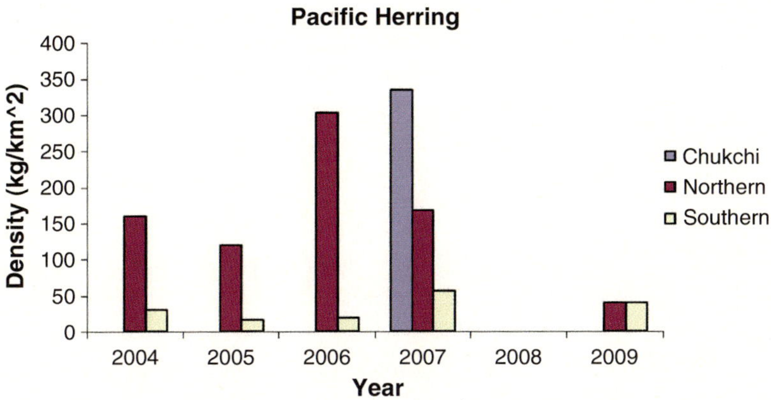

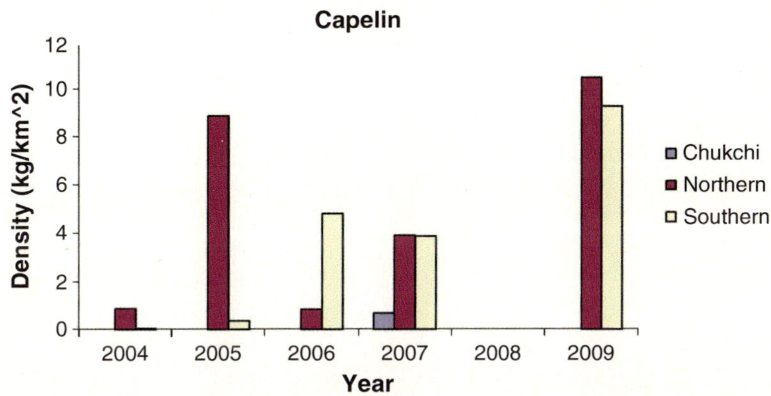

Fig. 11.3 Annual density of Pacific herring and capelin from 2004 to 2009 BASIS surveys in the Chukchi (2007 only), northern and southern Bering seas

They are important prey for seabirds and marine mammals and are in turn important consumers of zooplankton (Frost and Lowry 1981, 1983; Bradstreet et al. 1986; Jarvela and Thorsteinson 1999). As adults, Arctic cod are distributed across a broad range of depths and habitats, from nearshore bays to offshore shelves, both in benthic and pelagic waters (Craig et al. 1985; Benoit et al. 2008; Norcross et al. 2010; Rand and Logerwell 2010; Crawford et al. 2011). In the Chukchi Sea, Arctic cod densities have been reported to be higher in BSW compared to ACW, perhaps due to a higher abundance of zooplankton there (Gillispie et al. 1997). However, during BASIS 2007, juvenile Arctic cod abundances were higher in the ACW (Eisner et al. 2013), possibly due to changes in distribution associated with life history stages. Similarly, juvenile arctic cod dominated bottom trawl catches (56 % of fish caught) while capelin dominated beach seine catches (83 % of fish caught) during late-summer nearshore fish sampling in the northeastern Chukchi Sea (Thedinga et al. 2013). Annual variations in catches and species composition were attributed to fluctuations in water temperature, sea ice and wind conditions.

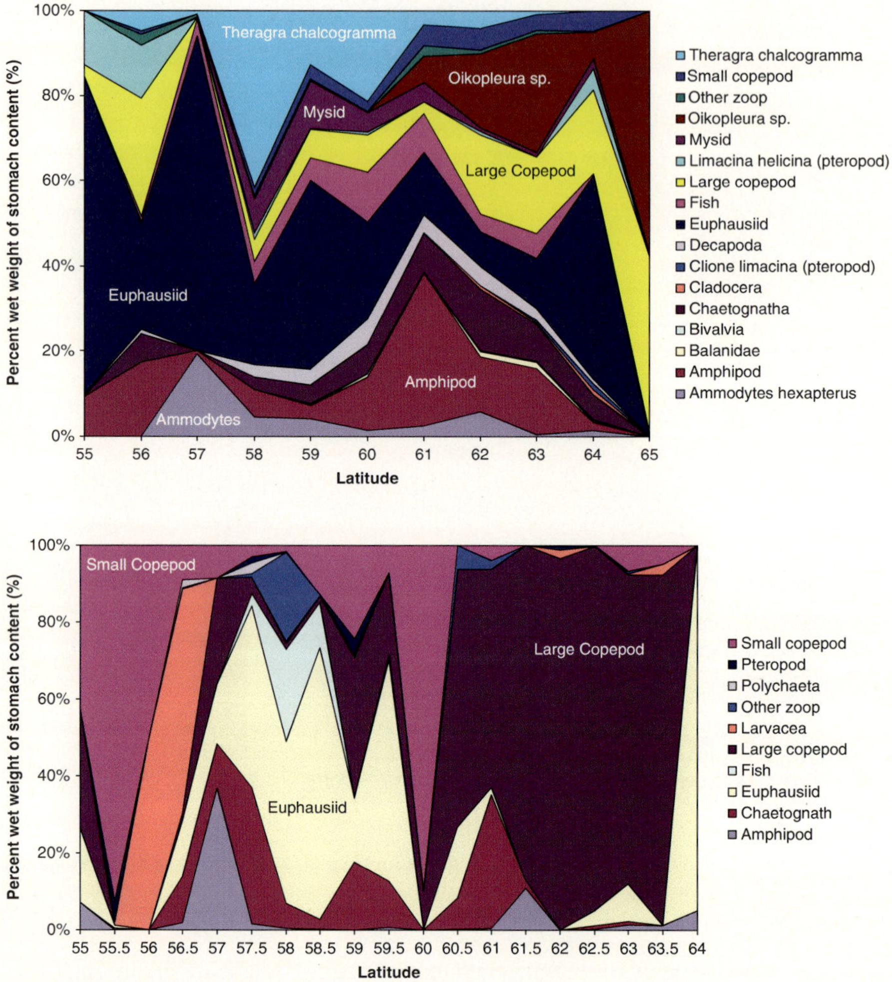

Fig. 11.4 Variability of Pacific herring (top) and capelin (bottom) diet with latitude

Strong linkages between primary production and the distribution and abundance of benthic macrofauna are typical of seasonally ice-covered systems (Wassmann 2011; Grebmeier 2012). The reduced numbers of pelagic grazers result in a large flux of uneaten phytoplankton to the benthos, which drives an abundant and diverse macrofaunal community (Dunton et al. 2005; Grebmeier et al. 2006a). Recent and on-going multidisciplinary investigations of the Chukchi Sea include the Shelf-Basin Interaction Study, the Russian-American Long-term Census of the Arctic (RUSALCA), the Chukchi Sea Offshore Monitoring in Drilling Area program and research supported by the oil and gas industry. The results of benthic ecology studies in 2008 and 2009 are typical of the area (Blanchard et al. 2010). Seventy percent of the epifaunal benthic biomass was brittle stars, 6 % was crabs (primarily snow crab, *Chionoecetes opilio*), 4–5 % was sea cucumbers and gastropods, 3 % was bivalves and colonial organisms such as ascidians, sponges, hydrozoans and bryozoans, 2 % was shrimp and sea anemones and 1 % was hermit crabs and sea stars. Snow crab is a circumpolar species for which there are substantial fisheries in the Atlantic and Pacific Oceans. Snow crabs in the Pacific occur in the northern Sea of Japan and the Bering, Chukchi and Beaufort Seas (Bluhm et al. 2009; Rand and Logerwell 2010). Paul et al. (1997) found snow crab at all depths (14–52 m) in the northeastern Chukchi Sea, with greatest abundance and biomass offshore south of 70 °N. They also reported that snow crabs tended to be smaller in the Chukchi Sea than those from the Bering Sea or northern Atlantic. All relatively large crabs (>34 mm carapace width females, and >35 mm males) were reproductively mature during the summer survey.

11.2.1.2 Beaufort Sea

Fish assemblages of the Beaufort Sea are poorly known compared to the northern Bering and Chukchi Seas. The most recent survey of offshore marine fishes and invertebrates in US waters was conducted in the western Beaufort Sea in August 2008 (Logerwell et al. 2011; Fig. 11.5a). Benthic fish and invertebrate species composition, distribution and abundance were assessed using standard Alaska Fisheries Science Center bottom trawl gear (Rand and Logerwell 2010). Fishes made up 6 % of the bottom trawl catch by weight, with invertebrates comprising the remaining 94 % of the catch. The four most abundant benthic fish were Arctic cod, eelpouts (*Lycodes* spp.), sculpin, and walleye pollock (Fig. 11.6a). The most abundant invertebrates were notched brittle stars (*Ophiura sarsi*), snow crab, mollusks (*Musculus* spp.), and the mudstar (*Ctenodiscus crispatus*).

Fish were also sampled in nearshore coastal waters of Cooper Island, a barrier island in the western Beaufort Sea, each August 2004–2007, and 2009, and in September 2009 (Fig. 11.5b). Capelin was the most abundant species captured by beach seine in the Beaufort Sea (Fig. 11.6b), whereas least cisco (*Coregonus sardinella*) was the most abundant species captured by beach seine in Elson Lagoon. Nearshore trawl catch in the Beaufort Sea was dominated by Arctic cod (Fig. 11.6b). The similarity in fish fauna near Cooper Island to other nearshore areas of the Beaufort Sea studied over 20 years ago (Craig et al. 1985; Jarvela and Thorsteinson 1999) indicates that species use of coastal waters has remained relatively unchanged. In Canadian waters, offshore sites in the Mackenzie River estuary and adjacent Beaufort Sea were surveyed in July and August from 2004 through 2007 (e.g., Majewski et al. 2011). Fishing gear included a multi-mesh gill net, mid-water trawl, and benthic trawl. Gill net catches were dominated by Pacific herring and midwater catches by Arctic cod. Benthic catches were dominated by Arctic cod, stout eelblenny (*Anisarchus medius*), Arctic alligator fish, Arctic staghorn sculpin and ribbed sculpin (*Triglops pingelii*).

The association of Arctic cod with sea ice is well documented in coastal and offshore Arctic locations across the globe (reviewed Bradstreet et al. 1986).

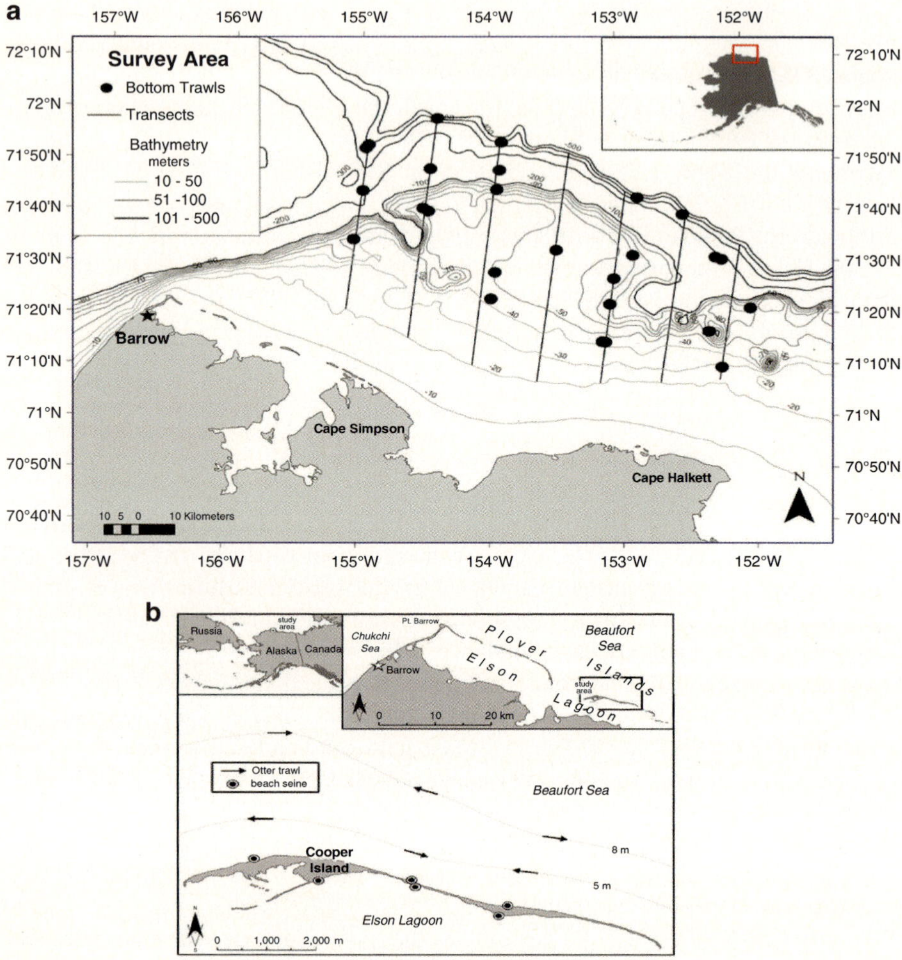

Fig. 11.5 Offshore (**a**: 2008 only) and coastal (**b**: 2004–2009) fish sampling sites in the western Beaufort Sea

Arctic cod not only dominate the benthic community in the Beaufort Sea, but also are abundant in the pelagic community, as indicated by a hydroacoustic survey conducted concurrently with the benthic survey discussed above (Parker-Stetter et al. 2011) and by hydroacoustic surveys in Lancaster Sound and Franklin Bay in the Canadian Beaufort Sea (Crawford and Jorgenson 1996; Benoit et al. 2008). Arctic cod use a wide variety of benthic and pelagic niches including the ice edge, ice-pocket refugia, demersal and near-bottom habitats (Crawford 2009; Crawford et al. 2011), and occur as both dense shoals (Benoit et al. 2008), and as non-schooling individuals (Bradstreet et al. 1986; Hop et al. 1997). Pelagic Arctic cod have been observed to aggregate in the warmest part of the water column in deep canyons

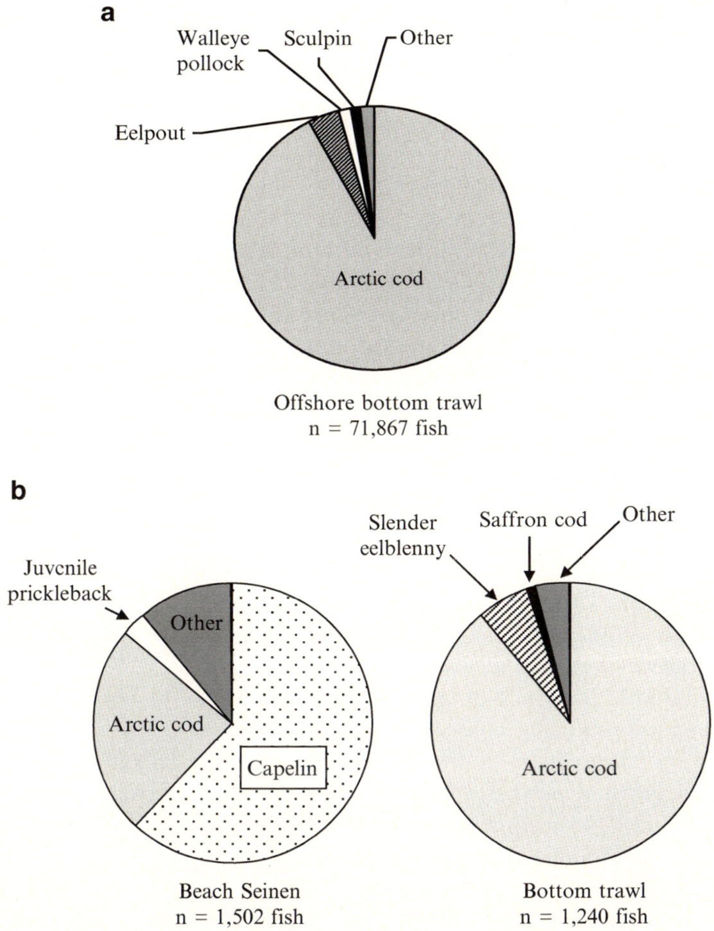

Fig. 11.6 Composition of fish species at offshore (**a**) and coastal (**b**) sampling sites in the western Beaufort Sea

(such as Barrow canyon and McKenzie canyon) and over the deep the Beaufort Sea slope, perhaps for energetic benefit, even if the temperature difference is only a few tenths of a degree (Crawford et al 2011). Examples of this have been found in both Lancaster Sound and in the Beaufort Sea (Crawford and Jorgenson 1996; Hop et al. 1997; Crawford 2009), and most recently in Franklin Bay where huge shoals of cod were found in the warmer-deeper areas of the Bay (Benoit et al. 2008). In contrast, benthic and pelagic Arctic cod at the Beaufort Sea shelf-break have been observed to be associated with cold, high salinity water derived from the Chukchi Sea, perhaps due to high secondary productivity expected to be found in these waters (Logerwell et al. 2011; Parker-Stetter et al. 2011).

11.2.2 Marine Birds

Marine birds include species commonly called 'seabirds' and others called 'seaducks', as well as phalaropes (*Phalaropus* spp.), which are shorebirds that feed in marine waters and can be found in pelagic waters during migration. In the PAR, an analysis of at-sea counts from ships (Sigler et al. 2011) categorized marine bird taxa into two groups: a NBS-Chukchi Sea cluster dominated by auklets (*Aethia* spp.) in the NBS and short-tailed shearwaters (*Puffinus tenuirostris*) in the Chukchi Sea; and a Beaufort Sea cluster, dominated by benthic-foraging eiders and long-tailed ducks (*Clangula hyemalis*), surface foraging glaucous gulls (*Larus hyperboreus*) and Arctic terns (*Sterna paradisaea*) and the pursuit diving black guillemot (*Cepphus grylle*). It is important to note that data from ship-based surveys conducted in mid-summer do not reflect the extensive use of the nearshore Chukchi Sea by sea ducks during spring migration and late-summer molt (see Sect. 11.3.2.2). Studies of marine birds have been conducted from ships at sea and on breeding colonies, as summarized below.

11.2.2.1 At-Sea Distribution

There are over 40 species of marine birds common to the PAR, based on pelagic surveys (Table 11.2). Much of what we know about marine birds at sea has been gathered opportunistically, but recent surveys on research-oriented icebreakers have improved our knowledge of marine bird distribution and relative abundance in arctic waters. The NBS, in particular the waters influenced by the oceanic AW, as well as the Bering Strait and shallow waters of the Chukchi Sea, support high densities of birds, comparable to the Bering Sea shelf edge (Fig. 11.7). Seabird distribution is often influenced by temperature and salinity gradients as well as other oceanographic features that promote productivity and concentrate prey (Piatt and Springer 2003; Piatt et al. 1991, 2007; Gall et al. 2013) and there are clear differences in distribution among species that prey on fish (piscivores), those that feed on zooplankton (planktivores) and those that feed on benthic prey. For example, in the northern Bering and southern Chukchi seas, piscivores are concentrated in the ACW, especially near headlands along the coasts, near islands and in areas where eddies are common (Piatt and Springer 2003). Conversely, planktivores are strongly associated with AW west of SLI, the Diomede islands, the AW-BSW in the central-southern Chukchi and are nearly absent from ACW (Piatt and Springer 2003) Planktonic prey amassed near convergent fronts of the ACW also attract foraging phalaropes (mainly red phalaropes *P. fulicarius*), terns and kittiwakes (*Rissa* spp.) (Piatt et al. 1991; Logerwell et al. 2011).

During summer, piscivorous seabirds from the mainland colonies of the southeastern Chukchi Sea feed mainly on Arctic cod, (Piatt et al. 1991), the most abundant pelagic fish in the area (Logerwell et al. 2011). The seabirds also consume sand lance, saffron cod, sculpins, and herring. Piatt et al. (1991) found that by August,

Table 11.2 Common marine bird species observed in the North Bering Sea, Chukchi Sea, and Beaufort Sea, based on pelagic surveys from 2006 to 2009 (Kuletz, unpublished data)

			Average at-sea densities		
Common name	Scientific name	Forage mode	N. Bering	Chukchi	Beaufort
Red-throated loon	*Gavia stellata*	Piscivore	<0.01	0.00	0.01
Pacific loon	*Gavia pacifica*	Piscivore	0.01	<0.01	<0.01
Laysan albatross	*Phoebastria immutabilis*	Piscivore	0.01	0.00	0.00
Northern fulmar	*Fulmarus glacialis*	Piscivore	4.88	0.69	<0.01
Short-tailed shearwater*	*Puffinus tenuirostris*	Piscivore	1.74	4.68*	0.18
Fork-tailed storm-petrel	*Oceanodroma furcata*	Planktivore	0.51	0.01	0.00
Pelagic cormorant	*Phalacrocorax pelagicus*	Piscivore	0.03	0.00	0.00
Red-faced cormorant	*Phalacrocorax urile*	Piscivore	0.01	0.00	0.00
Long-tailed duck	*Clangula hyemalis*	Benthic	0.01	0.04	0.09
Common eider	*Somateria mollissima*	Benthic	<0.01	0.37	0.25
King eider	*Somateria spectabilis*	Benthic	<0.01	0.01	0.07
Spectacled eider	*Somateria fischeri*	Benthic	86.13**	0.05	0.10
White-winged scoter	*Melanitta fusca*	Benthic	<0.01	0.00	0.01
Surf scoter	*Melanitta perspicillata*	Benthic	<0.01	0.00	0.01
Unidentified phalarope	*Phalaropus* spp.	Planktivore	0.04	0.49	0.18
Red phalarope	*Phalaropus fulicaria*	Planktivore	0.12	0.01	0.00
Red-necked phalarope	*Phalaropus lobatus*	Planktivore	0.05	0.22	0.00
Pomarine jaeger	*Stercorarius pomarinus*	Piscivore	0.04	0.06	0.00
Parasitic jaeger	*Stercorarius parasiticus*	Piscivore	0.01	0.01	0.01
Long-tailed jaeger	*Stercorarius longicaudus*	Piscivore	<0.01	0.01	0.00
Glaucous gull	*Larus hyperboreus*	Piscivore	0.31	0.24	0.21
Glaucous-winged gull	*Larus glaucescens*	Piscivore	0.03	0.00	0.00
Slaty-backed gull	*Larus schistisagus*	Piscivore	0.05	<0.01	0.00
Herring gull***	*Larus argentatus*	Piscivore	0.05	<0.01	0.00
Ivory gull	*Pagophila eburnea*	Piscivore	0.02	0.04	<0.01
Black-legged kittiwake	*Rissa tridactyla*	Piscivore	1.48	1.27	0.39
Red-legged kittiwake	*Rissa brevirostris*	Piscivore	0.01	0.00	0.00
Ross's gull	*Rhodostethia rosea*	Piscivore	0.01	0.14	0.27
Sabine's gull	*Xema sabini*	Planktivore	<0.01	0.03	<0.01
Arctic tern	*Sterna paradisaea*	Piscivore	<0.01	0.01	0.43
Unidentified murre	*Uria* spp.	Piscivore	2.25	0.39	0.03
Common murre	*Uria aalge*	Piscivore	0.47	0.14	0.00
Thick-billed murre	*Uria lomvia*	Piscivore	1.07	0.57	<0.01
Dovekie	*Alle alle*	Planktivore	<0.01	<0.01	0.01
Black guillemot	*Cepphus grylle*	Piscivore	0.03	0.02	0.10
Pigeon guillemot	*Cepphus columba*	Piscivore	0.01	0.00	0.00
Kittlitz's murrelet	*Brachyramphus brevirostris*	Piscivore	0.01	0.09	0.00
Ancient murrelet	*Synthliboramphus antiques*	Piscivore	0.05	0.07	0.00
Unidentified auklet spp	*Aethia* spp.	Planktivore	0.89	0.41	0.00
Parakeet auklet	*Aethia psittacula*	Planktivore	0.13	0.09	0.00
Crested auklet	*Aethia cristatella*	Planktivore	1.06	1.23	0.00

(continued)

Table 11.2 (continued)

Common name	Scientific name	Forage mode	Average at-sea densities		
			N. Bering	Chukchi	Beaufort
Least auklet	*Aethia pusilla*	Planktivore	2.00	0.07	0.00
Horned puffin	*Fratercula corniculata*	Piscivore	0.05	0.02	0.00
Tufted puffin	*Fratercula cirrhata*	Piscivore	0.21	0.05	0.00

The average densities are indices of relative abundance, derived from 3 km transect segments averaged within each region from March through October. Note that the average density for spectacled eiders is dominated by high densities encountered during spring surveys that targeted wintering areas for that species. Primary forage mode indicates the major prey types consumed by each species, although many marine birds typically use a variety of prey. Piscivore here can include a diet with a variety of fishes and squid

Note that survey effort in the Chukchi Sea was low, but occurred in fall when shearwaters were abundant

*Includes 'unidentified dark shearwaters', which were likely short-tailed shearwaters

**Includes observations of the main overwintering flocks of spectacled eiders

***Includes American and Vega subspecies

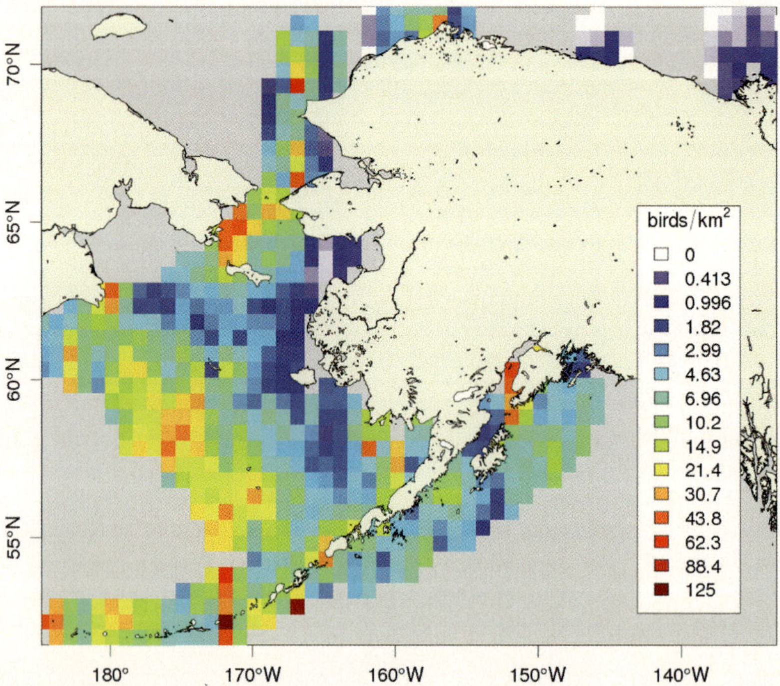

Fig. 11.7 Broad scale distribution of marine birds offshore Alaska based on USFWS at-sea surveys conducted 2006–2009; note high densities associated with the Anadyr Water and Chirikov Basin in the northern Bering Sea and near shoals in the Chukchi Sea

forage fish had apparently moved into the ACW from Bering/Anadyr waters, where birds took advantage of the late summer abundance of prey. Offshore, the most abundant birds in the southeastern Chukchi Sea in late summer are murres (*Uria* spp.), shearwaters, and kittiwakes (Springer et al 1984; Piatt et al. 1991, Kuletz unpublished data). In the oceanic Bering Sea/Anadyr waters, the avifauna is dominated by planktivorous auklets feeding on copepods (e.g. *Neocalanus plumchrus*) concentrated in the upper water column by stratification and upwelling (Gall et al. 2006; Kitaysky and Golubova 2000). Changes in stratification and salinity can dramatically alter seabird foraging distribution in the northeastern Chukchi Sea. During 2008–2010, seabird abundance, species composition, and distribution varied depending on the amount of Bering Sea water present, which influenced the strength of stratification and salinity (Gall et al. 2013). In this study, diving planktivores such as least and crested auklets associated with areas of strong stratification, while surface-feeding planktivores like phalaropes preferred well-mixed waters.

In addition to upwelling and fronts, another source of prey concentrated at the surface is provided by the plumes from feeding gray whales and seabirds respond to such ephemeral patches by altering their distribution at sea (Obst and Hunt 1990). In the Chirikov Basin, the main species observed taking advantage of whale activities were surface feeders (northern fulmar *Fulmar glacialis*, red phalarope, black-legged kittiwake *R. tridactyla*), and one diving species (thick-billed murre *U. lomvia*). When feeding at the plumes, the surface feeding birds were able to include benthic crustaceans in their diet, the most common being tube-dwelling *Ampelisca* amphipods (Obst and Hunt 1990).

11.2.2.2 Breeding Colonies

The abundant avifauna that breeds in the northern Bering and Chukchi seas is a reflection of the enormous productivity of this marine ecosystem. Of the roughly 29 million seabirds nesting among 1,714 colonies throughout Alaska, 21 million nest in Bering Sea. Half of Alaska's breeding birds are concentrated at 12 colonies, four of which occur in the NBS (Stephensen and Irons 2003). The largest seabird colonies are on St Matthew, St Lawrence and the Diomede and King islands, with smaller colonies along the coastline of the Bering Strait region (Fig. 11.8). Planktivorous seabirds predominate on the islands, the most numerous being the crested auklet (*Aethia cristatella*) and least auklet (*A. pusilla*). North of the Bering Strait, the largest seabird colonies in Alaska are at Cape Thompson and Cape Lisburne, each with ~500,000 birds. The predominate species at these mainland colonies are piscivorous seabirds such as thick-billed murre and common murres (*U. aalge*), black-legged kittiwakes, and tufted puffins (*Fratercula cirrhata*) and horned puffins (*F. corniculata*) (Springer et al. 1984). During summer, at least 25 seabird species regularly occur in the southeastern Chukchi Sea (Piatt et al. 1991). Farther north and east along the Chukchi, small colonies of piscivorous black guillemots, arctic terns and black-legged kittiwakes are scattered along bluffs and the low-lying barrier islands of the arctic coasts. Along the Beaufort coast, the kittiwakes and shorebirds are replaced

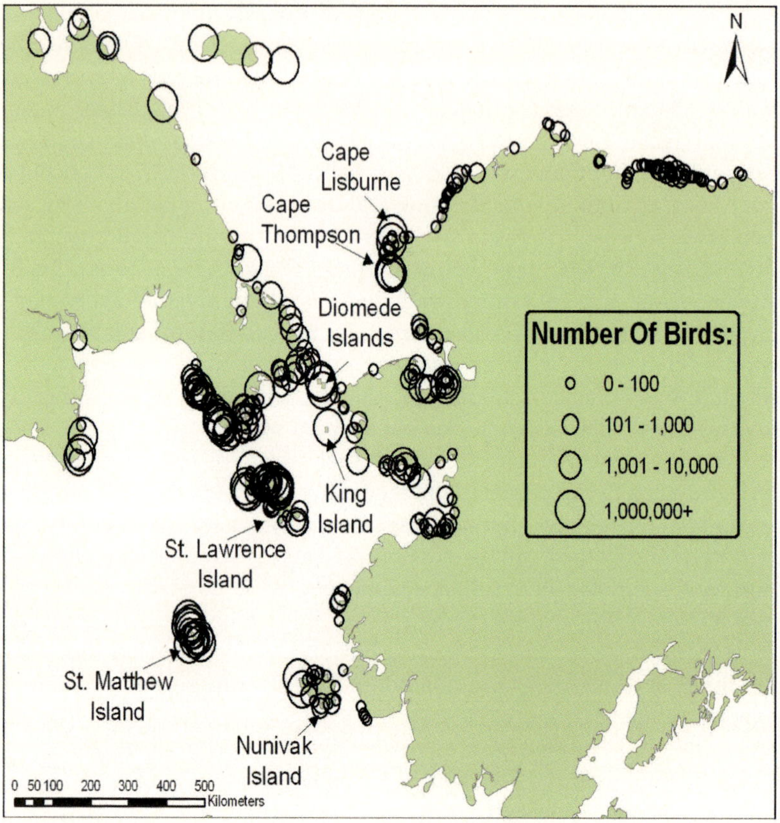

Fig. 11.8 Seabird colonies in the PAR, from North Pacific seabird colony database (http://alaska.fws.gov/mbsp/mbm/northpacificseabirds/colonies/)

by large numbers of benthic-feeding eiders, scoters, and long-tailed ducks (Fischer and Larned 2004).

Seabirds are obligate central place foragers during the breeding season, and the location of nesting habitat in relation to good forage grounds is critical to breeding success. The combination of upwelling along the continental shelf and the advection of prey to the central Chukchi Sea results in a constant flow of food past the colonies of the NBS (Piatt and Springer 2003). The concentration of auklets in the NBS is thus a combination of a rich food source (primarily copepods and euphausiids) within foraging range of rocky, volcanic islands. Because food appears to be abundant, Stephenson and Irons (2003) suggested that in the NBS auklets may only be limited by the availability of good nesting habitat.

Seabirds are affected by shifts in prey distribution, but the impact to breeding birds varies widely among species. Murres feed on a range of prey sizes, can dive down to 100 m and will travel up to 126 km to forage for their chicks (Mehlum et al. 1998); thus, they likely can adapt to shifts in prey type or distribution easier than can smaller birds or surface feeders. Kittiwakes are surface feeders (Baird 1994) that can forage up to 100 km from their nests, but often stay within 40 km. Terns are also surface feeders, but can only forage within 20 km from their nests and still successfully raise chicks (Hatch 2002). Furthermore, the relationship between nesting site and forage grounds can be altered by changes in ice cover. For example, in a 35-year study of black guillemots nesting northeast of Barrow, breeding success decreased and predation by polar bears increased coincident with the dramatic summertime retreats of sea ice in the Beaufort Sea (Divoky, in Moline et al. 2008). In this case, the guillemots lost access to their prey, zooplankton and fish associated with the rapidly retreating sea ice edge, which simultaneously stranded more polar bears along the coasts and resulted in increased nest predation. Changes in access to prey may be indirect, such as the extent and duration of ice cover affecting local productivity and subsequent prey composition or size, ultimately affecting seabird breeding success (Hunt et al. 2002; Gall et al. 2006; Kitaysky and Golubova 2000; Mueter and Litzow 2008; Sheffield-Guy et al. 2009).

11.2.2.3 Seasonal Dynamics

The marine avifauna of the PAR undergoes dramatic seasonal changes. Seabirds at the breeding colonies are at the northern edge of their species' ranges and often the timing of egg laying, chick hatching and fledging occur later than in populations of the same species at lower latitudes, as described by Day et al. (1999) for the Kittlit's murrelet (*Brachyramphus brevirostris*) nesting phenology throughout Alaska. During summer, breeding populations of birds swell into the millions, and are joined by non-breeding seabirds and post-breeding loons, seaducks, phalaropes, and shorebirds. By late summer, extreme migrators such as short-tailed shearwaters arrive by the millions after foraging in the southern Bering Sea and are often associated with (relatively) warm, saline water masses. Given their large numbers, the arrival of shearwaters in the PAR may have repercussions for resident birds (Shaffer et al. 2006), which could be acerbated by longer ice-free fall conditions.

In late summer and early autumn, seabirds use post-breeding staging areas for molting and feeding to replace body mass lost during egg production and chick-raising and, for some species, to prepare for long-distance migrations. Migratory routes in these areas are not well understood, but several regions stand out as major thoroughfares. In the fall, streams of birds move past Pt. Barrow, primarily heading west, where a variety of species have been observed feeding on massive swarms of euphausiids (e.g., Hatch 2002). Similarly, large numbers of birds move past Cape Thompson, Cape Lisburne, and the large islands in the NBS. Often, birds show sex-specific migratory patterns. For example, among common murres tagged in the Chukchi, males (likely with their flightless young) drifted west with the currents towards Siberia, whereas the females flew south toward the Bering Sea (Hatch 2002). Red phalarope females leave inland nesting areas prior to males in late summer, but by August both sexes and all age groups are at offshore staging sites (Tracy et al. 2002) and large groups move west past Pt. Barrow (Kuletz et al. 2008).

Of note, species not commonly associated with high latitudes, such as ancient murrelets (*Synthliboramphus antiquus*), have also shown evidence of northward migration through the NBS and the Bering Strait (Gall et al. 2013; Kuletz et al. 2008 and unpublished data), presumably to feed on abundant euphausiids and other prey. During years when water temperatures remain cold throughout the open water season, species that are more typically associated with ice, such as black guillemots and ivory gulls (*Pagophila eburnean*), are more abundant in the northeast Chukchi Sea (Gall et al. 2013).

As the sea ice extends south in late autumn, areas without open water leads become depauperate of avifauna and open water leads and polynyas become important habitats for overwintering marine birds, notably the spectacled eider (Lovvorn et al. 2009). Specifically, from March through May, the region of flaw leads between landfast and moving pack ice along the northeast Chukchi Sea coast is a critical migration corridor for essentially all sea ducks and loons destined for breeding sites along the U.S. and Canadian shorelines of the Beaufort and Chukchi seas. Ledyard Bay, along the Chukchi Sea coast, is designated a critical habitat due to its importance as a late-summer molting area for spectacled eiders and other sea duck species. In the early spring, two pagophilic seabirds that remain in the ice-covered regions are the black guillemot and the rare Kittlitz's murrelet. Although these murrelets are typically associated with coastal glaciers of southcentral Alaska, Day et al. (2011) estimate that the Kittlitz' murrelet population in the PAR is in the order of 5,200 birds. Little is known about its pelagic habitat or winter distribution, but recent spring surveys found Kittlitz's murrelets in open water leads and polynyas of the NBS (Kuletz, unpublished data). Kittlitz's murrelets and black guillemots were often in the same open lead, likely feeding on euphausiids, amphipods and arctic char found under and at the ice edge. Two other seabirds found in the ice are the ivory gull and Ross' gull (*Pagophila eburnean*), with the former known to feed on juvenile arctic cod and amphipods in addition to scavenging (Divoky, in Moline et al. 2008).

11.2.3 Marine Mammals

Sixteen species of marine mammals occur in the PAR, including eight that are year-round residents and considered core Arctic species. Eight others occur primarily in summer and autumn and are considered seasonally migrant species (Table 11.3).

11.2.3.1 Core Arctic Species

Polar bears (*Ursus maritimus*), Pacific walruses (*Odobenus rosmarus divergens*), bearded seals (*Erignathus barbatus*) and ringed seals (*Phoca hispida*) are considered 'ice-obligate' species, due to their reliance on sea ice for some part of their life history (e.g., hunting, resting, parturition and molting) (Kovacs et al. 2011; Laidre et al. 2008; Moore and Huntington 2008). Polar bears are pan-arctic in distribution, with

Table 11.3 Fifteen marine mammal species occur in the PAR, including eight **core Arctic** and seven seasonal migrants

Species	Population	Sea ice[a]	Diet
Polar bear (*Ursus maritimus*)	Chukchi and So. Beaufort (2 of 19 recognized)	Ice-obligate	Marine mammals, esp. ringed seals
PINNIPEDS			
Walrus (*Odobenus rosmarus divergens*)	Bering-Chukchi-Beaufort (1 of 3 subspecies)	Ice-obligate	Benthic fauna, esp. clams
Bearded seal (*Erignathus barbatus*)	*Bering-Chukchi-Beaufort	Ice-obligate	Benthic fauna, esp. clams and crabs
Ringed seal (*Phoca hispida*)	*Bering-Chukchi-Beaufort	Ice-obligate	Fish, crustaceans, large zooplankton
Ribbon seal (*Phoca fasciata*)	*Bering-Chukchi	Ice-associated	Fish, crustaceans, cephalopods
Spotted seal (*Phoca largha*)	*Bering-Chukchi	Ice-associated	Fish, crustaceans, cephalopods
Northern (Steller's) sea lion (*Eumetopias jubatus*)	*Bering-Chukchi	Seasonal migrant	Fish, crustaceans, cephalopods
CETACEANS			
Beluga (*Delphinapterus leucas*)	E. Bering, E. Chukchi and Beaufort (3 of 4, in PAR)	Ice-associated	Fish, crustaceans, cephalopods
Bowhead whale (*Balaena mysticetus*)	Bering- Chukchi-Beaufort (1 of 4 recognized)	Ice-associated	Zooplankton esp. copepods and euphausiids
Gray whale (*Eschrichtius robustus*)	Eastern North Pacific (1 of 2 recognized)	Seasonal migrant	Benthic and pelagic crustaceans
Humpback whale (*Megaptera novaeangliae*)	*Bering-Chukchi	Seasonal migrant	Fish, large zooplankton
Fin whale (*Balenoptera physalus*)	*Bering-Chukchi	Seasonal migrant	Fish, large zooplankton
Minke whale (*Baleanoptera acutorostrata*)	*Bering-Chukchi	Seasonal migrant	Fish, large zooplankton
Killer whale (*Orcinus orca*)	*Bering-Chukchi	Seasonal migrant	Fish and marine mammal ecotypes

(continued)

Table 11.3 (continued)

Species	Population	Sea ice[a]	Diet
Dall's porpoise (*Phocoenoides dalli*)	*Bering-Chukchi	Seasonal migrant	Fish, cephalopods
Harbor porpoise (*Phocoena phocoena*)	*Bering-Chukchi	Seasonal migrant	Fish, cephalopods

Population (# *vis.* pan-Arctic), or * = sub-population in PAR, relationship to sea ice and primary diet are listed for each species
[a]Moore and Huntington (2008)

two populations recognized in the PAR; the southern Beaufort Sea and the Chukchi Sea populations. Polar bears have annual movement patterns within individual home ranges, the size of which varies with ice type and seal availability (Amstrup et al. 2000). There is general fidelity of individuals to denning areas, although a shift from sea ice to land-based sites has occurred concomitant with the decline in sea ice in the PAR (Fischbach et al. 2007). Pacific walrus overwinter in the Bering Sea, but exhibit sex and age-specific summertime habitat preferences, with adult females, calves and juveniles using sea ice as a platform for resting and from which to feed in the Chukchi Sea, while adult males make feeding trips from terrestrial haul outs mostly in the Bering Sea (Jay et al. 2011). Since 2007, walruses have increasingly used coastal haul-outs and increased foraging in nearshore areas in response to the dramatic loss of sea ice in the Chukchi Sea (Jay et al. 2012). Compared to the clumped distribution of walruses, bearded seals are more evenly distributed throughout the PAR. Both species are typically associated with broken pack ice over shallow shelf waters (usually <100 m), and near open-water leads and polynyas (Ray et al. 2010). Ringed seals are abundant throughout the PAR and are closely associated with shore fast and pack ice habitats. During winter and spring, adult ringed seals maintain territories under the ice, and excavate cave-like structures on top of the ice called lairs. Lairs are used by both sexes, provide protection from weather and predators and are important for pupping and nursing (Smith and Stirling 1978; Smith et al. 1991; Kelly et al. 2010). Although ribbon and spotted seals also rest and give birth to their pups on sea ice, they are not considered ice-obligate, but rather 'ice-associated' species. Both species are primarily subarctic in distribution and occur along the marginal ice zone over the continental shelf of the central Bering Sea from November to April; the remainder of the year is spent in open-water and, in the case of spotted seals, on coastal haul outs (Kovacs et al. 2011; Laidre et al. 2008; Lowry et al. 2000).

There are two core Arctic cetacean species in the PAR, the beluga (*Delphinapterus leucas*) and bowhead whale (*Balaena mysticetus*), both of which overwinter in the Bering Sea. A third Arctic species, the narwhal (*Monodon monoceros*), is seen occasionally in the PAR, but these reports are generally considered extralimital. Three genetically-distinct beluga populations (stocks) occur in the PAR (O'Corry-Crowe et al. 1997), two of which (eastern Chukchi Sea and Beaufort Sea) migrate north

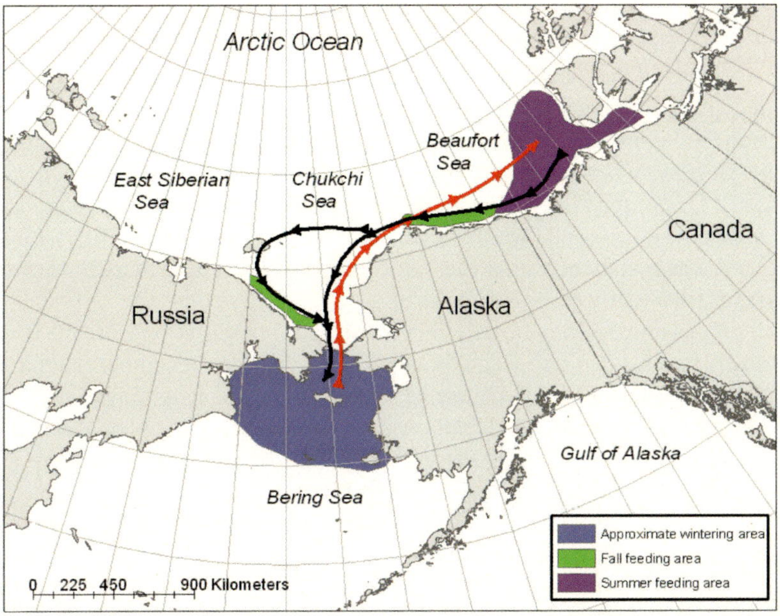

Fig. 11.9 Seasonal migratory cycle of bowhead whales in the PAR (From Moore and Laidre 2006); note that the Beaufort Sea population of beluga undertakes a similar seasonal migration

each spring, ultimately moving to the Chukchi and Beaufort seas during summer for feeding, calf-rearing and molting (Suydam et al. 2001). The third stock (eastern Bering Sea) uses coastal habitats in the NBS, with summer range identified as Norton Sound (O'Corry-Crowe et al. 1997). Bowhead whales typically migrate north through Bering Strait in April and May (Citta et al. 2012), using similar routes as the Beaufort Sea stock of belugas (Fig. 11.9). Visual surveys suggest most animals in both populations spend the summer months mainly in the Canadian Beaufort Sea and Amundsen Gulf region, although recent information from acoustic sampling suggests at least some beluga and bowhead whales summer near the Chukchi Plateau in the northern Chukchi Sea (Moore et al. 2012). During July, bowheads are widely distributed throughout the offshore Canadian Beaufort Sea, with sightings of at least a few whales near Point Barrow, Alaska in some years (Moore and Reeves 1993). By late August, oceanographic conditions concentrate bowhead prey on the continental shelf (Thomson et al. 1986; Ashjian et al. 2010) and whales aggregate to feed in specific areas across the Canadian and Alaskan summer range (Lowry et al. 2004; Moore et al. 2010). In Canadian waters, most bowhead whales feed in Amundsen Gulf in May and June, moving to continental shelf waters offshore of the Tuktoyaktuk Peninsula and other shelf and marine canyon locations from early August through late September (Richardson et al. 1987; Harwood et al. 2010). Prey may also concentrate far to the north. Recent satellite telemetry studies have shown bowhead whales from the PAR traveling north of Banks Island into the Northwest Passage to overlap with bowhead whales from the Atlantic (Heide-Jørgensen et al. 2011). The tagged bowheads presumably were attracted to the area for feeding and remained in the Viscount Melville Sound area for 2 weeks. In Alaskan waters, bowheads feed along the narrow continental shelf from Barter Island to waters east of Point Barrow, where large groups of feeding whales sometimes occur (Moore et al. 2010; Quakenbush et al. 2010). Bowheads also feed along the Chukotka coast in late summer and autumn (Moore et al. 1995; Quakenbush et al. 2010), with at least one tagged whale also migrating there in spring and remaining there throughout the summer of 2010 (Citta et al. 2012).

Both beluga and bowhead whales begin the autumn migration from the Canadian Beaufort by late August and reach the NBS in November-December (Richard et al. 2001; Quakenbush et al. 2010; Citta et al. 2012). Analysis of sightings from 10-years of aerial surveys showed bowheads selected shallow (<50 m) continental shelf habitat while beluga preferred slope (200–2,000 m) habitat during the autumn migration across the Alaskan Beaufort sea (Moore et al. 2000), which may reflect differences in the distribution of prey for each species. Limited sightings suggest bowheads winter in leads and polynyas that develop in the lee of islands and peninsulas (Moore and Reeves 1993). However, satellite telemetry data over two winter seasons showed that bowhead whales wintered in 90–100 % ice cover, far from the ice edge and polynyas (Citta et al. 2012). Information on the winter distribution of belugas is more limited, but telemetry data indicate that both the northern Bering and southern Chukchi seas are used (Richard et al. 2001, R. Suydam, personal communication).

11.2.3.2 Seasonally Migrant Species

One pinniped and seven cetacean species migrate seasonally to the PAR (Table 11.3). Steller sea lions (*Eumetopias jubatus*) routinely haul out on Saint Mathew Island and along the southern shore of SLI in summer. In addition, a recent study by Alaska Department of Fish and Game has documented their occurrence on SLI through December (L. Jemison, personal communication). Steller sea lions also haul out along the southern Chukotka coast, with reports of sub-adult animals along the northern Chukotka peninsula as far north as the east Siberian seas in summers (Boeskorov et al. 2011).

Gray whales feed extensively on pelagic, epi-benthic and benthic invertebrates in coastal central shelf and shoal areas in the NBS, Chukchi and western Beaufort seas during summer and autumn (Moore et al. 2000; Bluhm et al. 2007). Stafford et al. (2007) reported the detection of gray whale calls over-winter 2003–2004 on recorders located near 71° 30′N, 151° 50′W, demonstrating that some individuals can remain in the Beaufort Sea year-round. Although rare, there are occasional summertime reports of gray whales in the Beaufort Sea (Maher 1960), including animals feeding off the Tuktoyuktuk Peninsula in the Canadian Beaufort Sea (Rugh

Table 11.4 A summary of the number of sightings/number of cetaceans seen during marine mammal watches conducted on RUSALCA cruises, 2009 and 2010

	Rusalca 2009	**Rusalca 2009**	**Rusalca 2010**
	Wrangell Is.-Herald Valley	Southern Chukchi	Southern Chukchi
Species	10–13 September	4 September; **25–28 September**	1–9 August
Bowhead whale	2 sightings/6 whales	1 sighting/2 whales	5 sightings/15 whales
Gray whale	2 sightings/11 whales	7 sightings/85 whales **34 sightings/177 whales**	25 sightings/47 whales
Humpback whale		5 sightings/21 whales **8 sightings/28 whales**	11 sightings/18 whales
Minke whale			6 sightings/6 whales
Fin whale			1 sighting/1 whale
Harbor porpoise		1 sighting/1 porpoise	4 sightings/5 porpoise
Dall's porpoise			3 sightings/5 porpoise

Areas searched include: the Southern Chukchi = 65–69 °N, coast to coast, and Wrangell Island-Herald Valley = 70–73 °N, between 180 and 175 °W

and Fraker 1981). A female gray whale was tagged there in early September 2010, and remained there for a month before heading west to the Chukchi Sea (L. Quakenbush, unpublished data). The five other seasonally migrant cetacean species appear to be far less common in the PAR, although results from recent year-round passive acoustic sampling may soon change this view (e.g. Delarue et al. 2013). While hundreds of gray whales were seen in the southern Chukchi Sea during RUSALCA cruises in 2009 and 2010, humpback whales (*Megaptera novaeangliae*) numbered in the 10s of animals, with only six minke (*Balaenoptera acutorostrata*) and one fin whale (*B. physalus*) seen (Table 11.4). Nearly all of the cetacean sightings were on the rarely-surveyed western side of the International Date Line in the southern Chukchi Sea (Fig. 11.10), which corresponds to available records for fin, humpback, minke and gray whale distribution in the 1970s (e.g., Votrogov and Ivashin 1980) and the 1940–1950s (Nasu 1974). Killer whales (*Orcinus orca*) were also seen on the RUSALCA cruises; they are known to prey on gray whales in the NBS (Ljungblad and Moore 1983), but their residence time in the PAR is uncertain. Residents of SLI report seeing killer whales in May and June, while reports from Barrow usually occur in July and August (George et al. 1994), with no reported occurrence in the Canadian Beaufort Sea (Harwood and Smith 2002). Dall's porpoise (*Phocoenoides dalli*) were seen in the southern Chukchi Sea during the 2010 RUSALCA cruise (Fig. 11.10), a possible range-extension for this species, which is commonly found in the southeastern Bering Sea (e.g., Friday et al. 2012). Finally, harbor porpoise (*Phocoena phocoena*) are known to occur in the northern Bering and Chukchi seas, at least north to coastal waters near Barrow (Suydam and George 1992), although there are few details on numbers, habitat or prey preferences.

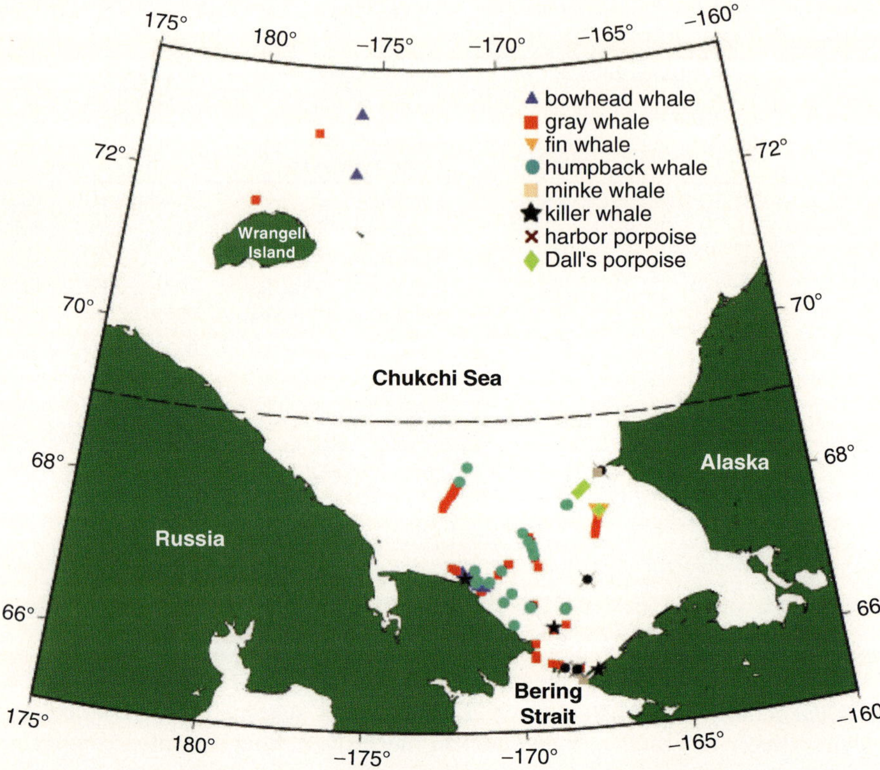

Fig. 11.10 Cetacean sightings during the RUSALCA cruises in 2009 and 2010; the cruise track extended north of 69 °N latitude (*dashed line*) only in 2009; counts provided in Table 11.4

11.3 Case Studies: Responses of UTL Species to Environmental Variability

Climate change is altering the habitats of UTL species in the PAR through biophysical forcing (see Overland et al. 2014; Frey et al. 2014; Nelson et al. 2014, all this volume). Responses of UTL species to altered habitats can be categorized as extrinsic, including shifts in range, timing (phenology) or regions of high abundance (hotspots); or intrinsic, including changes in diet, body condition, or body chemistry (isotopic, fatty acid and contaminant signatures). Responses are inter-related such that a shift in range, phenology or use of hotspots will be reflected in changes in diet, body condition and chemistry (Fig. 11.11). It is this connection that allows us to detect ecosystem reorganization by tracking changes in the ecology of UTL species. To date, there has been little or no coordination among researchers studying marine fishes, birds and mammals, and few studies have focused on responses to environmental variability. Here, we provide some examples of the responses of UTL species to environmental variability to explore how integrated research might better inform investigations of ecosystem reorganization related to climate change.

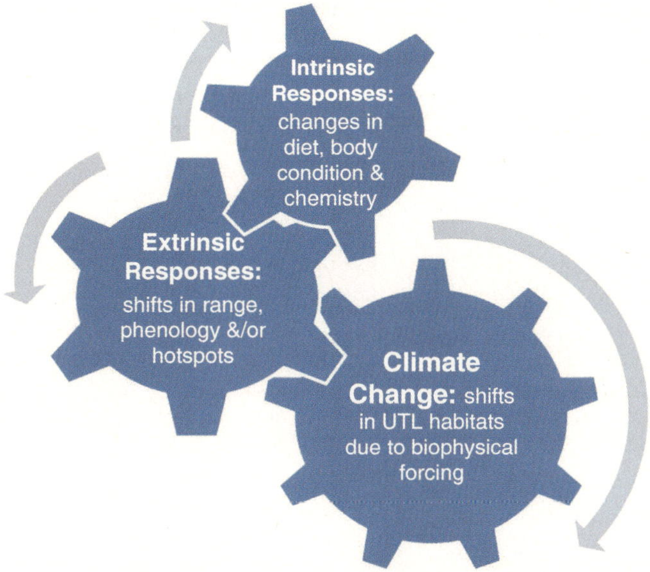

Fig. 11.11 Conceptual diagram of extrinsic and intrinsic responses of UTL species to alteration of habitats due to biophysical factors (sea ice, wind, trophic cascades) related to climate change

11.3.1 Fishes and Crabs

Environmental factors strongly influence fish distribution and abundance through changes in growth, survival, reproduction, and spawning distribution (Wood and McDonald 1997; Sundby and Nakken 2008). Rapid warming can exceed the ability of forage species to adapt, thereby causing a major restructuring of regional ecosystems as has been observed in the North and Baltic Sea ecosystems (Mackenzie and Schiedek 2007). It is likely that shifts in spatial distribution and northward range extensions are inevitable for fishes and crabs in the PAR and that marine community species composition will continue to change under a warming climate (Mueter et al. 2009). Examples of responses of fishes and crabs to environmental variability in the Bering, Chukchi and Beaufort seas are provided below.

11.3.1.1 Salmon and Forage Fish in the Northern Bering Sea

Climate and ocean conditions in the NBS varied annually during the BASIS surveys, providing an opportunity to examine the effects of ecosystem variability on pelagic fish. Specifically, 2004 and 2005 were characterized by a strong Aleutian Low pressure system centered over the Aleutian Islands during the winter, reduced sea ice concentrations and warm spring and summer ocean temperatures. In contrast, 2006 and 2007 were colder years, with a weak Aleutian Low and increased sea ice concentrations. This cooling trend continued through 2009 and into 2010. Andrews

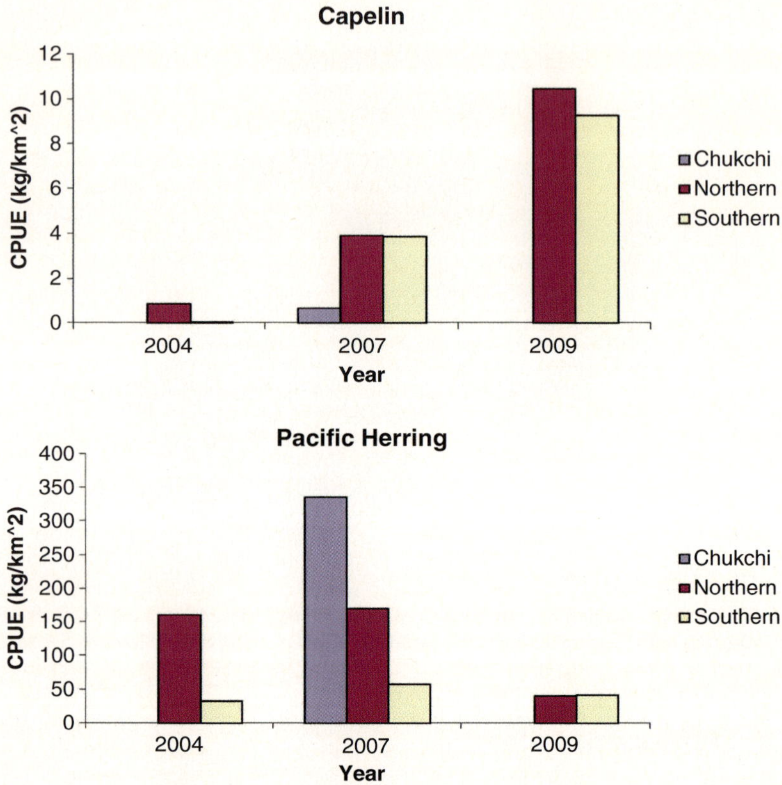

Fig. 11.12 Density of capelin and Pacific herring from BASIS surveys conducted in warm (2004) and cold (2007 and 2009) years in the Chukchi Sea (2007 only), northern Bering Sea (north of 60 °N) and southern Bering Sea

et al. (2009) compared the diet, length and whole body energy content of juvenile pink salmon between warm and cold years in the NBS. Fish were more abundant in juvenile salmon diet in warm years and zooplankton (amphipods, copepods, decapods, euphausiids and larvacea) more abundant in cold years. Juvenile pink salmon lengths were significantly longer in warm years reflecting increased growth rate, but had lower whole body energy. This result was unexpected because it suggests that salmon were consuming higher energy prey (fish) in warm years, but had lower body energy stores. Possible explanations are that prey quantity was more important than quality or that fish allocated more energy to storage during cold years.

Interannual variability in ocean conditions also impacted the abundance and diet composition of forage fish species in the NBS. Capelin density was lowest in the warm year, increased with cooling in 2007 and was even greater with continued cooling in 2009 (Fig. 11.12). In contrast, Pacific herring abundance was lowest in 2009. Capelin and Pacific herring diet in the NBS was also impacted by cooling from 2004 to 2009. During the warm years, herring diet was dominated by large euphausiids and large copepods, while in cold years diet composition shifted and

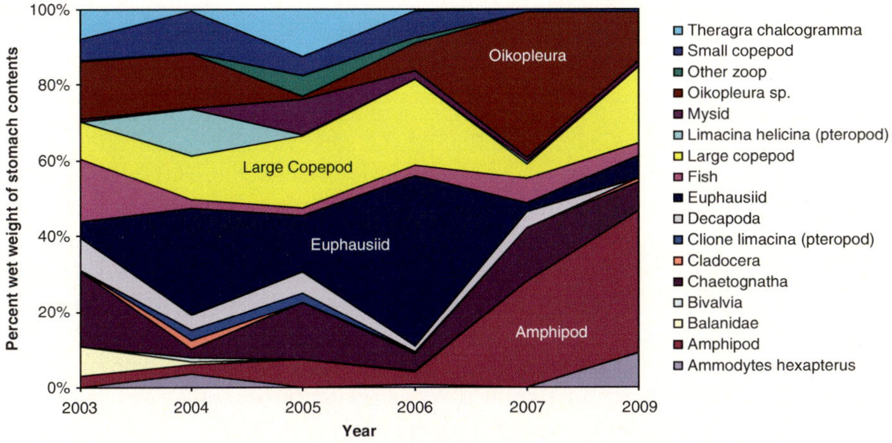

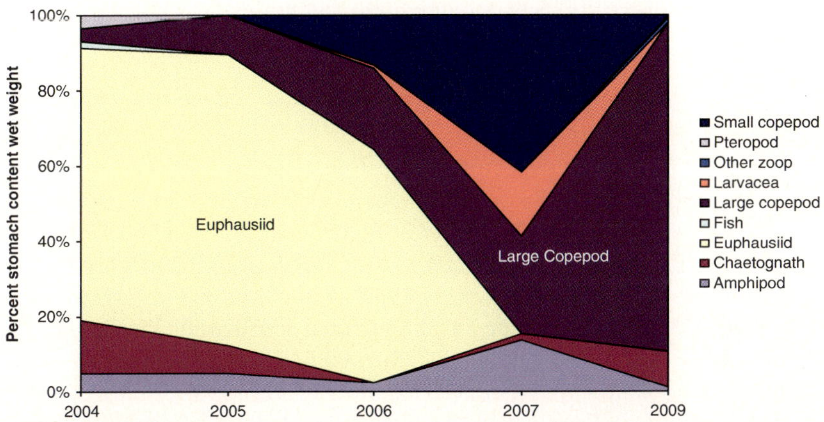

Fig. 11.13 Diet of Pacific herring (*top*) and capelin (*bottom*) in the northern Bering Sea in warm (2004, 2005), transition (2006) and cold (2007 and 2009) periods

herring were consuming mostly pelagic tunicates (*Oikopleura* spp.) and amphipods (Fig. 11.13). Capelin diets also shifted with ocean cooling. Capelin consumed predominantly euphausiids before 2007 and then switched to large and small copepods during the colder years of 2007 and 2009 (Fig. 11.13). The increase in large copepods in the diets of capelin was consistent with an increase in *Calanus* spp. in the BASIS zooplankton samples starting in 2006 (Eisner et al. 2014).

11.3.1.2 Snow Crab in the Chukchi Sea

A long-term comparison of the distribution, abundance and size of snow crab, a potentially important commercial species, is provided by data from the RUSALCA surveys of the 2000s, and data from a similar survey conducted in 1976 (Feder et al. 2005). The community structure of epibenthic megafauna was quantified from 44 stations sampled during RUSALCA surveys in 2004, 2007 and 2008 (Bluhm et al. 2009). Crustaceans were the second most abundant and biomass-rich group, with samples comprised predominantly of snow crab. The records of snow crab caught in Herald Canyon during the RUSALCA surveys represent an extension of the northern range limit of the species in the PAR (Bluhm et al. 2009). In addition, comparisons of the 2004 estimates of snow crab biomass with estimates from a similar 1976 survey (Feder et al. 2005) indicate that snow crab abundance has increased by one to two orders of magnitude (Bluhm et al. 2009, 2010). However, body size of crab does not appear to have increased in the Chukchi Sea over the past three decades, and very few large and reproductively mature female crab (over 40 mm carapace width) were found (Bluhm et al. 2009). The largest male snow crab caught in the Chukchi Sea in 1976 was 75 mm wide (Frost and Lowry 1983) and only a few males larger than that were caught in the early 1990s (Barber et al. 1997). Thus, it appears snow crab have responded to the sea ice reduction and warmer ocean temperatures of the 2000s by moving north, but not by growing larger.

11.3.1.3 Demersal Fish and Crab in the Beaufort Sea

A survey of offshore marine fishes of the western Beaufort Sea was conducted in 1977 (Frost and Lowry 1983) and not again until August 2008 (Rand and Logerwell 2010; Logerwell et al. 2011; Parker-Stetter et al. 2011). The 1977 survey was comprised of 33 stations sampled across the Chukchi and Beaufort seas from longitude 164° to 141°W (Frost and Lowry 1983), while the 2008 survey extended from longitude 155° to 152°W. Of the 34 taxa captured and identified from the 2008 Beaufort Sea survey, 17 had also been documented in the 1977 survey. The 2008 results document or confirm apparent extensions to the known ranges of four species: walleye pollock, Pacific cod, festive snailfish (*Liparis marmoratus*), and eyeshade sculpin (*Nautichthys pribilovius*). A Chukchi Sea survey in 1990 (Barber et al. 1997) reported Pacific cod at three stations located between 68 °N and 69 °N. Festive snailfish are a relatively rare species, likely due to difficulty in correctly identifying it. One specimen was recorded in the NE Bering Sea near SLI at 63°00'N, 169°20'W (Busby and Chernova 2001). Most recently, 19 confirmed festive snailfish were caught in ten tows during the 2010 AFSC bottom trawl survey of the Bering Sea, at latitudes as far north as 65 ° N (J. Orr, personal communication). Previous to these records, the species had only been documented in the Sea of Okhotsk. The northernmost record of the eyeshade sculpin previous to the 2008 survey was in the northern Chukchi Sea, west of Point Barrow (Barber et al. 1997). Given the scarcity of fish survey data in the Arctic, it is possible that these apparent range extensions are due to lack of sampling or taxanomic uncertainty. Further monitoring is necessary to document that the distributions of fish are changing as a result of climate change. In addition to the potential range extensions of pollock and Pacific cod, Bering flounder (*Hippoglossoides robustus*) were caught in the 2008 survey, but not the 1977 survey. Pollock and Pacific cod are abundant in the Bering Sea and are commercially valuable. Analysis of pollock otoliths showed that most of the fish caught during the 2008 survey were sub-adults, age-2.

In 1990, an ichthyoplankton survey in the Chukchi Sea found juvenile walleye pollock northwest of Barrow (Wyllie-Echeverria 1995). During the 2004 RUSALCA survey of the Chukchi Sea, (Mecklenburg et al. 2007) captured pollock ranging from 102 to 168 mm total length, indicating that these fish were likely sub-adults. So, although pollock are occurring in Arctic seas, fish of spawning age or size have not yet been documented and the origins of the juvenile fish are not known. Of note, the pollock caught in the western Beaufort Sea in 2008 were smaller at age than pollock in the Bering Sea that year (Rand and Logerwell 2010). This suggests that the fish were spawned in cold Arctic waters or were transported to cold waters shortly after spawning. The size difference is manifested first at age-2; age-1 pollock from the Bering and Beaufort seas were similar in size. This lends support to the latter hypothesis, that fish were spawned in north Pacific waters and transported into the Arctic sometime during their first year of life.

11.3.2 Marine Birds

At hemispheric scales, decadal trends in murre populations appear to respond to climatic shifts in SST, with evidence of synchrony in trends among colonies within an ocean basin, and opposite trends between the circumpolar Pacific and Atlantic populations (Irons et al. 2008). Longer and more extensive ice-free periods in the Chukchi and Beaufort seas could open avenues of migration between Atlantic and Pacific species. An example might be the dovekie (*Alle alle*), one of the most abundant Atlantic seabirds. This small alcid was documented as having a few hundred, possibly nesting birds in the NBS (Day et al. 1988). However, at-sea surveys between 2006 and 2010 found higher numbers (Kuletz et al. 2008 and unpublished data), suggesting an increase in the NBS breeding populations of dovekie, or increased immigration from the eastern Arctic, or both.

At regional scales within the Pacific Arctic and sub-Arctic, a number of studies have related changes in breeding success of planktivorous or piscivorous birds to presumed changes in the dispersion or abundance of their prey (Table 11.5; Springer et al. 1984, 1986; Kitaysky and Golubova 2000; Gall et al. 2006; Sheffield-Guy et al. 2009). Such changes typically reflect shifts in SST or in the position of fronts between currents, water masses, or thermal strata (Haney 1991; Piatt et al. 1991; Russell et al. 1999). For sea ducks that generally do not feed in marine habitats during the breeding season, linkages between the marine environment and breeding appear to depend mainly on reserves of fat and protein acquired during winter or migration (Lovvorn et al. 2003 and references therein, Table 11.5). Although the role of reserve levels in the success of breeders has been well studied in some sea duck species, effects of reserves on their propensity to initiate breeding is poorly understood but potentially important (Coulson 1984).

11.3.2.1 Nesting Auklets and the Anadyr Current

For least auklets (*Aethia pusilla*) and crested auklets (*A. cristatella*) nesting on the northern coast of SLI, chick survival varied among years depending on diet (Gall et al. 2006; Sheffield-Guy et al. 2009). Nestlings of both auklet species hatched earlier and had higher survival rates when they were fed the large, oceanic copepod *Neocalanus cristatus*. Conversely, the small copepod *Calanus marshallae* was common in the diet of least auklets, and the mid-sized copepod *N. flemingeri* was prevalent in the diet of crested auklets in the year of lowest chick survival. More recently, a 5-year study (2000–2004) in the same colonies found that the presence of *N. cristatus* in the diet was not alone the key to higher chick survival for either species; rather, high survival rates depended on a large proportion of euphausiids (*Thysanoessa* spp.) in chick diets (Sheffield-Guy et al. 2009). The availability of euphausiids appeared to compensate for years with relatively lower abundance of *N. cristatus*.

The large *N. cristatus*, which cannot breed over the shallow Bering Sea shelf, is in fact transported into this region from the shelf break 500 km away by the Anadyr Current. Access to *N. cristatus* by these breeding auklets appears to depend on the strength and position of the Anadyr current north of the island, which can vary substantially with the strength and direction of local winds. Northwesterly winds can shift this current tens of kilometers to the southeast in only 2–4 days (Gawarkiewicz et al. 1994), reducing its distance from these auklet breeding colonies. In contrast, southeasterly winds may push this current far enough away from the colony to reduce its accessibility and force the birds to feed on smaller copepods nearer the island (but see Obst et al. 1995). The auklets nesting on SLI may forage ~50 km from their colonies, thus shifts in prey availability and quality of this magnitude will have greater impact on the birds abilities to raise chicks, compared to auklets nesting closer to important foraging areas (Sheffield-Guy et al. 2009). Because this region is so shallow (mostly <50 m deep), local winds can have dramatic effects on hydrographic patterns (Woodgate et al. 2005). Long-term climatic changes in dominant wind direction, by controlling access to valuable foods, could thereby have important effects on the population dynamics of these two auklet species in this area. This example demonstrates that although changing ice conditions are often emphasized in projecting effects of climate change, altered wind patterns and resulting shifts in hydrography can also have critical effects on trophic relationships.

11.3.3 Eiders During Winter and Migration

A number of sea duck species winter in or migrate through the PAR. Of these, spectacled eiders (*Somateria fischeri*) and steller's eiders (*Polysticta stelleri*) are listed as threatened under the U.S. Endangered Species Act, king eiders (*S. spectabilis*) and common eiders (*S. mollissima*) declined by over 50 % from 1976 to 1996, and long-tailed ducks (*Clangula hyemalis*) decreased by 75 % from 1977 to 1994 (Hodges et al. 1996; Suydam et al. 2000). Reasons for these declines are mostly unknown.

All these species have been studied on their breeding areas. When at sea, the timing of movements and occupancy of different areas by king eiders migrating along the Beaufort and Chukchi Sea coasts were documented by satellite telemetry (Phillips et al. 2007; Oppel et al. 2008). However, functional habitat relationships in marine waters of the PAR have been studied only for spectacled eiders.

Research on spectacled eiders has focused on delineating areas that are critical to this species during winter in open-water leads south of SLI in the northern Bering Sea. Of prey in this area, bivalves have greatest biomass (Grebmeier and Cooper 1995); the diet of eiders collected at one site consisted only of the most abundant bivalve species (*Nuculana radiata*), and included no other prey that were present at lower densities (Lovvorn et al. 2003). An energetics model was used to estimate threshold prey densities above which spectacled eiders could balance their energy budget (Lovvorn et al. 2009). Prey densities from a regular sampling grid in 1999–2001 identified areas that exceeded that density, the corollary being that areas with densities below the threshold were non-essential to the eiders. However, analyses of past benthic surveys revealed that the locations of areas with high enough density of the dominant prey to support eiders in 1971–1974 were completely different from the locations of such areas in 1993–1994. Moreover, areas that could still support the eiders in 1999–2001 were only a small fraction of the suitable area only 5–8 years earlier (Lovvorn et al. 2009). In 1971–1974, by far the dominant bivalve prey species was *Macoma calcarea* while *N. radiata* was quite rare, whereas by 1993–1994 *N. radiata* had become superabundant and *M. calcarea* had declined dramatically. By 1999–2001, *N. radiata* had also shown major decline, while *M. calcarea* has remained at low densities. The dispersion and abundance of both bivalve species show strong positive correlations with sediment organic matter content (Grebmeier et al. 2006a, b). Thus, it is likely that changes in wind-driven hydrographic patterns that redistribute settled organic matter and young bivalves have an important role in these habitat shifts for eiders.

In addition to these changes in the abundance, dispersion, and species of prey, the eiders' access to prey can also be affected by ice conditions. In years of very low ice cover for most of winter such as 2001, inability of the eiders to rest on ice, where energy costs of thermoregulation are much lower than in water, could substantially reduce the area of habitat with prey densities high enough to meet the eiders' energy needs (Lovvorn et al. 2009). Conversely, in winters such as 2009 and 2010, when there was dense ice cover with few leads that allowed access to water, eiders can be excluded from the best feeding areas (Cooper et al. 2013). Decreased fat stores during 2009 suggested that in years when ice cover restricts access to the better feeding areas, spectacled eiders may have difficulty acquiring adequate reserves for migration and breeding (Lovvorn, unpublished data).

During warm years of the early 2000s, northward range shifts of important commercial fishes suggested that bottom trawl fisheries might expand north of St. Matthew Island into the wintering area of spectacled eiders. If the need arises to exclude habitat critical to eiders from possible impacts of intensive trawling, it is important to consider that suitable habitat for eiders may shift greatly in location and extent over time, and in annual accessibility depending on ice conditions. A much larger area must be set aside than is indicated by sampling prey dispersion over the span of only a few years or even a decade.

In the spring, spectacled eiders migrating to nesting areas along the Beaufort Sea coast move along the eastern coast of the Chukchi Sea. This narrow migration corridor is only 10–30 m deep, and is accessible in the spring only through open-water leads between land-fast and moving pack ice. As this migration path is also used by a number of other species, including king and common eiders, long-tailed ducks, and red-throated and yellow-billed loons, it is among the most important habitats for sea ducks and loons in the western Arctic (Oppel et al. 2009). During spring migration, ice conditions are highly variable there, resulting in unpredictable access to benthic foods. Rare instances in which ice completely excluded the birds from such habitats along the Beaufort Sea coast led to mortality of tens of thousands of eiders (Barry 1968; Fournier and Hines 1994). Even without high direct mortality, failure of females to achieve or maintain the minimum body reserves needed for breeding (Lovvorn et al. 2003), or to complete courtship and pair bonding which may occur mostly during migration (Lovvorn et al. 2012), might have important population impacts over a series of years with unfavorable weather. With earlier breakup of ice as predicted by climate trends, access to limited areas of habitat might be restricted if northwesterly winds drive drifting ice to become densely packed in those areas. Conversely, reduced extent or duration of land-fast ice, allowing greater wind-driven resuspension of shallow sediments, might alter sediment quality and the structure of benthic communities. Long-term studies of ice conditions and prey assemblages in that nearshore migration corridor (cf. Feder et al. 1994) may be important to understanding future population trends of diving birds that nest along the Arctic Coast.

11.3.4 Marine Mammals

The extreme sea ice retreats, common in late-summer since 2002 (Perovich 2011), served as a unifying theme for a special volume of papers focused on the effects of climate change on Arctic marine mammals (Huntington and Moore 2008). The impact of sea ice loss was described by species-specific sensitivity indices where a reliance on sea ice as a physical structure was one of nine factors considered (Laidre et al. 2008). Using the sensitivity indices, a conceptual model was developed that catalogued marine mammal species by their ecological relationship to sea ice (Moore and Huntington 2008). Here we provide empirical examples of three instances where marine mammals have exhibited responses to environmental variability, especially seasonal reduction in sea ice, in the PAR.

11.3.4.1 Timing and Relative Abundance of Bowhead Whales Feeding in the Canadian Beaufort Sea

Aerial surveys in the Canadian Beaufort Sea during late summer 2007–2009 found feeding aggregations of bowhead whales in areas similar to those of the 1980s (Richardson et al. 1987; Harwood and Smith 2002; Harwood et al. 2010). However, the feeding aggregations occurred roughly 2 weeks earlier in the season and the

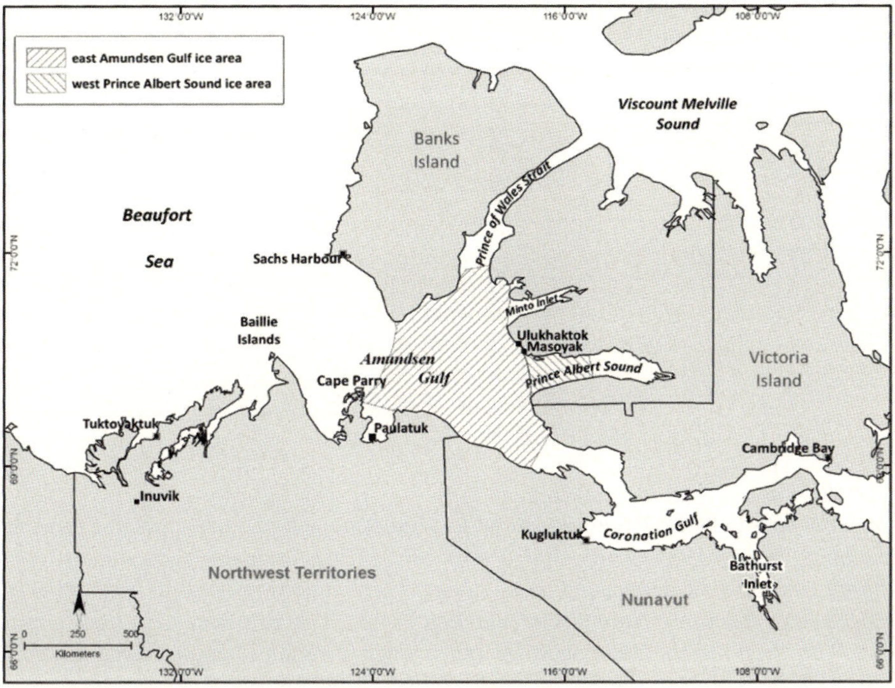

Fig. 11.14 Location of seal harvest, sampling and ice analysis areas in east Amundsen Gulf (Adapted from Harwood et al. 2012)

number of surfaced whales counted was consistently higher in the 2000s than in the 1980s. The study area, design, and survey methods were comparable across years, so the earlier timing and increased relative abundance of bowheads during the 2007–2009 surveys likely reflects an increase population size since the 1980s (George et al. 2004), an increase in the summer use of the Canadian Beaufort Sea shelf by bowheads, or both.

An examination of June sea ice cover in the aerial survey study area found that open water increased by 65 % per year for the period 1979–2002 (Moore and Laidre 2006). If both primary production and upwelling of prey on the summer feeding range has increased coincident with the loss of sea ice, as predicted by biophysical models of the Canadian Beaufort shelf (Arrigo and van Dijken 2011; Carmack et al. 2006), this may be a contributing factor to observed changes in timing and relative abundance as bowhead whales respond to earlier and better feeding opportunities.

11.3.4.2 Body Condition of Ringed Seals in the Western Canadian Arctic

Prime ringed seal breeding habitat, such as the large bays of eastern Amundsen Gulf (Fig. 11.14), combine the qualities of readily available food, as well as stable fast ice with adequate surface deformation to provide suitable sites for the construction

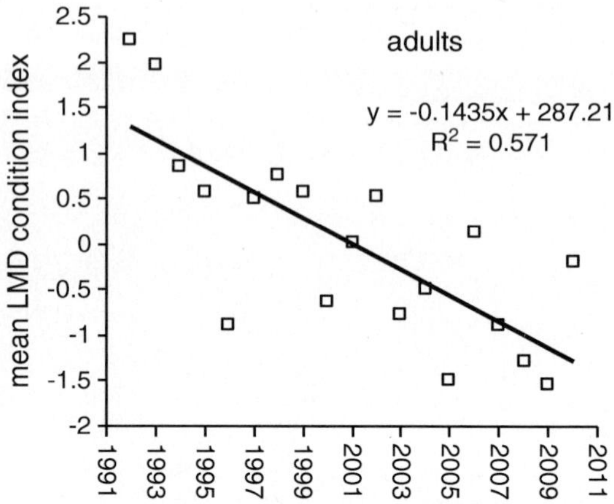

Fig. 11.15 Residual mean annual body condition of adult ringed seals sampled over 20 years in east Amundsen Gulf, 1992–2010 (Harwood et al. 2012). Body condition was calculated as the Length-Mass-blubber Depth (LMD) index, according to Ryg et al. (1990)

of subnivean lairs, which are essential to protect pups from both predators and the elements (Smith 1987; Kelly et al. 2010). Since 1992, there has been a statistically significant trend of decreasing mean annual body condition of adult and sub-adult ringed seals in eastern Amundsen Gulf (Fig. 11.15). That the changes in body condition were well matched among years and sex/age groupings, suggests that the trend can be linked to changes in the seal's diet, possibly from shifts in fish species composition or availability. This temporal trend of declining seal body condition was also apparent in ringed seals in Hudson Bay during the same period (Chambellant et al. 2012).

A parallel result in the Amundsen Gulf study was that body condition of seals was negatively correlated with the timing of fast ice clearance in spring (Fig. 11.16). The decline in condition was apparent in all sex/age groupings, and statistically significant in sub adults (Harwood et al. 2012) The mechanism is thought to be linked, at least in part, to how the timing of spring break up influences the amount of time there is for seals to feed, when they can start feeding, and the quality and/or quantity of prey (Wu et al. 2007; Smith 1987). There was also variation in reproductive output (reduced ovulation rates, percent pups in the harvest) following extreme heavy ice years, particularly in 1974 and 2005 when break-up of the fast ice in the breeding area was unusually late (Smith 1987; Harwood et al. 2012). Conversely, in 1998, break-up was 6 weeks early which led to higher body condition indices in all age and sex classes except for unweaned pups. In that year, break-up was so early that it shortened the lactation period by 3 weeks, and this resulted in pups being in poor condition and increased pup mortality (Smith and Harwood 2001).

While the seal population in this core habitat appears to recover from natural, extreme-year fluctuations (Stirling 1997; Smith 1987; Harwood et al. 2012), the possible magnified effect of several consecutive extreme ice years, compounded by the simultaneous occurrence of the temporal decline in seal body condition, is of particular concern. Seals residing in non-core and less populated habitats (e.g., offshore of the North Slope of Alaska and the west coast of Banks Island) may be more prone to environmentally-induced decline in body condition and reproductive failure than

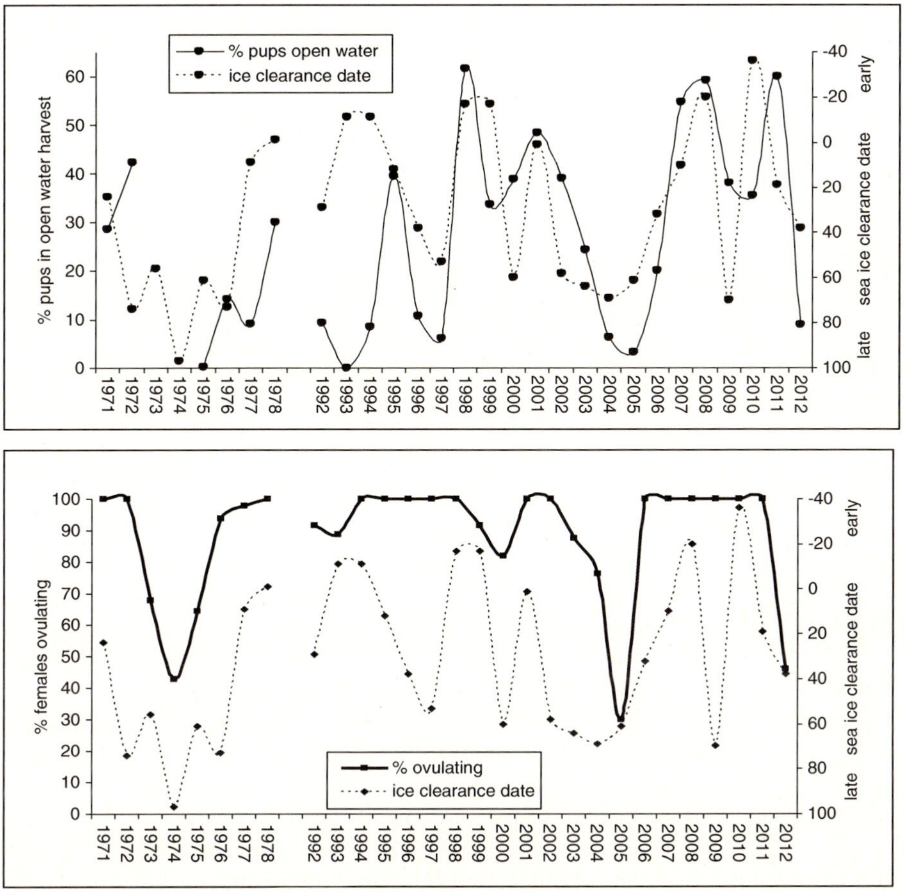

Fig. 11.16 Annual percent pups in the open water harvest and ovulation rates for ringed seals sampled during harvest-based monitoring in east Amundsen Gulf, 1971–1978 (Smith 1987) and 1992–2012 (Harwood et al. 2012)

seals in core, prime habitats such as Amundsen Gulf. Longterm studies are required to reveal trends, particularly in ecologically important areas. Identification of the types of prey taken by the seals is also needed to elucidate changes in diet and thus point to the links to broader environmental changes which are taking place.

11.3.4.3 Changes in Life-History and Diet of Walruses and Seals in the Northern Bering and Chukchi Seas

An analysis of five decades (1950s–2000s) of walrus harvest records in the Bering Strait region, showed that the age at harvest for males and females increased and the pregnancy rate decreased over time (Garlich-Miller et al. 2006). The age at first reproduction increased from 8 years in the 1950s and 1960s to 10 years in the late 1970s to early 1980s, suggesting that females were energetically stressed during the later period and the population overall was less productive. More recently, tens of thousands of walruses have shifted from resting on sea ice to resting on land along the northwest coast of Alaska and the northeastern coast of the Chukotka Peninsula coincident with the extreme seasonal retreats of sea ice since 2007 (Jay et al. 2011). In addition to increases in mortality (especially of calves) due to trampling, it is anticipated that increased use of land haul outs may result in increased energy expenditure during feeding bouts and reduced access to preferred feeding grounds. In modeling projected status of the Pacific walrus, the availability of sea ice habitat in summer and fall, along with harvest levels had the greatest influence on future population outcomes (Jay et al. 2011). Specifically, projections indicated that an increase in mortality of walrus calves at land haul outs had a greater effect on the population than an equivalent increase in harvest-related mortality distributed among all age classes (Udevitz et al. 2013).

An analysis of four decades (1960s–2000s) of body condition, growth and ovulation rate data from seals harvested for subsistence in Alaskan Bering and Chukchi sea communities indicated that for spotted and ribbon seals, the environmental conditions in the 1960s and 2000s produced similar levels of body condition and similar rates of growth and ovulation. However, environmental conditions in the 1970s were better for ribbon seals, which had higher indices of body condition, growth, and reproduction than in the 1960s and 2000s. Conversely, the conditions in the 1970s were worse for spotted seals according to the same indices (Quakenbush et al. 2009; Quakenbush and Citta 2008). For ice-obligate species, bearded and ringed seals, environmental conditions in the 2000s produced average or better body condition and rates of growth and ovulation compared to the past (Quakenbush et al. 2011a, b). Therefore, for all seal species, recent environmental conditions, although extreme regarding summer sea ice extent, have not affected indices of life history parameters beyond what has been measured in the past. Overall, the results for walruses and seals demonstrate that a change in the environment is not always negative and that the same change can affect species differently.

In addition to life history parameters such as body condition and productivity, investigations of seal stomach contents show how diet has changed in the Bering and Chukchi seas over four decades. For example, spotted and bearded seals ate more invertebrate prey and less fish in the 1960s and 1970s than they did in the late 1990s and 2000s (Fig. 11.17; Quakenbush et al. 2009, 2011b). Examination of the fish species eaten by spotted seals showed that there was an increase in all species (herring, rainbow smelt, Arctic cod, saffron cod) or group of species (all cod, flatfish) eaten, except sclupin. For bearded seals, the breakdown was similar except that sculpin increased along with the other fish. This combination of results suggests that sculpin were also likely increasing, but were not a preferred prey for spotted seals, at least when other fish species were available. Knowing how prey species are shifting in marine mammal diets can help us identify changes in the ecosystem. The increase in fish and concomitant decrease in invertebrates may be evidence that the productivity of the benthos is decreasing, as reported for some areas by Grebmeier (2012) and

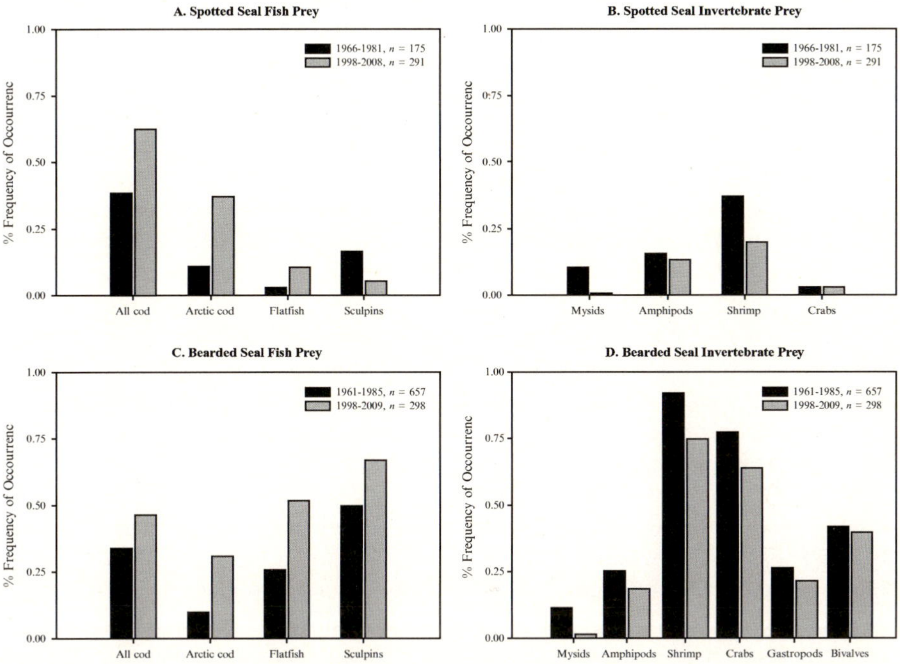

Fig. 11.17 Percent frequency of occurrence (% FO) of fish and invertebrate prey identified in the stomachs of spotted and bearded seals in the Bering and Chukchi seas, 1961–2009. % FO of all prey items were significantly different ($p<0.05$) between periods

predicted by models. The situation is likely more complicated, however, because several of the fish species that increased in the diet were benthic fish (e.g., flatfish and sculpin). In addition, the change in diet was not associated with a negative change in body condition or productivity indicating flexibility in diet for these seals.

11.4 UTL Species as Ecosystem Sentinels

Climate warming is altering marine ecosystems (Hoegh-Guldberg and Bruno 2010), with some of the clearest examples described in polar regions (Schofield et al. 2010; Wassmann 2011). Whether influenced by physical (bottom-up) or biological (top-down) forces, it appears that ecosystems reorganize in steps, not clines (Hare and Mantua 2000, and references therein). Such steps often follow abrupt changes physical forcing. For example, Scott and Marshall (2010) describe a 12-day abrupt change in the timing of sea ice breakup that occurred between 1988 and 1989 in western Hudson Bay, with no significant trend before or after the step. An increase in regional southwesterly winds in early June and a corresponding increase in surface temperature appeared to be contributing factors to this abrupt change in break-up timing, which has proven detrimental to polar bears that require a longer period to hunt seals from sea ice. A similar step-loss in seasonal sea ice cover in the PAR occurred in 2007, linked to a preconditioning phase of ice thinning (2000–2006) and to 'anomalous' wind conditions that year (Overland 2009; Overland et al. 2014, this issue). A step-change in top-down forcing can also reorganize ecosystems. For example, after the collapse of commercial fisheries off southwest Africa in the late 1960s, the trophic structure shifted, jellyfish biomass increased and one of its predators, the bearded goby (*Sufflogobius bibarbatus*), became the predominant prey for UTL species (Pennisi 2010). Of note, the reorganization of this ecosystem was detected by analyzing goby diet using stable isotope ratios of body tissues.

In a review of Arctic ecological dynamics associated with recent climate change, Post et al. (2009) list seven urgent research priorities. Three of the seven include studies of: (i) trophic interactions, (ii) scale dependence of climate responses and (iii) the importance of extreme events (i.e. tipping points) to species composition and ecosystem restructuring. The extreme loss of sea ice in late-summer 2007 and 2012 is an example of a physical 'tipping point' for the PAR; a threshold beyond which sea ice is not expected to return to its former late-summer state (Perovich 2011). This event provides an opportunity to develop a threshold-based research approach to study the effects that a dramatic shift in a defining physical element has on trophic interactions across the ecological scale of UTL species (see Fig. 11.1). Given that sea ice is forecast to continue to decline to an ice-free summer by about 2040 (Wang and Overland 2009), a research focus on the responses of UTL species could be more informative than studies of sea ice alone (see Hare and Mantua 2000).

11.4.1 UTL-Focused Research Framework

The example case studies demonstrate that marine fishes, birds and mammals respond in measurable ways to variability in sea ice, water temperature and salinity, sediment type and available prey. The challenge now is to propose how UTL species can best serve as sentinels for broad-scale ecosystem shifts and reorganization. As a starting point, we suggest that UTL species could be used to investigate shifts within three ecosystem processes in the PAR: (i) trophic interactions, focused on the relative importance of arctic cod, forage fishes and zooplankton in UTL diets, (ii) foraging dynamics, focused on prey delivery and aggregation via pelagic-benthic coupling, upwelling, advection and the formation of fronts and eddies, and (iii) species composition, focused on shifts in UTL species phenology and range expansion (Fig. 11.18 and Table 11.5). Note that trophic interactions focus on intrinsic responses, while investigations of ecological scale of prey delivery and species composition focus on extrinsic responses to climate change (see Fig. 11.11). Further, tracking responses using a suite of UTL species as ecosystem 'samplers' would provide data over a broad range of temporal and spatial scales (see Fig. 11.1). By combining information across taxa we may be able to recognize and perhaps even predict steps, or thresholds, that could result in cascading changes in ecosystem structure.

Table 11.5 Three ecosystem processes that could be investigated in the PAR by tracking responses in focal (and alternate) UTL species

Ecosystem process	Focal species	Potential responses[related ref.]
Trophic interactions: shifts in UTL species' diet among arctic cod, other forage fishes and various zooplankton spp.	Black guillemots, (loons)	Shift in diet, colony location or nesting failure[a]
	Shearwaters and auklets	Shifts in distribution and at-sea abundance[j]
	Ringed seals, (beluga)	Shift in distribution, diet, body condition[b]
	Bearded, ringed and spotted seals	Shift in diet and body condition[c]
	Bowhead whales	Shift in diet and body condition[d]
Foraging dynamics: shifts in UTL prey delivery and concentration via pelagic-benthic coupling, upwelling, advection and the formation of fronts and eddies	Spectacled eiders	Winter survival and pre-breeding reserves[e, f]
	Least and crested auklets	Shift chick survival[g]
	Bowhead whales (ringed seals)	Shift in feeding aggregations and availability to hunters[h]
Species composition: shifts in UTL species phenology, range expansion and/or relative abundance	Salmon, pollock and snow crab capelin and Pacific herring	Shift in distribution, size, diet[i]
	Auklets, shearwaters	Shifts in feeding dispersion related to changes major currents, stratification or mixing and oceanographic fronts[j]
	Murres, puffins, kittiwakes	Shifts in feeding dispersion related to stratification or mixing[j] and distribution shifts of fishes[k]
	Gray, humpback, fin and killer whales	Shift in phenology, distribution, diet[l, m]

[a]Divoky *in* Moline et al. (2008); [b]Harwood et al. (2012); [c]Quakenbush et al. (2009, 2011a, b); [d]Lowry et al. 2004; [e]Lovvorn et al. (2003); [f]Lovvorn et al. (2009); Clarke et al. (2013); [g]Gall et al. (2006); [h]Ashjian et al. (2010); [i]Logerwell et al. (2011); [j]Gall et al. (2013); [k]Springer et al. (1984); [l]Moore (2008); [m]Delarue et al. (2013) and Ferguson et al. (2010)

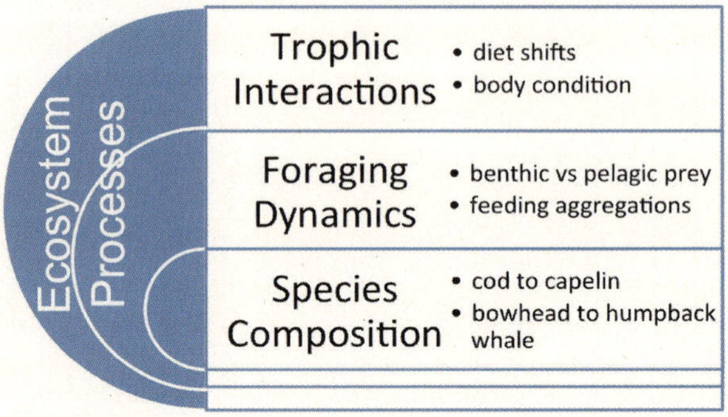

Fig. 11.18 Three ecosystem processes that could be investigated in the PAR by research focused on the trophic interactions, foraging dynamics and/or species composition of UTL species. Sampling across a broad range of spatial and temporal scales (see Fig. 11.1) would be achieved by including research on a spectrum of UTL species in multidisciplinary oceanographic programs

11.4.1.1 Trophic Interactions

Arctic cod are a vital trophic link between plankton and UTL species in the PAR marine ecosystem (e.g. Bradstreet et al. 1986; Frost and Lowry 1984; Ainley and DeMaster 1990), yet little is known about the distribution, abundance and responses to ecosystem variability of this key species. Observations of and models based on UTL stomach analyses suggest that cod are a conduit for up to 93 % of the energy flow between zooplankton and seabirds/marine mammals (Frost and Lowry 1984; Welch et al. 1992). While the functional response of Arctic cod populations to the seasonal loss of sea ice is unknown, a shift away from cod as a key trophic link can be tracked via seabird and marine mammal diets, especially in black guillemots and ringed seals. Reliance of these two UTL species on Arctic cod has been studied in the PAR over the past three to four decades (e.g., Divoky *in* Moline et al. 2008; Frost and Lowry 1981, 1984; Harwood et al. 2012) providing a long-term baseline against which to assess shifts in diet, body condition and reproduction rates. Recent hydro-acoustic surveys have identified dense shoals of Arctic cod, which likely play an important role in structuring the Beaufort Sea ecosystem (e.g., Benoit et al. 2008; Parker-Stetter et al. 2011; Crawford et al. 2011). Using UTL species to investigate how the reduction of sea ice affects the stability of arctic cod availability seems the best option to discover if, when, and how this fundamental trophic relationship is altered.

Shifts in the availability of other forage fish and zooplankton species could also be evaluated by tracking the diet and body condition of various UTL species (Table 11.5). For example, the biology of ringed, bearded and spotted seals has recently been reviewed (Quakenbush et al. 2009, 2011a, b), providing information on their distribution, diet and body condition dating back to the 1960s. Similarly, bowhead whale diet and body condition have been studied, via collaboration with Alaskan Native hunters, since the late 1970s (Lowry et al. 2004; Moore et al. 2010). These studies demonstrate that bowhead stomach contents vary, principally between endemic-copepods and advected-euphausiids, between seasons and across years. These longitudinal observations provide a foundation against which current diet and body condition can be compared and evaluated in light of physical changes to the PAR ecosystem over the past 40–60 years. A concerted effort to standardize and combine observations of diet and body condition from seabirds, to seals to whales would provide insights to ecosystem variability and responses to physical forcing unavailable through other means.

11.4.1.2 Foraging Dynamics

The northern Bering and Chukchi seas are considered benthic-dominated systems because productivity related to the presence of sea ice is generated faster than it can be consumed in the water column and it falls to the sea floor resulting in regions of very high benthic productivity (Hunt et al. 2013; Grebmeier 2012; Nelson et al. 2014, this volume). Conversely, the southern Bering Sea is considered a pelagic-dominated system because the earlier seasonal retreat of sea ice is followed by a phytoplankton bloom that is harvested by zooplankton and fishes consuming much of the productivity the water column before it reaches the sea floor (e.g., Hunt et al. 2002). Therefore, benthic-feeding UTL species should thrive where there is persistent sea ice, but pelagic-feeding UTL species may be favored with the loss of sea ice. While this simplified conceptual model does not include the importance of prey advection (Springer et al. 1989), nor the dynamics of upwelling that bring deep-dwelling prey onto the continental shelves (Carmack et al. 2006), it does offer a conceptual model where in benthic-foraging UTL species can act as sentinels to ecosystem restructuring as sea ice declines.

The long-term study of spectacled eiders has revealed how dynamic the benthic prey assemblages associated with NBS polynyas are (Lovvorn et al. 2003, 2009), and continued research in this area would provide detail of future changes at a regional scale. Over broader spatial scales, integration of long-term research on benthic-feeding bearded seal, walrus and gray whales would provide a window into how the benthic prey assemblages on which these species depend are altered with changing physical conditions (e.g. Quakenbush et al. 2011b; Jay et al. 2011; Moore 2008). With regard to the pelagic vs. benthic scenario, ringed and spotted seals should fare better than bearded seals in a reduced-ice environment because their diet consists of more pelagic fishes. However, if the bearded seal diet is flexible, we may see an increase in pelagic fish species in the diet of all seals. By looking for "steps," like reduced benthic feeding, via continued monitoring of walrus and seal diet and body condition, we may be able to determine when a 'tip' between benthic and pelagic trophic pathways occurs. Interpretation of these data would be greatly enhanced if information on UTL species movements relative to sea ice conditions and measures of productivity are also available.

The simplified pelagic-benthic coupling model omits the fundamental processes of production, advection, upwelling and concentration of prey that are essential to the foraging success of UTL species (e.g., Benoit-Bird et al. 2013). Indeed, pelagic foraging seabirds are influenced by changes in transport of zooplankton from the Bering into the Chukchi Sea (e.g. Springer et al. 1984). Specifically, Gall et al. (2013) found that seabird species composition and total abundance changed over 3 years (2008–2010), depending on the degree of Bering Sea water intrusion onto the northeastern Chukchi shelf. In particular, planktivorous auklets were associated with regions of strong stratification and high salinity that occurred where Bering Sea water was in the upper layer. While copepods and other arctic-endemic zooplankton are an essential component in the diet of many UTL species (Nelson et al. 2014, this volume) it is noteworthy that euphausiids are common in the diet of fishes (see Fig. 11.13), seabirds and marine mammals, demonstrating that a Pacific-derived prey plays a prominent role in UTL foraging in the PAR. Investigation of step-changes to in foraging dynamics could be achieved by tracking feeding aggregations of bowhead whales in the Beaufort Sea coupled with continued sampling of stomach contents and body condition (see Lowry et al. 2004). Recent work pairing such sampling with multidisciplinary oceanographic studies near Barrow resulted in the development of a 'krill-trap' conceptual model, a first step in elucidating the effects of winds and tides on the availability and concentration of bowhead whale prey (Ashjian et al. 2010; Okkonen et al. 2011). Of note, the dominance of euphausiids in bowhead whale stomachs was first discovered in the 1980s (Lowry et al. 2004), through research partnerships between scientists and local hunters; without this partnership, it would likely be assumed that bowheads were feeding solely on copepods, as broadly reported in the literature.

11.4.1.3 Species Composition

There are numerous examples in the literature of temperate species moving to more northern or alpine habitats as the climate warms, as well as examples of changes in migration timing and breeding seasons (e.g., Root et al. 2003; Tewksbury et al. 2011). Changes in UTL species composition due to range expansions and shifts in phenology are anticipated in response to warming in the PAR, which in turn will alter the trophic dynamics of the system. Indeed, shifts in phenology and range for several Arctic species appear to be well underway (e.g. Post et al. 2009). Salmon, snow crab and several species of whales may be good candidates for tracking shifts in species composition in the PAR, due to commercial importance and protected-species status, respectively. Densities of pink and chum salmon near Bering Strait (see Fig. 11.2), and their comparatively large size in the Chukchi Sea, will be research foci as the BASIS study continues to track the status of these species. Research partnerships with coastal residents could augment information on salmon species composition, size and diet, as well as track the occurrence of more temperate fish species such as capelin (e.g. Thedinga et al. 2013), a common prey of humpback whales. Similarly, snow crab of commercial size have already been reported in the western Beaufort Sea (Logerwell et al. 2011), with continued monitoring of this commercially important species is anticipated.

With regard to cetaceans, gray whales have demonstrated responses to environmental changes in the North Pacific (Moore 2008), including detection of their calls in the Beaufort Sea over winter (Stafford et al. 2007). Humpback whales are seen routinely in the southwestern Chukchi waters (Fig. 11.10; Clarke et al. 2013), with at least one recent report of a feeding whale as far north as Barrow, while fin whale calls are commonly recorded in the northeastern Chukchi Sea from July-October (Delarue et al. 2013). Although these reports may indicate a northward shift of seasonally migrant cetaceans, they should be interpreted within the context of decades-earlier summaries of fin, humpback and gray whale distribution in the southern Chukchi Sea, especially the productive zone just northwest of Bering Strait (Maher 1960; Nasu 1974; Votrogov and Ivashin 1980). Populations of these seasonally migrant cetacean species are rebounding from commercial harvests that continued through the mid-twentieth century, so recent sightings may reflect both a response to climate warming and a re-occupation of their historic summer range by expanding populations. While an increase in the occurrence of killer whales has not been consistently reported in the PAR, they have been reported seen more often in the Hudson Bay region of the Canadian Arctic concomitant with sea ice loss related to climate warming (Ferguson et al. 2010). Their role as top-predators of both fishes and marine mammals (e.g. Springer et al. 2008 and references therein) suggest they could dramatically alter the PAR marine ecosystem should they expand their range or remain in northern waters longer in response to the recent loss of sea ice.

11.5 Summary

Marine fishes, birds and mammals are formidable oceanographers (Fedak 2004; Smith 2001) and sample the PAR ecosystem at spatial and temporal scales unavailable through standard means. To survive, each species must adapt to environmental variability and in doing so can identify ecosystem alterations. In this chapter, we have provided an overview of ecology for UTL species, followed by examples where particular species in each guild have shown measurable responses to ecosystem variability. At broad spatial and temporal scales, fishes and snow crab appear to be moving north in response to climate warming; but, there are insufficient data on the quintessential species, arctic cod, to evaluate its reaction to systemic change. At more regional scales, marine birds and mammals have changed aspects of their phenology, body condition, productivity and diet, as they respond to sea ice variability and prey availability. Because these species must find dense assemblages of prey to forage effectively, they are particularly adept at identifying regions of high secondary productivity and will continue to do so if and when those regions change with climate warming.

11.5.1 Tracking Biological Responses in an Era of Rapid Change and Extreme Events

The intensity and frequency of extreme events has emerged as a common facet of climate change (Jentsch et al. 2007) and it seems likely that the next 10–50 years will be marked by tremendous variability in physical conditions in the PAR. The ranges between sea ice seasonal maxima and minima extents are now record-setting, with extensive late-summer retreats in the Chukchi and Beaufort seas (2007, 2011 and 2012) followed by broad advances in spring (2008, 2012) in the Bering Sea (http://nsidc.org). Sea ice is also thinning at a rapid rate, with ice volume in 2012 roughly half that of 2007. In addition, with increased heat stored in the ocean, storm events north of Bering Strait are now common, bringing novel forcing conditions to seasonally open-water areas and unprecedented erosion to coastal areas. The PAR appears to be responding to all these physical changes with increased primary production (Arrigo and van Dijken 2011; Arrigo et al. 2012) and possibly new pathways of zooplankton prey delivery to UTL species via advection and upwelling dynamics (Carmack et al. 2006).

It is in this context that we urge the development of a UTL-focused research approach, using marine fishes, birds and mammals as sentinels or guides to reorganization in ecosystem structure. Just as Grebmeier et al. (2010) proposed focused-sampling at highly productive regions along a latitudinal gradient of the PAR as a means of tracking biological responses to physical forcing, here we suggest focused-sampling on species that reflect ecosystem shifts by measurable changes in their migration and movement patterns, distribution and relative abundance, diet, reproductive output and body condition. Linking the results of each of these focused-sampling approaches would strengthen both. As in the launch of the Distributed Biological Observatory (http://www.arctic.noaa.gov/dbo), using extant long-term data sets to identify study sites and species would accelerate a program of research based on marine fishes, birds and mammals as monitors of step-changes in the PAR ecosystem.

11.5.2 Integration of Science and Local Knowledge

Several of the marine mammal studies cited in this chapter are based upon partnerships among researchers trained in the 'western science' and indigenous people, holders of local or 'traditional' knowledge (e.g. Ashjian et al. 2010; Moore et al. 2010; Quakenbush et al. 2010; Citta et al. 2012; Harwood et al. 2012). Partnerships between local residents and scientists in arctic research are becoming more common place (Weatherhead et al. 2010 and references therein). One example, the Sea Ice for Walrus Outlook (SIWO), combines weather forecasting and satellite observations of sea ice with local observations by walrus hunters from April through June in the Bering Strait region (http://www.arcus.org/search/siwo). This combination of indigenous observations with scientific approaches has resulted in a more comprehensive understanding of springtime biophysics in this dynamic region. To be successful, it is essential for the science and indigenous partners to respect and trust each other's skills, knowledge and abilities, that regular communication is fostered at all steps, and that provisions are in place to monitor progress (Bell and Harwood 2012). The credible blending of indigenous and scientific views and skills improves the likelihood of ultimately understanding the resource, its habitats and its inherent ecological relationships.

Indigenous residents of the PAR rely on marine fishes, birds and mammals both for sustenance and cultural identity. Scientists and policy makers study UTL species' ecology, in some cases to act upon regulatory statutes for their conservation as people that live outside the Arctic relate to UTL species, especially birds and mammals, for their 'charismatic' nature. This strong link between UTL species and humans makes them a natural nexus, or connection, between people (Arctic residents, scientists, policy makers and the public) and ecosystem science. Observing, understanding and predicting future states in Arctic ecosystems will require the combined focus and will of all these human resources.

11.6 Personal Communications

C. George, North Slope Borough, Dept of Wildlife Mgt, Barrow, AK USA<craig.george@north-slope.org>

L. Jemison, Alaska Dept of Fish & Game, Juneau, AK USA<lauri.jemison@alaska.gov>

J. Orr, Alaska Fisheries Science Center<james.orr@noaa.gov>

Acknowledgements We thank our many colleagues for their years of work in the field and laboratory resulting in the published accounts upon which much of this chapter is based. We also thank Kate Stafford (APL/UW) for providing cetacean sightings from the 2010 RUSALCA cruise, as well as the Principal Investigators and crew of that program. Finally, we thank three anonymous reviewers for helpful comments on the initial version of this chapter.

References

Aagaard K, Roach AT, Schumacher JD (1985) On the wind-driven variability of the flow through Bering strait. J Geophys Res 90:7213–7221

Ainley DG, DeMaster DP (1990) The upper trophic levels in polar marine ecosystems. In: Smith WO Jr (ed) Polar oceanography, Part B Chemistry, biology and geology. Academic, San Diego

Allen MJ, Smith GB (1988) Atlas and zoogeography of common fishes in the Bering Sea and northeastern Pacific. NOAA Technical Report NMFS66: 151 p

Amstrup SC, Durner GM, Stirling I, Lunn NN, Messier F (2000) Movements and distribution of polar bears in the Beaufort Sea. Can J Zoo 78:948–966

Anderson PJ, Piatt JF (1999) Community reorganization in the Gulf of Alaska following ocean climate regime shift. Mar Ecol Prog Ser 189:117–123

Andrews AG, Farley EV, Moss JH, Murphy JM, Husoe EF (2009) Energy density and length of juvenile pink salmon (*Oncorhynchus gorbuscha*) in the eastern Bering Sea from 2004 to 2007: a period of relatively warm and cool sea surface temperatures. N Pac Anadromous Fish Comm Bull 5:183–189

Arrigo KR, van Dijken GL (2011) Secular trends in Arctic Ocean net primary production. J Geophys Res 116:C09011. doi:10.1029/2011 JC007151

Arrigo KR, Perovich DK, Pickart RS, +28 co-authors (2012) Massive phytoplankton blooms under arctic sea ice. Science 336:1408

Ashjian CJ, Braund SR, Campbell RG, George JC, Kruse Maslowski W, Moore SE, Nicolson CR, Okkonen SR, Sherr BF, Sherr EB, Spitz Y (2010) Climate variability, oceanography, bowhead whale distribution and Inupiat subsistence whaling near Barrow, Alaska. Arctic 63:179–194

Baird PH (1994) Black-legged Kittiwake (*Rissa tridactyla*). In: Poole A, Gill F (eds) The birds of North America No. 92. Academy of Natural Sciences/American Ornithologists' Union, Philadelphia/Washington, DC

Barber WE, Smith RL, Vallarino M, Meyer RM (1997) Demersal fish assemblages of the northeastern Chukchi Sea, Alaska. Fish Bull 95:195–209

Barry TW (1968) Observations on natural mortality and native use of eider ducks along the Beaufort Sea coast. Can Field Nat 82:141–144

Bell RK, Harwood LA (2012) Harvest-based monitoring in the inuvialuit settlement region: steps for success. Arctic 65:421–432

Benoit D, Simard Y, Fortier l (2008) Hydroacoustic detection of large winter aggregations of Arctic cod (*Boreogadus saida*) at depth in ice-covered Franklin Bay (Beaufort Sea). J Geophys Res 113. doi:10.1029/2007JC004276

Benoit-Bird KJ, Battaile BC, Heppell SA, +9 co-authors (2013) Prey patch patterns predict habitat use by top marine predators with diverse foraging strategies. PLoS One. doi:10.1371/journal.pone.0053348

Blanchard AL, Parris C, Nichols H (2010) 2009 environmental studies program in the northeastern Chukchi Sea: benthic ecology of the Burger and Klondike survey areas. University of Alaska, Fairbanks

Bluhm BA, Coyle KO, Konar B, Highsmith R (2007) High gray whale relative abundances associated with an oceanographic front in the south-central Chukchi Sea. Deep Sea Res II 54:2919–2933

Bluhm BA, Iken K, Hardy SM, Sirenko BI, Holladay BA (2009) Community structure of epibenthic megafauna in the Chukchi Sea. Aquat Biol 7:269–293

Bluhm BA, Iken K, Hopcorft RR (2010) Observations and exploration of the Arctic's Canada Basin and the Chukchi Sea: the hidden ocean and RUSALCA expeditions. Deep Sea Res II 57:1–4

Boeskorov GG, Davydov SP, Kochnev AA, Lang EM (2011) Penetration of the Steller's sea lion (*Eumetopias jubatus*) into waters of the Chukchi and East-Siberian Seas. Zool Zhurnal 90:123–128 [in Russian, English Abstract]

Boetius A, Albrecht S, Bakker K, + 14 co-authors (2013) Export of algal biomass from the melting Arctic sea ice. Science. doi:10.1126/science.1231346

Bradstreet MSW, Finley KJ, Sekerak AD, Griffiths WB, Evans CR, Fabijan FF, Stallard HE (1986) Aspects of the biology of Arctic cod (*Boreogadus saida*) in arctic marine food chains. Can Tech Rep Fish Aquat Sci 1491: 1–193

Busby MS, Chernova NV (2001) Redescription of the festive snailfish, *Liparis marmoratus* (Scorpaeniformes: Liparidae), with a new record from the northern Bering Sea. Ichthyol Res 48:187–191

Carmack E, Barber D, Christensen J, Macdonald R, Rudels B, Sakshaug E (2006) Climate variability and physical forcing of the food webs and the carbon budget on panarctic shelves. Prog Oceanogr 71:145–182

Chambellant M, Stirling I, Gough WA, Ferguson SH (2012) Temporal variations in Hudson Bay ringed seal (*Phoca hispida*) life-history parameters in relation to environment. J Mamm 93:267–281

Citta JJ, Quakenbush LT, George JC, Small RJ, Heide-Jørgensen MP, Brower H, Adams B, Brower L (2012) Winter movements of bowhead whales (*Balaena mysticetus*) in the Bering Sea. Arctic 65:13–34

Clarke J, Stafford K, Moore SE, Rone B, Aerts L, Crance J (2013) Subarctic cetaceans in the southern Chukchi Sea: evidence of recovery or response to a changing ecosystem. Oceanography 26:136–149

Cooper LW, Sexson MG, Grebmeier JM, Gradinger R, Mordy CW, Lovvorn JR (2013) Linkages between sea ice coverage, pelagic-benthic coupling and the distribution of spectacled eiders: observations in March 2008, 2009 and 2010 from the northern Bering Sea. Deep Sea Res II 94:31–43

Coulson JC (1984) The population dynamics of the Eider duck *Somateria mollissima* and evidence of extensive non-breeding by adult ducks. Ibis 126:525–543

Craig PC, Griffiths L, Haldorson L, McElderry H (1985) Distributional patterns of fishes in an Alaskan Arctic lagoon. Polar Biol 4:9–18

Crawford RE (2009) Forage fish habitat distribution near the Alaskan coastal shelf areas of the Beaufort and Chukchi seas. Final Report, oil spill recovery institute, Cordova. November 2009. Available: www.pws-osri.org

Crawford RE, Jorgenson JK (1996) Quantitative studies of Arctic cod (*Boreogadus saida*) schools: important energy stores in the Arctic food web. Arctic 49:181–193

Crawford RE, Vagle S, Carmack E (2011) Water mass and bathymetric characteristics of polar cod habitat along the continental shelf and slope of the Beaufort and Chukchi seas. Polar Biol. doi:10.1007/s00300-011-1051-9

Cui X, Grebmeier JM, Cooper LW, Lovvorn JR, North CA, Seaver WL, Kolts JM (2009) Spatial distribution of groundfish in the northern Bering Sea in relation to environmental variation. Mar Ecol Prog Ser 393:147–160

Day RH, DeGange AR, Divoky GJ, Troy DM (1988) Distribution and subspecies of the Dovekie in Alaska. Condor 90:712–714

Day RH, Kuletz KJ, Nigro DA (1999) Kittlitz's murrelet (*Brachyramphus brevirostris*). In: Poole A, Gill F (eds) The birds of North America, No. 435. Academy of Natural Sciences/American Ornithologists' Union, Philadelphia/Washington, DC

Day RH, Gall AE, Prichard AK, Divoky GJ, Rojek NA (2011) The status and distribution of Kittlitz's Murrelet *Brachyramphus brevirostris* in northern Alaska. Mar Ornithol 39:53–63

Delarue J, Martin B, Hannay D, Berchok C (2013) Acoustic occurrence and affiliations of fin whales detected in the northeaster Chukchi Sea, July to October 2007–2010. Arctic 66 (in press)

Dunton K, Goodall J, Schonberg SV, Grebmeier JM, Maidment DR (2005) Multi-decadal synthesis of benthic-pelagic coupling in the western Arctic: role of cross-shelf advective processes. Deep Sea Res II 52:3462–3477

Eisner L, Hillgruber N, Martinson E, Maselko J (2013) Pelagic fish and zooplankton species assemblages in relation to water mass characteristics in the northern Bering and southeast Chukchi seas. Polar Biol 36:87–113

Eisner L, Napp J, Mier K, Pinchuk A, Andrews A (2014) Climate-mediated changes in zooplankton community structure for the eastern Bering Sea. Deep Sea Res II. doi:10.1016/j.dsr2.2014.03.004

Fedak MA (2004) Marine mammals as platforms for oceanographic sampling: a "win/win" situation for biological operational oceanography. Mem Natl Inst Polar Res Spec Iss 58:133–147

Feder HM, Naidu AS, Jewett SC, Hameedi JM, Johnson WR, Whitledge TE (1994) The northeastern Chukchi Sea: benthos-environmental interactions. Mar Ecol Prog Ser 111:171–190

Feder HM, Jewett SC, Blanchard A (2005) Southeastern Chukchi Sea (Alaska) epibenthos. Polar Biol 28:402–421

Ferguson SH, Higdon JW, Chmelnitsky EG (2010) The rise of killer whales as a major Arctic predator. In: Ferguson SH et al (eds) A little less arctic: top predators in the world's largest Northern Inland Sea, Hudson Bay, 117. Springer Science+Business Media B.V. doi:10.1007/978-90-481-9121-5_6

Fischbach AS, Amstrup SC, Douglas DC (2007) Landward and eastward shift of Alaskan polar bear denning associated with recent sea ice changes. Polar Biol 30:1395–1405

Fischer JB, Larned WW (2004) Summer distribution of marine birds in the western Beaufort Sea. Arctic 57:143–159

Fournier MA, Hines JE (1994) Effects of starvation on muscle and organ mass of king eiders, *Somateria spectabilis*, and the ecological and management implications. Wildfowl 45:188–197

Francis RC, Hare SR, Hollowed AB, Wooster WS (1998) Effects of interdecadal climate variability on the oceanic ecosystems of the NE Pacific. Fish Ocean 7:1–21

Frey KE, Maslanik JA, Clement Kinney J, Maslowski W (2014) Chapter 3: Recent variability in sea ice cover, age, and thickness in the Pacific Arctic region. In: Grebmeier JM, Maslowski W (eds) The Pacific Arctic region: ecosystem status and trends in a rapidly changing environment. Springer, Dordrecht, pp 31–64

Friday NA, Waite JM, Zerbini A, Moore SE (2012) Cetacean distribution and abundance in relation to oceanographic domains on the eastern Bering Sea shelf: 1999–2004. Deep Sea Res 65–70:260–272

Frost KJ, Lowry LF (1981) Trophic importance of some marine gadids in northern Alaska and their body-otolith size relationships. Fish Bull 79:187–192

Frost KJ, Lowry LF (1983) Demersal fishes and invertebrates trawled in the northeastern Chukchi and western Beaufort Seas 1976–1977. NOAA Technical Report NMFS-SSRF-764. U.S. Department of Commerce. 22p

Frost KJ, Lowry LF (1984) Trophic relationships of vertebrate consumers in the Alaskan Beaufort Sea. In: Barnes PW, Schell DM, Reimnitz E (eds) The Alaskan Beaufort Sea: ecosystems and environments. Academic, San Diego

Gall AE, Roby DD, Irons DB, Rose IC (2006) Differential response in chick survival to diet in least and crested auklets. Mar Ecol Prog Ser 308:279–291

Gall AE, Day RH, Weingartner TJ (2013) Structure and variability of the marine-bird community in the northeastern Chukchi Sea. Continental Shelf Res 67:96–115

Garlich-Miller JL, Quakenbush LT, Bromaghin JF (2006) Trends in age structure and productivity of Pacific walruses harvested in the Bering Strait region of Alaska, 1952–2002. Mar Mamm Sci 22:880–896

Gawarkiewicz G, Haney JC, Caruso MJ (1994) Summertime synoptic variability of frontal systems in the northern Bering Sea. J Geophys Res 99:7617–7625

George JC, Philo M, Hazard K, Withrow D, Carroll GM, Suydam R (1994) Frequency of killer whale (*Orcinus orca*) attacks and ship collisions based on scarring on bowhead whales (*Balaena mysticetus*) of the Bering-Chukchi-Beaufort sea stock. Arctic 47(3):247–255

George JC, Zeh J, Suydam R, Clark C (2004) Abundance and population trend (1978–2001) of western Arctic bowhead whales surveyed near Barrow, Alaska. Mar Mamm Sci 20:755–773

Gillispie JG, Smith RL, Barbour E, Barber WE (1997) Distribution, abundance and growth of arctic cod in the northeastern Chukchi Sea. Am Fish Soc Symp 19:81–89

Grebmeier JM (2012) Shifting patterns of life in the Pacific Arctic and sub-Arctic Seas. Ann Rev Mar Sci 4:16.1–16.16

Grebmeier JM, Cooper LW (1995) Influence of the St. Lawrence Island polynyas upon the Bering Sea benthos. J Geophys Res 11:4439–4460

Grebmeier JM, Cooper LW, Feder HM, Sirenko BI (2006a) Ecosystem dynamics of the Pacific-influenced Northern Bering and Chukchi Seas in the Amerasian Arctic. Prog Oceanogr 71:331–361

Grebmeier JM, Overland JE, Moore SE, Farley EV, Carmack EC, Cooper LW, Frey KE, Helle JH, McLoughlin FA, McNutt L (2006b) A major ecosystem shift observed in the northern Bering Sea. Science 311:1461–1464

Grebmeier JM, Moore SE, Overland JE, Frey KE, Gradinger R (2010) Biological-response to recent Pacific Arctic Sea ice retreats. EOS T Am Geophys Un 91(18):161–162

Haney JC (1991) Influence of pycnocline topography and water-column structure on marine distributions of alcids (Aves: Alcidae) in Anadyr Strait, northern Bering Sea, Alaska. Mar Biol 110:419–435

Hare SR, Mantua NJ (2000) Empirical evidence for North Pacific regime shifts in 1977 and 1989. Prog Oceanogr 47:103–145

Harwood LA, Smith TG (2002) Whales of the Beaufort Sea: an overview and outlook. Arctic 55(Supp 1):77–79

Harwood LA, Auld J, Joynt A, Moore SE (2010) Distribution of bowhead whales in the SE Beaufort Sea during late summer, 2007–2009. DFO Can Sci Advis Sec Res Doc 2009/111 22 pp

Harwood LA, Smith TG, Melling H, Alikamik J, Kingsley MCSK (2012) Ringed seals and sea ice in Canada's Western Arctic: harvest-based monitoring 1992–2011. Arctic 65:377–390

Hatch JJ (2002) Arctic Tern (*Sterna paradisaea*). In: Poole A, Gill F (eds) The birds of North America, No. 707. Wiley and Colleagues: Research Publications, Philadelphia

Heide-Jørgensen MP, Laidre KL, Quakenbush LT, Citta J (2011) Northwest Passage opens for bowhead whales. Biol Lett. doi:10.1098/rsbl.2011.0731

Helle JH (2009) Birth of Bering-Aleutian Salmon International Survey (BASIS). N Pac Anadromous Fish Comm Bull 5:vii–viii

Hodges JI, King JG, Conant B, Hanson HA (1996) Aerial surveys of waterbirds in Alaska 1957–94: population trends and observer variability. US Dep Int, Nat Bio Serv Info Techical Report 49

Hoegh-Guldberg O, Bruno JF (2010) The impact of climate change on the world's marine ecosystems. Science 328:1523–1528

Hop H, Welch HE, Crawford RE (1997) Population structure and feeding ecology of Arctic cod schools in the Canadian High Arctic. Am Fish Soc Symp 19:68–80

Hopcroft RR, Kosobokova KN, Pinchuk AI (2010) Zooplankton community patterns in the Chukchi Sea during summer 2004. Deep Sea Res II 57:27–39

Hunt GL Jr, Stabeno P, Walters G, Sinclair E, Brodeur RD, Napp JM, Bond NA (2002) Climate change and control of the southeastern Bering Sea pelagic ecosystem. Deep Sea Res II 49:5821–5953

Hunt GL Jr, Blanchard AL, Boveng P, + 12 co-authors (2013) The Barents and Chukchi Seas: comparison of two Arctic shelf ecosystems. J Mar Syst 109–110:43–68

Huntington HP, Moore SE (2008) Assessing the impacts of climate change on Arctic marine mammals. Ecol App 18:S1–S2

Ianelli JN, Honkalehto T, Barbeaux S, Kotwicki S, Aydin K, Williamson N (2012) Assessment of the walleye pollock stock in the Eastern Bering Sea. Stock assessment and fishery evaluation reports for 2013. North Pacific Fishery Management Council, 605 West 4th Ave., Suite 306, Anchorage, 99501

Irons DB, Anker-Nilssen T, Gaston AJ, Byrd GV, Falk K, Gilchrist G, Hario M, Hjernquist M, Krasnov YV, Mosbech A, Olsen B, Petersen A, Reid JB, Robertson GJ, Strøm H, Wohl KD (2008) Fluctuations in circumpolar seabird populations linked to climate oscillations. Glob Chg Biol 14:1455–1463

Irvine JR, Macdonald RW, Brown RJ, Godbout L, Reist JD, Carmack EC (2009) Salmon in the Arctic and how they avoid lethal low temperatures. N Pac Anadromous Fish Comm Bull 5:39–50

Jarvela LE, Thorsteinson LK (1999) The epipelagic fish community of Beaufort Sea coastal waters, Alaska. Arctic 52:80–94

Jay CV, Marcot BG, Douglas DC (2011) Projected status of the Pacific walrus (*Odobenus rosmarus divergens*) in the twenty-first century. Polar Biol 34:1065–1084

Jay CV, Fischbach AS, Kochnev AA (2012) Walrus areas of use in the Chukchi Sea during sparse sea ice cover. Mar Ecol Prog Ser 468:1–13

Jentsch A, Kreyling J, Beierkuhnlein C (2007) A new generation of climate-change experiments: events, not trends. Front Ecol Environ 5:365–374

Kelly BP, Badajos OH, Kunnasranta M, Moran JR, Martinez-Baker M, Wartzok D, Boveng P (2010) Seasonal home ranges and fidelity to breeding sites among ringed seals. Polar Biol. doi:10.1007/s00300-010-0796

Kitaysky AS, Golubova EG (2000) Climate change causes contrasting trends in reproductive performance of planktivorous and planktivorous alcids. J Anim Ecol 69:248–262

Kondzela C, Garvin M, Riley R, Murphy J, Moss J, Fuller SA, Gharrett A (2009) Preliminary genetic analysis of juvenile chum salmon from the Chukchi Sea and Bering Strait. N Pac Anadromous Fish Comm Bull 5:25–27

Kovacs KM, Lydersen C, Overland JE, Moore SE (2011) Impacts of changing sea-ice conditions on Arctic marine mammals. Mar Biodiv 41:181–194

Krebs CJ, Berteaux D (2006) Problems and pitfalls in relating climate variability to population dynamics. Clim Res 32:143–149

Kuletz KJ, Labunski EA, Renner M, Irons DB (2008) The North Pacific pelagic seabird observer program. North Pacific Research Board Final Report, Project No. 637

Laidre KL, Stirling I, Lowry LF, Wiig O, Heide-Jorgensen MP, Ferguson SH (2008) Quantifying the sensitivity of arctic marine mammals to climate-induced habitat change. Ecol App 18(2): S97–S125

Levin SA (1992) The problem of pattern and scale in ecology. Ecology 73:1943–1967

Ljungblad DK, Moore SE (1983) Killer whales (*Orcinus orca*) chasing gray whales (*Eschrichtiusrobustus*) in the northern Bering Sea. Arctic 26:361–364

Logerwell E, Rand K, Weingartner T (2011) Oceanographic characteristics of the habitat of benthic fish and invertebrates in the Beaufort Sea. Polar Biol 34:1783–1796

Lovvorn JR, Richman SE, Grebmeier JM, Cooper LW (2003) Diet and body condition of spectacled eiders wintering in pack ice of the Bering Sea. Polar Biol 26:259–267

Lovvorn JR, Grebmeier JM, Cooper LW, Bump JK, Richman SE (2009) Modeling marine protected areas for threatened eiders in a climatically changing Bering Sea. Ecol App 19:1596–1613

Lovvorn JR, Mossotti RH, Wilson JJ, McKay D (2012) Eiders in offshore pack ice show previously unknown courtship behavior: acceleration of readiness for a constrained breeding period? Polar Biol 35:1087–1095

Lowry LF, Burkanov VN, Frost KJ, Simpkins MA, Davis R, DeMaster DP, Suydam R, Springer A (2000) Habitat use and habitat selection by spotted seals (*Phoca largha*) in the Bering Sea. Can J Zoo 78:1959–1971

Lowry LF, Sheffield G, George JC (2004) Bowhead whale feeding in the Alaskan Beaufort Sea, based on stomach content analyses. J Cet Res Manage 6:215–223

Mackenzie BR, Schiedek D (2007) Daily ocean monitoring since the 1860s shows record warming of northern European seas. Glob Chge Biol 13:1335–1347

Maher WJ (1960) Recent records of the California gray whale (*Eschrichtius glaucus*) along the north coast of Alaska. Arctic 13:257–265

Majewski AR, Lowdon MK, Reist JD, Park BJ (2011) Fish catch data from Herschel Island, Yukon Territory, and other offshore sites in the Canadian Beaufort Sea, July and August 2007, aboard the CCGS *Nahidik*. Can Data Rep Fish Aquat Sci 1231: vi+50 p

Maslanik J, Stroeve J, Fowler C, Emery W (2011) Distribution and trends in Arctic sea ice age through spring 2011. Geophys Res Lett 38:L13502. doi:10.1029/2011GL047735

Mecklenburg CW, Stein DL, Sheiko BA, Chernova NV, Mecklenburg TA, Holladay BA (2007) Russian-American long term census of the Arctic: benthic fishes trawled in the Chukchi Sea and Bering Strait, August 2004. Northwest Nat 88:168–187

Mehlum F, Hunt GL, Decker MB, Norlund N (1998) Hydrographic features, cetaceans and the foraging of thick-billed murres and other marine birds in the northwestern Barents Sea. Arctic 51:243–252

Moline MA, Karnovsky NJ, Brown Z, Divoky GJ, Frazer TK, Jacopy CA, Torres JJ, Fraser WR (2008) High latitude changes in ice dynamics and their impact on polar marine ecosystems. Ann NY Acad Sci 1134:267–319

Moore SE (2005) Long-term environmental change and marine mammals. In: Reynolds JE III, Perrin WF, Reeves RR, Montgomery S, Ragen TJ (eds) Marine mammal research: conservation beyond crisis. John Hopkins University Press, Baltimore

Moore SE (2008) Marine mammals as ecosystem sentinels. J Mamm 89(3):534–540

Moore SE, Huntington HP (2008) Arctic marine mammals and climate change: impacts and resilience. Ecol App 18:S157–S165

Moore SE, Laidre KL (2006) Trends in sea ice cover within habitats used by bowhead whales in the western Arctic. Ecol App 16:932–944

Moore SE, Reeves RR (1993) Distribution and movements. In: Burns JJ, Montague JJ, Cowles CJ (eds) The bowhead whale. Allen Press, Lawrence

Moore SE, George JC, Coyle KO, Weingartner TJ (1995) Bowhead whales along the Chukotka coast in autumn. Arctic 48:155–160

Moore SE, DeMaster DP, Dayton PK (2000) Cetacean habitat selection in the Alaskan Arctic in summer and autumn. Arctic 53:432–447

Moore SE, George JC, Sheffield G, Bacon J, Ashjian CJ (2010) Bowhead whale distribution and feeding near Barrow, Alaska in late summer 2005–06. Arctic 63:195–205

Moore SE, Stafford KM, Melling H, Berchok C, Wiig Ø, Kovacs KM, Lydersen C, Richter-Menge J (2012) Comparing marine mammal acoustic habitats in the Atlantic and Pacific sectors of the High Arctic: year-long records from Fram Strait and the Chukchi Plateau. Polar Biol 35:475–480

Moss JH, Murphy JM, Farley EV, Eisner LB, Andrews AG (2009) Juvenile pink and chum salmon distribution, diet, and growth in the northern Bering and Chukchi Seas. N Pac Anadr Fish Comm Bull 5:191–196

Mueter FJ, Litzow MA (2008) Sea ice retreat alters the biogeography of the Bering Sea continental shelf. Ecol App 18:309–320

Mueter FJ, Broms C, Drinkwater KF, Friedland KD, Hare JA, Hunt GR, Melle W, Taylor M (2009) Ecosystem response to recent oceanographic variability in high-latitude Northern Hemisphere ecosystems. Prog Oceanogr 81:93–110

Murphy JM, Templin WD, Farley EV, Seeb JE (2009) Stock-structured distribution of western Alaska and Yukon juvenile chinook salmon (*Oncorhynchus tshawytscha*) from United States BASIS surveys, 2002–2007. N Pac Anadr Fish Comm Bull 5:51–59

Nasu K (1974) Movement of baleen whales in relation to hydrographic conditions in the northern part of the North Pacific Ocean and Bering Sea. In: Hood DW, Kelley EJ (eds) Oceanography of the Bering Sea, with emphasis on renewable resources. Institute of Marine Science, University of Alaska, pp 345–361

Nelson RJ, Ashjian C, Bluhm B, Conlan K, Gradinger R, Grebmeier J, Hill V, Hopcroft R, Hunt B, Joo H, Kirchman D, Kosobokova K, Lee S, Li WKW, Lovejoy C, Poulin M, Sherr E, Young K (2014) Chapter 10: Biodiversity and biogeography of the lower trophic taxa of the Pacific Arctic region: sensitivities to climate change. In: Grebmeier JM, Maslowski W (eds) The Pacific Arctic region: ecosystem status and trends in a rapidly changing environment. Springer, Dordrecht, pp 269–336

Niebauer HJ, Schell DM (1993) Physical environment of the Bering Sea population. In: Burns JJ, Montague JJ, Cowles CJ (eds) The bowhead whale. Allen Press, Lawrence

Norcross BL, Holladay BA, Busby MS, Mier K (2010) Demersal and larval fish assemblages in the Chukchi Sea. Deep Sea Res II 57:57–70

Norris S (2003) Neutral theory: a new unified model for ecology. Bio Sci 53(2):124–129

O'Corry-Crowe GM, Suydam RS, Rosenberg A, Frost KF, Dizon AE (1997) Phylogeography, population structure and dispersal patterns of the beluga whale *Delphinapterus leucas* in the western Neararctic revealed my mitochondrial DNA. Molec Ecol 6:955–970

Obst BS, Hunt GL (1990) Marine birds feed at gray whale mud plumes in the Bering Sea. Auk 107:678–688

Obst BS, Russell RW, Hunt GL, Eppley ZA, Harrison NM (1995) Foraging radii and energetic of least auklets (*Aethia pusilla*) breeding on three Bering Sea islands. Physiol Zool 68:647–672

Okkonen S, Ashjian CA, Campbell R, Clarke JT, Moore SE, Taylor KD (2011) Satellite observations of circulation features associated with bowhead whale feeding hotspot near Barrow, Alaska. Remote Sens Environ 115:2168–2174

Oppel S, Powel AM, Dickson DL (2008) Timing and distance of king eider migration and winter movements. Condor 110:296–305

Oppel S, Dickson DL, Powell AN (2009) International importance of the eastern Chukchi Sea as a staging area for migrating king eiders. Polar Biol 32:775–783

Overland JE (2009) Meteorology of the Beaufort Sea. J Geophys Res 114:C00A07

Overland JE, Wang J, Pickart RS, Wang M (2014) Chapter 2: Recent and future changes in the meteorology of the Pacific Arctic. In: Grebmeier JM, Maslowski W (eds) The Pacific Arctic region: ecosystem status and trends in a rapidly changing environment. Springer, Dordrecht, pp 17–30

Parker-Stetter SL, Horne JK, Weingartner TJ (2011) Distribution of Arctic cod and age-0 fish in the U.S. Beaufort Sea. Polar Biol 34:1543–1557

Paul JM, Paul AJ, Barber WE (1997) Reproductive biology and distribution of the snow crab from the northeastern Chukchi Sea. Am Fish Soc Sym 19:287–294

Pennisi E (2010) How a little fish keeps overfished ecosystem productive. Science 329:268

Perovich DK (2011) The changing arctic sea ice cover. Oceanography 24:162–173
Phillips LM, Powell AN, Taylor EJ, Rexstad EA (2007) Use of the Beaufort Sea by king eiders breeding on the North Slope of Alaska. J Wild Mgt 71:1892–1898
Piatt JF, Springer AM (2003) Advection, pelagic food webs and the biogeography of seabirds in Beringia. Mar Ornith 31:141–154
Piatt JF, Wells JL, Maccharles A, Fadely B (1991) The distribution of seabirds and their prey in relation to ocean currents in the southeastern Chukchi Sea. Can Wild Serv Occ Paper 68:21–31
Piatt JF, Sydeman WJ, Wiese F (2007) Introduction: a modern role for seabirds as indicators. Mar Ecol Prog Ser 352:199–204
Pickart RS, Pratt LF, Torres DJ, Whitledge TE, Proshutinsky AY, Aagaard K, Agnew TA, Moore GWK, Dail HJ (2010) Evolution and dynamics of the flow through Herald Canyon in the western Chukchi Sea. Deep Sea Res II 57:5–26
Post E, Forchhammer MC, Bret-Harte MS, +22 co-authors (2009) Ecological dynamics across the Arctic associated with recent climate change. Science 325:1355–1358
Proshutinsky A, Krishfield R, Timmermans ML, Toole J, Carmack E, McLaughlin F, Williams WJ, Zimmermann S, Itoh M, Shimada K (2009) Beaufort gyre freshwater reservoir: state and variability from observations. J Geophys Res 114:C00A10
Quakenbush L, Citta J (2008) Biology of the ribbon seal in Alaska. Report to the NMFS, US Department of Commerce. Available at: http://www.fakr.noaa.gov/protectedresources/seals/ice/ribbon/rsbiology2008.pdf
Quakenbush, L, Citta J, Crawford J (2009) Biology of the spotted seal (*Phoca largha*) in Alaska from 1962 to 2008. Report to the NMFS, US Department of Commerce. Available at: http://www.fakr.noaa.gov/protectedresources/seals/ice/spotted/ssbiology072509.pdf
Quakenbush LT, Citta JJ, George JC, Small RJ, Heide-Jorgensen MP (2010) Fall and winter movements of bowhead whales (*Balaena mysticetus*) in the Chukchi Sea and within a potential petroleum development area. Arctic 63(3):289–307
Quakenbush L, Citta J, Crawford J (2011a) Biology of the ringed seal (*Phoca hispida*) in Alaska, 1960–2010. Report to the NMFS, US Department of Commerce. Available at: http://www.fakr.noaa.gov/protectedresources/seals/ice/ringed/biologyofringedseals1960-2010.pdf
Quakenbush L, Citta J, Crawford J (2011b) Biology of the bearded seal (*Erignathus barbatus*) in Alaska, 1961–2009. Report to the NMFS, US Department of Commerce. Available at: http://www.fakr.noaa.gov/protectedresources/seals/ice/bearded/biologyofbeardedseals1961-2009.pdf
Rand K, Logerwell EA (2010) The first survey of the abundance of benthic fish and invertebrates in the offshore marine waters of the Beaufort Sea since the late 1970s. Polar Biol. doi:10.1007/s00300-110-0900-2
Ray GC, Overland JE, Hufford GL (2010) Seascape as an organizing principle for evaluating walrus and seal sea-ice habitat in Beringia. Geophys Res Lett 37:L20504. doi:10.1029/2010GL044452
Richard PR, Heide-Jørgensen MP, Orr J, Dietz R, Smith TG (2001) Summer and autumn movements and habitat use by belugas in the Canadian High Arctic and adjacent waters. Arctic 54:207–222
Richardson WJ, Davis RA, Evans CR, Lujunglad DK, Norton P (1987) Summer distribution of bowhead whales, *Balaena mysticetus*, relative to oil industry activities in the Canadian Beaufort Sea, 1980–84. Arctic 40(2):93–104
Root TL, Price JT, Hall KR, Schneilder SH, Rosenzweig C, Pounds JA (2003) Fingerprints of global warming on wild animals and plants. Nature 42:57–60
Rugh DJ, Fraker MA (1981) Gray whale sightings in the eastern Beaufort Sea. Arctic 34:135
Russell RW, Harrison NM, Hunt GL (1999) Foraging at a front: hydrography, zooplankton, and avian planktivory in the northern Bering Sea. Mar Ecol Prog Ser 182:77–93139
Ryg M, Lydersen C, Markussen NH, Smith TG, Oritsland NA (1990) Estimating the blubber content of phocid seals. Can J Fish Aquat Sci 47:1223–1227
Schofield O, Ducklow HW, Martinson DG, Meredith MP, Moline MA, Fraser WR (2010) How do polar marine ecosystems respond to rapid climate change? Science 328:1520–1523
Scott JBT, Marshall GJ (2010) A step-change in date of sea-ice breakup in western Hudson Bay. Arctic 63:155–164

Shaffer SA, Tremblay Y, Weimerskirch H, Scott D, Thompson DR, Sagar PM, Moller H, Taylor GA, Foley DG, Block BA, Costa DP (2006) Migratory shearwaters integrate oceanic resources across the Pacific Ocean in an endless summer. Proc Natl Acad Sci U S A 103:12799–12802
Sheffield-Guy LM, Roby DD, Gall AI, Irons DB, Rose IC (2009) The influence of diet and ocean conditions on productivity of auklets on St Lawrence Island, Alaska. Mar Ornithol 37:227–236
Shimada K, Kamoshida T, Itoh M, Nishino S, Carmack E, McLaughlin F, Zimmermann S, Proshutinsky A (2006) Pacific Ocean inflow: influence on catastrophic reduction of sea ice cover in the Arctic Ocean. Geophys Res Lett 33:L08605
Sigler MF, Renner M, Danielson SL, Eisner LB, Lauth RR, Kuletz KJ, Logerwell EA, Hunt GL (2011) Fluxes, fins and feathers: relationships among the Bering, Chukchi and Beaufort Seas in a time of climate change. Oceanography 24:250–265
Smith TG (1987) The ringed seal, *Phoca hispida*, of the Canadian western Arctic. Can Bull Fish Aquat Sci 216:1–81
Smith TG (2001) Marine mammals as 'educated' oceanographic platforms. Arctic 54:350–356
Smith TG, Harwood LA (2001) Observations of neonate ringed seals, Phoca hispida, after early break-up of the sea ice in Prince Albert Sound, N.T. Canada, spring 1998. Polar Biol 24:215–219
Smith TG, Stirling I (1978) Variation in the density of ringed seal (*Phoca hispida*) birth lairs in the Amundsen Gulf, Northwest Territories. Can J Zoo 56:1066–1071
Smith TG, Hammill MO, Taugbol G (1991) A review of the developmental, behavioural and physiological adaptations of the ringed seal, *Phoca hisipda*, to life in the arctic winter. Arctic 44:124–131
Springer AM, Roseneau DG, Murphy EC, Springer MI (1984) Environmental controls of marine food webs: food habits of seabirds in the eastern Chukchi Sea. Can J Fish Aquat Sci 41:1202–1215
Springer AM, Roseneau DG, Lloyd DS, McRoy CP, Murphy EC (1986) Seabird responses to fluctuating prey abundance in the eastern Bering Sea. Mar Ecol Prog Ser 32:1–12
Springer AM, McRoy CP, Turco KR (1989) The paradox of pelagic food webs in the northern Bering Sea II. Zooplankton communities. Cont Shelf Res 9:359–386
Springer AM, Estes JA, van Vliet GB, Williams TM, Doak DF, Danner EM, Pfister B (2008) Mammal-eating killer whales, industrial whaling, and the sequential megafaunal collapse in the North Pacific Ocean: a reply to critics of Springer *et al.* 2003. Mar Mamm Sci 24:414–442
Stafford KM, Moore SE, Spillane M, Wiggins S (2007) Gray whale calls recorded near Barrow, Alaska throughout the winter of 2003–04. Arctic 60:167–172
Stephensen SW, Irons DB (2003) A comparison of colonial breeding seabirds in the eastern Bering Sea and Gulf of Alaska. Mar Ornithol 31:167–173
Stephenson SA (2006) A review of the occurrence of Pacific salmon (*Oncorhynchus* spp.) in the Canadian western Arctic. Arctic 59:37–46
Stevenson DE, Lauth RR (2012) Latitudinal trends and temporal shifts in the catch composition of bottom trawls conducted on the eastern Bering Sea Shelf. Deep Sea Res II 65–70:251–259
Stirling I (1997) The importance of polynyas, ice edges and leads to marine mammals and birds. J Mar Syst 10:9–21
Stroeve JC, Maslanik J, Serreze MC, Rigor I, Meier W, Fowler C (2011) Sea ice response to an extreme negative phase of the Arctic Oscillation during winter 2009/2010. Geophys Res Lett 38:L02502. doi:10.1029/2010GL045662
Sundby S, Nakken O (2008) Spatial shifts in spawning habitats of Arcto-Norwegian cod related to multidecadal climate oscillations and climate change. ICES J Mar Sci 65:953–962
Suydam RS, George JC (1992) Recent sightings of harbor porpoise, *Phocoena phocoena*, near Point Barrow, Alaska. Can Field Nat 106:489–492
Suydam RS, Dickson DL, Fadely JB, Quakenbush LT (2000) Population declines of king and common eiders of the Beafurot Sea. Condor 102:219–222
Suydam RS, Lowry LF, Frost KJ, O'Corry-Crowe GM, Pikok D (2001) Satellite tracking of eastern Chukchi Sea beluga whales into the Arctic Ocean. Arctic 54:237–243
Tewksbury JJ, Sheldon KS, Ettinger AK (2011) Moving farther and faster. Nat Clim Change 1:396–397

Thedinga JF, Johnson SW, Neff AD, Hoffman CA, Maselko JM (2013) Nearshore fish assemblages of the northeastern Chukchi Sea, Alaska. Arctic 66 (in press)

Thomson DH, Fissel DB, Marko JR, Davis RA, Borstad GA (1986) Distribution of bowhead whales in relation to hydrometerological events in the Beaufort Sea. Environ Studies Rev Funds Rep No. 028

Tracy DM, Schamel D, Dale J (2002) Red Phalarope (*Phalaropus fulicarius*). In: Poole A, Gill F (eds) The birds of North America, No. 698. Academy of Natural Sciences/American Ornithologists' Union, Philadelphia/Washington, DC

Udevitz MS, Taylor RL, Garlich-Miller JL, Quakenbush LT, Snyder JA (2013) Potential population-level effects of increased haulout-related mortality of Pacific walrus calves. Polar Biol 36:291–298

Votrogov LM, Ivashin MV (1980) Sighting of fin and humpback whales in the Bering and Chukchi seas. Rep Int Whal Commn 30:247–251

Wang M, Overland JE (2009) A sea ice free summer Arctic within 30 years? Geophys Res Lett 36:L07502. doi:10.1029/2009GL037820

Wassmann P (2011) Arctic marine ecosystems in an era of rapid climate change. Prog Oceanogr 90:1–17

Weatherhead E, Gearheard S, Barry RG (2010) Changes in weather persistence: insight from Inuit knowledge. Global Environ Change 20:523–528

Weingartner T, Aagaard K, Woodgate R, Danielson S, Sasaki Y, Cavalieri D (2005) Circulation on the north central Chukchi Sea shelf. Deep Sea Res II 52:3150–3174

Welch HE, Bergman MA, Siferd TD, Martin KA, Curtis MF, Crawford RE, Conover RJ, Hop H (1992) Energy flow through the marine ecosystem of the Lancaster sound region, Arctic Canada. Arctic 45:343–357

Wood CM, McDonald DG (1997) Global warming: implications for freshwater and marine fish. Cambridge University Press, Cambridge

Woodgate RA, Aagaard K, Weingartner TJ (2005) A year in the physical oceanography of the Chukchi Sea: moored measurements from autumn 1990–1001. Deep Sea Res II 52:3116–3149

Woodgate RA, Weingartner TJ, Lindsay R (2012) Observed increases in Bering strait oceanic fluxes from the Pacific to the Arctic from 2001 to 2011 and their impacts on the Arctic Ocean water column. Geophys Res Lett 39. doi:10.1029/2012GL05092

Wu Y, Peterson IK, Tang CDL, Platt T, Sathyendranath S, Fuentes-Yaco C (2007) The impact of sea ice on the initiation of the spring bloom on the Newfoundland and Labrador Shelves. J Plankton Res 29:509–514

Wyllie-Echeverria T (1995) Sea ice conditions and the distribution of walleye Pollock (*Theragra chalcogramma*) on the Bering and Chukchi Shelf. In: Beamish RV (ed) Climate change and northern fish populations. Canadian Special Pub Fish Aquatic Science, Nanaimo

Chapter 12
Progress and Challenges in Biogeochemical Modeling of the Pacific Arctic Region

Clara J. Deal*, Nadja Steiner*, Jim Christian, Jaclyn Clement Kinney, Ken L. Denman, Scott M. Elliott, Georgina Gibson, Meibing Jin, Diane Lavoie, Sang H. Lee, Warren Lee, Wieslaw Maslowski, Jia Wang, and Eiji Watanabe

Abstract At this early stage of modeling marine ecosystems and biogeochemical cycles in the Pacific Arctic Region (PAR), numerous challenges lie ahead. Observational data used for model development and validation remain sparse, especially across seasons and under a variety of environmental conditions. Field data are becoming more available, but at the same time PAR is rapidly changing. Biogeochemical models can provide the means to capture some of these changes. This study introduces and synthesizes ecosystem modeling in PAR by discussing differences in complexity and application of one-dimensional, regional, and global earth system models. Topics

*Both authors contributed equally.

C.J. Deal (✉) • G. Gibson • M. Jin
International Arctic Research Center, University of Alaska Fairbanks, PO Box 757340, Fairbanks, AK 99775-7340, USA
e-mail: deal@iarc.uaf.edu

N. Steiner • J. Christian
Institute of Ocean Sciences, Fisheries and Oceans Canada, Sidney, BC, Canada

Canadian Centre for Climate Modeling and Analysis, Environment Canada, Victoria, BC, Canada

J. Clement Kinney • W. Maslowski
Department of Oceanography, Graduate School of Engineering and Applied Sciences, Naval Postgraduate School, Dyer Road, Bldg. SP339B, Monterey, CA 93943, USA
e-mail: jlclemen@nps.edu; maslowsk@nps.edu

K.L. Denman
School of Earth and Ocean Sciences, University of Victoria, Victoria, BC, Canada

S.M. Elliott
Climate Ocean Sea Ice Modeling, Los Alamos National Laboratory, Los Alamos, NM, USA

D. Lavoie
Maurice Lamontagne Institute, Fisheries and Oceans Canada, Mont Joli, QC, Canada

S.H. Lee
Department of Oceanography, Pusan National University, Busan, South Korea

J.M. Grebmeier and W. Maslowski (eds.), *The Pacific Arctic Region: Ecosystem Status and Trends in a Rapidly Changing Environment*, DOI 10.1007/978-94-017-8863-2_12,

include the general structure of ecosystem models and specifics of the combined benthic, pelagic, and ice PAR ecosystems, the importance of model validation, model responses to climate influences (e.g. diminishing sea ice, ocean acidification), and the impacts of circulation and stratification changes on PAR ecosystems and biogeochemical cycling. Examples of modeling studies that help place the region within the context of the Pan-Arctic System are also discussed. We synthesize past and ongoing PAR biogeochemical modeling efforts and briefly touch on decision makers' use of ecosystem models and on necessary future developments.

Keywords Biogeochemical models • Ecosystem modeling • Arctic • Sea ice biogeochemistry • Carbon cycling modeling

12.1 Introduction

Rapid climate change occurring in the Pacific Arctic Region (PAR) and questions about the impact on marine ecosystems and biogeochemical cycles have led to the development of models to interpret observations and predict future changes and responses. Large uncertainties exist regarding the ways in which the biology of the Arctic, including PAR, is responding to climate change. Biogeochemical models can help elucidate these changes, synthesize observations, and guide resource-limited field campaigns. Models are particularly useful in PAR due to limited access, incomplete satellite coverage, and poor spatial resolution of field data. To comprehend the impact of climate change on the ecosystem and on biogeochemical cycling, modeling based upon measures of key processes and a range of in situ validation data is required.

The PAR ecosystem is a dynamic system. Biogeochemical models attempt to capture the dynamics and cycling of the biochemical (i.e., living) and geochemical (i.e., non-living) parts of an ecosystem. These dynamic models are formulated as a combination of a set of differential equations, some algebraic equations, and a parameter list (Box 12.1). To qualify for the term "ecosystem model", a model must

W. Lee
Canadian Centre for Climate Modeling and Analysis, Environment Canada, Victoria, BC, Canada

J. Wang
Great Lakes Environmental Research Laboratory (GLERL), National Oceanic and Atmospheric Administration, Ann Arbor, MI, USA
e-mail: jia.wang@noaa.gov

E. Watanabe
International Arctic Research Center, University of Alaska Fairbanks, Fairbanks, AK, USA

Research and Development Center for Global Change, Japan Agency for Marine-Earth Science and Technology, Yokohama, Japan

Box 12.1: Ecosystem Model Basics

The simplest marine ecosystem models usually represent a planktonic food web with one or more nutrient (N), phytoplankton (P), zooplankton (Z) and often detritus (D) compartments (i.e., state variables). More complex models may contain more components (e.g., bacteria, zooplankton life stages, DOM), additional biomes (e.g. sea ice, benthos), and gases, or they may extend to fish and higher trophic levels (HTLs) in order to more readily address human dimensions (e.g., fisheries, lifestyle, and habitat changes). Models that include socioeconomic interactions and responses are beginning to be introduced but are as yet unavailable for PAR and thus lie beyond the scope of this chapter.

Figure 12.1 shows a schematic of compartments and processes represented in ecosystem models with multiple layers of complexity. The basic NPZD model

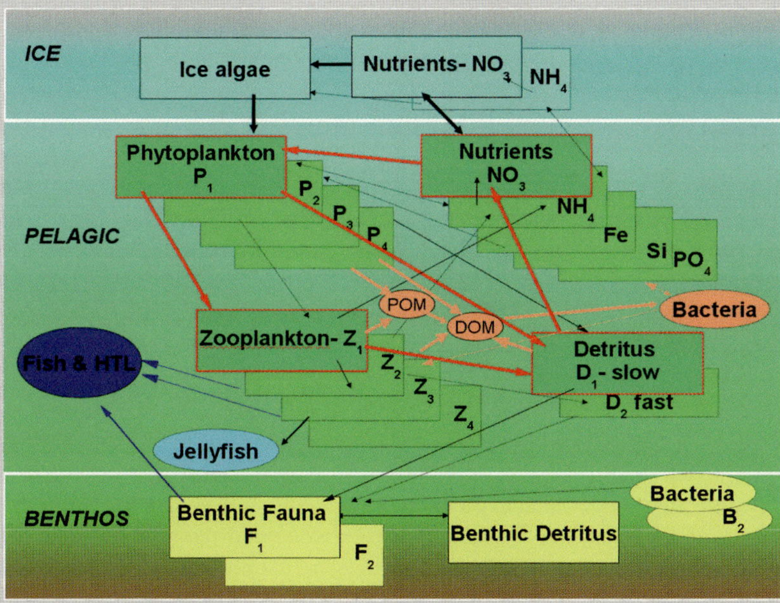

Fig. 12.1 Simplified schematic of ecosystem model components with multiple layers of complexity. The simplest NPZD (Nutrient, Phytoplankton, Zooplankton, Detritus) model as used in some ESMs; e.g., The Canadian Earth System CanESM1.5 (discussed in Sect. 12.10) is indicated by red frames and arrows which show processes linking compartments. Increased complexity is displayed by multiple nutrients (NO_3, NH_4, Fe, Si, PO_4), phytoplankton (P_1, P_2, P_3, P_4) and zooplankton (Z_1, Z_2, Z_3, Z_4) functional types, and detritus (D_1, D_2), as well as additional components for ice algae and benthic communities, fish and higher trophic levels (HTLs), dissolved and particulate organic matter (DOM, POM), and bacteria. Links within levels of higher complexity are only drawn indicatively and do not represent the full extent necessary to model a real ecosystem

(continued)

Box 12.1: (continued)

is indicated by red frames and arrows that show processes linking compartments. Increased complexity is indicated by multiple boxes for N, P, Z, and D, as well as multiple arrows representing interactions and dependencies. (In text, subscripts are usually used to denote the number of compartments; e.g., $N_3P_2ZD_3$ denotes three N, two P, one Z, and three D compartments).

Numerically, temporal changes in model compartments are expressed by conservation equations fixed in space. There are several basic transfer functions (or processes) to consider: phytoplankton response to light, phytoplankton uptake of nutrients, zooplankton grazing, and loss terms due to death, excretion, predation, and sinking. Several functional forms for each process have been used, ranging from simple linear responses to nonlinear forms. A thorough description of modeling microbial foodwebs would extend over many pages and go beyond the scope of this chapter. Instead, we refer the reader to literature reviews that discuss different aspects and numerics of biogeochemical modeling, model components, and equations, as well as parameterizations for ecosystem processes and biogeochemical interactions (e.g. Tett 1987; Davidson 1996; Gentleman 2002; Vichi et al. 2007; Hood and Christian 2008).

Model differential equations require the definition of an initial state, which is based on available observational data. The basic model unit used is typically mmol N m^{-3} since nitrogen is limiting to PP in the oceans, and in the Arctic specifically (Tremblay and Gagnon 2009).

include explicit phytoplankton and heterotrophic organisms; the simplest such model is a nutrient-phytoplankton-zooplankton (NPZ) model. Biogeochemical models have been used for several decades in an attempt to explain, analyze, and predict what we can only sparsely measure. With their ability to extend over spatial and temporal scales far beyond what we can observe, models are useful tools in scientific research.

In this chapter, we present a synthesis of biogeochemical model development within PAR ranging from one-dimensional (1-D) and high-resolution regional, to global earth system modeling. The choice of an appropriate model is determined by the questions being asked and the data available. Validation is a large part of model development and evaluation, and is discussed in more detail in Box 12.2. With this biogeochemical model synthesis, we address the ability of dynamic modeling tools to help understand the impacts of climate change and to explore physical-chemical-biological interactions within PAR. We will discuss several examples of PAR modeling studies and will point to knowledge gaps and future research needs.

Box 12.2: Validation of Biogeochemical Models

In view of the growing human effect on the ocean and the increased interest in understanding ocean biogeochemistry (e.g., Doney 2010), the purpose of coupled biological/physical models goes beyond information synthesis and numerical experimentation. Marine system models are increasingly used to enable cost analyses, to predict outcomes of management choices, and eventually to support high-stakes decision making (e.g., Stow et al. 2009). Naturally, a model's sensitivity and skill must be tested and measured to enable us to understand model capabilities and to be confident in ecosystem projections. With the increasing complexity of physical and biogeochemical models, this is no easy task.

Model intercomparisons show that model skill is not necessarily associated with model complexity or model type, nor is it possible to identify one single model that is most skilled according to all criteria (Friedrichs et al. 2009). Different statistical quantities (i.e., skill metrics) may capture different aspects of model performance, and a thorough assessment of model skill may require use of multiple types of skill metrics simultaneously (Stow et al. 2009). The equations within a model represent hypotheses about how the system works; they might not be correct. Similarly, observations provide an approximation of the truth, limited by measurement uncertainties as well us under-sampling in space and time. Stow et al. (2009) state that a model starts to exhibit skill when the observational and predictive uncertainty halos overlap; the ideal case is a complete overlap.

More specifically for ecosystem applications, Fennel (2009) writes that the skill of physical-biogeochemical ocean models, which include truncated food webs, depends on how effectively unresolved processes are parameterized in terms of resolved processes. An important aspect is the mixture of time scales at different trophic levels of the food web. Since truncated models are still the only available choice for many biogeochemical model studies, we require a thorough understanding of potential error sources.

Stow et al. (2009) highlight several graphical techniques that are useful for assessing model skill, including simple time series plots of observations and model predictions, bivariate plots, and misfit representations as well as quantitative metrics like correlation coefficients, root mean squared error, bias, model efficiency, etc. Diagrams summarizing multiple aspects of model performance in a single figure have begun to appear, especially in the coupled model literature, as a convenient way to quantify and communicate model performance. Most common is likely the Taylor diagram, which provides summary information about how the linear correlation coefficient and the

(continued)

Box 12.2: (continued)

variance comparisons each contribute to the unbiased root mean square difference (RMSD) (Taylor 2001). Jolliff et al. (2009) point out that in certain cases the Taylor diagram provides an incomplete picture, because it often yields no information about other aspects of model performance such as the bias (the comparison of mean values) or the total RMSD (a metric for overall model and data agreement). Hence they introduce the target diagram, a Cartesian coordinate plot providing summary information about how the magnitude and sign of the bias and the pattern agreement (unbiased RMSD) each contribute to the total RMSD magnitude. While the purpose of both the Taylor and target diagrams is to compactly summarize statistical quantities to aid in model skill assessment, Jolliff et al. (2009) suggest that target diagrams may be better suited for ocean ecosystem modeling because they better summarize the overall agreement between model and data.

Since many ecosystem models are highly complex, the outputs from such models are highly multivariate. Allen and Somerfield (2009) explore multivariate approaches for skill assessment (e.g., principal component analysis or vector correlations). These allow the simultaneous examination of the ways in which numerous variables vary in relation to each other, both spatially and temporally. Doney et al. (2009) point out that the multivariate analyses of model dynamics may be particularly useful when the model data skill assessments are applied to fully-coupled climate or earth system models (e.g., Doney et al. 2006; Schneider et al. 2008; Orr et al. 2005). In global models, it is especially challenging to replicate observations from local time-series because there are many subgridscale processes and representation issues that tend to confound the comparison. Persistent physical biases in coupled models propagate into the ecological/biogeochemical mean state and seasonal cycle. Because the coupled ocean–atmosphere models generate their own internal climate variability, assessment of simulated interannual-to-decadal variability can only be done statistically, not directly (Doney et al. 2009). It has been stated repeatedly that the success of the biological and chemical simulations are highly dependent on the hydrodynamic fields and hence require a high-quality underlying physical circulation model (e.g., Matear and Holloway 1995; Doney et al. 2009; Smith et al. 2009). It is, therefore, essential to include physical model biases and errors in ecological and biogeochemical model data assessment.

Cross-validation and Monte Carlo simulation have been suggested as other options to test model skill and uncertainty or, more specifically, interpolative and extrapolative skill of marine ecosystem models (e.g., Hemmings et al. 2004; Friedrichs et al. 2007; Smith et al. 2009; Wallhead et al. 2009). These methods

(continued)

Box 12.2: (continued)

allow predictive objectives to be specified, including how much interpolation and extrapolation is required of the models. Skill assessment is then focused on estimating the skill function using skill metrics. Monte Carlo type approaches have also been used to assess ecosystem model sensitivity. Such approaches usually involve defining a domain of possible model input parameters, randomly selecting parameter inputs given the specified domain, performing deterministic computations using these inputs (running the model), and finally aggregating and analyzing the results to assess model variance and determine a statistical relationship between the model input parameters and output (e.g., Megry and Hinckley 2001; Gibson and Spitz 2011). Gibson and Spitz (2011) suggest that sensitivity analysis should be considered a prerequisite for any new marine ecosystem model. In addition to providing a measure of uncertainty to ecosystem projections, it provides helpful information during model tuning as well as insight into expected behavior of the ecosystem model under alternative physical conditions or in different regions. Sensitivity analysis can also indicate where further field and laboratory studies should be focused. This is especially true in regions where data access is limited and hence skill assessment via model data intercomparison is biased due to seasonally- or regionally-focused measurements.

In all skill assessment approaches, a major requirement is data accuracy and availability. It is essential that observations represent the true state of the system with acceptable accuracy and precision, and provide a reasonable representation of model time and space scales. Unfortunately, suitable coherent datasets are still very limited, especially in polar regions, which are within limited satellite range and frequently inaccessible for a large part of the year. We hope that the current expansion of ocean and coastal observing systems, along with improvements in satellite data product quality and advanced data management, will help to address this issue. For the PAR region, the BEST/BSIRP and the DBO (Grebmeier et al. 2010) are good examples of such improvements.

12.2 PAR Characteristics Particularly Relevant for Biogeochemical Modeling

Seasonal ice cover, wide shallow shelves, extreme environmental conditions, and extensive advection create unique biophysical relationships in the spatially and temporally dynamic PAR system. Sea ice extent can vary by hundreds of kilometers from year to year; differences of up to about 25 % have been observed based on

Scanning Multichannel Microwave Radiometer and Special Sensor Microwave/ Imager (SSM/I) data in March and September, respectively (Frey et al. 2014, this volume). As the ice melts in spring, increased light and salinity-induced stabilization of the water column initiate the phytoplankton bloom (Sakshaug 2004) that follows the retreat of the marginal ice zone. These waters may be too cold for zooplankton to be abundant, resulting in low levels of grazing, and much of the phytoplankton biomass is left to sink to the benthos undigested (Bluhm and Gradinger 2008). Phytoplankton biomass from an open water and later-occurring bloom is more often recycled by grazers because the water column is warmer then, thus, the food source for the benthos is reduced. Recent reductions in sea ice have resulted in a shift of the subarctic pelagic-dominated ecosystem typical of the southeastern Bering Sea northward into the northern Bering Sea (Grebmeier et al. 2006, 2010). Are existing ecosystem models capable of simulating these temporal and spatial shifts that determine benthic or pelagic dominance? Capturing the timing of ice-associated and open water blooms in a model is critical for simulating the resulting food web structure.

More than half of PAR is composed of very shallow shelf seas. On these arctic shelves, sea ice, benthic, and pelagic systems are intimately linked. More than 60 % of Chukchi Sea primary production (PP) is available for local export to the benthos or offshore transport to the adjacent basin (Campbell et al. 2009). Nutrients regenerated from the benthos can fuel pelagic production when distributed upward through re-suspension processes and mixing events. A holistic modeling study of arctic warming impacts on biological productivity and biogeochemical cycling in PAR thus calls for a benthic component, including grazing by the organisms that feed on the benthos.

Implementation of a sea ice ecosystem component is also important for PAR, especially in the Chukchi Sea, where sea ice is the prime source of production and algal biomass during the spring (Gradinger 2009). Jump-starting of the phytoplankton bloom by sea ice algae is another potential link between sea ice and pelagic systems, just as entrainment of phytoplankton into the sea ice in autumn may "seed" the spring ice algal bloom (McRoy, personal communication 2006; Werner et al. 2007). Model studies can help to elucidate these significant roles and assist us to zero in on the short window of opportunity that is available for observing these key connections.

Much of the PAR system is governed by advection (Carmack and Wassmann 2006). Circulation, currents, and flow structures (e.g., eddies, fronts) can transport and/or concentrate biomass and nutrients and cause *hot spots* for grazers at all trophic levels. For example, elevated Euphausiid abundance in the fall at Barrow, Alaska has been explained by transport from the northern Bering Sea and by southerly winds (Berline et al. 2008). Movements and other changes in flows due to changes in climate may exert a disproportionate influence on ecosystems, species distributions, and community structure. Hence, accurate high-resolution modeling of physical properties is highly important in modeling the biogeochemistry of PAR and will be addressed in Sect. 12.9.

12.3 A Brief History of PAR Biogeochemical Models

One of the first numerical modeling studies of the planktonic ecosystem in PAR focused on the Bering Strait and was performed as part of the Inner Shelf Transfer and Recycling (ISHTAR) program. The three-dimensional (3-D) ecohydrodynamic model developed at the University of Liege was adapted to the northern Bering Sea (Nihoul et al. 1993) to assess system variability in summer. Reconstructions of the first available data fields from ISHTAR by inverse modeling showed the strong influence of major circulation features on the biogeochemical cycles and ecosystems (Brasseur and Haus 1990). Nihoul et al. (1993) found seasonal patterns of primary and secondary production to be dominated by the main water mass transports, upwelling, and fronts (e.g., the strong upwelling of nutrients along the east Siberian coast was shown to sustain the biology of the region). Maxima in their simulated PP patterns appeared first along the Anadyr Stream front, and later farther north in the Chukchi Sea. The measured PP pattern is similar, although the absolute production appears to be lower in recent years compared to previous decades (Lee et al. 2007; Lee unpublished data).

Using 3-D circulation, plankton, and benthos models, Walsh et al. (2004) explored the effects of interannual changes in ice cover and water motion on carbon and nitrogen cycling within the Chukchi and Beaufort seas. Fall case studies were simulated with the analysis focusing on carbon dioxide (CO_2) cycling. Walsh et al. (2005) used the same ecological model to explore the relative roles of light and nutrients in controlling year-round production within the Chukchi and Beaufort seas. They developed seasonal carbon budgets by making use of extensive field data obtained during 2002, a year signified by the lowest ice cover in 20 years. (Since then the ice cover has decreased even further, with a minimum in 2012). Their analysis of biogeochemical processes in light-limited and nutrient-limited regimes (i.e., end of summer in open seas, and under extensive ice cover in the nutrient-poor upper waters of the Canadian Basin) provided insights into the possible consequences of future climate changes at these high latitudes. They concluded that increased light availability from future ice retreat would not benefit the shelf food web or increase carbon sequestration in the Chukchi and Beaufort seas unless additional nutrients became available.

Sea ice ecosystem models have been applied in PAR only recently. They have been used to investigate environmental controls and variability of sea ice PP and ice algal biomass. For a few western Arctic locations where sea ice algal dynamics and habitats have been rather well characterized, ice algal production has been simulated in 1-D (Lavoie et al. 2005); most often coupled to biological production models in the water column (Jin et al. 2006a, 2007, 2009; Lavoie et al. 2009, 2010; Lee et al. 2010; Pogson et al. 2011). The Jin et al. sea ice biological model has been coupled to a global dynamic sea ice model to assess large-scale variability of ice PP and algal biomass (Deal et al. 2011). All of these sea ice ecosystem models are based on the first ice algae model developed by Arrigo et al. (1993) for sea ice in Antarctica. The Arrigo et al. model contains a detailed treatment of internal ice light and nutrient fluxes. However, ice in the Arctic is different (Dieckmann and Hellmer 2003). For example, there is more multi-year ice (MYI) and less extensive flooding in arctic ice; therefore, nutrients are more limited for algal communities close to the ice surface, while the bottom 2–3 cm are the most favorable sub-habitat for ice algal blooms (Gradinger et al. 2005 and references therein; Lee et al. 2010). The primary source of nutrients for ice algae is the mixed layer, and algae accumulation at the base of the ice will thus be greatly affected by factors that govern this nutrient exchange.

During the most recent International Polar Year (IPY March 2007 to March 2009), a large effort was expended to improve simulations of PAR ecosystem dynamics by adding ice and benthic biology, achieving higher resolution to better resolve features important to the biology, looking at longer timescales to understand long-term dynamics, examining coupling within earth system models (ESMs), and working on future projections.

12.4 Modeling PAR in 1-D: Introduction and Locations

Before proceeding to fully 3-D ecosystem models, it is often useful to first develop and test a coupled ice-ocean ecosystem model in a 1-D context. Generally, 1-D models are used for detailed diagnostic and prognostic modeling and process studies. In PAR they have been applied at specific locations to investigate the impacts of shrinking sea ice cover, increasing freshwater flux, and shifting climate regime on primary and export production. The vertical 1-D model structure simplifies analysis of model results and allows us to examine the relative importance of physical processes (i.e., water column stratification, mixing, snow melt, snow and sea ice thickness, freshwater runoff). A fast-running 1-D model can be a testing tool for developing parameterizations of complex biogeochemical or physical processes that affect ecosystem dynamics. Time series observations at specific sites allow for extensive testing and validation of 1-D models. Observations might include laboratory studies or mesocosm experiments, which are by design 1-D. Well-designed 1-D model experiments can provide a first take on addressing science questions about ecosystem responses to changes in temperature, stratification, seasonal ice zones (SIZs), and acidification associated with current environmental transitions.

Figure 12.2 includes PAR locations that have worked well for 1-D ice-ocean ecosystem modeling investigations: the outer shelf region of the Canadian Beaufort Sea (Lavoie et al. 2009, 2010), the National Oceanic and Atmospheric Administration (NOAA)/Pacific Marine Environmental Laboratory M2 mooring site in the Bering Sea SIZ (Jin et al. 2007, 2009; Gibson and Spitz 2011), the multi-year pack ice in the Canadian Basin (Lee et al. 2010), and the landfast ice zone in the Chukchi shelf (Jin et al. 2006a). Ideally, 1-D modeling sites are chosen because (1) they are representative of a particular region, (2) intensive biophysical time series measurements are available, and (3) it can be reasonably assumed that horizontal advection and diffusion effects are negligible.

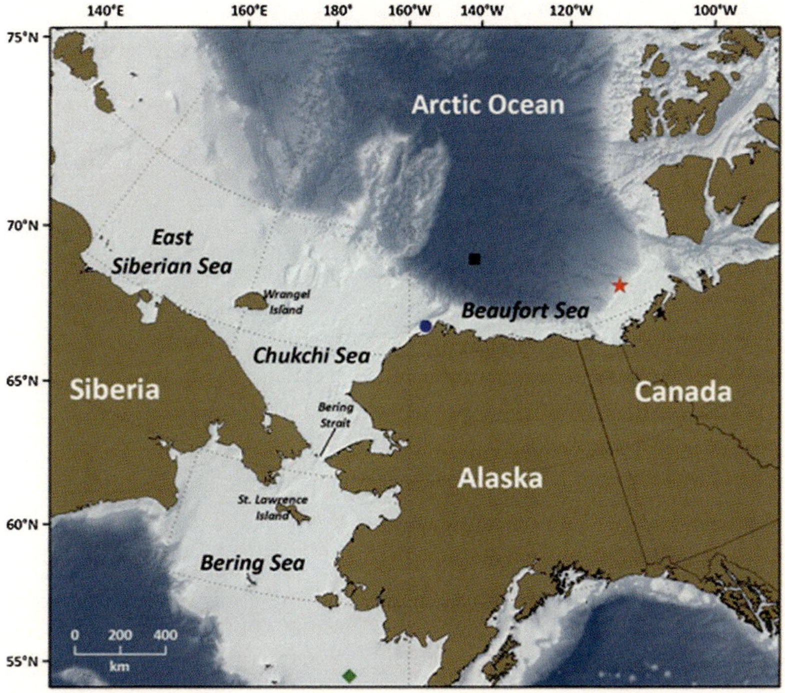

Fig. 12.2 Map of PAR, defined as the marine region extending from north of St. Matthews Island to the Beaufort Sea and the Arctic Ocean. Symbols denote locations where 1-D ice-ocean ecosystem models have been applied (*green diamond*: M2 mooring site in SIZ of the Bering Sea; *blue dot*: landfast ice zone of the Chukchi shelf; *black square*: multi-year pack ice in the Canadian Basin; *red star*: outer shelf region of the Canadian Beaufort Sea)

12.5 Results from 1-D Modeling Studies

Lavoie et al. (2005) used a detailed treatment of the boundary layer between the upper ocean and the sea ice bottom to understand the temporal patterns observed in ice algae. In the model ice algae are assumed to be limited by silicic acid. When melting occurs at the ice bottom, ice algal cells are expelled from the ice into the mixed layer at a rate proportional to the ice melt rate. Bottom ice growth and melt, which are affected by heat and salt fluxes between the ice and the ocean, were shown to be important factors for ice algae biomass accumulation at the base of the ice.

Lavoie et al. (2009) coupled the ice algae model (Lavoie et al. 2005) to a pelagic N_3PZD_3 model (see Box 12.1) to simulate the seasonal cycle and to study the physical processes controlling primary and export production on the Beaufort Shelf. In a continuing study, output from the Canadian Centre for Climate Modeling and Analysis (CCCma) coupled global climate model (CGCM2) was used to force this 1-D model to project future PP for the years 2041–2060 and 2081–2100 (Lavoie et al. 2010) (Fig. 12.3a). Their goal was to investigate the

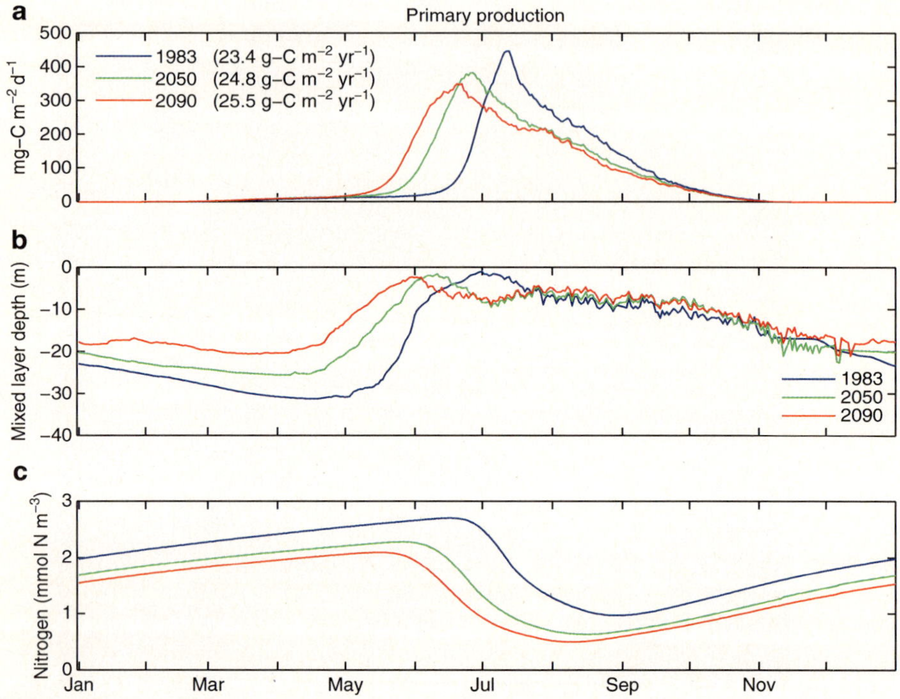

Fig. 12.3 Simulated average annual cycle for each 18-year period for the outer shelf of the Canadian Beaufort Sea: (**a**) depth-integrated primary production (phytoplankton and ice algae), (**b**) mixed layer depth, and (**c**) nitrogen concentration in the top 40 m (Figures from Lavoie et al. 2010)

impacts on PP of increased freshwater runoff and longer ice-free season. It was found that the subsurface bloom is the most important current contributor to total annual PP, although the contribution of the surface bloom can be relatively important in some years, when conditions in the previous fall lead to higher nutrient replenishment in the mixed layer by wind and winter convective mixing (Melling and Moore 1995). The simulations suggest that by mid century, the relative importance of the phytoplankton surface spring bloom compared with the subsurface bloom will have decreased further. In the model this is shown to be a result of increased stratification (due to increased runoff) that reduces the depth of winter convective mixing (Fig. 12.3b, c). Overall, the projected PP increase is modest (<10 %) by 2090 (Fig. 12.3a) and results essentially from a lengthening of the subsurface bloom. Lavoie et al. (2010) suggest that the increasing importance of the subsurface versus surface blooms could lead to a greater underestimation of chlorophyll a (chl a) concentration and PP by satellite sensors, which cannot detect subsurface chl a. Their simulations also project a decrease in the future relative importance of the ice algae contribution to total PP due to the decrease in the length of the growing season linked to earlier snow and ice melt.

The effect of stratification on the projected Beaufort Sea shelf PP could be representative of other arctic shelf seas affected by high freshwater runoff such as the Laptev and Siberian seas. However, each shelf is unique, and extrapolation of results from one shelf to another requires caution (e.g., Carmack et al. 2006). In the Bering Sea a decrease in PP is projected due to a stratification-dependent decrease in nutrient supply (see Sect. 12.9.2). However, this might not hold true for shelf seas where production is already limited more by nutrients than by light. A comparison of surface PP in the Beaufort Sea between the years 1998–2002 and 2007, a year of very low ice extent, shows little variation compared to other arctic regions (Fig. 12.2e in Arrigo et al. 2008). However, entrainment of nutrients into the surface layer through ice edge upwelling (Mundy et al. 2005), eddies (Tremblay et al. 2008), or increased wind mixing could play a more important role in the future.

Results from Jin et al. (2009) multi-species model indicate that the ecosystem in the Bering Sea SIZ responds to climate regime shifts by a switch in the dominant phytoplankton group. Their physical model was forced by sea surface wind, heat and salt fluxes, oceanic tides, and sea ice data from Hadley Center (monthly) before 1978 and SSM/I (daily) after 1979. Their sea ice algae model has four compartments, including ice algae and three nutrients, nitrate + nitrite, ammonium, and silicate. The ocean ecosystem model is $N_3P_3Z_3D$. Remotely-sensed sea ice concentration data were used to approximate the sea ice thickness in the model, as in Jin et al. (2007). Simulated dominant phytoplankton switched from ice-associated cold water species to warm water species, and zooplankton abundance increased coinciding with changes in the timing and vertical distribution of lower trophic level production after the Pacific Decadal Oscillation index reversal in 1977. Before 1977, most of the simulated annual PP was from early blooms in icy cold water. Because grazing was light, much of the assimilated carbon was transferred to the benthos. After 1977 open water phytoplankton blooms prevailed in warmer water, supporting high pelagic secondary production and zooplankton biomass.

Lee et al. (2010) applied the Jin et al. (2007) model to three sites in PAR where ice core observations were conducted (Lee et al. 2010). The sites included one clean sediment-free site (Chukchi) and one sediment-rich site (Beaufort), both in landfast sea ice offshore Barrow, plus one oceanic Ice Exercise site. Comparisons among model sites revealed that strong light attenuation by trapped sediments controlled ice PP. Thicker ice at the oceanic site also attenuated light, delaying the peak in ice algal biomass, and maximum production reached only 10 % of that at the Chukchi site because of nutrient limitation.

A 1-D version of the Bering Ecosystem Study (BEST)-NPZ ecosystem model (Sect. 12.8) was used for a sensitivity analysis to gain insights into the relative importance of biological vs. physical model parameters (Gibson and Spitz 2011). Sensitivity studies are critically important because reliable model projections require model validation, an understanding of model sensitivities, and a measure of model uncertainty. Box 12.2 introduces many of the currently applied methods. Working with the BEST-NPZ model, Gibson and Spitz (2011) used a Monte Carlo-type approach to address uncertainty in model forecasts. Their 1-D model was run iteratively with randomly selected biological inputs and environmental conditions (including temperature, salinity, ice cover, and light) that were based on triangular

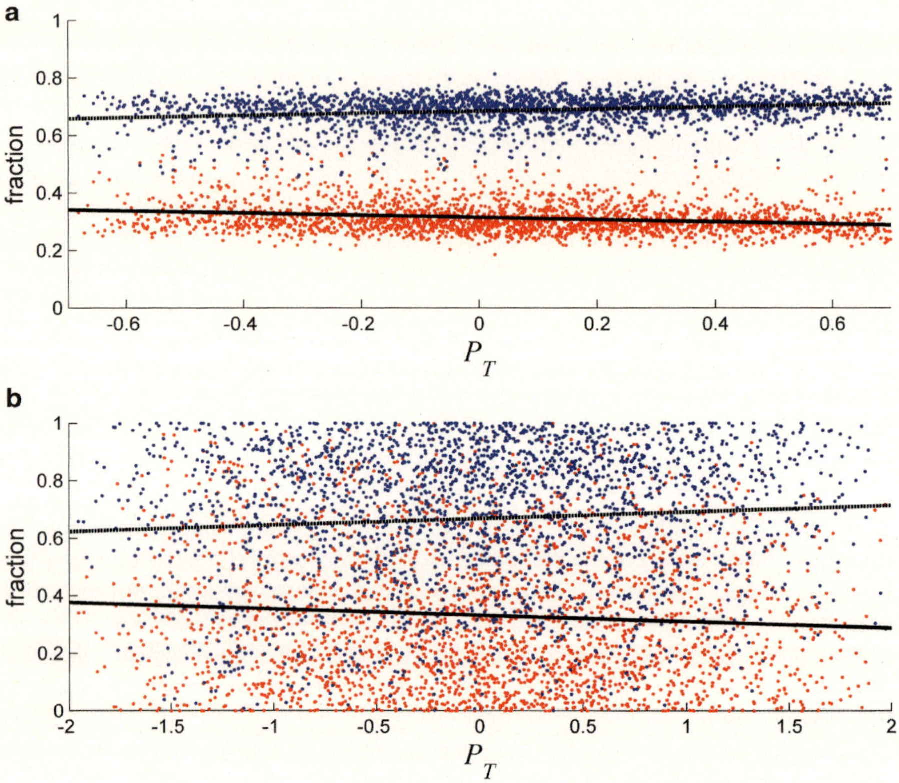

Fig. 12.4 Fraction of secondary production generated by small zooplankton (microzooplankton and small copepods; red) versus large zooplankton (large copepods and euphausiids; *blue*) for (**a**) Ex. I, in which biological parameters were varied by ±10 % of the baseline values and physical parameters were varied within −0.7–1 SD and (**b**) Ex. IV, in which biological parameters were varied within best guesses for their actual range, and physical parameters were varied within ±2.0 SD. P_T is temperature (Figure from Gibson and Spitz 2011)

probability distributions. Baseline values for each biological parameter were estimated based on literature reviews and on input from BEST-Bering Sea Integrated Ecosystem Research Program (BSIERP) field scientists. In their first experiment (Ex. I), the biological parameters varied within ±10 % of the baseline values. In their fourth experiment (Ex. IV), the parameter range expanded to within their "best guess" of minimum and maximum values or, if the range was unknown, to ±60 % of the baseline value. Physical parameters varied within −0.7 and 1.0 SD in Ex. I and within ±2.0 SD in Ex. IV. Results suggest that the fraction of total secondary production attributable to small zooplankton (small copepods and microzooplankton) vs. large zooplankton (large copepods and euphausiids) will increase with increasing water temperature, indicating a shift to a longer pelagic food web (Fig. 12.4a). The relationship holds even with a broader range of parameter variability (Fig. 12.4b).

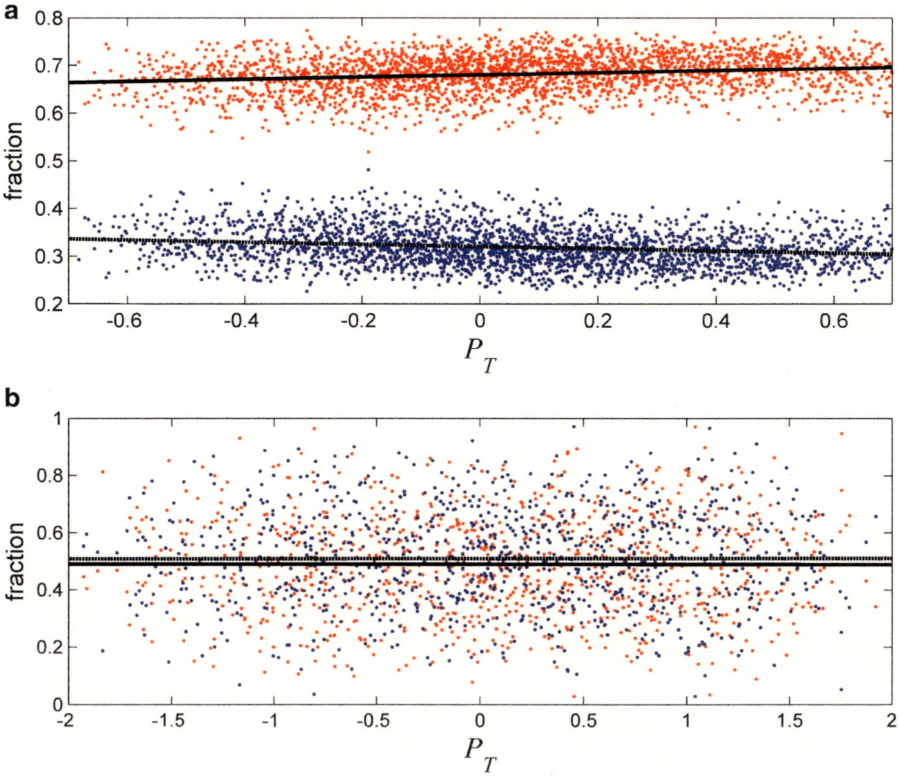

Fig. 12.5 Fraction of secondary production by mesozooplankton (small copepods and large copepods and euphasiids; *red*) versus benthic infauna (*blue*) for (**a**) Ex. I and (**b**) Ex. IV. For Ex. descriptions see Fig. 12.4 caption. P_T is temperature (Figure from Gibson and Spitz 2011)

Gibson and Spitz (2011) also ranked parameters using a least squares linearization, a multiple regression between the parameters' deviation from the mean and the model output. When they varied each parameter by only a small amount, they found that zooplankton grazing rate and efficiency, light, temperature, and ice thickness and timing were as important as benthic biomass and benthic production. But when they varied parameters within their "best guess", benthic grazing rate, benthic half saturation uptake rate, benthic assimilation efficiency, and benthic respiration rate were the most important parameters. The range implemented for the benthic parameters was relatively broad due to limited information in the literature. As a result, while benthos appears sensitive to temperature increases (Fig. 12.5a) when the parameter range is small, the relationship is not clear when the biological range is broader (Fig. 12.5b). The biological variability swamps the physical signal, highlighting the importance of narrowing down some of the benthic parameters.

12.6 Introduction to High-Resolution Regional and Pan-Arctic Models

High-resolution models provide us with tools that we can use to address science questions about biogeochemical responses to transitions in terrestrial, gateway, and shelf-to-basin fluxes, as well as to changes in ice cover. They also allow us to make first estimates of the consequences of these responses for higher trophic levels (HTLs) and socioeconomic activities. We include here models with grid cells of ¼ degree (~25 km) resolution or less. A 25 km resolution model barely resolves the Bering Strait, which is 50 km wide at its narrowest and provides heat, nutrients, Pacific fauna, and ~40 % of the total freshwater input to the Arctic Ocean (Woodgate and Aagaard 2005). Arctic models that have a boundary at the Bering Strait usually prescribe boundary conditions from climatology or, if available, coarser-resolution global models. Coarse-resolution global circulation models (GCMs) usually artificially widen the strait to allow two-way flow. The main advantage of high resolution is the ability to resolve advective processes and spatial details such as smaller-scale circulation, currents, orographic features, flow structures, and ocean eddies that contribute to mixing and transport.

Below, recent high-resolution modeling studies in the PAR are presented, examining how climate-induced changes in physical forcing (e.g., temperature) and sea ice loss may alter productivity. These modeling studies usually involve evaluating mean fields, integrals, seasonal cycles, and interannual variability. In addition, sensitivity studies provide insights into ecosystem behavior in a changing climate.

12.7 Primary Productivity: Modeling the Present and Recent Past

The coupled 3-D pan-Arctic biology/sea ice/ocean model by Zhang et al. (2010) was developed to investigate the impact of declining arctic sea ice on the marine planktonic ecosystem from 1988 to 2007. Their model's horizontal resolution averages 22 km and there are 30 vertical levels of different thicknesses as small as 5 m in the top 30 m. The biological model is a modified version of the Kishi et al. (2007) 11-component lower-trophic-level model adapted to the Arctic Ocean. The physical model results are in good agreement with observed downward trends in summer sea ice extent and thickness (Rothrock et al. 1999, 2008; Stroeve et al. 2008). In the model, these changes in sea ice led to an increase in both surface and subsurface PP, mainly in the seasonally and permanently ice-covered Arctic Ocean. The general upward trend in PP was attributed to increasing photosynthetically active radiation and nutrient availability (mainly due to strengthened upwelling), and surface water temperature in the Arctic Ocean peripheral seas. Compared to in situ observations, the model was basically able to capture the timing of the spring bloom in the seasonally ice covered waters of the Chukchi/Beaufort seas. As mentioned earlier, Lavoie et al. (2010) suggest that only the subsurface bloom in the Beaufort Sea will increase due to the reduction of nutrient replenishment to the mixed layer during winter. Models often

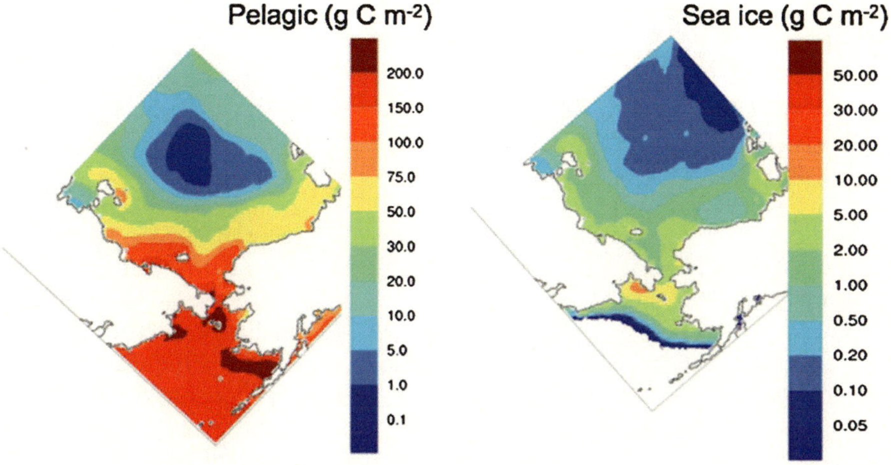

Fig. 12.6 Modeled annual primary production in the ocean upper 100 m (*top left*) and sea ice (*top right*), averaged over 1995–2006, from the study of Jin et al. (2012)

disagree and the differences (e.g., processes included, spatial domain, time period) need to be kept in mind when interpreting, contrasting, and utilizing model results.

Jin et al. (2012) coupled an ice ecosystem model to the global POP-CICE (Parallel Ocean Program-Los Alamos National Laboratory Sea Ice Model) with an open-ocean pelagic ecosystem model (Moore et al. 2004). The global domain avoids open boundary settings in the Bering Strait. Horizontal resolution ranges from 20 to 85 km, averaging 40 km north of 70°N. There are 40 vertical layers ranging from 10 m in the surface to 250 m below 2,000 m. The sea ice algal ecosystem component was coded into CICE (Hunke and Lipscomb 2008; Hunke and Bitz 2009) and first tested in CICE-standalone with a mixed-layer ocean model (Deal et al. 2011). Jin et al. (2012) examined the simulated PP in arctic sea ice and ocean waters for the model period 1992–2007 and found that annual PP increased from 305 Tg C year^{-1} in 1998 to 550 Tg C year^{-1} in 2007. Annual mean ice PP was highest during low ice years from 2005 to 2007. In the Zhang et al. (2010) modeling study, total PP increased by 50 %, from 456 Tg C year^{-1} in 1988 to 682 Tg C year^{-1} in 2007. These PP values are comparable to satellite derived estimates of 375 Tg C year^{-1} in 1998 and 485 Tg C year^{-1} in 2006 (Pabi et al. 2008).

Simulated annual PP in the ocean (upper 100 m) and sea ice from 1995 to 2006 in PAR by Jin et al. (2012) are shown in Fig. 12.6. The ocean production (top left) reveals high productivity in the Bering Sea (>150 g C m^{-2} year^{-1}). The Chukchi Sea production downstream of the Bering Strait ranges from 150 g C m^{-2} year^{-1} at the Bering Strait down to about 50 g C m^{-2} year^{-1} along the shelf break, similar to the range of 55–145 g C m^{-2} year^{-1} observed during 2002–2004 (Lee et al. 2007; Rho and Whitledge 2007). Arrigo et al. (2008) suggest an annual production rate (mean 1998–2002) of about 40 g C m^{-2} year^{-1} for the Chukchi Sea. The production in the perennial-ice-covered central Arctic is much lower; less than 1 g C m^{-2} year^{-1} in some regions. The modeled annual ice algal production (Fig. 12.6, top right) is around 2 g C m^{-2} $year^{-1}$ within the observed ranges of 0.7–5 g C m^{-2} $year^{-1}$ offshore Barrow in the Chukchi Sea (Jin et al. 2006a). To investigate the decadal change of ocean PP from 1990s to 2007, the average annual PP in the ocean upper 100 m for low ice years (2002, 2003, 2005–2007) was subtracted from model results for high ice years (1998–2001, 2004). The differences indicate a northward shift in ocean PP with a decrease in the Bering Sea and an increase in the more northern shelf and PAR SIZ (Jin et al. 2012).

The recent Arctic Ocean biogeochemical model intercomparison by Popova et al. (2012) includes the Jin et al. (2012), Zhang et al. (2010), and three other coupled physical-biological models: Yool et al. (2010), Dupont (in review); and Popova et al. (2010). The study compares and contrasts the substantial variability in biogeochemical model structure and complexity, and the differences between the physical models (e.g., resolution, initial and boundary conditions, sea ice submodel). Results show that the models broadly reproduce the present-day large-scale distribution pattern of total annual PP across the Arctic Ocean in agreement with satellite-derived (Arrigo et al. 2008) and in situ data, including the high values of observed PP in the Chukchi Sea (Carmack et al. 2006). However, the models disagree when it comes to the physical factors controlling PP. The study focuses on winter mixing as the main mechanism controlling basin-scale patterns of nutrient supply. An important conclusion is that the models' vertical mixing, in particular winter mixing, needs to be improved before they can predict future annual Arctic Ocean PP under continued sea ice retreat.

In addition to these pan-Arctic biogeochemical models, Slagstad et al. (2011) have recently extended their Barents Sea ecosystem model (Wassmann et al. 2006) to the Arctic domain. For many years, their ecosystem model consisting of 13 state variables stood out as the most validated and comprehensive physical-biological model applied within the Arctic. As with the models above, high PP in the Chukchi Sea is reproduced, but they also test the effect of retreating sea ice in the forthcoming century on primary and secondary production. One experimental outcome projects the expansion of the typical Arctic zooplankton species *C. glacialis* into the Chukchi and East Siberian Seas.

12.8 Regional Modeling Focusing on the Bering and Chukchi Seas

In the western Arctic, the most comprehensive marine ecosystem numerical modeling effort to date is part of the National Science Foundation (NSF)-supported BEST and the North Pacific Research Board-supported BSIERP. The BEST-BSIERP partnership is working to develop and implement a suite of integrated climate, physical oceanography, lower and upper trophic level ecosystem, and economic models. The ultimate objective is to produce a series of multi-year forecasts for the eastern Bering Sea ecosystem using alternate climate projections. The 15-component BEST-NPZ ecosystem model is a modified version of the Global

Ocean Ecosystems Dynamics model (Hinckley et al. 2009). Its domain includes the entire Bering Sea. The model has been designed for coupling with a HTL fish model (Forage-Euphausiid Abundance in Space and Time) under development through the BSIERP research program and has both a sea ice ecosystem (modified from Jin et al. 2006a) and a benthic component. Inclusion of the benthic submodel with explicit benthic processes, infauna, and detritus makes this model unique among PAR ecosystem models, thus enabling studies of the partitioning of organic carbon between pelagic and benthic systems.

Using the Coupled Ice Ocean Model (CIOM; Wang et al. 2009 and references therein; Hu and Wang 2010) and the Physical Ecosystem Model (PhEcoM; Wang et al. 2003; Jin et al. 2006b), a 3-D coupled CIOM-PhEcoM was developed and implemented in the Bering and Chukchi seas. Figure 12.7 shows CIOM-PhEcoM simulated chl a, and remotely-derived chl a (SeaWifs) on April 29, 2008 and June 28, 2008. The model simulations are in general consistent with observed features, indicating that their ice-ocean-ecosystem model captures some of the important dynamics of this complex coupled system. For example, relatively high chl a concentrations (proxy for phytoplankton biomass) are apparent along the Bering Slope Current (BSC) in the daily snapshots of model results and observations. And, in the deep western Aleutian Basin, eddies or gyres (Hu and Wang 2010) entrain phytoplankton, in particular along the Kamchatka Current (Fig. 12.7a, b). Due to cloud cover, only a trace of high values is seen in the remotely-derived chl a observations along the BSC and Kamchatka Current. Chl a (Fig. 12.7c, d) and nitrate (not shown), indicate that after the sea ice has melted in late June, phytoplankton blooms in the Gulf of Anadyr and western Bering Strait are sustained by the nutrient-rich water of the Anadyr Current transiting northward into the Chukchi Sea.

12.9 Advancements in the Simulation of Physical Features

While ecosystem models are implemented in existing, usually well-established physical models, modelers are aiming for even higher resolution and more accurate representation of small-scale (<10 km horizontal) processes. So far, these kinds of improvements have mostly been achieved without a coupled ecosystem module. Even though adding ecosystem variables can help to pinpoint problems with physical parameterizations, the increase in computational resources often makes it more feasible to develop the physical model first. Eventually, the appropriateness of the physical model and its temporal and spatial scale will depend on the goal of the biogeochemical modeling effort.

A limited number of 3-D physical modeling studies have focused on the physical oceanography of the northern Bering and Chukchi seas (e.g., Overland and Roach 1987; Spaulding et al. 1987; Nihoul et al. 1993; Hermann et al. 2002; Hu and Wang 2008). These studies have proven useful in simulating the major circulation features of the region. However, they have been limited by low spatial resolution and/or small domains with lateral boundary conditions prescribed in close proximity to the

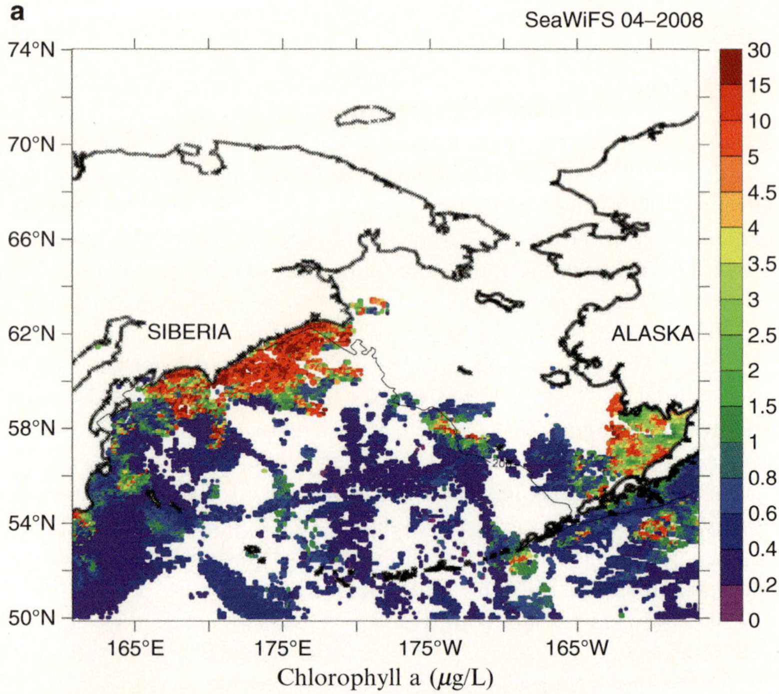

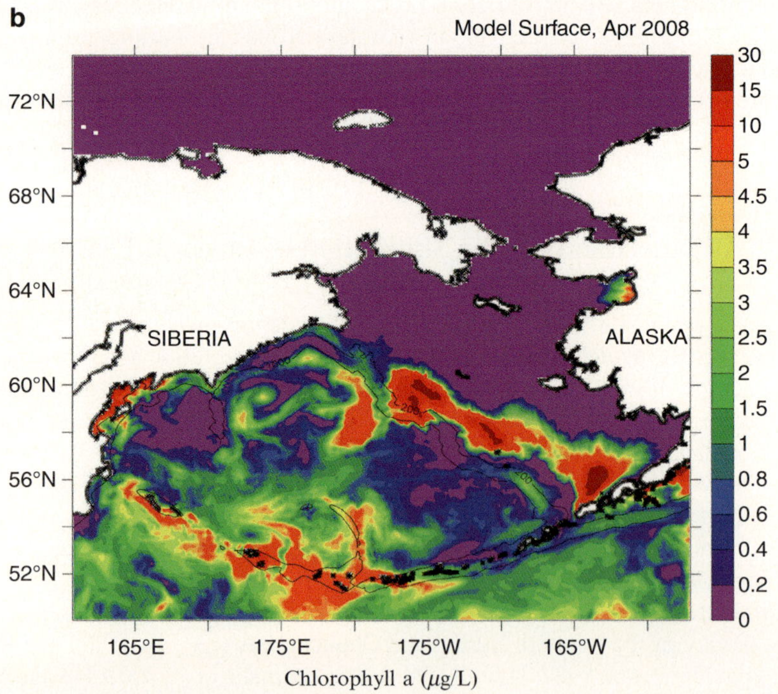

Fig. 12.7 (**a**) CIOM-PhEcoM simulated chl a, and (**b**) remotely-derived chl a (from the SeaWifs) for April 29, 2008; and (**c**) CIOM-PhEcoM-simulated chl a, and (**d**) remotely-derived chl a for June 28, 2008. Units for chl a (*color bar*) are μg l^{-1}

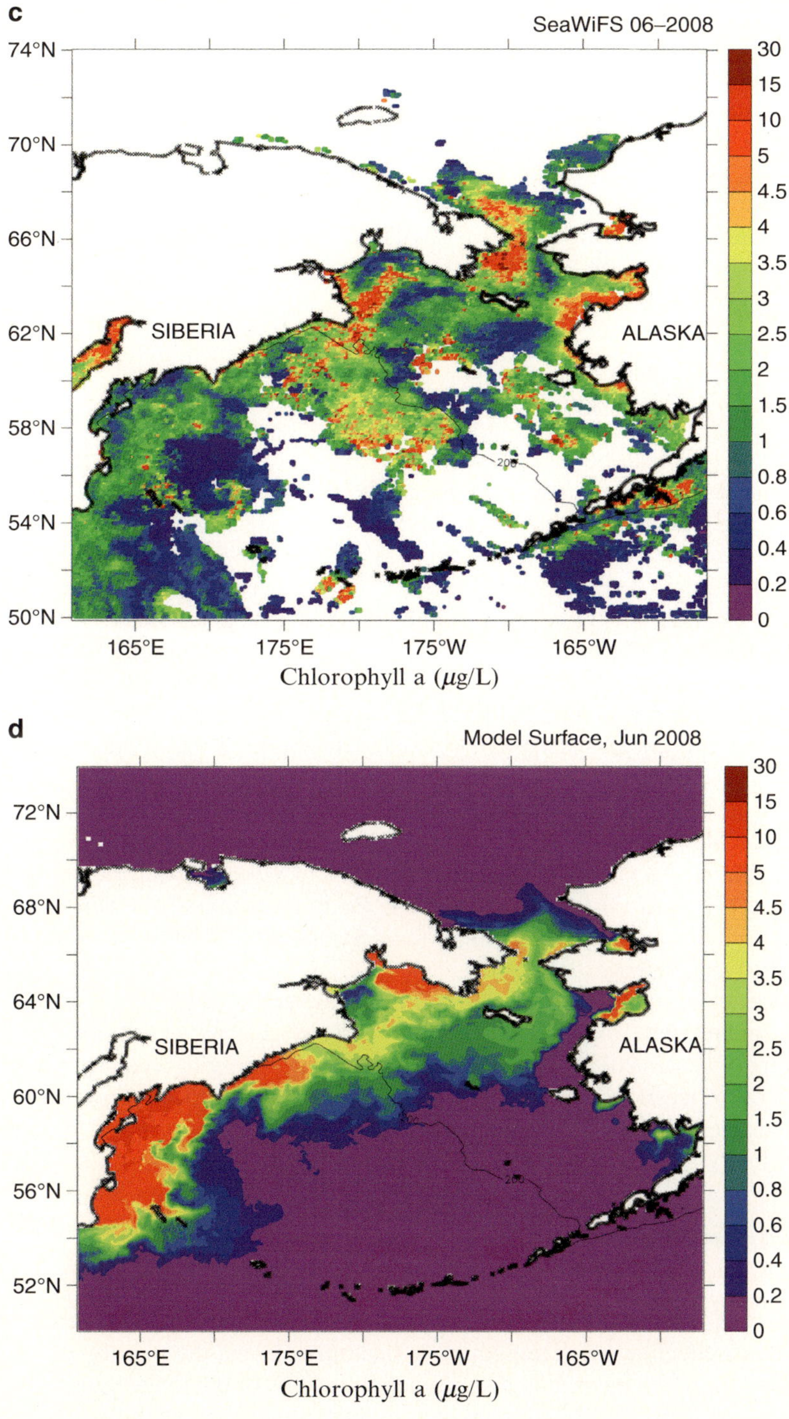

Fig. 12.7 (continued)

region of interest. In addition, these studies were integrated for only a short period of time (1 month to 1 year). Recent advancements in computing capability and updated bathymetry information have allowed the development of relatively high-resolution models, with pan-Arctic domains. In Chap. 7 of this book Clement Kinney et al. (2014) discuss volume transport and property fluxes through the Bering Strait, comparing several high-resolution models. The following subsections highlight examples from only one of these models (the Naval Postgraduate School Arctic Modeling Effort, NAME) and from an additional model for the Chukchi-Beaufort seas illustrating high-resolution capabilities related to biogeochemical modeling.

12.9.1 Gateway Fluxes

Flow through restricted passages such as the Bering Strait and in narrow currents such as the Anadyr and Alaska Coastal currents can significantly impact ecosystem dynamics, but are challenging to adequately resolve. Model resolutions on the order of kilometers are required. NAME uses a ~9 km ocean model that permits flow through the narrow straits and passages of PAR and realistically represents the coastal currents (Clement Kinney et al. 2009, 2014, this volume). NAME also resolves flow reversals that affect ice conditions in the central and eastern Bering Sea, which in turn affect nutrient and biomass transport through the Bering Strait.

The NAME 26-year (1979–2004) mean circulation in the northern Bering and Chukchi seas is shown in Fig. 12.8. Representing the narrow and shallow flow through the Shpanberg, Anadyr, and Bering straits is critical to the overall flow field and to the mass and property exchanges across this region. The strongest velocities occur within the Anadyr Current, along the coast of Siberia, and in the Bering Strait. A strong northeastward flow also exists in Barrow Canyon. This mean circulation agrees with the available observations (e.g., Stabeno et al. 1999), and it is presented primarily for reference.

Evidence of major impacts on the PAR ecosystem through the redistribution of ocean boundaries and habitats due to climate change are already observed (Grebmeier et al. 2010). A model not capable of resolving changes in gateway fluxes (also ocean fronts, currents, and general transport pathways) could easily miss changes in species transport and habitat shifts.

12.9.2 Vertical Structure

The vertical structure of the water column—i.e., the depth of any permanent or seasonal pycnoclines, the strength of vertical mixing, and the stability of the water column—plays a large role in determining the access of phytoplankton to light and nutrients, and thus PP. Although not yet coupled to an ecosystem model, NAME has

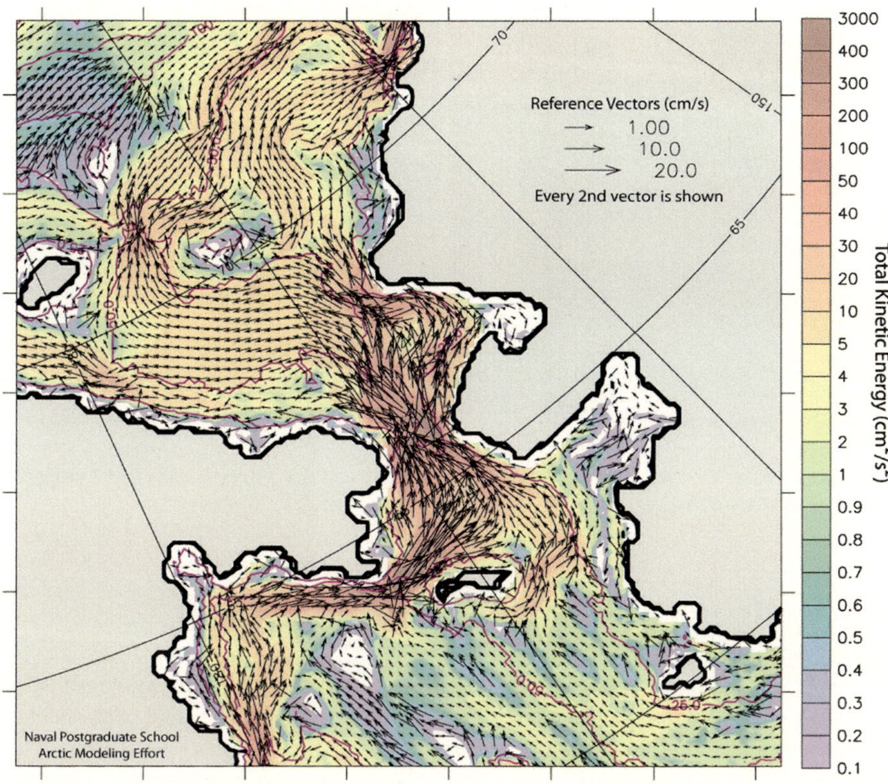

Fig. 12.8 Twenty-six-year mean (1979–2004) circulation in the upper 100 m from NAME. Every 2nd vector in each direction is shown. Shading represents total kinetic energy. *Magenta lines* denote bathymetry contours

demonstrated the ability to capture the influence on stratification of surface warming, sea-ice melt, and river discharge interacting with ocean circulation over large areas. The 2.5 m vertical resolution is currently state-of-the-art for a pan-Arctic model (Clement Kinney et al. 2014, this volume).

Eddy kinetic energy (EKE) is an important model parameter; it commonly represents regions of active mixing, including in the vertical. Vertical mixing is believed to lead to increased biological productivity in areas that become nutrient-limited in the euphotic zone. Simulated seasonally-averaged EKE in the surface layer for a typical summer and autumn is shown in Fig. 12.9. North of St. Lawrence Island, in the Chirikov Basin, and north of Bering Strait, there is a distinct decrease in EKE. Clement et al. (2005) propose that the high primary and secondary (benthic) production often found in this region (Grebmeier et al. 1988; Springer and McRoy 1993; Grebmeier and Dunton 2000) might be at least partially attributed to the high EKE and associated vertical mixing providing nutrient supply just upstream.

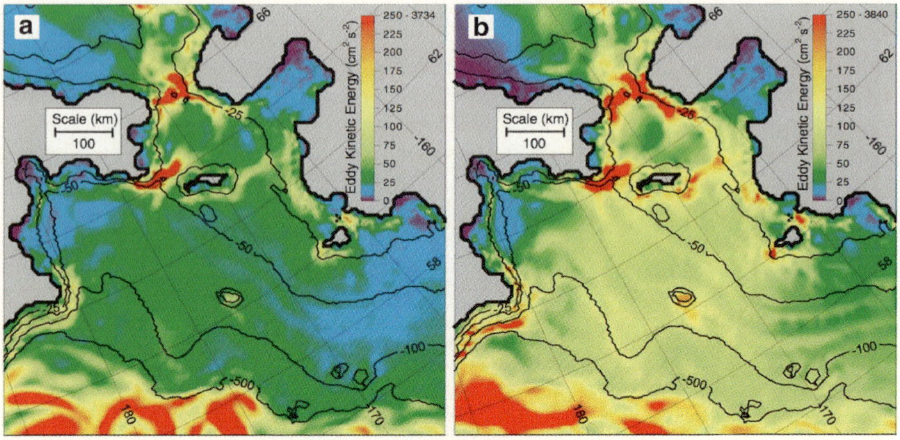

Fig. 12.9 Seasonally-averaged EKE at the surface calculated from NAME daily 1987 snapshots (**a**) Summer (J–A–S) average, and (**b**) Autumn (O–N–D) average

12.9.3 Mesoscale Eddies and Shelf-Basin Exchange

Eddies generated along the Chukchi-Beaufort shelf slope contribute to the transport of Pacific water into the basin interior and to the shelf-basin exchange of particulate organic carbon (POC), including zooplankton (e.g., Llinas et al. 2009). Zooplankton effectively link the lower and upper trophic levels, supporting large populations of mammals, seabirds, and species targeted by local fisheries (Nelson et al. 2014, this volume). Thus it is important to be able to model the impacts of the changes in these current patterns and flow structures on the ecosystem.

A number of anti-cyclonic eddies have been captured in the surface and halocline layers above 300 m depth in the Canadian Basin by in situ and drifting buoy observations (Manley and Hunkins 1985; Mathis et al. 2007). The Moderate-Resolution Imaging Spectroradiometer (MODIS) sea surface temperature has also detected a few warm-core eddies along the Beaufort shelf break (Fig. 12.10). Detailed physical and chemical properties of a cold-core eddy observed on the Chukchi Sea continental slope have been surveyed in Mathis et al. (2007) and Kadko et al. (2008). Both studies mention that the bottom-intensified shelf-edge current forms such mesoscale eddies, which play a significant role in the transport of carbon, oxygen, and nutrients associated with the Pacific winter water into the upper halocline of the Canadian Basin.

Numerical experiments using the eddy-resolving coupled sea ice-ocean Center for Climate System Research (CCSR) Ocean Component Model (COCO; Hasumi 2006), with a horizontal resolution of about 2.5 km, reveal that the interannual variations in mesoscale eddy activities and the shelf-basin exchange of Pacific water in the western Arctic Ocean depend on sea ice cover and shelf-wide surface wind conditions (Watanabe and Hasumi 2009; Watanabe 2011) (Figs. 12.10 and 12.11). Mesoscale eddies were also identified by the NAME model (Maslowski et al. 2008)

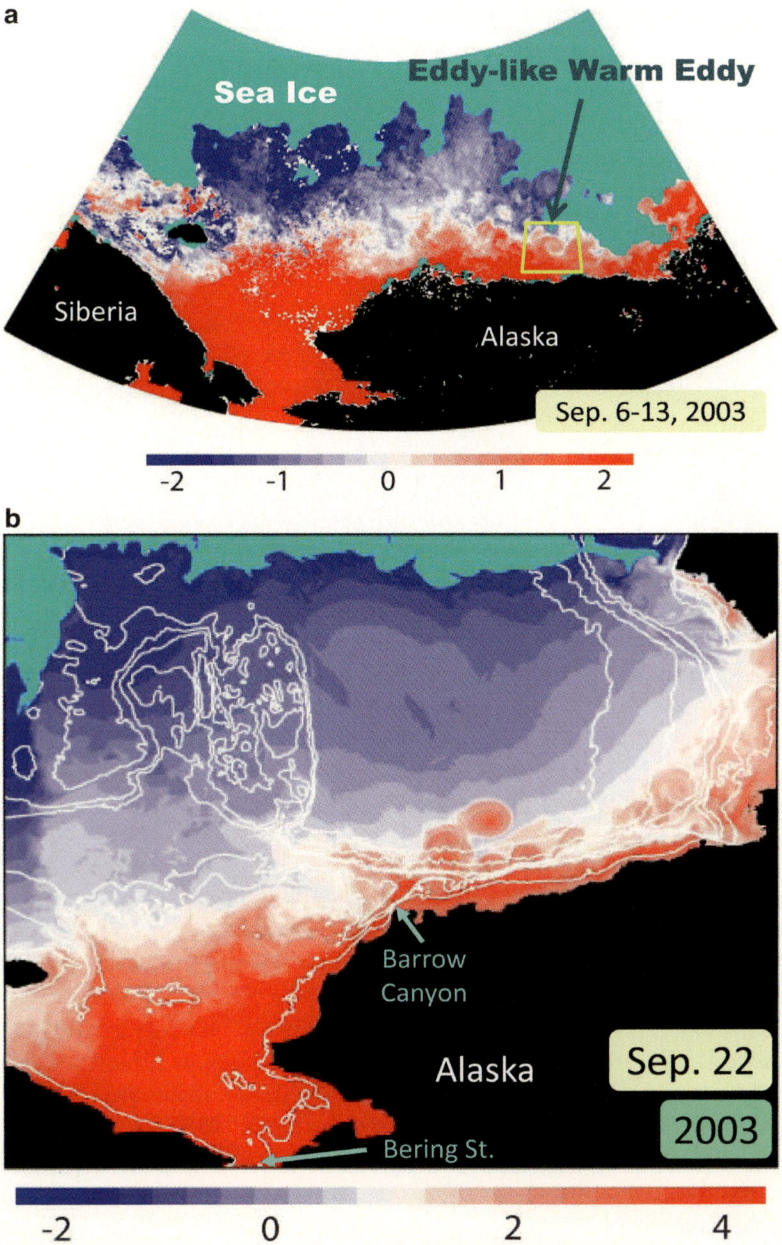

Fig. 12.10 (*Top*) MODIS 8-day-composite sea surface temperature during September 6th–13th, 2003 [deg C; color bar]. AMSR-E sea ice area is overlaid by *sky-blue shade*. (*bottom*) Sea surface temperature [deg C; color bar] and sea ice area in September simulated by the eddy-resolving version of the coupled sea ice-ocean model CCSR-COCO. *White contours* denote bottom topography (Figures from Watanabe 2011)

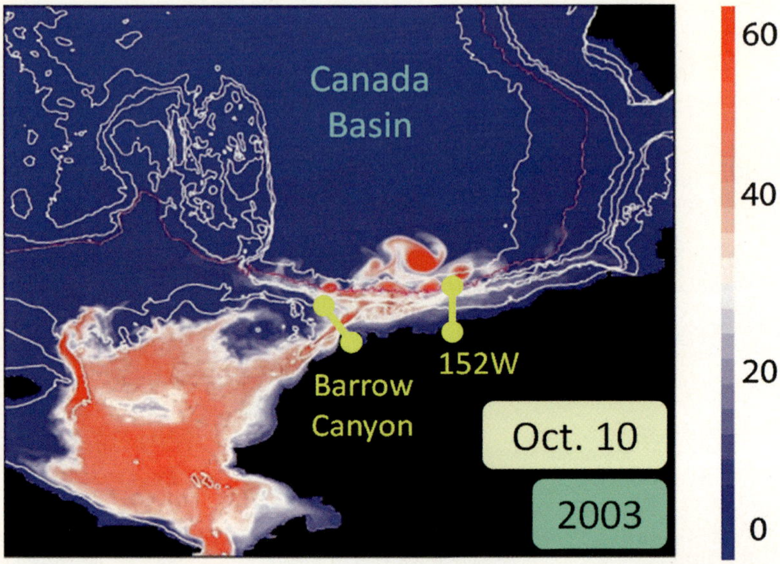

Fig. 12.11 Pacific water content in October 2003 (m), simulated by the eddy-resolving COCO with atmospheric forcing. The Pacific water content is calculated by the vertical integration of the Pacific water tracer concentration from the ocean surface to sea bottom in each grid (Figure from Watanabe 2011)

in the Beaufort Sea and Canadian Basin throughout the years. Mesoscale activity has the potential to significantly influence PP as illustrated in the northern Gulf of Alaska (Okkonen et al. 2003), where mesoscale eddies produce tongue-like features with maximum chl a concentrations. PP is enhanced in this region via the transport of phytoplankton and nutrients (including iron) from the outer shelf into the deep basin. In the Barents Sea, Wassmann et al. (2006) found considerably increased PP in their eddy-resolving 4-km-resolution model compared to an earlier 20-km-resolution model (Slagstad and Wassmann 1997).

High-resolution modeling is indispensable to resolve not only mesoscale eddies but also ice-edge and shelf-break upwelling. The ability of a model to simulate localized mechanisms of nutrient supply, on the order of ~1 km, depends on model resolution. In addition to more accurately predicting PP, high model resolution can help to clarify the underlying physical mechanisms responsible for spatial and temporal variations in marine ecosystems.

12.10 Modeling the PAR Ecosystem from an ESM Perspective

ESMs are coupled atmosphere-ocean-land-biosphere models designed to run for time periods of centuries to thousands of years and forced by changes in atmospheric CO_2 and other greenhouse gases, aerosols, solar variability, volcanism, ozone, and land use change. Model adjustment to its forcing (spin up) is usually done in a preindustrial control mode with forcing levels fixed to preindustrial times. The spin-up time is determined by deep ocean overturning timescales, which can be thousands of years. Upon reaching a stable state, model runs can be executed in historical mode, forced by known CO_2 concentrations or emissions from preindustrial to present. Based on current states, the models are then run for future scenarios. Up to now simulations have mostly been based on the scenarios outlined in the Special Report on Emission Scenarios (SRES, Nakicenovic et al. 2000). For the Intergovernmental Panel on Climate Change (IPCC) 5th Assessment Report (AR5), new scenarios known as Representative Concentration Pathways (RCPs) have been created (Moss et al. 2010).

The development of biogeochemical components in ESMs has been advanced significantly over recent years. Results from the first generation of ESMs which incorporated an interactive carbon cycle are discussed, e.g., by Friedlingstein et al. (2006) and Denman et al. (2007). The next generation of models is currently being archived for the 5th Coupled Model Intercomparison Experiment (CMIP5) in support of the IPCC AR5. This is the first CMIP to include ocean biogeochemical fields.

Generally ESMs have fairly coarse resolution, both horizontally (1–3.5°) and vertically (~50 m at the surface for earlier models and ~10 m for more recent models). This makes it difficult to resolve biological or chemical processes happening in the euphotic zone as well as small-scale physical processes, which are important for the biogeochemistry (see previous section). ESMs simulate the overall spatial and seasonal pattern with fair accuracy. However, based on the nature of those models and current data availability, it is not surprising that model skill (Box 12.2) for biogeochemical variables in global ESMs (which are usually based on limited local data sources) are rather low (e.g., Schneider et al. 2008; Steinacher et al. 2010). This is especially true in areas like the Arctic, which is characterized by few and seasonally-biased data and which includes vast shelf areas requiring high vertical model resolution.

Keeping those limitations in mind, we can focus on the uses of global ESMs and discuss their applicability for specific regions such as the PAR. ESMs provide insight into global connections and interactions between ocean basins, which might be important for the specific regions of interest. Moreover they are able to provide an estimate of changes in response to rising atmospheric CO_2 and other greenhouse gas levels and climate warming. They allow us to put recent changes into perspective with past and projected climates, and to study influences on and feedbacks from energy flows, carbon cycling, and biological productivity. More specifically, they can be used to study how biogeochemical cycling will respond to transitions in temperature, vertical stratification, SIZs, and acidification. This is a first step toward addressing consequences for socioeconomic activities. Guided by the outcomes from ESMs, we then must look to higher-resolution regional models of particular ecosystems to address more specific questions.

12.10.1 Ocean Acidification and Primary Production (PP) Changes Projected Over the Twenty-First Century

The Arctic, encompassing PAR, is one of the regions most sensitive to recent and expected future climate changes. Climate models predict a significant retreat in sea ice cover (e.g., Holland et al. 2006; Solomon et al. 2007), signs of which are already being observed in PAR (e.g. Stroeve et al. 2011). Recently the scientific community has paid increasing attention to ocean acidification (lower pH) and the particular vulnerability of arctic ecosystems (e.g., Orr et al. 2005; Steinacher et al. 2009; Denman et al. 2011; Gattuso and Hansson 2011). Acidification can significantly affect the growth, metabolism, and life cycle of marine organisms (see Box 12.3). The ocean's acidity is affected by increased atmospheric CO_2 and increased freshwater inputs from river runoff and ice melt. The combined effect renders arctic waters especially vulnerable to decreased saturation states of calcium carbonate ($CaCO_3$) minerals such as aragonite and calcite (Yamamoto-Kawai et al. 2009, 2011; Denman et al. 2011). Signs of aragonite undersaturation in surface waters in the Arctic have been reported by Chierici and Fransson (2009), Bates et al. (2009) and Yamamoto-Kawai et al. (2009, 2011). The Arctic Monitoring and Assessment Programme highlighted ocean acidification as an important issue and an Arctic Ocean Acidification assessment report is now in preparation.

Steinacher et al. (2009), using the Climate System Model (CSM) 1.4-Carbon of the US National Center for Atmospheric Research (NCAR), and Denman et al. (2011), using the Canadian ESM CanESM1, report model projections of aragonite and calcite saturation, carbonate concentration, and pH, based on SRES scenarios; all project significant acidification in the Arctic over the next century. CMIP5 results are now publicly available, allowing for a more thorough analysis. As an early example we present results for the PAR from the CanESM1.5, which is a higher-resolution version of CanESM1.0 (see Arora et al. 2009 and Christian et al. 2010 for model details). CanESM1.5 is a precursor to the AR5 model version CanESM2 (e.g. Arora et al. 2011), which includes an improved ocean model and a new atmosphere but was not ready in time for this synthesis. The ocean component includes a NPZD ecosystem with simple parameterization of N_2 fixation, calcification, and iron limitation (Zahariev et al. 2008). In CanESM1.5 and 2 the ocean has more vertical levels (40) and greater vertical resolution (~10 m in the top 107 m) than in CanESM1.0 (Arora et al. 2011). The horizontal resolution is 1.41° in longitude and 0.94° in latitude. The CanESM1.5 atmosphere is the same as in CanESM1.0 but with slightly higher resolution (~2.8°). Results are presented from a historical run (1850 to 2005) and a scenario run (2006–2100). The latter is based on RCP 8.5 with aerosol optical depth as in the SRES A2 scenario. Both are carbon-intense scenarios with RCP8.5 utilizing the highest total CO_2 emission scenario that has been used in AR5 experiments.

Figure 12.12 shows annually-averaged surface pH, aragonite saturation, and depth-integrated PP simulated for the end of the century (2096–2100, left hand side) and differences (end of century minus current) projected for the next 100 years.

Box 12.3: Ecosystem Response to Acidification and Multiple Stressors: A Challenge for Ecosystem Models

In recent years, laboratory studies testing the response of certain ecosystem components to acidification have revealed signs of dissolution in existing shells (e.g., Riebesell et al. 2000; Fabry et al. 2008), altered rates of essential metabolic processes and efficiency (e.g., Dupont and Thorndyke 2009; Barcelos e Ramos et al. 2010), impaired ability to sense chemical cues from predators (Munday et al. 2010), inhibited calcification, lower growth rates, degradation of mechanical integrity (e.g., Comeau et al. 2009; Gaylord et al. 2011), and underdevelopment of essential functions (e.g., Dupont and Thorndyke 2009). The observed responses have been found to be species-specific and to affect different life stages in different ways. Melzner et al. (2009) studied the physiological basis for high CO_2 tolerance and found that species with a higher metabolic rate and a powerful ion regulatory apparatus, or with more buffer fluids (e.g., blood, egg fluid) surrounding their cells, are able to cope more easily with high CO_2. Early developmental stages that expose cells directly to the surroundings seem to be a bottleneck for species survival.

Pörtner and Farrell (2008) associate the ability to cope with climate change with an aquatic lifeform's aerobic thermal window (the temperature range within which the species is able to execute vital functions). The thermal window varies with life stages; eggs, early larvae, and spawning animals are most vulnerable because their thermal windows are narrow. This window is expected to decrease with increasing partial pressure of CO_2 (pCO_2) and decreasing oxygen (O_2; Pörtner and Farrell 2008; Pörtner 2010).

As pCO_2 rises, species distribution and abundance will change according to each organism's tolerance of and ability to adapt to the increase. The extinction of one keystone species creates an opportunity for another species with a more flexible and robust regulatory system or a different thermal window. Species that are able to do so might migrate to more suitable habitats. Hence, trophic structure and biodiversity are key to the resistance and resilience of marine ecosystems to future perturbations (e.g., Dupont and Thorndyke 2009). Cheung et al. (2011), using a bioclimatic envelope model, found that ocean acidification and reduced O_2 content lead to reduced growth performance, an increase in the rate of range shifts, and lower estimated fisheries catch potentials. Beyond the individual species response, species interaction is an important factor. Many ecological processes are synchronized; hence, the timing of processes and developmental stages is crucial.

The future cost of acidification and other climate effects (such as warming) on ecosystems (e.g., biogeographic shifts, decreases in marine harvests, loss of coastline protection, and shoreline destabilization) will depend on the marine ecosystem response and on changes in human uses of marine resources

(continued)

Box 12.3: (continued)

(Cooley et al. 2009; Sumaila and Cheung 2010) and the ability of community and infrastructure planning to adapt successfully to change. Coupled ecosystem models and bioclimatic envelope models will help to address those issues, although socioeconomic models might be required to compare probable outcomes of different management choices (Cooley et al. 2009). How can we reliably calculate species/ecosystem tipping points in coupled ecosystem models and understand how the structure and function of whole ecosystems will adapt to multi-decadal environmental change? Recommendations include studying a wide range of species, mechanisms of physiological response, complete life cycles (over several generations), organisms that are less vulnerable (to detect the physiological traits crucial for ecological success in a future ocean), species already experiencing stress (from, e.g., high pCO_2), and multi-stressor environments (e.g., Dupont and Thorndyke 2009; Melzner et al. 2009; Ridgwell et al. 2009; Denman et al. 2011). Lynch et al. (2009) point to the need to focus on system integration of model theory and observation as an overarching goal. Much of what is mentioned above is described in much more detail in the book on ocean acidification by Gattuso and Hansson (2011).

Surface pH in the PAR has decreased from ~8.2 in preindustrial times to ~8.0 at present. In the RCP8.5 scenario pH is projected to further decrease to ~7.9 by mid-century and below 7.7 by the end of the century (Fig. 12.12a), which corresponds with the results in Steinacher et al. (2009). While pH change highlights the increased ocean acidification to be expected over the next century, aragonite and calcite saturation also depend on salinity and are significantly affected by increased freshwater contributions from river inflow, precipitation, and ice melt (Yamamoto-Kawai et al. 2009; Denman et al. 2011). Hence the surface ocean north of Bering Strait, which receives higher freshwater contributions, is predicted to become undersaturated (below 1.0) in aragonite before the southern PAR (Fig. 12.12b). From now to the end of the century, CanESM1.5 results project the aragonite saturation state to change by about 0.8–1.0 in most of the PAR with somewhat smaller changes north and northeast of the Bering Strait, where surface waters have already become undersaturated. Increased PP within and north of Bering Strait (Fig. 12.12f), which acts to increase aragonite (and calcite) saturation states in the surface layer, might contribute to somewhat smaller changes in this area. Increased PP could also be responsible for rendering simulated summer surface saturations in the northern Bering Strait slightly higher than in the subsurface (not shown).

The freshwater influence at the surface can lead to a subsurface maximum in the mean saturation state, with undersaturation in the deep ocean. The switch from surface supersaturation to deeper ocean undersaturation is commonly called the saturation horizon or saturation depth. The latter has been slowly moving upward

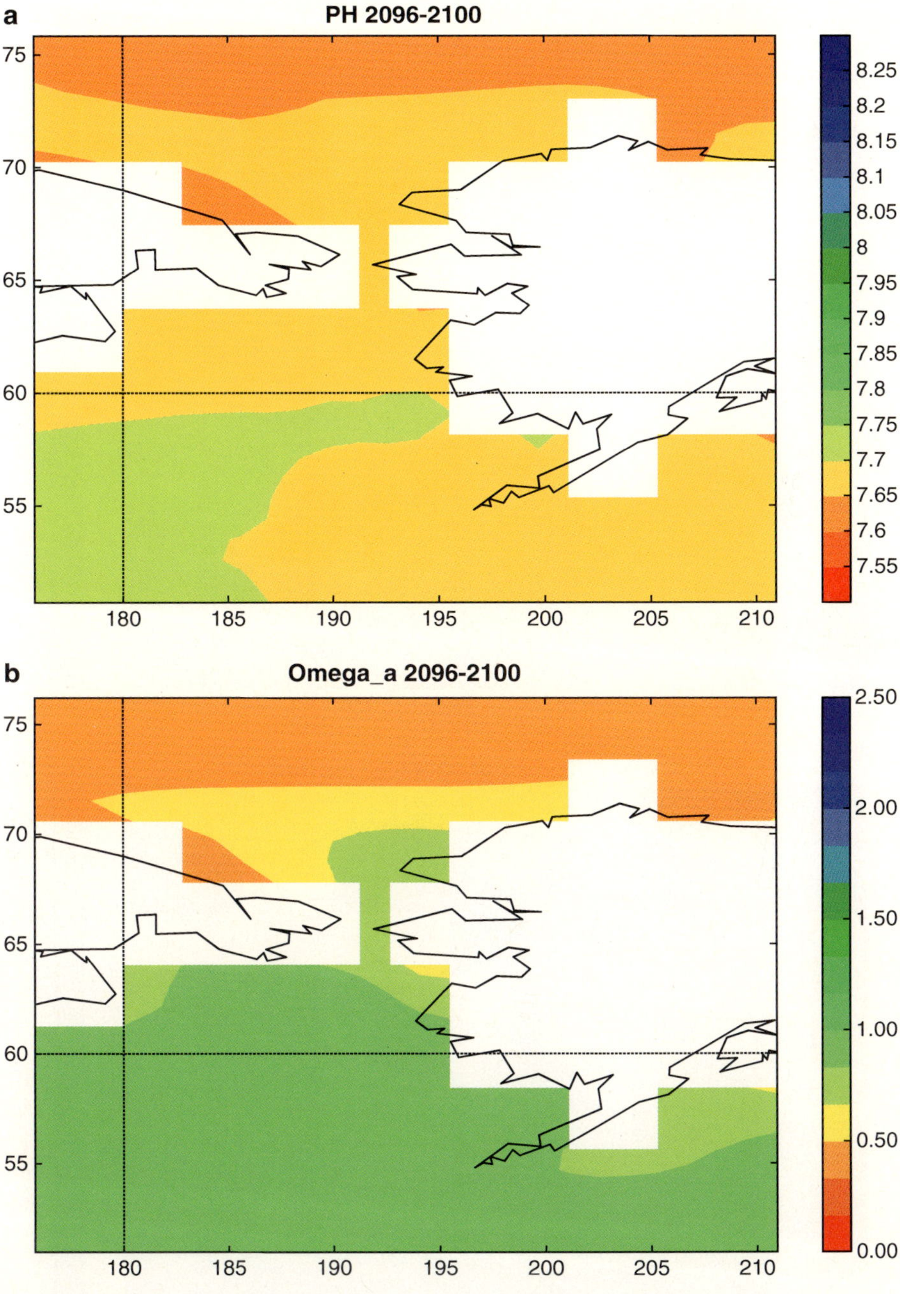

Fig. 12.12 Model predictions for the end of the century (2096–2100, **a**, **b**, **c**) and differences between the end of the century and current times (1996–2005) (**d**, **e**, **f**), simulated with CanESM1.5 and using the IPCC AR5 RCP8.5 scenario (see text). (**a**, **d**) annually-averaged pH, and (**b**, **e**) aragonite saturation state in the surface layer (top 10 m). (**c**, **f**) annually-averaged and vertically-integrated primary production in g C m^{-2} $year^{-1}$

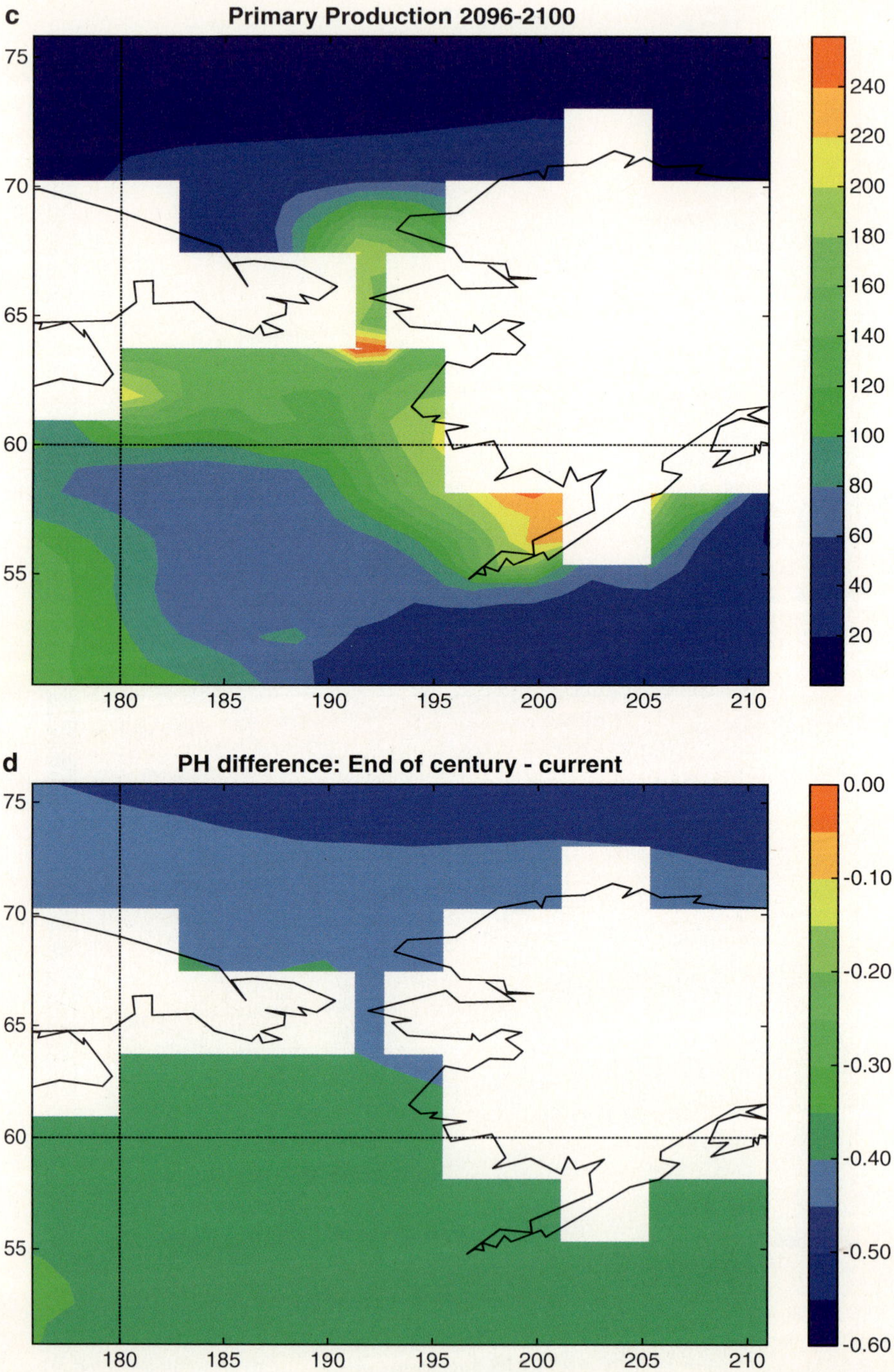

Fig. 12.12 (continued)

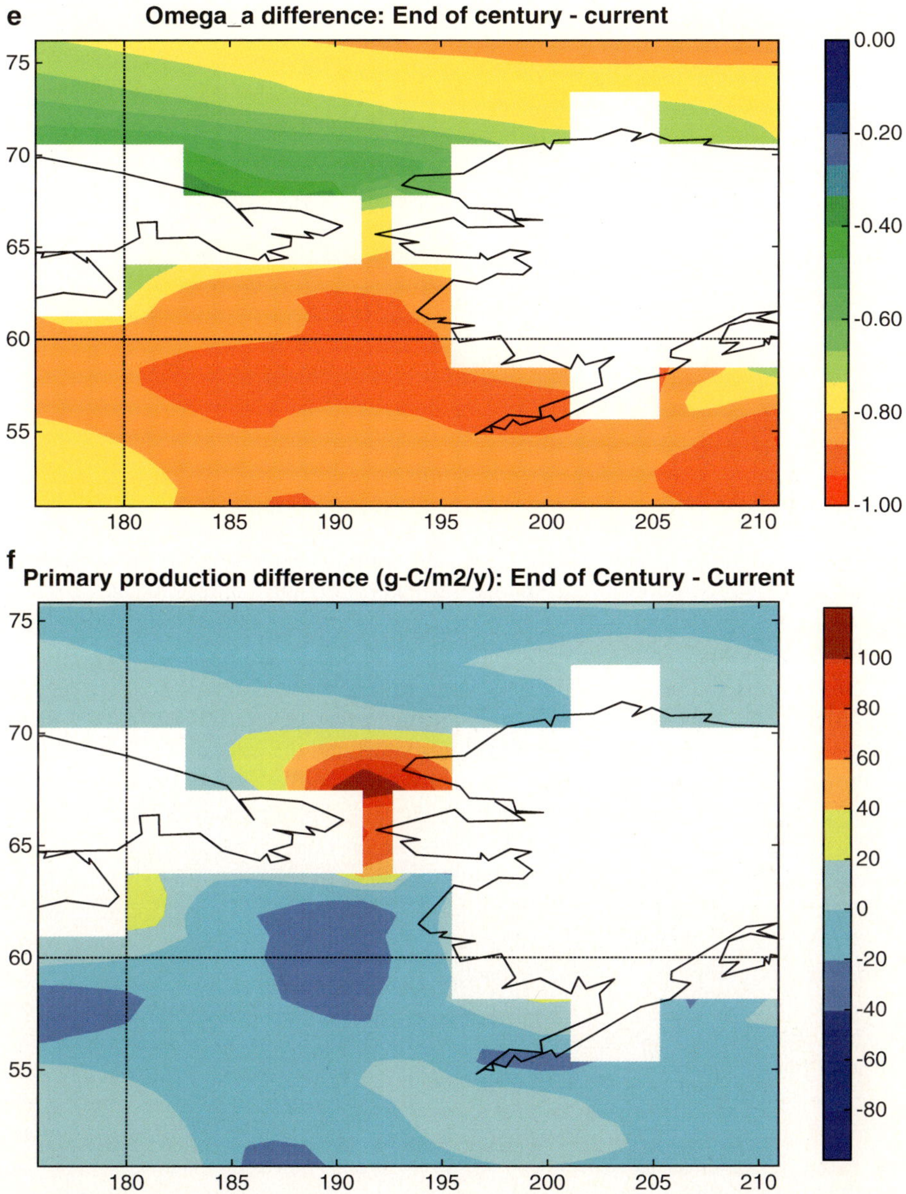

Fig. 12.12 (continued)

since preindustrial times and is projected to continue to do so in the future (e.g., Steinacher et al. 2009; Denman et al. 2011). The aragonite saturation horizon in the PAR is currently still below shelf level. CanESM1.5 projects the aragonite saturation horizon to rise to 200 m in almost all of the PAR by mid-century and the entire PAR, from surface to bottom, to be undersaturated by the end of the century. This could have severe consequences for the pelagic and benthic communities as well as HTLs and fisheries in the area.

The simulated pattern and magnitude of PP in current times is very similar to the simulations of pelagic production by Jin et al. 2012 (Fig. 12.4, top left). Figure 12.12c shows annually-averaged depth-integrated PP projected for the end of the century with changes from current conditions in Fig. 12.12f. CanESM1.5 projects an increase in Bering Strait and Chukchi Sea PP due to the increased light and temperature that accompanies retreating sea ice. The increases range from 10 to 50 g C m^{-2} $year^{-1}$ by mid-century and 20–110 g C m^{-2} $year^{-1}$ by the end of the century with the highest increase within the Bering Strait. At the same time, the model shows a decrease of about 15 g C m^{-2} $year^{-1}$ by mid-century and up to 40 g C m^{-2} $year^{-1}$ by the end of the century in the central and southern Bering Sea. This decrease can likely be attributed to changes in stratification. A scenario discussed in Mathis et al. (2010) suggests that a decrease in sea ice causes increased availability of solar radiation and a more stratified water column. The latter would limit production through limited nutrient fluxes. A model analysis confirms that suggestion. The CanESM1.5 results show intensified warming in the area over the second part of the century leading to temperatures about 3 °C higher than current. Nutrient levels are decreased by about 30–60 % at the surface and subsurface, while nutrient levels in deeper layers are higher by about 15–40 % compared to present time values at the same depth and corresponding to reduced mixed layer depths (MLDs).

An analysis of the multimodel mean from 17 of the previous CMIP3 models (W. Merryfield, CCCma, pers. Comm.) indicated that, while the multimodel mean shows shallower MLDs under anthropogenic warming, the primary influence differs with region. In the northeast Pacific, increased near-surface stratification was most important. Steinacher et al. (2010) compare global fields of projected PP in the IPSL-CM4-LOOP model from the Institut Pierre Simon Laplace (IPSL), the COSMOS ESM from the Max Planck Institute for Meteorology (MPIM), and two versions of the NCAR Community Climate System Model (CSM1.4-Carbon and CCSM3-BEC) for the SRES A2 scenario (a high-CO_2 scenario similar to RCP8.5). A robust result from the intercomparison is a decreasing trend in global net PP. However, the magnitude differs among models and regions. Two processes are primarily responsible for changes in PP. First, a reduction in nutrient supply to the surface in a more stratified ocean leads to decreased production; second, an increase in light and temperature with retreating sea-ice and sufficient nutrients supports increased production. While the models are generally consistent in their responses, IPSL projects a decrease in Arctic PP, related to a reduced supply of macro nutrients, whereas CSM1.4, MPIM, and CCSM3 project an increase due to reduced temperature and light limitation (Note that absolute changes in the Arctic are small). In the PAR (Steinacher et al. 2010, their Fig. 12.3) all models simulate present values that are too low; the most realistic values are produced by the IPSL model. Projections show increases in PP for all models but the IPSL, which shows a decrease in the Bering Sea and an increase north of the Bering Strait, similar to the CanESM model results. In a multi-model average weighted by skill score, the IPSL model dominates in the PAR, although the skill scores are generally very low in that area (<0.2).

Discussing discrepancies between results from empirical approaches and process-based ESMs, Steinacher et al. (2010) point out the importance of a realistic representation of nutrient cycling and distribution in order to project changes in PP with some realism.

However, as stated earlier, ESMs are computationally expensive. Constant tension exists between simple grid resolution adjustments, which can improve circulation physics, and the tracer/reaction-intensive biology and chemistry. Some processes might not be resolved and parameterized adequately, especially in areas of narrow straits and shallow shelves, both of which are characteristic of the PAR. Also, ESMs so far do not account for changing carbon fluxes via rivers and coastal erosion, which can be significant (Cai et al. 2014, this volume). Fabry et al. (2009) point out that the polar seas constitute ecosystem laboratories for diverse studies in ocean acidification, including assessing potential acclimation and adaptation and modeling future impacts at population, community, and regional scales. The PAR is already an area with an intensive research program and it can serve as a test bed for developing detailed biological and biogeochemical models. It will be reasonable to perform simulations at a variety of scales ranging from local (point or column) to regional, and then to refine and transfer appropriate parameterizations to global ESMs.

12.10.2 Projections of Fisheries and Biodiversity Impacts with Bioclimatic Envelope Models

While of intense interest to managers and policy makers, projections of HTLs in the Earth system modeling framework are not yet possible. Bioclimatic envelope models are an intermediate step to allow projections of global marine biodiversity impacts (e.g. Cheung et al. 2009) and fisheries catch potential (e.g. Cheung et al. 2010, 2011) under climate change. In these models, shifts in species distribution can be predicted by evaluating changes in physical and biological conditions relative to those suitable for a given species. Suitability is determined by correlating current environmental conditions with maps of current species abundances. Future environmental conditions are provided from climate models or ESMs (e.g. Cheung et al. 2009, 2011). Cheung et al. (2009) predict numerous local species extinctions in the sub-polar regions and intense species invasion in the Arctic and southern oceans. Cheung et al. (2010) project 30–70 % increase in maximum fisheries catch potential in high-latitude regions with the largest changes in the Pacific. They note, however, that their model does not yet account for the effects of changes in ocean biogeochemistry (e.g. ocean acidification, oxygen reductions, and thermal window limitations, see Box 12.3) and phytoplankton community structure. Cheung et al. (2011) introduced a model including these new factors, but have not yet applied it to the Arctic or the Pacific oceans.

12.10.3 Marine Methane Emissions

Methane (CH_4) from subsea permafrost (Shakhova et al. 2010) and trapped in the form of clathrate hydrates is expected to leak from shallow arctic shelves, partly driven by penetration of the North Atlantic warming signal into central ocean layers (Reagan and Moridis 2008). Archer (2007) concluded that destabilizations will be particularly important around the rim of the central arctic basin because the Atlantic layer is the warmest source of water to the Arctic and both the Chukchi and Beaufort shelf zones will be directly affected.

Modeling arctic marine methane chemistry is in its infancy, and simulating methane emissions from the ocean have so far been afforded low priority in ESMs. For the most part they are represented as a constant flux equally distributed over the global ocean. The IPCC AR3 and AR4 (Prather et al. 2001; Denman et al. 2007) cite global emissions of 10–15 Tg methane year^{-1}, although the most recent measurements in the Pacific (Bates et al. 1996) and Atlantic (Rhee et al. 2009) suggest these values may constitute strong overestimates. The arctic continental shelves are believed to be a large potential source to the global atmosphere. Macdonald et al. (2009) estimate that they provide up to four times the annual flux estimated for all other coastal seas combined. New studies further suggest that aerobic marine methane production will be sensitive to changes in water-column stratification and also to the nutrient limitation likely to result from greenhouse-gas-induced warming (Karl et al. 2008). An additional source of methane may be organic carbon buried in sediments over glacial timescales and then transformed into methane under anaerobic conditions (Reeburgh 2007).

The premise of a recent modeling study by Elliott et al. (2010) is that CH_4 gas emanating from such systems will be consumed along the periphery of the PAR by methanotrophs and anaerobic oxidizers of methane (Reeburgh 2007). In the model, consumer organisms are pulled out of the sediments along with bubbles and fluids flowing into regional-scale plumes. As metabolism progresses using the new methane sources, hypoxia and nutrient depletion appear at scales of tens to hundreds of kilometers. Elliott et al. (2010) conclude that these altered water masses will form a new geochemical component of the central Arctic Ocean milieu; different regions will display distinct responses to future methane throughput.

12.10.4 Ocean Biogeochemistry-Aerosol Connections

Another important factor for consideration in modeling climate with earth system simulations is the generation of aerosols over the oceans. While current atmospheric models treat aerosols as produced mostly from sea salt and volatile reduced sulfur (Jones et al. 2001), some authors suggest a strong but poorly-quantified link between the particulate organic matter of polar surface waters and aerosol and arctic cloud systems (e.g., Leck and Bigg 2005b, 2008; Matrai et al. 2008), and a possibly overestimated role of sea salt (e.g., Leck and Bigg 2005a; Bigg and Leck 2008).

The role of volatile dimethylsulfide (DMS) is still unclear as well. DMS is formed via a complex network of food web interactions from its precursor, dimethylsulphoniopropionate (DMSP), produced by many marine phytoplankton and sea ice algae. In the atmosphere, DMS is oxidized to sulfate aerosols, which serve as cloud condensation nuclei (CCN). Shaw (1983) was the first to suggest that biogenic sulfur gases may regulate climate through generation of sulfate aerosols that scatter sunlight back to space. However, quantitative descriptions of key processes and direct links are still lacking (e.g., Ayers and Cainey 2007). The relative importance of sea salt, sulfate, and organics in determining the CCN number concentration is still unclear, as are the responses of plankton community structure and/or planktonic DMSP production to climate change. The underlying marine biogeochemistry has not yet been modeled adequately at any scale with respect to recent findings regarding the organic-to-CCN connection.

Global process models for the marine sulfur cycle constitute the first real attempt by the earth system simulation community to examine climatic effects of ocean biogeochemistry by mechanisms other than greenhouse gases (Six and Maier-Reimer 2006; Elliott 2009; Le Clainche et al. 2010). Several authors on this review are now involved in implementing generic sulfur reaction schemes into the CICE-POP-CCSM family of codes (Hunke and Lipscomb 2008; Elliott et al. 2012). Despite the development of 1-D and high-resolution regional biogeochemical models for northern high latitudes that include sea ice algae (see Sects. 12.5 and 12.6) and ocean DMS cycling (Jodwalis et al. 2000), polar biogeochemical cycling and especially the ice algae remain underdeveloped in GCMs (Elliott 2009).

12.11 Gaps and Needs: Introduction

PAR biogeochemical modelers agree that more observations are needed, specifically observations sustained throughout the year and biological process and rate measurements. For example, there is little known about the biology of marine organisms such as zooplankton at depth under the ice, and the wintertime distribution and physiological state of phytoplankton. Also, biogeochemical transfer processes within and through sea-ice have yet to be elucidated (e.g. Loose et al. 2011). Advances in modeling sea ice are critical overall: Directly and indirectly, sea ice influences nutrient availability and atmospheric composition (e.g., trace gas exchange, aerosol sources) and, along with snow cover, controls the light available to PAR ecosystems for much of the year. Many recent and on-going efforts are directed toward improving modeling of biochemical and geochemical processes, and of physical transport within the sea ice matrix. Regional downscaling from GCMs is recognized as an important tool in producing regional climate information for impact and adaptation studies. Relevant studies are at an early stage but have great potential for future research.

12.12 Availability of Observations

Wassmann et al. (2011) studied footprints of climate change in the Arctic ecosystem and point out the significant lack of reliable baseline data from which change can be identified, particularly for planktonic and benthic systems. The sparseness and heterogeneity of the available biological and biogeochemical observations, for instance seawater nutrient concentrations and plankton biomass, also creates problems for 3-D model initialization and validation. An ideal dataset would cover all seasons and most of the model domain. The World Ocean Atlas (WOA) monthly climatology http://www.nodc.noaa.gov/OC5WOA05/pr_woa05.html is the only gridded dataset available for model initialization of seawater nutrient concentrations in PAR. However, a few new arctic databases are being developed with the goal of improved understanding of Arctic Ocean PP and its changing physical controls such as light, nutrients, and stratification. The NSF Arctic Primary Productivity Observational Synthesis Project has perhaps the largest nutrient database as a result of this effort. A merged-calibrated file for the Arctic is now also available from the Carbon dioxide in the Atlantic Ocean (CARINA) data synthesis project, which is accessible via the Carbon Dioxide Information Analysis Center (CDIAC, http://cdiac.esd.ornl.gov/oceans/CARINA/Carina_inv.html). The quality of the data in this merged product is significantly higher than data in the WOA, but the measurements are patchily distributed and insufficient to grid, especially by season and in shallow water. Carbon data from the Pacific Ocean are being synthesized in the North Pacific Marine Science Organization effort. Additional datasets from large arctic projects exist that have not yet been added to the WOA, such as the NSF Shelf Basin Interactions (SBI) and the NOAA IPY project: Russian American Long-Term Census of the Arctic (RUSALCA). Observationally-driven IPY projects, such as Canada's Three Oceans, measured spatial variability along ship transects and recorded temporal variability using year-round moorings. Many of the recent measurement programs within the Pacific Arctic are part of the Pacific Arctic Group's (PAG) Distributed Biological Observatory (DBO) which aims at linking physics and biology (Grebmeier et al. 2010). Efforts for data management and sharing of DBO data, e.g. within The Advanced Cooperative Arctic Data and Information Service (ACADIS) of the Arctic Observing Network (AON), are underway. ACADIS provides data management support for all projects funded by NSF's Arctic Science Program.

Long-term observational networks that address spatial scaling issues (e.g., fine-scale observations vs. coarse scale models) and link observations to processes influencing biogeochemical dynamics are needed for modeling efforts. Time series observations over very broad spatial domains will aid in validating 3-D biogeochemical models. Synthesis products from IPY efforts, such as maps of benthic biomass, provide models with more realistic initial conditions across the model domain. Remote-sensing data provide synoptic biological information, but these are snapshots of the surface ocean layer under clear sky conditions and are not indicative of the hydrography below the surface. Biological rate measurement and process-oriented datasets needed for parameterizing models are rare for PAR.

New modeling approaches and data assimilation procedures that directly incorporate observations from a wide range of measurement systems including biological and biogeochemical observations are yet to be developed. The demand for marine system models to support policy and management decisions is on the rise. The focus needs to be set on better enabling diagnostic modeling of the present-day and more accurate modeling for real-time forecasts and management projects. For now PAR models could be designed to take advantage of ocean color products that are routinely produced and distributed, and expand from there as arctic biological databases emerge. To build up these databases, sufficient support and effort needs to be invested in comprehensive sampling of the Arctic.

12.13 Modeling Brine Processes and Tracer Transport in Sea Ice

Current ice algae models usually apply simple parameterizations for nutrient supply, salt, and material release. Oceanic emissions are essentially prohibited because sea ice forms a barrier to gas exchange in models. However, laboratory and field studies show that sea ice can be highly permeable to CO_2 (Gosink et al. 1976; Kelley and Gosink 1979; Golden et al. 1998) and that it controls and contributes to air-sea flux of climatically-active trace gases, such as CO_2 and DMS (Semiletov et al. 2004; Delille et al. 2007; Zemmelink et al. 2008; Miller et al. 2011). A simplification helpful for modelers is the "rule of fives": At temperatures below about–5 °C, corresponding to a brine volume fraction of about 5 % and a typical bulk sea ice salinity of 5 parts per thousand, brine inclusions are mostly disconnected, and sea ice is effectively impermeable to fluid flow. For temperatures above –5 °C, the brine inclusions become connected over larger scales, allowing fluid and gas transport through the sea ice (e.g. Golden et al. 1998; Petrich and Eicken 2009) Modelers are now looking into more accurate representations of brine dynamics to describe permeability and salinity profiles within ice, which affect physical properties like heat transport and melting (e.g., Golden et al. 2007; Petrich and Eicken 2009; Vancoppenolle et al. 2009) as well as gas fluxes and nutrient and material cycling (Vancoppenolle et al. 2010). Notz and Worster (2009) revisited sea ice desalination processes and concluded that gravity drainage in winter and flushing during summer are the only significant contributors to salt loss from sea ice. They point out that while models of flushing are fairly realistic, the lack of a physically-based quantitative description of gravity drainage is a significant shortcoming of sea ice models. From the biogeochemical perspective, improved brine process modeling will allow us to assess the importance of nitrate versus silicic acid or other nutrient limitations, the production of dissolved organic carbon (DOC) and extracellular polymeric substances (EPS) by ice algae, and their subsequent recycling by bacteria. The latter appears to play an important role in organic carbon cycling within the sea ice (Krembs et al. 2002; Riedel et al. 2008). EPS are sticky and play an important role in cell aggregation and sinking. Although ice algae do not represent a major fraction of annual PP, their actual influence on the export of organic carbon could be strong.

Improved knowledge of gas exchange through sea ice has the potential to alter our understanding of the seasonal amplitude, and localized sources and sinks of atmospheric CO_2 and other gases in PAR. However, little is known about the complex dynamic processes involved. Rysgaard et al. (2007) suggest that $CaCO_3$ precipitation in sea ice might act as a sink for CO_2. Recent observations confirm the presence of the mineral ikaite likely caused by the precipitation of $CaCO_3$ within brine channels in the Arctic (Dieckmann et al. 2010). CO_2 is pumped in and out of central ice layers through processes that are partly dominated by chemical thermodynamics but that also involve complex brine dynamics that affect alkalinity and pH within the bulk ice structure.

12.14 Snow Component

While most Arctic snow melts completely by Spring, it has a significant effect on physical and biological processes within the sea ice matrix and below. Snow remains one of the most important variables needing improvement in sea ice models as well as ice ecosystem models. Snow cover controls ice growth rate and the amount and timing of light availability at the underside of the ice, which in turn regulate ice algal bloom onset, biomass accumulation, and termination of the ice algal bloom (Lavoie et al. 2005, 2009; Mundy et al. 2005; Jin et al. 2006a). Snow cover also affects the onset and magnitude of the phytoplankton spring bloom, through its effect on ice melt, timing of ice disappearance and release of nutrients and ice algae. The highly variable processes in the evolution of the snow (and ice) cover, especially during the melting season (e.g., formation of melt ponds, snow-ice formation during flooding), which affect both light conditions and habitat for sea-ice ecosystems, are still insufficiently described in models and will need further attention in future model development.

12.15 Microbial Loop

Heterotrophic bacteria as well as the rest of the microbial loop need attention in PAR biogeochemical models. Contributions of the microbial loop to carbon cycling (e.g., PP and export production) are substantial and are discussed in a previous chapter (Mathis et al. 2014, this volume). Important bacterial processes such as decomposition of organic matter and remineralization of nutrients are typically included in PAR biogeochemical models through implicit representation of bacteria. In this way, Lavoie et al. (2009) were able to simulate detrital flux comparable to the mean annual POC flux estimated from sequential sediment traps, although microplankton, dissolved organic matter (DOM), ammonium, and bacteria compartments would have enhanced the model. DOM concentrations in arctic rivers are among the highest in the world (Dittmar and Kattner 2003). Cai et al. (2014, this volume) summarize riverine DOC fluxes from many studies in the Pacific region of the Arctic Ocean. DOC supplied by these rivers has traditionally been considered refractory, but new studies question this assumption and suggest that DOC delivered by rivers during the spring freshet is labile (Holmes et al. 2008, and references therein). Similarly, Garneau et al. (2008) found that some areas of the Mackenzie shelf environment could be a source of CO_2 due to high bacterial production, and they suggest that the bacterial-to-PP ratio could increase in the future due to the increase in allochthonous substrate supplied by rivers. To represent these processes realistically in biogeochemical models, an explicit description of bacteria and/or related compounds need to be included.

12.16 Modeling Adaptation and Synergistic Effects

At this point most ecosystem models, especially those incorporated into global ESMs, are fairly simple with either constant or temperature-dependent parameterizations. However, to adequately simulate both functional and structural ecosystem changes in a changing climate, this might not be sufficient. It is not known how Arctic ecosystems will adapt to the drastic changes happening in such short time spans. Laboratory studies show significant impact of, for example, ocean acidification and other stressors on marine life (Box 12.3); however, they also show that, while some species will fail to survive, others will thrive under the same conditions. Many open questions wait to be resolved: Will the less fortunate simply be replaced by more suitable species? Do community structures stay the same and just shift their location? Or, if creating longer-term projections, do species adapt genetically and/or behaviorally or die out? How do timings of migration, growth and reproduction patterns and predator–prey interactions change?

From a modeling perspective, we ask: Are our coarsely-grouped phytoplankton and zooplankton compartments representative of those found in today's climate, and will they still be representative tomorrow? Do current parameterizations provide accurate predictions because better adapted species will simply replace less competitive/poorly adapted species? Or do we need to think about genetically or physiologically adaptive processes occurring over generations represented on timescales of global ESMs? While species replacement and migration might not necessarily require changes in parameterizations, evolutionary adaption will. How can this be achieved if the adaptation potential for individual species varies? Laboratory studies show that for some species one generation is enough to cope with the environmental changes like e.g. impacts of ocean acidification (Dupont and Thorndyke 2008), however many more generations might be needed for genetic mutations to manifest. The synergism of effects is an important factor. Organisms might be able to adjust to one effect, but perhaps not to multiple stressors including warming, ocean acidification, hypoxia, habitat destruction, overfishing, introduced species, etc. One type of modeling that would allow projections on how these changes might occur requires a complex adaptive systems approach (e.g. Norberg 2004). In the approach of Norberg, the ability of the ecosystem to evolve with a changing environment/climate is a function of the existing biodiversity within species or functional groups of species. In a related approach, Pahlow et al. (2008) allow the parameters within the model to change or adapt in response to changes in environmental variables, but the essential structure of the ecosystem remains unchanged.

12.17 Dynamical Downscaling

Dynamical downscaling, where the output from one (or more) GCM is used to force a finer-resolution regional climate model (e.g., Liang et al. 2008) could be applied to PAR. The statistical downscaling method used in Lavoie et al. (2010) to provide forcing for future projections is limited by the ability of the GCM to represent important mesoscale features and processes. Regional climate downscaling (RCD) techniques are increasingly being utilized to produce regional climate information for impact and adaptation studies. Hence it is critical that the potentials, limitations, and uncertainties of RCD-based information are well understood by the modeling and user communities. The World Climate Research Program has formed a Task Force on Regional Climate Downscaling (TFRCD) to address those issues and foster international collaboration between global climate modelers, the downscaling community, and end-users to better support impact and adaptation activities. The TFRCD initiated a framework called the Coordinated Regional Climate Downscaling Experiment (CORDEX), which among other integrative approaches defines a common set of Regional Climate Model domains for dynamical downscaling (http://copes.ipsl.jussieu.fr/RCD_CORDEX.html). The CORDEX domains include a setup for the Arctic, however so far the program has been limited to the atmosphere. To further advance modeling of arctic marine biogeochemistry Arctic System Models are needed (Roberts et al. 2010) and are now starting to become feasible (e.g., Maslowski et al. 2012). Both dynamical and statistical downscaling methods will benefit from the continuously improving presentation of the Arctic Ocean in GCMs. Marine biogeochemical modeling has received a significant boost in recent years and will provide an important addition to RCD models, especially with respect to future fisheries management.

12.18 Summary

Models are useful and necessary tools for studying the impact of climate related changes on the complex interactions that influence PAR ecosystems. We expect that continuing advances in computing power and resources combined with increasing attention to this sensitive region will rapidly advance PAR biogeochemical modeling in coming years. The goal of this model synthesis is to provide an overview, to evaluate the ability of current PAR ecosystem models to address interactions within the ecosystem and between physical and biogeochemical properties, and to discuss model suitability to simulate ecosystem changes in a changing climate. The specifics of PAR with its seasonal sea ice zone and shallow shelves call for coupled ecosystem models including benthic, pelagic, and ice communities. Some models already incorporate all three components and are able to represent observed shifts from a benthic-dominated system to a pelagic-dominated system in response to sea ice retreat, as well as seeding of spring algae blooms by ice algae. Existing models can (1) simulate annual ice algal and pelagic production within the ranges of the observations, (2) represent hot spots in production and differences in regional distribution of biomass, (3) explore the effects of interannual changes in ice cover and water motion on carbon and nitrogen cycling, (4) analyze biogeochemical processes under alternate light- and nutrient-limited regimes, and (5) study seasonal cycles and the physical processes controlling primary and export production under both current conditions and future scenarios. These results provide valuable insights into possible future ecosystem dynamics in a changing Arctic, enabling first estimates of consequences for HTLs and socioeconomic activities.

The subsurface phytoplankton bloom is a major contributor of fixed carbon, and we caution that satellite-derived estimates of phytoplankton production, which are not able to track subsurface blooms, might underestimate actual concentrations. This shortcoming must be taken into account when validating ecosystem models, which provide vertical structure and depth-integrated PP. Light attenuation is a major factor limiting PP, and factors like ice-entrapped sediments, thicker ice, and/or snow depth have been shown to affect the amount of light available to marine algae. We point out that ecosystem structures and physical environments in different regions are quite specific and that ecosystem models from other regions (e.g., the southern ocean or even the Barents Sea) cannot readily be applied to PAR.

We highlight the importance of capturing the timing and location of the ice edge and open water blooms for simulating the food web structure that follows. An accurate representation thereof requires high-resolution modeling. Since much of the PAR system is governed by advection, proper representation of currents, circulation, frontal systems, and vertical stratification as well as potential future changes in those systems is important. The very-high-resolution physical models that exist are able to represent relevant features (e.g., eddies). However, these models are not yet coupled to a marine ecosystem component.

Both 1-D and regional models forced by climate model output as well as climate models themselves provide first estimates of future and current PAR changes. Model results suggest not only a PP increase in the northern PAR due to the ice retreat, but also some reduction in the southern PAR attributed to changes in the upper ocean stratification. They also show a reduction in the importance of the ice algae contribution due to the earlier snow and ice melt, while other models suggest even more ice algal production in low ice years. Models also suggest switches in the dominant phytoplankton groups due to changes in sea ice cover (i.e., timing of sea ice retreat), which might support either pelagic- or benthic-dominated systems.

Projections with ESMs highlight imminent acidification, which renders PAR waters undersaturated in aragonite and calcite from top to bottom by the end of the century. This acidification could carry severe consequences for pelagic and benthic communities. It is important to note that ESMs do not yet incorporate any physiological or behavioral changes beyond regional shifts due to acidification or other environmental influences. Many of the possible physiological and behavioral changes are not yet understood and must be explored in laboratory and field studies.

We emphasize model validation as a procedure of major importance in developing model parameterizations, highlighting the need for observational data from within PAR. Consistent and continuous interaction between modelers and field scientists, preferably starting in the planning stage of a project, is essential. These interactions have proven beneficial from an observational standpoint as well. While data are essential for validating models, models in turn can be used to interpret and extrapolate data, identify hot spots for observations, test hypotheses, and guide the design of field programs.

In closing, biogeochemical modeling in PAR has advanced significantly over recent years. Limitations still exist due partly to the lack of gridded datasets and the limited understanding of links between individual ecosystem components and possible adaptation processes in a changing climate. We expect integrated observing and modeling approaches for PAR to increase over the coming years, advancing our understanding of the underlying systems and our ability to accurately represent those systems in model equations.

Acknowledgments We thank the two anonymous reviewers for their helpful comments and Dr. Candace O'Connor for editorial assistance. The National Science Foundation (NSF) provided funds for meetings as part of the International Collaboration to Achieve Circumpolar Synthesis and Integration Project (ARC-0653838). NOAA provided travel support to C. Deal through the University Corporation for Atmosphere Research. The departments of Fisheries and Oceans and Environment Canada supported the contributions of N. Steiner, J. Christian, K. Denman, D. Lavoie, and W. Lee. The IARC/JAMSTEC Cooperative Agreement and NSF ARC-0652838 supported contributions by C. Deal and M. Jin. Funding support for contributions by J. Clement Kinney and W. Maslowski was provided by multiple grants from the Climate Change Prediction Program of the Department of Energy, the Arctic System Science (ARCSS) Program of the NSF, and the Office of Naval Research. Funding support for contributions by G. Gibson was provided by a grant from NSF (ARC-0732538). Funding support for S.H. Lee was provided by the MOF, Korea (K-PORT, PM12020). J. Wang was supported by the NOAA CPO Office of Arctic Research on ice-ocean-ecosystem modeling under the RUSALCA and PAG projects. This is GLERL Contribution No. 1704. E. Watanabe was funded by the IARC/JAXA Contract JFY2008-2010 and the Grants-in-Aid for Scientific Research (S) of Japan Society for the Promotion of Science (JSPS), No. 22221003

References

Allen J, Somerfield P (2009) A multivariate approach to model skill assessment. J Mar Syst 76:83–94. doi:10.1016/j.jmarsys.2008.05.0

Archer D (2007) Methane hydrate stability and anthropogenic climate change. Biogeosciences 4:521–544

Arora V, Boer G, Christian J et al (2009) The effect of terrestrial photosynthesis down-regulation on the twentieth century carbon budget simulated with the CCCma earth system model. J Climate 22:6066–6088

Arora V, Scinocca J, Boer G et al (2011) Carbon emission limits required to satisfy future representative concentration pathways of greenhouse gases. Geophys Res Lett 38, L05805. doi:10.1029/2010GL046270

Arrigo K, Kremer J, Sullivan CW (1993) A simulated Antarctic fast ice ecosystem. J Geophys Res 98:6929–6946

Arrigo KR, van Dijken G, Pabi S (2008) Impact of a shrinking Arctic ice cover on marine primary production. Geophys Res Lett 35, L19603. doi:10.1029/2008GL035028

Ayers GP, Cainey JM (2007) The claw hypothesis: a review of the major developments. Environ Chem 4:366–374. doi:10.1071/EN07080

Barcelos e Ramos J, Müller M, Riebesell U (2010) Short-term response of the coccolithophore Emiliana huxleyi to an abrupt change in seawater carbon dioxide concentrations. Biogeosciences 7:177–186

Bates N, Michaels A, Knap A (1996) Seasonal and interannual variability of the oceanic carbon dioxide system at the US'JGOFS Bermuda Atlantic time-series site. Deep Sea Res II 43(2–3): 347–383. doi:10.1016/0967-0645(95)00093-3

Bates N, Mathis J, Cooper L (2009) The effect of ocean acidification on biologically induced seasonality of carbonate mineral saturation states in the western Arctic Ocean. J Geophys Res 114:C11,007. doi:10.1029/2008JC004862

Berline L, Spitz YH, Ashjian CJ, Campbell RG, Maslowski W, Moore SE (2008) Euphaussid transport in the western Arctic Ocean. Mar Ecol Prog Ser 360:163–178

Bigg E, Leck C (2008) The composition of fragments of bubbles bursting at the ocean surface. J Geophys Res 113:D11209. doi:10.1029/2007JD009078

Bluhm BA, Gradinger R (2008) Regional variability in food availability for Arctic marine mammals. Ecol Appl 18:577–596

Brasseur P, Haus J (1990) Application of a three-dimensional variational inverse model to the analysis of ecohydrodynamic data in the Bering and Chukchi seas. J Mar Syst 1:383–401

Cai W-J, Bates NR, Guo L, Anderson L et al (2014) Chapter 8: Carbon fluxes across boundaries in the Pacific Arctic region in a changing environment. In: Grebmeier JM, Maslowski W (eds) The Pacific Arctic region: ecosystem status and trends in a rapidly changing environment. Springer, Dordrecht, pp 199–222

Campbell RG, Sherr EB, Ashijian CJ et al (2009) Mesoplankton prey preference and grazing impact in the western Arctic Ocean. Deep Sea Res II 56:1274–1289

Carmack EC, Wassmann P (2006) Food webs and physical-biological coupling on pan-Arctic shelves: unifying concepts and comprehensive perspectives. Prog Oceanogr 71:446–477

Carmack EC, Barber D, Christensen J et al (2006) Climate variability and physical forcing of the food webs and the carbon budget on pan-Arctic shelves. Prog Oceanogr 71:145–181

Cheung W, Lam VW, Sarmiento J et al (2009) Projecting global marine biodiversity impacts under climate change scenarios. Fish Fish 68. doi:10.1111/j.1467-2979.2008.00315.x

Cheung W, Lam VW, Sarmiento J et al (2010) Large-scale redistribution of maximum fisheries catch potential in the global ocean under climate change. Glob Change Biol 16:24–35. doi:10.1111/j.1365-2486.2009.01995.x

Cheung W, Dunne J, Sarmiento J et al (2011) Integrating ecophysiology and plankton dynamics into projected maximum fisheries catch potential under climate change in the Northeast Atlantic. ICES J Mar Sci 68:1008–1018

Chierici M, Fransson A (2009) Calcium carbonate saturation in the surface water of the Arctic Ocean: undersaturation in freshwater influenced shelves. Biogeosciences 6(11):2421–2431

Christian J, Arora V, Boer G et al (2010) The global carbon cycle in the Canadian Earth System Model (CanESM1): preindustrial control simulation. J Geophys Res 115. doi:10.1029/2008JG000920

Clement JL, Maslowski W, Cooper L, Grebmeier J, Walczowski W (2005) Ocean circulation and exchanges through the northern Bering Sea—1979–2001 model results. Deep Sea Res II 52:3509–3540

Clement Kinney J, Maslowski W, Okkonen S (2009) On the processes controlling shelf-basin exchange and outer shelf dynamics in the Bering Sea. Deep Sea Res II 56:1351–1362. doi:10.1016/j.dsr2.2008.10.023

Clement Kinney J, Maslowski W, Aksenov Y, de Cuevas B, Jakacki J, Nguyen A, Osinski R, Steele M, Woodgate RA, Zhang J (2014) Chapter 7: On the flow through Bering strait: a synthesis of model results and observations. In: Grebmeier JM, Maslowski W (eds) The Pacific Arctic region: ecosystem status and trends in a rapidly changing environment. Springer, Dordrecht, pp 167–198

Comeau S, Gorsky G, Jeffree R, Teyssié JL, Gattuso JP (2009) Impact of ocean acidification on a key arctic pelagic mollusc (Limacina helicina). Biogeosciences 6:1877–1882

Cooley S, Kite-Powell H, Doney S (2009) Ocean acidification's potential to alter global marine ecosystem services. Oceanography 22(4):172–181

Davidson K (1996) Modelling microbial foodwebs. Mar Ecol Prog Ser 145:279–296

Deal C, Elliott S, Jin M, Hunke E, Maltrud M, Jeffery N (2011) Large scale modeling of primary production and ice algal biomass within arctic sea ice in 1992. J Geophys Res 116, C07004. doi:10.1029/2010JC006409

Delille B, Jourdain B, Borges A, Tison JL, Delille D (2007) Biogas (CO_2, O_2, dimethylsulfide) dynamics in spring antarctic fast ice. Limnol Oceanogr 52(4):1367–1379

Denman KL, Brasseur G, Chidthaisong A et al (2007) Couplings between changes in the climate system and biogeochemistry. In: Solomon S, Qin D, Manning M et al (eds) Climate change 2007: the physical science basis. Contribution of working group I to the fourth assessment report of the Intergovernmental Panel on Climate Change. Cambridge University Press, Cambridge/New York

Denman K, Christian J, Steiner N, Pörtner HO, Nojiri Y (2011) Potential impacts of future ocean acidification on marine ecosystems and fisheries. ICES J Mar Sci. doi:10.1093/icesjms/fsr074

Dieckmann GS, Hellmer HH (2003) The importance of sea ice: an overview. In: Thomas DN, Dieckmann GS (eds) Sea ice: an introduction to its physics, chemistry, biology, and geology. Blackwell, Oxford, pp 10–13

Dieckmann GS, Nehrke G, Uhlig C et al (2010) Brief communication : ikaite ($CaCO_3.6H_2O$) discovered in Arctic sea ice. Cryosphere 4:227–230

Dittmar T, Kattner G (2003) The biogeochemistry of the river and shelf ecosystem of the Arctic Ocean: a review. Mar Chem 83:103–120

Doney S (2010) The growing human footprint on coastal and open-ocean biogeochemistry. Science 328(5985):1512–1516

Doney S, Lindsay K, Fung I, John J (2006) Natural variability in a stable 1,000 year coupled climate-carbon cycle simulation. J Climate 19(13):3033–3054

Doney SC, Lima I, Moore JK et al (2009) Skill metrics for confronting global upper ocean ecosystem-biogeochemistry models against field and remote sensing data. J Mar Syst 76:95–112. doi:10.1016/j.jmarsys.2008.05.0

Dupont S, Thorndyke M (2008) Ocean acidification and its impact on the early life-history stages of marine animals. In: Briand F (ed) Impact of acidification on biological, chemical and physical systems in the Mediterranean and Black Seas, vol 36, Monograph. CIESM, Monaco, pp 89–97

Dupont S, Thorndyke M (2009) Impact of CO_2-driven ocean acidification on invertebrates early life history – what we know, what we need to know and what we can do. Biogeosci Discuss 6:3109–3131. doi:10.5194/bgd-6-3109-2009

Elliott S (2009) Dependence of DMS global sea-air flux distribution on transfer velocity and concentration field type. J Geophys Res 114. doi:10.1029/2008JG000710

Elliott S, Reagan M, Moridis G, Smith P (2010) Geochemistry of clathrate-derived methane in Arctic Ocean waters. Geophys Res Lett. doi:10.1029/2010GL043369

Elliott S, Deal J, Humphries G et al (2012) Pan-Arctic simulation of coupled nutrient-sulfur cycling due to sea ice biology. J Geophys Res 117, G01016. doi:10.1029/2011JG001649

Fabry V, Seibel B, Feely R, Orr J (2008) Impacts of ocean acidification on marine fauna and ecosystem processes. ICES J Mar Sci 65(3):414–432

Fabry V, Mclintock J, Mathis J, Grebmeier J (2009) Ocean acidification at high latitudes: the bellwether. Oceanography 22(4):160–171

Fennel W (2009) Parameterizations of truncated food web models from the perspective of an end-to-end model approach. J Mar Syst 76:171–185. doi:10.1016/j.jmarsys.2008.05.0

Frey KE, Masalanik JA, Clement Kinney J, Maslowski W (2014) Chapter 3: Recent variability of sea ice cover, age, and thickness in the Pacific Arctic region. In: Grebmeier JM, Maslowski W (eds) The Pacific Arctic region: ecosystem status and trends in a rapidly changing environment. Springer, Dordrecht, pp 31–64

Friedlingstein P, Cox P, Betts R et al (2006) Climate-carbon cycle feedback analysis: results from the C4MIP model intercomparison. J Climate 19:3337–3353. doi:10.1175/JCLI3800.1

Friedrichs M, Dusenberry J, Anderson L et al (2007) Assessment of skill and portability in regional marine biogeochemical models: the role of multiple planktonic groups. J Geophys Res 112:C08001. doi:10.1029/2006JC003852

Friedrichs MA, Carr ME, Barber RT et al (2009) Assessing the uncertainties of model estimates of primary productivity in the tropical Pacific Ocean. J Mar Syst 76:113133. doi:10.1016/j.jmarsys.2008.05.0

Garneau M-E, Roy S, Lovejoy C et al (2008) Seasonal dynamics of bacterial biomass and production in a coastal arctic ecosystem: Franklin Bay, western Canadian Arctic. J Geophys Res 113:C07S91

Gattuso J-P, Hansson L (2011) Ocean acidification. Oxford University Press, New York

Gaylord B, Hill TM, Sanford E et al (2011) Functional impacts of ocean acidification in an ecologically critical foundation species. J Exp Biol 214(15):2586–2594. doi:10.1242/eb.055939

Gentleman W (2002) A chronology of plankton dynamics in silico: how computer models have been used to study marine ecosystems. Hydrobiologia 480:69–85

Gibson GA, Spitz Y (2011) Impacts of biological parameterization, initial conditions and environmental forcing on parameter sensitivity and uncertainty in a marine ecosystem model for the Bering Sea. J Mar Syst 88:214–231

Golden KM, Ackley SF, Lytle VI (1998) The percolation phase transition in sea ice. Science 282:2238–2241

Golden KM, Eicken H, Heaton AL, Miner J, Pringle D, Zhu J (2007) Thermal evolution of permeability and microstructure in sea ice. Geophys Res Lett 34:L16,501. doi:10.1029/2007GL030447

Gosink TA, Pearson JG, Kelley JJ (1976) Gas movement through sea-ice. Nature 263:41–42

Gradinger R (2009) Sea-ice algae: major contributors to primary production and algal biomass in the Chukchi and Beaufort seas during May/June 2002. Deep Sea Res II 44(8):1623–1644. doi:10.1016/j.dsr2.2008.10016

Gradinger R, Meiners K, Plumley G, Zhang Q, Bluhm B (2005) Abundance and composition of the sea ice meiofauna in off-shore pack ice of the Beaufort Gyre in summer 2002 and 2003. Polar Biol 28:171–181

Grebmeier JM, Dunton KH (2000) Benthic processes in the northern Bering/Chukchi seas: status and global change. In: Huntington HP (ed) Impacts of changes in sea-ice and other environmental parameters in the Arctic. Report of the Marine Mammal Commission workshop, Girdwood, 2000

Grebmeier JM, McRoy CP, Feder HM (1988) Pelagic-benthic coupling on the shelf of the northern Bering and Chukchi seas. I. Food supply source and benthic biomass. Mar Ecol Prog Ser 48:57–67

Grebmeier JM, Overland J, Moore SE (2006) A major ecosystem shift in the northern Bering Sea. Science 311:1461–1464

Grebmeier JM, Moore SE, Overland JW, Frey KE, Gradinger R (2010) Biological response to recent Pacific Arctic sea ice retreats. EOS 91:161–168

Hasumi H (2006) CCSR Ocean Component Model (COCO) version 4.0. Center for Climate System Research report, University of Tokyo

Hemmings J, Srokosz M, Challenor P, Fasham M (2004) Split-domain calibration of an ecosystem model using satellite ocean colour data. J Mar Syst 50(3–4):141–179

Hermann AJ, Stabeno PJ, Haidvogel DB (2002) A regional tidal/subtidal circulation model of the southeastern Bering Sea: development, sensitivity analyses and hindcasting. Deep Sea Res II 49:5495–5967

Hinckley S, Coyle KO, Gibson G (2009) A biophysical NPZ model with iron for the Gulf of Alaska: reproducing the differences between an oceanic HNLC ecosystem and a classical northern temperate shelf ecosystem. Deep Sea Res II 56:2520–2536

Holland M, Bitz C, Tremblay B (2006) Future abrupt reductions in the summer arctic sea ice. Geophys Res Lett 33(23):L23503. doi:10.1029/2006GL028024

Holmes RM et al (2008) Lability of DOC transported by Alaskan rivers to the Arctic Ocean. Geophys Res Lett 35:L03402

Hood R, Christian J (2008) Ocean nitrogen cycle modelling. In: Capone D (ed) Nitrogen in the marine environment, 2nd edn. Academic Press, San Diego CA

Hu H, Wang J (2008) Modeling the ocean circulation in the Bering Sea. Chinese J Polar Sci 19:193–211

Hu H, Wang J (2010) Modeling effects of tidal and wave mixing on circulation and thermohaline structures in the Bering Sea: process studies. J Geophys Res 115:C01006. doi:10.1029/2008JC005175

Hunke EC, Bitz CM (2009) Age characteristics in multidecadal Arctic sea ice simulation. J Geophys Res 114:C08013. doi:10.1029/2008JC005186

Hunke E, Lipscomb W (2008) CICE: the Los Alamos sea ice model documentation and software user's manual (Version 4.0). Los Alamos National Laboratory, Los Alamos LA-CC-06-012 edn

Jin M, Deal CJ, Wang J, Shin K-H, Tanaka N, Whitledge TE, Lee SH, Gradinger R (2006a) Controls of the land fast ice-ocean ecosystem offshore Barrow, Alaska. Ann Glaciol 44:63–72

Jin M, Deal CJ, Wang J, Tanaka N, Ikeda M (2006b) Vertical mixing effects on the phytoplankton bloom in the southeastern Bering Sea mid-shelf. J Geophys Res 111:C03002. doi:10.1029/2005JC002994

Jin M, Deal CJ, Wang J et al (2007) Ice-associated phytoplankton blooms in the southeastern Bering Sea. Geophys Res Lett 34:L06612. doi:10.1029/2006GL028849

Jin M, Deal CJ, Wang J et al (2009) Response of lower trophic level production to long-term climate change in the southeastern Bering Sea. J Geophys Res 114:C04010. doi:10.1029/2008JC005105

Jin M, Deal CJ, Lee S et al (2012) Investigation of Arctic sea ice and ocean primary production for the period 1992 to 2007 using a 3-D global ice-ocean ecosystem model. Deep Sea Res II. doi:10.1016/j.dsr2.2011.06.003

Jodwalis CM, Benner RL, Eslinger DL (2000) Modeling of dimethyl sulfide ocean mixing, biological production and sea-to-air flux for high latitudes. J Geophys Res 105:14387–14399

Jolliff JK, Kindle JC, Shulman I, Penta B, Friedrichs MA, Helber R, Arnone RA (2009) Summary diagrams for coupled hydrodynamic-ecosystem model skill assessment. J Mar Syst 76:64–82. doi:10.1016/j.jmarsys.2008.05.014

Jones A, Roberts D, Woodage M, Johnson C (2001) Indirect sulphate aerosol forcing in a climate model with an interactive sulphur cycle. J Geophys Res 106:20293–20310

Kadko D, Pickart RS, Mathis J (2008) Age characteristics of a shelf-break eddy in the western Arctic and implications for shelf-basin exchange. J Geophys Res 113:C02018. doi:10.1029/2007JC0044292241

Karl DM, Beversdorf L, Björkman KM, Church MJ, Martinez A, Delong EF (2008) Aerobic production of methane in the sea. Nat Geosci 1:473–478

Kelley J, Gosink T (1979) Gases in sea ice: 1975–1979. Technical report. Office of Naval Research, Arlington, 104 pp

Kishi MJM et al (2007) NEMURO—a lower trophic level model for the North Pacific marine ecosystem. Ecol Model 202:12–25

Krembs C, Eicken H, Junge K et al (2002) High concentrations of exopolymeric substances in Arctic winter sea ice: implications for the polar ocean carbon cycle and cryoprotection of diatoms. Deep Sea Res I 49:2163–2181

Lavoie D, Denman K, Michel C (2005) Modeling ice algal growth and decline in a seasonally ice-covered region of the Arctic (resolute passage, Canadian archipelago). J Geophys Res 110:C11009. doi:10.1029/2005JC002922

Lavoie D, Macdonald RW, Denman KL (2009) Primary productivity and export fluxes on the Canadian shelf of the Beaufort Sea: a modelling study. J Mar Syst 75:17–32

Lavoie D, Denman KL, Macdonald RW (2010) Effects of future climate change on primary productivity and export fluxes in the Beaufort Sea. J Geophys Res 115:C04018. doi:10.1029/2009JC005493

Le Clainche Y, Vezina A, Levasseur M et al (2010) A first appraisal of prognostic ocean DMS models and prospects for their use in climate models. Global Biogeochem Cycles 24:GB3021. doi:10.1029/2009GB003721

Leck C, Bigg E (2005a) Biogenic particles in the surface microlayer and overlaying atmosphere in the central Arctic Ocean during summer. Tellus 57B(4):305–316. doi:10.1111/j.1600-0889.2005.00148.x

Leck C, Bigg E (2005b) Source and evolution of the marine aerosol – a new perspective. Geophys Res Lett 32:L19803. doi:10.1029/2005GL023651

Leck C, Bigg E (2008) Comparison of sources and nature of the tropical aerosol with the summer high Arctic aerosol. Tellus 60B(1):118–126. doi:10.1111/j.1600-0889.2007.00315.x

Lee S, Whitledge TE, Kang S (2007) Recent carbon and nitrogen uptake rates of phytoplankton in Bering Strait and the Chukchi Sea. Cont Shelf Res 27:2231–2249

Lee S, Jin M, Whitledge TE (2010) Comparison of bottom sea-ice algal characteristics from coastal and offshore regions in the Arctic Ocean. Polar Biol. doi:10.1007/s00300-010-0820-1

Liang X-Z, Kunkel KE, Meehl GA (2008) Regional climate models downscaling analysis of general circulation models present climate biases propagation into future change projections. Geophys Res Lett 35:L08709

Llinas L, Pickart RS, Mathis JT, Smith SL (2009) Zooplankton inside an Arctic Ocean cold-core eddy: probable origin and fate. Deep Sea Res II 56:1290–1304

Loose B, Miller LA, Elliott S, Papakyriakou T (2011) Sea ice biogeochemistry and material transport across the frozen interface. Oceanography 24(3):202–218

Lynch DR, McGillicuddy DJ Jr, Werner FE (2009) Skill assessment for coupled biological/physical models of marine systems. J Mar Syst 76:1–3. doi:10.1016/j.jmarsys.2008.05.002

Macdonald R, Anderson LG, Christensen JP, Miller LA, Semiletov IP, Stein R (2009) The Arctic Ocean: budgets and fluxes. In: Liu K-K, Atkinson L, Quinones R, Talaue-McManus L (eds) Carbon and nutrient fluxes in continental margins: a global synthesis. Springer, Berlin

Manley TO, Hunkins K (1985) Mesoscale eddies of the Arctic Ocean. J Geophys Res 90:4911–4930

Maslowski W, Clement Kinney J, Marble D (2008) Towards eddy-resolving models of the Arctic Ocean. In: Hecht MW, Hasumi H (eds) Ocean modeling in an eddying regime, vol 177, Geophysical monograph series, Washington DC

Maslowski W, Clement Kinney J, Higgins M, Roberts A (2012) The future of Arctic sea ice. Annu Rev Earth Planet Sc 40:625–654

Matear R, Holloway G (1995) Modeling the inorganic phosphorus cycle of the North Pacific using an adjoint data assimilation model to assess the role of dissolved organic phosphorus. Global Biogeochem Cycles 9:101–119

Mathis JT, Pickart RS, Hansell DA (2007) Eddy transport of organic carbon and nutrients from the Chukchi shelf: impact of the upper halocline of the western Arctic Ocean. J Geophys Res 112:C05011. doi:10.1029/2006JC003899

Mathis J, Cross J, Bates N, Moran SB, Lomas M, Mordy C, Stabeno P (2010) Seasonal distribution of dissolved inorganic carbon and net community production on the Bering Sea shelf. Biogeosciences 7:1769–1787. doi:10.5194/bg-7-1769-2010

Mathis JT, Grebmeier JM, Hansell DA, Hopcroft RR, Kirchman DL, Lee SH, Moran SB, Bates NR, VanLaningham S, Cross JN, Cai W-J (2014) Chapter 9: Carbon biogeochemistry of the Western Arctic: primary production, carbon export and the controls on ocean acidification. In: Grebmeier JM, Maslowski W (eds) The Pacific Arctic region: ecosystem status and trends in a rapidly changing environment. Springer, Dordrecht, pp 223–268

Matrai P, Tranvik L, Leck C, Knulst J (2008) Are high arctic surface microlayers a potential source of aerosol organic precursors? Mar Chem 108:109–122

Megry BA, Hinckley S (2001) The effect of turbulence on feeding of larval fishes: a sensitivity analysis using an individual-based model. ICES J Mar Sci 58:1015–1029

Melling H, Moore RM (1995) Modification of halocline source waters during freezing on the Beaufort Sea shelf: evidence from oxygen isotopes and dissolved nutrients. Cont Shelf Res 15:89–113

Melzner F, Gutowska MA, Langenbuch M et al (2009) Physiological basis for high CO_2 tolerance in marine ectothermic animals: pre-adaptation through lifestyle and ontogeny? Biogeosciences 6:2313–2331

Miller L, Papakyriakou TN, Collins RE, Deming JW, Ehn JK, Macdonald RW, Mucci A, Owens O, Raudsepp M, Sutherland N (2011) Carbon dynamics in sea ice: a winter flux time series. J Geophys Res 116:C02028. doi:10.1029/2009JC006058

Moore JK, Doney SC, Lindsay K (2004) Upper ocean ecosystem dynamics and iron cycling in a global three-dimensional model. Global Biogeochem Cycles 18:GB4028. doi:10.1029/2004GB002220

Moss RH, Edmonds JA, Hibbard KA et al (2010) The next generation of scenarios for climate change research and assessment. Nature 463:747–756. doi:10.1038/nature08823

Munday P, Dixson D, Cormick M, Meekan M, Ferrari M, Chivers DP (2010) Replenishment of fish populations is threatened by ocean acidification. Proc Natl Acad Sci U S A. doi:10.1073/pnas.1004519107

Mundy CJ, Barber DG, Michel C (2005) Variability of snow and ice thermal, physical and optical properties pertinent to sea ice algae biomass during spring. J Mar Syst 58:107–120

Nakicenovic G, Alcamo J, Davis G, de Vries J Fenhann B, Gaffin S et al (2000) IPCC special report on emissions scenarios. Technical report. Cambridge University Press, United Kingdom/New York

Nelson RJ, Ashjian C, Bluhm B, Conlan K, Gradinger R, Grebmeier J, Hill V, Hopcroft R, Hunt B, Joo H, Kirchman D, Kosobokova K, Lee S, Li WKW, Lovejoy C, Poulin M, Sherr E, Young K (2014) Chapter 10: Biodiversity and biogeography of the lower trophic taxa of the Pacific Arctic region: sensitivities to climate change. In: Grebmeier JM, Maslowski W (eds) The Pacific Arctic region: ecosystem status and trends in a rapidly changing environment. Springer, Dordrecht, pp 269–336

Nihoul JC, Adam P, Brasseur P et al (1993) Three-dimensional general circulation model of the northern Bering Sea's summer ecohydrodynamics. Cont Shelf Res 13:509–542

Norberg J (2004) Biodiversity and ecosystem functioning: a complex adaptive systems approach. Limnol Oceanogr 49:1269–1277

Notz D, Worster MG (2009) Desalination processes of sea ice revisited. J Geophys Res 114:C05006. doi:10.1029/2008JC004885

Okkonen SR, Weingartner TJ, Danielson SL et al (2003) Satellite and hydrographic observations of eddy-induced shelf-slope exchange in the northwestern Gulf of Alaska. J Geophys Res 108:C23033. doi:10.1029/2002JC001342

Orr J, Fabry V, Aumont O et al (2005) Anthropogenic ocean acidification over the twenty-first century and its impacts on calcifying organisms. Nature 437:681–686

Overland JE, Roach AT (1987) Northward flow in the Bering and Chukchi seas. J Geophys Res 92:7097–7105

Pabi S, van Dijken G, Arrigo KR (2008) Primary production in the Arctic Ocean, 1998–2006. J Geophys Res 113:C08005. doi:10.1029/2007JC004578

Pahlow M, Vezina AF, Casault B, Maass H, Malloch L, Wright DG, Lu Y (2008) Adaptive model of plankton dynamics for the North Atlantic. Prog Oceanogr 76:151–191

Petrich C, Eicken H (2009) Growth, structure and properties of sea ice. In: Thomas DN and Dieckmann GS (eds) Sea ice. Wiley-Blackwell, Iowa, pp 23–77

Pogson L, Tremblay B, Lavoie D, Michel C, Vancoppenolle M (2011) Development and validation of a one-dimensional snow-ice algae model against observations in resolute passage, Canadian Arctic archipelago. J Geophys Res 116:C04010. doi:10.1029/2010JC006119

Popova EE, Yool A, Coward AC et al (2010) Control of primary production in the Arctic by nutrients and light: insights from a high resolution ocean general circulation model. Biogeosciences 7:3569–3591. doi:10.5194/bg-7-3569-2010

Popova EE, Yool A, Coward AC et al (2012) What controls primary production in the Arctic Ocean? Results from an intercomparison of five general circulation models with biogeochemistry. J Geophys Res 117:C00D12. doi:10.1029/2011JC007112

Pörtner H (2010) Oxygen- and capacity-limitation of thermal tolerance: a matrix for integrating climate-related stressor effects in marine ecosystems. J Exp Biol 216(6):881–893

Pörtner H, Farrell A (2008) Physiology and climate change. Science 322:690–692

Prather M, Ehhalt D, Dentener F et al (2001) Atmospheric chemistry and greenhouse gases. In: Climate change 2001: the scientific basis. Cambridge University Press, Cambridge, England, pp 239–287

Reagan M, Moridis G (2008) Dynamic response of oceanic hydrate deposits to ocean temperature change. J Geophys Res 113. doi:10.1029/2008JC004938

Reeburgh W (2007) Oceanic methane biogeochemistry. Chem Rev 107:486–513

Rhee T, Kettle AJ, Andreae MO (2009) Methane and nitrous oxide emissions from the ocean: a reassessment using basin-wide observations in the Atlantic. J Geophys Res 114:D12304. doi: 10.1029/2008JD011662

Rho T, Whitledge TE (2007) Characteristics of seasonal and spatial variability of primary production over the southeastern Bering Sea shelf. Cont Shelf Res 27:2556–2569

Ridgwell A, Schmidt DN, Turley C, Brownlee C, Maldonado MT, Tortell P, Young JR (2009) From laboratory manipulations to earth system models: scaling calcification impacts of ocean acidification. Biogeosciences 6:2611–2623

Riebesell U, Zondervan I, Rost B, Tortell P, Zeebe R, Morel F (2000) Reduced calcification of marine plankton in response to increased atmospheric CO_2. Nature 407:364–367

Riedel A, Michel C, Gosselin M et al (2008) Winter-spring dynamics in sea-ice carbon cycling in the coastal Arctic Ocean. J Mar Syst 74:918–932

Roberts A, Cassano J, Döscher R et al (2010) A science plan for regional Arctic system modeling, a report to the national science foundation from the international Arctic science community, international Arctic research center technical papers 10–0001, International Arctic Research Center, University of Alaska Fairbanks, 47 p

Rothrock DA, Yu Y, Maykut GA (1999) Thinning of the Arctic Sea ice cover. Geophys Res Lett 26:3469–3472

Rothrock DA, Percival DB, Wensnahan M (2008) The decline in arctic sea-ice thickness: separating the spatial, annual, and interannual variability in a quarter century of submarine data. J Geophys Res 113:C05003. doi:10.1029/2007JC004252

Rysgaard S, Glud R, Sejr M et al (2007) Inorganic carbon transport during sea ice growth and decay: a carbon pump in polar seas. J Geophys Res 112:C03016. doi:10.1029/2006JC003572

Sakshaug E (2004) Primary and secondary production in the Arctic Seas. In: Stein R, MacDonald RW (eds) The organic carbon cycle in the Arctic Ocean. Springer, Berlin, pp 57–81

Schneider B, Bopp L, Gehlen M, Segschneider J, Frölicher TL, Cadule P, Friedlingstein P, Doney SC, Behrenfeld MJ, Joos F (2008) Climate-induced interannual variability of marine primary and export production in three global coupled climate carbon cycle models. Biogeoscience 5:597–614

Semiletov I, Makshtas A, Akasofu SI, Andreas E (2004) Atmospheric CO_2 balance: the role of arctic sea ice. Geophys Res Lett 31(5):L05,121. doi:10.1029/2003GL017996

Shakhova N, Semiletov I, Salyuk A, Yusupov V, Kosmach D, Gustafsson Ô (2010) Extensive methane venting to the atmosphere from sediments of the east Siberian Arctic shelf. Science 237:1246–1250

Shaw GE (1983) Bio-controlled thermostasis involving the sulfur cycle. Clim Change 5(297): 297–303. doi:10.1007/BF02423524

Six KD, Maier-Reimer E (2006) What controls the oceanic dimethylsulfide (DMS) cycle. Global Biogeochem Cycles 20. doi:10.1029/2005GB002674

Slagstad D, Wassmann P (1997) Climatic change and carbon flux in the Barents Sea: 3-D simulations of ice-distribution, primary production and vertical export of particulate organic carbon. Mem Natl Inst Polar Res 51(Special Issue):119–141

Slagstad D, Ellingsen IH, Wassmann P (2011) Evaluating primary and secondary production in an Arctic Ocean void of summer sea ice: an experimental simulation approach. Prog Oceanogr 90:117–131

Smith KW, McGillicuddy DJ, Lynch DR (2009) Parameter estimation using an ensemble smoother: the effect of the circulation in biological estimation. J Mar Syst 76:162–170

Solomon S, Qin D, Manning M, Chen Z et al (eds) (2007) The physical science basis. Contribution of working group I to the fourth assessment report of the IPCC. Cambridge University Press

Spaulding M, Isaji T, Mendelsohn D et al (1987) Numerical simulation of wind-driven flow through the Bering Strait. J Phys Oceanogr 17:1799–1816

Springer AM, McRoy CP (1993) The paradox of pelagic food webs in the northern Bering Sea. III. Patterns of primary production. Cont Shelf Res 13:575–599

Stabeno PJ, Schumacher JD, Ohtani K (1999) The physical oceanography of the Bering Sea. In: Loughlin TR, Ohtani K (eds) Dynamics of the Bering Sea. University of Alaska Sea Grant, Fairbanks, pp 1–28

Steinacher M, Joos F, Frölicher T, Plattner GK, Doney S (2009) Imminent ocean acidification of the arctic projected with the NCAR global coupled carbon-cycle climate model. Biogeosciences 6(4):515–533

Steinacher M, Joos F, Frölicher T et al (2010) Projected 21st century decrease in marine productivity: a multi-model analysis. Biogeosciences 7:979–1005

Stow CA, Jolliff J, McGillicuddy DJ, Doney SC, Allen JI, Friedrichs MA, Rose KA, Wallhead P (2009) Skill assessment for coupled biological/physical models of marine systems. J Mar Syst 76:4–15. doi:10.1016/j.jmarsys.2008.05.006

Stroeve J, Holland MM, Meier W et al (2008) Arctic sea ice decline: faster than forecast. Geophys Res Lett 34:L09501

Stroeve J, Serreze M, Holland M, Kay J, Maslanik J, Barrett A (2011) The Arctic's rapidly shrinking sea ice cover: a research synthesis. Clim Change. doi:10.1007/s10584-011-0101-1

Sumaila R, Cheung W (2010) Cost of adapting fisheries to climate change. The World Bank discussion paper 5(8):1496–1501

Taylor K (2001) Summarizing multiple aspects of model performance in a single diagram. J Geophys Res 106:7183–7192

Tett P (1987) Modelling the growth and distribution of marine microplankton. In: Fletcher M, Joe J, Gray T (eds) Ecology of microbial communities. Cambridge University Press, Cambridge

Tremblay J-E, Gagnon G (2009) The effects of irradiance and nutrient supply on the productivity of Arctic waters: a perspective on climate change. In: Nihoul JCJ, Kostianoy AG (eds) Influence of climate change on the changing Arctic and sub-Arctic conditions. Proceedings of the NATO advanced research workshop, Liège, May 2008. Springer, Dordrecht, pp 73–94

Tremblay J-E et al (2008) Vertical stability and annual dynamics of nutrients and chlorophyll fluorescence in the coastal, southeast Beaufort Sea. J Geophys Res 113:C07S90. doi:10.1029/2007JC004547

Vancoppenolle M, Fichefet T, Goosse H (2009) Simulating the mass balance and salinity of arctic and antarctic sea ice. 2. Importance of sea ice salinity variations. Ocean Model 27:54–69

Vancoppenolle M, Goosse H, de Montety A, Fichefet T, Tremblay B, Tison JL (2010) Modeling brine and nutrient dynamics in antarctic sea ice: the case of dissolved silica. J Geophys Res 115: C02005. doi:10.1029/2009JC005369

Vichi M, Pinardi N, Masina S (2007) A generalized model of pelagic biogeochemistry for the global ocean ecosystem. Part I: theory. J Mar Syst 64:89–109

Wallhead PJ, Martin AP, Srokosz MA, Franks PJ (2009) Skill assessment via cross-validation and Monte Carlo simulation: an application to Georges bank plankton models. J Mar Syst 76:134–150. doi:10.1016/j.jmarsys.2008.05.0

Walsh JJ, Dieterle DA, Maslowski W et al (2004) Decadal shifts in biophysical forcing of Arctic marine food webs: numerical consequences. J Geophys Res 109:C05031

Walsh JJ, Dieterle DA, Maslowski W et al (2005) A numerical model of seasonal primary production within the Chukchi/Beaufort seas. Deep Sea Res II 52:3541–3576. doi:10.1016/j.dsr2.2005.09.009

Wang J, Deal C, Wan Z, Jin M, Tanaka N. Ikeda M (2003) User's guide for a physical ecosystem model in the subpolar and polar oceans. IARC FRSGC technical report 03–01. 75 pp

Wang J, Hu H, Mizobata K, Saitoh S (2009) Seasonal variations of sea ice and ocean circulation in the Bering Sea: a model-data fusion study. J Geophys Res 114:C02011. doi:10.1029/2008JC004727

Wassmann P, Slagstad D, Riser CW, Reigstad M (2006) Modelling the ecosystem dynamics of the Barents Sea including the marginal ice zone II. Carbon flux and interannual variability. J Mar Syst 59:1–24

Wassmann P, Duarte CM, Agusti S et al (2011) Footprints of climate change in the Arctic marine ecosystem. Global Change Biol 17:1235–1249. doi:10.1111/j.1365-2486.2010.02311

Watanabe E (2011) Beaufort shelfbreak eddies and shelf-basin exchange of Pacific summer water in the western Arctic Ocean detected by satellite and modeling analyses. J Geophys Res 116:C08034. doi:10.1029/2010JC006259

Watanabe E, Hasumi H (2009) Pacific water transport in the western Arctic Ocean simulated by an eddy-resolving coupled sea ice-ocean model. J Phys Oceanogr 39:2194–2211

Werner I, Ikävalko J, Schünemann H (2007) Sea-ice algae in Arctic pack ice during late winter. Polar Biol 30:1493–1504

Woodgate RA, Aagaard K (2005) Revising the Bering Strait freshwater flux into the Arctic Ocean. Geophys Res Lett 32:L02602. doi:10.1029/2004GL021747

Yamamoto-Kawai M, McLaughlin F, Carmack E, Nishino S, Shimada K (2009) Aragonite undersaturation in the Arctic Ocean: effects of ocean acidification and sea ice melt. Science 326(5956):1098–1100

Yamamoto-Kawai M, McLaughlin F, Carmack E (2011) Effects of ocean acidification, warming and melting of sea ice on aragonite saturation of the Canada Basin surface water. Geophys Res Lett 38:L03601. doi:10.1029/2010GL045501

Yool A, Oschlies A, Nurser AJG, Gruber N (2010) A model-based assessment of the TrOCA approach for estimating anthropogenic carbon in the ocean. Biogeosciences 7:723–751. doi:10.5194/bg-7-723-2010

Zahariev K, Christian J, Denman K (2008) A global ocean carbon model with parameterizations of iron limitation, calcification and N_2 fixation: preindustrial, historical and fertilization simulations. Prog Oceanogr 77(1):56–82

Zemmelink H, Houghton L, Dacey J et al (2008) Stratification and the distribution of phytoplankton, nutrients, inorganic carbon, and sulfur in the surface waters of Weddell Sea leads. Deep Sea Res II 55:988–999

Zhang J, Spitz Y, Steele M et al (2010) Modeling the impact of declining sea ice on the Arctic marine planktonic ecosystem. J Geophys Res 115:C10015. doi:10.1029/209JC005387

Index

J.M. Grebmeier and W. Maslowski (eds.), *The Pacific Arctic Region: Ecosystem Status and Trends in a Rapidly Changing Environment*, DOI 10.1007/978-94-017-8863-2,

Printed by Printforce, the Netherlands